INTEGRATIVE CLUSTER ANALYSIS IN BIOINFORMATICS

INTEGRATIVE CLUSTER ANALYSIS IN BIOINFORMATICS

Basel Abu-Jamous, Rui Fa and Asoke K. Nandi
Brunel University London, UK

This edition first published 2015
© 2015 John Wiley & Sons, Ltd

Registered Office
John Wiley & Sons, Ltd, The Atrium, Southern Gate, Chichester, West Sussex, PO19 8SQ, United Kingdom

For details of our global editorial offices, for customer services and for information about how to apply
for permission to reuse the copyright material in this book please see our website at www.wiley.com.

Library of Congress Cataloging-in-Publication Data

Abu-Jamous, Basel.
Integrative cluster analysis in bioinformatics / Basel Abu-Jamous, Dr Rui Fa, and Prof. Asoke K. Nandi.
    pages   cm
  Includes bibliographical references and index.
  ISBN 978-1-118-90653-8 (cloth)
1. Bioinformatics–Mathematics.   2. Cluster analysis.   I. Fa, Rui.   II. Nandi, Asoke Kumar.   III. Title.
  QH324.2.A24 2015
  519.5′3–dc23

                                              2014032428

A catalogue record for this book is available from the British Library.

Set in 10/12pt Times by SPi Publisher Services, Pondicherry, India
Printed and bound in Singapore by Markono Print Media Pte Ltd

1   2015

To
Wael Abu Jamous and Eman Arafat
Hui Jin, Yvonne and Molly Fa
Marion, Robin, David and Anita Nandi

Brief Contents

Contents

Preface

Bioinformatics is a new and interdisciplinary field which is concerned with the development of methods for recording, organising, and analysing biological data. With the advent of human genome-sequencing, microarrays and high-performance computing, developing software tools to generate useful biological knowledge has been a major activity.

Clustering techniques have been increasingly used in the analysis of high-throughput biological datasets. Although many generic clustering methods have been used successfully to analyse biological datasets, many specific properties of those datasets require customised methods which are specifically designed to meet such properties. Therefore, both aspects – the design and the application of clustering methods in this field – have been under active investigations and considerations by many research groups around the world. Indeed, there have been many activities in assimilating the work in this field, especially by designing state-of-the-art methods which uniquely address various issues that are particularly relevant in biological data.

This book attempts to outline the complete pathway from the basics of molecular biology to the generation of biological knowledge. It is supplied with an introductory part to molecular biology at a level which can be understood by researchers coming from a numeric background, such as computer scientists and information engineers. The introductory part helps those readers to get introduced to the basic biological knowledge needed to appreciate the specific applications of the methods in this book. The book also explains the structure and properties of many types of high-throughput datasets commonly found in biological studies, including public repositories/databases, pre-processing like normalisation and the identification of differentially expressed genes. A major part of the book will cover various clustering methods and clustering-validation techniques as well as their applications to biological datasets, representing an integrative analysis. It should be remarked that not all clustering methods have been utilised in bioinformatics yet. Some of the most recent state-of-the-art clustering methods, which can deal with specific problems that appear in biological datasets, are paid much attention, especially in how they, and their possible successors, could be used to enhance the pace of biological discoveries in the future. Although proposed in the context of bioinformatics, some specialised sophisticated methods can also be used by other researchers to apply them to other analogous problems. Therefore, the general community of researchers in the field of machine learning are also targeted by most of the contents of this book.

There are books mainly focusing on various aspects related to microarrays, such as biological experimental design, image processing, identification of differentially expressed genes, supervised classification, etc., while covering clustering analysis at a less thorough level. In any case, these tend to focus on clustering of microarray datasets rather than considering a wider range of biological datasets. Yet, other books provide more thorough coverage of clustering than the aforementioned books, but they do not provide sufficient background in the field of molecular biology for researchers coming from numeric backgrounds to appreciate and understand the origins of the available datasets. These kinds of book tend to be mainly data-clustering books and naturally belong to the computational side of bioinformatics rather than to the interface between the computational and the biological sides. This book does sit at this interface and goes far beyond those currently available, in that it presents some biological preliminaries as well as some state-of-the-art clustering methods that are specifically designed to suit specific issues which appear in biological datasets.

As the field is still developing, such a book cannot be definitive or complete. This book is designed to target researchers in bioinformatics ranging from entry level researchers (e.g. senior bachelor students and master students) to the most senior researchers (e.g. heads of research groups). It is hoped that graduate students should be able to learn enough basics before studying journal papers; researchers in related fields should be able to get a broad perspective on what has been achieved; and current researchers in this field should be able to use it as a reference.

Further to the material provided in the book, a companion website hosting a selected collection of software and links to publicly available datasets can be accessed by using the following URL: https://code.google.com/p/integrative-cluster-bioinformatics/.

A work of this magnitude will unfortunately contain errors and omissions. We would like to take this opportunity to apologise unreservedly for all such indiscretions in advance. We would welcome comments and corrections; please send them by email to a.k.nandi@ieee.org or by any other means.

BASEL ABU-JAMOUS, RUI FA AND ASOKE K. NANDI
London, UK
Feb 2015

List of Symbols

X	Data matrix
x_n	The nth data object
S	Similarity matrix
N	Number of data objects
M	Number of features or samples
K	Number of clusters
$D(\cdot,\cdot)$	Dissimilarity between two input data objects
$S(\cdot,\cdot)$	Similarity between two input data objects
Z	Partition (vector sequence with the length of N)
U	Partition matrix (with N rows and K columns)
$u_{k,n}$	Entry in the partition matrix, indicating the membership of the nth data object in the kth cluster
C	Cluster set, containing K clusters $\{C_1, ..., C_K\}$
C_k	The kth cluster, containing the indices of those data objects belonging to it
c_k	The centroid of the kth cluster
$\kappa(\cdot,\cdot)$	Kernel function
m	The fuzzifier
μ	The mean vector
Σ	The covariance matrix
Σ	The within-cluster covariance matrix
$(\cdot)^T$	The matrix transpose operator
$(\cdot)^{-1}$	The matrix inverse
$\det(\cdot)$	The determinant operator
$\exp(\cdot)$	The exponential function
$|\cdot|$	The cardinality of a set
$\|\cdot\|$	The Euclidean norm
Θ	The total parameter set

G	Number of groups
τ	The mixing parameter
$\mathcal{L}(\cdot)$	The likelihood function
$\boldsymbol{A}$	The adjacency matrix
$\boldsymbol{B}$	The incident matrix
$\boldsymbol{D}$	The degree matrix
$\boldsymbol{L}$	The Laplacian matrix
G	A graph
V	The set of vertices
E	The set of edges
$\boldsymbol{M}$	The modularity matrix
Q	The modularity value
N_V	Number of vertices
N_E	Number of edges
$\boldsymbol{\theta}$	A smaller parameter set

About the Authors

Basel Abu-Jamous received his BSc degree in computer engineering from the University of Jordan, Amman, Jordan, in 2010. He received his MSc degree in information and intelligence engineering from the University of Liverpool, Liverpool, UK, in 2011. He was awarded the Sir Robin Saxby Prize for the 2010/2011 academic year based on his performance in his MSc degree. Mr Abu-Jamous started his PhD degree in electrical engineering and electronics at the University of Liverpool in 2011 and then transferred with his supervisor, Professor Asoke K. Nandi, to complete his PhD degree at Brunel University London, UK. He was awarded a prize in the posters section of the 6th Annual Student Research Conference 2013 at Brunel University London. Currently, he is a research assistant at the Department of Electronic and Computer Engineering at Brunel University London, a position to which he was appointed in January 2015. Mr Abu-Jamous has authored or co-authored several journal and international conference papers. His research interests include bioinformatics, computational biology, and the broader areas of information engineering and machine learning.

Dr Rui Fa received his PhD degree in electrical engineering and information systems from the University of Newcastle, UK, in 2007. From January 2008 to September 2010, he held research positions in the University of York and the University of Leeds working in radar signal processing and wireless communication projects. From October 2010, Dr Fa extended his research fields to bioinformatics and computational biology, and joined the University of Liverpool involving in a collaborative research project with the Universities of Oxford, Cambridge, and Bristol, which is funded by the National Institute for Health Research (NIHR). Dr Fa joined Brunel University London as a senior research fellow in 2013. His current research interests include bioinformatics and computational biology, systems biology, machine learning, Bayesian statistics, statistical signal processing, and network science. Dr Fa has authored and co-authored more than 60 peer-reviewed journal and conference papers.

Professor Asoke K. Nandi joined Brunel University London in April 2013 as the Head of Electronic and Computer Engineering. He received a PhD from the University of Cambridge, and since then has worked in many institutions, including CERN, Geneva, Switzerland; University of Oxford, Oxford, UK; Imperial College London, UK; University of Strathclyde, Glasgow, UK; and University of Liverpool, Liverpool, UK. His research spans many different topics, including bioinformatics, communications, machine learning (feature selection, feature generation, classification, clustering, and pattern recognition), and signal processing.

In 1983 Professor Nandi was a member of the UA1 team at CERN that discovered the three fundamental particles known as W^+, W^- and Z^0, providing the evidence for the unification of the electromagnetic and weak forces, which was recognized by the Nobel Committee for Physics in 1984. He has been honoured with the Fellowships of the Royal Academy of Engineering (UK) and the Institute of Electrical and Electronics Engineers (USA). He is a Fellow of five other professional institutions, including the Institute of Physics (UK), the Institute of Mathematics and its Applications (UK), the British Computer Society (UK), the Institution of Mechanical Engineers (UK), and the Institution of Engineering and Technology (UK). His publications have been cited more than 16 000 times and his h-index is 60 (Google Scholar).

Part One

Introduction

1

Introduction to Bioinformatics

1.1 Introduction

Interesting research fields emerge through the collaboration of researchers from different, sometimes distant, disciplines. Examples include biochemistry, biophysics, quantum information science, systems engineering, mechatronics, business information systems, management information systems, geophysics, biomedical engineering, cybernetics, art history, media technology and others. This marriage between disciplines yields findings which blend the views of different areas over the same subject or set of data.

The stimuli leading to such collaborations are numerous. For example, one discipline may develop tools that generate types of data that require another discipline to analyse. In other cases, one field scratches a layer of unknowns to discover that significant parts of its scope are actually based on the principles of another field, such as the low-level biological studies of the chemical interactions in the cells, which delivered biochemistry as an interdisciplinary field. Other interdisciplinary fields emerged because of their complementary involvement in building different parts of the same target system or in understanding different sides of the same research question; for example, mechatronics engineering aims at building systems which have both mechanical and electronic parts, such as all modern automobiles. Interdisciplinary areas like business information systems and management information systems have emerged due to the high demand for information systems which target business and management aspects; although generic information systems would meet many of those requirements, a customised field focusing on such applications is indeed more efficient given such high demand.

The interdisciplinary field of this book's focus is *bioinformatics*. The motive behind this field's emergence is the increasingly expanding generation of massive raw biological data following the developments in high-throughput techniques in the last couple of decades. The scale of this high-throughput data is orders of magnitude higher than what can be efficiently analysed

Integrative Cluster Analysis in Bioinformatics, First Edition. Basel Abu-Jamous, Rui Fa and Asoke K. Nandi.
© 2015 John Wiley & Sons, Ltd. Published 2015 by John Wiley & Sons, Ltd.

in a manual fashion. Consequently, information engineers were recruited in order to contribute to data analysis by employing their computational methods. Cycles of computational analysis, sharing of results, interdisciplinary discussions and abstractions have led, and are still leading, to many key discoveries in biology and medicine. This success has attracted many information engineers towards biology and many biologists towards information engineering to meet in a potentially rich intersection area, which itself has grown in size to establish the field of *bioinformatics.*

1.2   The "Omics" Era

A new suffix has been introduced to the English language in this era of high-throughput data expansion; that is "-omics", and its relatives "-ome" and "-omic". This started in the 1930s when the entire set of genes carried by a chromosome was called the *genome,* blending the words "gene" and "chromosome" (OED, 2014). Consequently, the analysis of the entire genome was called *genomics,* and many known research journals carried the term "genome" or "genomics" in their titles such as Genomics, Genome Research, Genome Biology, BMC Genomics, Genome Medicine, the Journal of Genetics and Genomics (JGG), and others.

The -ome suffix was not kept exclusive for the genome; it has been rather generalised to indicate the complete set of some type of molecule or object. The proteome is the complete set of proteins in a cell, tissue or organism. Similarly are the transcriptome, metabolome, glycome and lipidome for the complete sets of transcripts, metabolites, glycans (carbohydrates) and lipids. In a respective order, large-scale studies of those complete sets are known as proteomics, transcriptomics, metabolomics, glycomics and lipidomics. The -ome suffix was further generalised to include the complete sets of objects other than basic molecules. For example, the microbiome is the complete set of microorganisms (e.g. bacteria, microscopic fungi, etc.) in a given environment such as a building, a sample of soil or the human gut (Kembel *et al.,* 2014). More *omic* fields have also emerged such as *agrigenomics* (the application of genomics in agriculture), pharmacogenomics and pharmacoproteomics (the application of genomics and proteomics to pharmacology), and others.

All of those biological fields of omics involve high-throughput datasets which are subject to information engineering involvements, and therefore reside at the core focus of bioinformatic research. An even higher level of omics analysis involves integrative analysis of many types of omic datasets. *OMICS: a Journal of Integrative Biology* is a journal which targets research studies that consider such collective analysis at different levels from single cells to societies.

More types of high-throughput omic datasets are expected to emerge. The role of bioinformatics as an interdisciplinary field will be more important. This is not only because each of those omic datasets is massive in size when considered individually; it is also because of the size of information hidden in the relations between those generally heterogeneous datasets, which requires more sophisticated computational methods to analyse.

1.3   The Scope of Bioinformatics

The scope of bioinformatics includes the development of methods, techniques and tools which target storage, retrieval, organisation, analysis and presentation of high-throughput biological data.

1.3.1 Areas of Molecular Biology Subject to Bioinformatics Analysis

In a very general statement, each part of molecular biology which produces high-throughput data is subject to bioinformatics analysis. On the other hand, low-throughput data which can be manually analysed do not represent subjects for bioinformatics. The omics fields described in the previous section are indeed included in bioinformatics analysis. This includes aspects of DNA, RNA and protein sequence analysis, gene and protein expression, genetics of diseases including cancers and special phenotypes, analysis of gene regulation, chemical interaction regulation, enzymatic regulation, other types of regulation, analysis of flowing signals in cells, networks of genetic, protein and other molecular interactions, comparable analysis of the diversity of genomes between individuals or organisms in an environment or across different environments, and others.

1.3.2 Data Storage, Retrieval and Organisation

The human genome is a linear thread of more than three billion base-pairs (letters). In 2012, and after more than 4 years after its starting point, the 1000 Genomes Project Consortium announced the completion of sequencing of the complete genomes of 1092 individuals from fourteen different populations (The 1000 Genomes Project Consortium, 2014). Moreover, the genomes of thousands of organisms, other than humans, have been sequenced and stored during the last two to three decades. As for gene expression data, tens of thousands of massive microarray datasets have been generated in the last two decades. Add to that the increasing amounts of data generated for protein expression, DNA binding and other types of high-throughput data. Data generation has not stopped and is expected to increase rapidly due to the massive advances in technologies and cost reduction. Therefore, it is crucial to store such amounts of datasets in an efficient manner which allows for quick and efficient access by large numbers of researchers from different parts of the world simultaneously.

Given the current trend, which is to offer most of the generated high-throughput datasets for public use in centralised databases, it becomes essential to standardise the way in which data are organised, annotated and labelled. This enhances information exchange and mutual understanding between different research groups in the world.

Taken together, the scope of bioinformatics indeed includes designing and implementing appropriate databases for high-throughput biological data storage, building means of data access to those databases such as web services and network applications, organising different levels of data pieces by standard formats and annotations, and, undoubtedly, maintaining and enhancing the availability and the scalability of these data repositories.

1.3.3 Data Analysis

Elaine Mardis, the Professor of Genetics in the Genome Institute at Washington University, and a collaborator in the 1000 Genomes Project, titled her "musing" published in *Genome Medicine* in 2010 as "the $1,000 genome, the $100,000 analysis?" (Mardis, 2010). Mardis discussed the tremendous drop in the cost of sequencing the complete genome of an individual human from hundreds of millions of dollars to a few thousands, and that it is expected to reach the line of $1,000. She mused, based on many facts and observations, that the cost of data analysis, which does not seem to be dropping, will constitute the major part of the total cost, rather than the cost of data generation.

A small proportion of human genes, out of 20 000–25 000, has been well described and understood, while many gaps in our understanding of the vast majority of them do still exist. Identifying the sequence of a gene from a thousand individuals and measuring its expression profile under many conditions are not sufficient to understand its function. What increases the level of complexity is the fact that genes are highly interrelated in terms of their functions and regulation. Many genes' products work in concert to achieve a common objective; many others perform different related or unrelated tasks in different parts of the cell; many genes' products, if met within the same location, would have conflicts resulting in them negatively affecting each other's function; moreover, many genes' products control, directly or indirectly, the expression of other genes. These are examples of complexities that are not directly seen in raw sequence or expression datasets.

The quest to answer such questions and to unveil more regarding the unknowns is being carried out by large interdisciplinary collaborations, which when blended belong to the field of bioinformatics. Computational methods already existing in the field of machine learning were borrowed to be employed in biological data analysis. However, owing to the high demand, enormous size and various special characteristics, various computational methods have been designed specifically within the area of bioinformatics. Recruiting appropriate existing computational methods and/or designing customised methods for more efficient analysis of the biological data represents a large aspect of bioinformatics.

Furthermore, the gap between the existing methods in bioinformatics and the amount of information hidden in the existing data is large. This calls for more innovative out-of-the-box methods which have the ability to capture the diverse types of hidden information. A key feature of the desired methods is the ability to analyse multiple heterogeneous datasets, which can be of one or multiple types, in order to fetch high-level and low-level comprehensive and collective conclusions (Abu-Jamous *et al.*, 2013). It is expected that such methods would have a level of complexity and sophistication which enables them to delve into the inherently embodied complexities of the biological systems.

1.3.4   Statistical Analysis

In order to rely on the measurements offered by a high-throughput dataset or the results provided by a computational method, quantitative quality measures are required. Owing to the large-scale nature of analysis in bioinformatics, which typically involves large numbers of objects or samples with many stochastic variables, tests and techniques based on statistics represent the most intuitive choice for quality assessment and significance identification.

1.3.5   Presentation

A table with tens of thousands of numbers, a list of thousands of gene names, a network of tens of thousands of gene relations, a string of billions of characters representing the genome of an individual, a list of scores assigned to thousands of genes based on some computational method, a table of gene clusters produced by a partitioning algorithm, and others are examples of ways of data presentation that are not normally comprehensible by human researchers. Thus, effective techniques for comprehensible data and results presentations are required; designing such techniques is certainly part of the roles of researchers in bioinformatics.

Presentation takes the forms of figures, tables, and text passages. Successful choice of the type of figure or table to be used depends on the ability of that figure to highlight the aspects of interest in a large amount of data in a comprehensible and conclusive way. Colour-coding, symbol-coding, lines, arrows, labels, shapes and others are examples of pieces that need to be brought together cleverly to produce a powerful figure. If such figures can be produced to visualise data or results, they are more desirable than tables, which are in their turn more desirable than text passages. Indeed, supporting text is usually required to describe the figure and explain how it should be read.

Bioinformaticians, as part of their scope of research, have been designing different types of figures and tables that suite the nature of biological high-throughput data. Many of those figures belong to conventional families of figures that are used in various areas of research, and many others are novel or customised versions of the conventional ones.

1.4 What Do Information Engineers and Biologists Need to Know?

Most fruitful bioinformatic studies usually involve collaborations between both biologists and information engineers rather than being carried out solely by one of the two parties. Though, for a successful collaboration, researchers from both sides need to take steps towards each other to reside at the interface between the two fields.

Information engineers are experts in machine learning and related areas, and in many cases, in parts of statistics. They have the ability to design, implement and apply computational methods. An information engineer who works in bioinformatics needs to learn the principles of molecular biology such as the main components and processes of the living cell at its molecular level, and what is known as the *central dogma of molecular biology*. By learning such introductory amounts of molecular biology, the information engineer will be able to comprehend terms like gene, protein, DNA, RNA, messenger RNA, non-coding RNA, chromatin, transcription, translation, gene regulation, gene expression, ribosome, genetic interaction, protein physical interaction, pathway, transport, cellular signalling and others. The information engineer will also be able to understand the structure of the commonly used high-throughput biological datasets such as nucleic acid sequences, microarrays and genetic interaction networks. Moreover, it is important, at least within the specific targeted application, to be aware of the major biological processes taking place and the main questions of research requiring answers. Additionally, familiarity with the statistical properties lying beneath the raw measurements provided by the high-throughput datasets, such as the levels and types of noise, is indeed crucial for reliable analysis. We provide most of what an information engineer needs to know in molecular biology in Parts Two and Three of this book.

In contrast, molecular biologists are familiar with the living cell and its processes. They may be familiar with the structure of high-throughput datasets, but this becomes essential for the type of dataset considered if they are involved in a bioinformatic collaboration. As for the computational methods, they usually need to know them at the *black-box* level of abstraction, that is they need to understand the structure of the raw data provided to the method as an input, the structure of the result generated by the method as an output, and the semantic relation between the input and the output. In most of the cases, they do not need to delve into the mathematical and logical details that are carried out within the course of the method's application. However, sometimes a more detailed understanding of the method, deeper than the black-box level of abstraction, is very useful for better comprehension of its results.

It can be seen in the preceding paragraphs that both parties need to understand to a good level of detail in the structures of the datasets and the results. This is because the datasets and the results respectively represent the cargoes transferred from biologists to information engineers and from information engineers to biologists across the interface between the two fields. Accompanied with the results, and sometimes with the datasets, statistical measures are provided. Good understanding of the proper interpretation and indications of such measures is certainly required from both parties.

1.5 Discussion and Summary

The massive amounts of high-throughput biological datasets generated in the last two to three decades have motivated biologists to collaborate with information engineers in order to be able to analyse such manually-incomprehensible data. The collaborations have grown considerably, leading to the identification of the interface between biology and informatics as a standalone interdisciplinary field named as *bioinformatics*.

The scope of bioinformatics includes designing, implementing and applying methods for storing, retrieving, organising, analysing and presenting biological high-throughput data. Many of these methods were borrowed from other applications of information engineering, while many others were specially designed for bioinformatics applications. *Omics* datasets, which measure certain types of biological molecules or objects at a large scale, reside at the core focus of bioinformatics. Examples of omics include genomics, proteomics and glycomics, which are large-scale analyses of genes, proteins and glycans (carbohydrates), respectively.

In a successful collaboration, both information engineers and biologists need to have certain levels of knowledge and understanding of each other's fields. Information engineers need to understand the principles of molecular biology and the main biological questions in the context under consideration, while biologists need to learn abstract levels of description of the computational methods involved in the analysis. Both parties need to be familiar with the structure of the raw datasets and the results, as well as the correct indications and consequences of the adopted statistical measures.

References

Abu-Jamous, B., Fa, R., Roberts, D.J. and Nandi, A.K. (2013). Paradigm of tunable clustering using binarization of consensus partition matrices (Bi- CoPaM) for gene discovery. *Plos One*, **8**(2), e56432.

Kembel, S.W., Meadow, J.F., O'Connor, T.K. *et al.* (2014). Architectural design drives the biogeography of indoor bacterial communities. *Plos One*, **9**(1), e87093.

Mardis, E.R. (2010). The $1,000 genome, the $100,000 analysis? *Genome Medicine*, **2**, p. 84.

OED (2014). *Oxford English Dictionary Online*, Oxford University Press, Oxford.

The 1000 Genomes Project Consortium (2012). An integrated map of genetic variation from 1,092 human genomes. *Nature*, **491**, pp. 56–65.

2

Computational Methods in Bioinformatics

2.1 Introduction

Owing to the diversity in the types of high-throughput biological data as well as the diversity in the objectives of bioinformatic research studies, various classes of computational methods have been utilised in bioinformatics. Algorithms belonging to the fields of machine learning, data mining, optimisation, template matching, image processing, soft computing, modelling, simulation and string manipulation have been employed in bioinformatics. Moreover, many biological datasets have been modelled and analysed in the light of principles from network theory, control theory, information theory and statistics. Additionally, the implementation of such methods has been carried out by using different tools and technologies such as Structured Query Language (SQL) databases, web protocols and services, high-level programming languages, mark-up languages, statistical and modelling software applications, and sophisticated toolboxes, packages, and tools customised for bioinformatics.

This book focuses on *clustering* algorithms, which represent one of the most widespread classes of methods recruited in bioinformatics. For integrative cluster analysis, other classes of methods need to be adopted in a cascade of steps starting from raw datasets and ending at final mature sets of results; we have focused on such classes of methods, such as normalisation and visualisation, in this book as well.

For a more complete view of the field, we devote this chapter as a brief introduction to computational methods used in bioinformatics in general. This will position cluster analysis in its place within the wider context of bioinformatics, and shall establish a base on which the reader can build by referring to the relevant resources in the literature.

Integrative Cluster Analysis in Bioinformatics, First Edition. Basel Abu-Jamous, Rui Fa and Asoke K. Nandi.
© 2015 John Wiley & Sons, Ltd. Published 2015 by John Wiley & Sons, Ltd.

2.2 Machine Learning and Data Mining

Machine learning is an area in artificial intelligence which is concerned about building algorithms and methods that can learn from empirical data. Machine learning is divided into two main branches, supervised learning and unsupervised learning. In *supervised learning,* a representative set of known, labelled, data is learned in order to be able to predict unknown data by generalisation. In contrast, unsupervised learning aims at learning hidden structures in a given set of unlabelled data. Other branches in machine learning, including semi-supervised learning, also exist.

Data mining aims at discovering patterns and useful pieces of information in a large set of data. The focus in data mining is the final objective and not the method. Therefore, methods from machine learning are commonly utilised by data mining in addition to other methods and techniques from statistics, database systems, discrete mathematics and optimisation. The key aspect which allows data mining analysis to employ machine learning methods is the ability to format and represent the mined data in a structure that is accessible by machine learning methods. Such data is generally organised as a set of objects with quantitative features.

2.2.1 Supervised Learning

The given set of known and labelled data that is subjected to supervised learning is known as *learning data.* Learning data generally includes a set of observed examples whose *targets* are known. The target of an example object represents the "correct answer" to the question under investigation given that particular object. An illustrative example is a set of known personal photographs which are used to train a supervised learning method to answer the question "is the given photograph of a male or a female?" The training set is labelled, that is the correct gender of the person in each of the given photographs is known and is fed to the method to be trained on. If training was successful, the supervised method will be able to identify, with high accuracy, the gender of a previously unseen personal photograph other than the ones provided in the training set. The problem targeted by this illustrative example, known as *supervised classification,* constitutes a major part of the area of supervised learning. Classification methods learn how to classify sets of objects into two or more classes after being trained on a set of objects whose correct classes are known.

Numerous methods for linear and non-linear classification exist. Examples include k-nearest neighbour (KNN) (Altman, 1992), artificial neural networks (ANN) (Haykin, 1999), support vector machines (SVM) (Cortes and Vapnik, 1995), Fisher's linear discriminant (Duda, Hart and Stork, 2000), Bayes classifier (Devroye, Györfi and Lugosi, 1997), decision trees (Quinlan, 1986), regression-based classifiers (Naseem, Togneri and Bennamoun, 2010), and genetic programming (GP) (Liu and Xu, 2009). Many linear methods, like the SVM, can be made non-linear by applying what is known as the kernel trick, which non-linearly transforms the data into another space, in which the originally linear method is applied (Shawe-Taylor and Cristianini, 2004). Moreover, and owing to the differences between those methods in their points of strength and weakness, ensembles of classifiers are used in which many methods vote for the answer, and the answer which obtains the majority of votes is considered to be the final consensus answer (Breiman, 1996).

Supervised classification methods are commonly used in bioinformatics. For example, a given set of gene expression samples taken from patients and healthy individuals can be provided to a classifier to learn the gene expression patterns that distinguish between the two classes. Two main consequences follow from successful training of such classifier – first, it can assist the diagnosis of a gene expression sample taken from a new individual whose medical condition is unknown; secondly, the learnt pattern of gene expression which distinguishes between the two medical conditions can be used to advance our understanding of the genetic causes and relations to the disease under consideration. Cancers, owing to their largely genetic dependency and our incomplete knowledge about them, are widely investigated through supervised classification analysis.

2.2.2   Unsupervised Learning

In contrast to supervised learning, the data investigated by unsupervised learning methods are unlabelled. A dominating class of unsupervised learning methods is *unsupervised clustering*, which, when applied in bioinformatics, represents the focus of this book. In clustering, a set of observed objects with measured features is partitioned into a number of clusters of objects such that those objects which are included in the same cluster are similar to each other while being dissimilar to the objects included in the other clusters based on a predefined similarity/ dissimilarity criterion (Abu-Jamous *et al.*, 2013). One of the challenges that appear in unsupervised learning but not in supervised learning is the identification of the correct number of classes/clusters. This might be estimated based on *a priori* field-specific knowledge, or might need to be learnt as part of the learning process. Owing to the massive amounts of unknowns in molecular biology, the unsupervised nature of clustering, that is its lack of requirements of *a priori* known labels, makes it widely applied in bioinformatics, especially in gene clustering (Eisen, 1999; Salem, Jack and Nandi, 2008; Abu-Jamous *et al.*, 2013; Fa *et al.*, 2014).

Numerous families of clustering methods exist in the literature, and they are thoroughly explained in Part Four of this book. These families of methods include partitional clustering (e.g. *k*-means (Pena, Lozano and Larranaga, 1999)), neural network-based clustering (e.g. self-organising maps (SOMs) (Xiao *et al.*, 2003) and self-organising oscillator networks (SOON) (Salem, Jack and Nandi, 2008)), mixture model clustering (e.g. finite mixture model (Bailey and Elkan, 1994)), hierarchical clustering (divisive and agglomerative) (Eisen, 1999), fuzzy clustering (e.g. fuzzy c-means) (Dimitriadou, Weingessel and Hornik, 2002), consensus clustering (Avogadri and Valentini, 2008; Vega-Pons and Ruiz-Shulcloper, 2011; Abu-Jamous *et al.*, 2013), graph clustering (e.g. spectral clustering algorithm (Ng *et al.*, 2001)), biclustering (Cheng and Church, 2000), and others.

2.3   Optimisation

Optimisation is the process of finding the best object or solution in a given set of objects or possible solutions based on a predefined criterion (Nemhauser, 1989). The criterion is normally formulated as a *fitness function* of the numerically represented features of the objects. A fitness function must be strictly monotonically increasing or decreasing with the quality of the objects as serves the problem under investigation. Therefore, the problem of optimisation can be

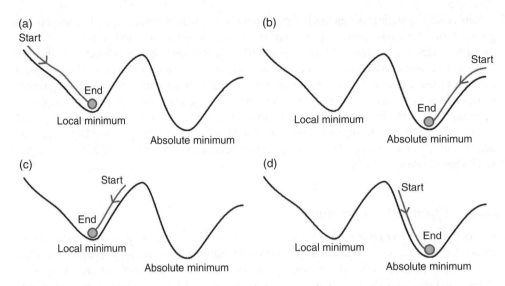

Figure 2.1 Hill climbing optimisation is highly dependent on the starting position. Absolute (global) optimum was found in (b) and (d) while the algorithm converged to local optima in (a) and (c)

formulated as the problem of finding the object at which the fitness function has its absolute maximum or absolute minimum value as it strictly monotonically increases or decreases with the quality, respectively. Such a point is known as the *global optimum* or the *absolute optimum*.

A major problem of optimisation is the problem of being stuck in *local optima* (plural of optimum). A local optimum is a point in the fitness function at which the function's value is better than its value at all of the surrounding points in the feature space.

Hill climbing is an intuitive and fast algorithm for optimisation, but is well known for having the problem of convergence to local optima. Figure 2.1 illustrates the process of minimisation by hill climbing while considering four different starting points. The starting point is an initial solution that is picked randomly or based on *a priori* knowledge. Then, the points locally surrounding that point are examined by the fitness function. The neighbouring point which has the minimum value is chosen as the next point. Similar tests are performed over the neighbours of that point, and this step is repeated iteratively. When the current point has the minimum fitness compared with all of its direct neighbours, the algorithm terminates and this final point is considered the final solution. The path of the points selected by the algorithm is shown in the four examples in Figure 2.1, and the final solution at which it converges is shown as a solid circle. It can be clearly seen in the figure that the starting point greatly influences the path of optimisation by hill climbing as well as the point of convergence.

Brute force optimisation, also known as *exhaustive search*, is theoretically guaranteed to arrive at the global (absolute) optimum solution, but it is completely infeasible in most real problems. In brute force, the fitness values of absolutely all of the possible points in the search space are calculated, and then the best of them is simply selected. The common infeasibility of this method is due to the size of the search space in most real problems. For example, the problem of identifying the best *permutation* of N objects has a search space of $(N!)$, that is the number of possible permutations of those N objects is the factorial of N. If there is a very simple fitness function which can be evaluated in 1 ns (10^{-9} s) by a modern processor, the amount of time needed to evaluate the fitness function for all of the possible solutions is ($10^{-9} \times N!$) seconds. Therefore, CPU time will be 0.12 μs (10^{-6}) for five objects ($N = 5$), 3.6 ms (10^{-3}) for ten objects ($N = 10$), 22 min for fifteen objects ($N = 15$), 77 years for twenty objects ($N = 20$), and 500 million years for twenty-five objects ($N = 25$). Practical problems would include tens to tens of thousands of objects, and it is very clear that brute force solutions are infeasible in such cases.

More sophisticated optimisation methods exist such as Newton's method (Nocedal and Wright, 2006), quadratic programming (Nocedal and Wright, 2006), numerical analysis, gradient descent (Snyman, 2005), and heuristic methods such as simulated annealing (Kirkpatrick, Gelatt and Vecchi, 1983), tabu search (Glover, 1989; Glover, 1990), particle swarm optimisation (Xiao et al., 2003), and genetic algorithm (GA) (Mitchell, 1998). Furthermore, many of these methods have many variants and hybrid approaches.

Optimisation techniques are widely adopted in machine learning methods applied in bioinformatics and in other fields. This is by structuring the machine learning problem or part of it in the form of an optimisation problem. Additionally, optimisation is recruited to solve many other problems in bioinformatics such as determining proteins' three-dimensional secondary and ternary structures, template matching in DNA and RNA sequences, regression-based techniques such as the locally weighted scatter-plot smoothing (lowess) normalisation technique and others.

2.4   Image Processing: Bioimage Informatics

Various types of biological high-throughput data involve images at one point or another within their pipelines of steps. Examples include microarrays, next generation sequencing (NGS) and fluorescence microscopy. In fluorescence microscopy, the molecules of interest in cells (e.g. specific protein(s)) are tagged with fluorescent labels, which can be detected relatively easily when images for those cells are captured. Image pre-processing and processing techniques are involved to quantify and qualify the pieces of information embedded in those biological images such as the level of abundance of that molecule and its localisation within the cell.

Owing to recent advancements in microscopic optics, fluorescent tagging and robotics, automation of microscopic imaging has become feasible. Consequently, fluorescent microscopy has started to be used to produce tens of thousands to millions of images in an automated fashion at large scales (e.g. genome wide) over many time points, that is high-throughput filming of large-scale biological parameters. Key parameters that can be measured in this high-throughput manner include macromolecule (e.g. protein) diffusion, protein–protein interactions and subcellular localisation and concentration. The analysis of such types of high-throughput biological data is an emerging branch of bioinformatics known as *bioimage informatics* (Pepperkok and Ellenberg, 2006; Peng, 2008; Li et al., 2013).

There are many challenges in bioimage informatics. Storing and organising the many terabytes of data produced by a single high-throughput microscopy experiment is one. Another major challenge is pre-processing the produced images as they are produced in an efficient way. Pre-processing normally includes segmentation and filtering of non-interesting areas. Also, classification of the types of images and sub-image components would be necessary; for example, the types of cells or sub-cellular components within an image can greatly influence downstream analysis. Then, features, preferably quantitative, are extracted from the images to produce numeric matrices that can be exposed to machine learning, data mining and statistical analysis. Pre-processing, classification and quantification are most preferably carried out online while image acquisition and storing are in progress. This is intensively demanding and would most likely require access to advanced computing infrastructures such as computing clusters and massive storage disks. Indeed, developments in image-processing algorithms that can handle such a vast scale of data efficiently are required to progress in this area.

2.5   Network Analysis

Modelling the massive amounts of biological data that are and will be available has taken many forms, which have consequently led to many approaches of analysis. The network form is an increasingly considered form of representing relations between biological objects (e.g. proteins, genes, etc.). As known in network theory, a network consists of nodes (vertices) and edges (links). Therefore, the provided biological objects can be represented as nodes, while the relations that link between those objects can be represented as edges (Képès, 2007).

Many properties characterise different types of networks. For example, the edges may be undirected or directed (e.g. arrows), networks may allow or not allow cycles, that is cyclic or acyclic (sequences of edges which terminate at the same node from which they originate), and networks may allow or not allow self-connected nodes, where a node is connected to itself. Different existing network analysis methods are only applicable to networks with specific types of edges and nodes; thus, bioinformatic network analysis requires proper network modelling and sufficient understanding of its properties.

Examples of biological networks include protein–protein interaction networks, genetic interaction networks and gene regulatory networks. Protein–protein interactions are physical interactions between protein molecules; consequently, protein–protein interaction networks model proteins as nodes, and physical interactions as undirected edges (Chen et al., 2014). Genetic interaction networks represent genes as nodes and interactions as edges; a genetic interaction is considered to exist between two genes if the effect of perturbing both genes together is not equivalent to the added effects of perturbing each gene individually, and therefore there are no self-connected nodes in genetic interaction networks (Tong et al., 2004; Costanzo et al., 2010). Gene regulatory networks model the regulatory roles of the products of some genes over other genes; a gene is said to regulate another gene if its product activates or represses the expression of that target gene; therefore, the network edges which represent such regulatory roles are not only directed, but also of one of two different types, activation or repression (Maetschke et al., 2014). Moreover, a gene may regulate its own expression by positive or negative feedback loops, which allows for self-connected nodes to be formed in such networks. Other types of biological networks which are also commonly considered include signalling networks, metabolic networks, hybrid networks with more than one type of nodes and/or edges, and others.

Biological network analysis can enhance our understanding of the interactive roles of genes, proteins, metabolites and others. For that to happen, more advancements in the computational methods that are capable of analysing biological networks are required (Horton, 2014). Indeed, such methods need to suit the different sets of properties that characterise their target networks of investigation.

2.6   Statistical Analysis

Principles and techniques from statistics have been heavily used in bioinformatics either directly or by constituting influential parts of other classes of methods such as machine learning and optimisation. Moreover, statistical measures, like p-values, E-values, radii of confidence, false-discovery rates (FDR) and others have been used to quantify the significance and the reliability of the results produced by various bioinformatic analysis experiments.

At the high-throughput data generation step, it is common to produce many replicates for the same condition in order to increase the statistical reliability of the measurements. The arithmetic mean, the geometric mean or the median can be taken for those measurements as a representative value (Churchill, 2002; Quackenbush, 2002; Wu and Irizarry, 2007; Wu, 2009). After data generation, data correction and normalisation techniques consider statistical techniques to format the data in a way which is suitable for downstream analysis. For example, quantile normalisation of microarray datasets is based on the assumption that all microarray samples produced in a single experiment at a genome-wide scale should have the same statistical distribution, and consequently they are all statistically manipulated to meet this pre-justified assumption before downstream comparisons and investigations take place (Bolstad *et al.*, 2003).

As datasets are normalised, they undergo thorough analysis, which is usually carried out by employing machine learning, data mining, optimisation and other classes of sophisticated methods. Although mere statistical analysis is not common at this step, it is most likely that statistical techniques are involved in critical parts of such sophisticated methods. Supervised classification methods like Fisher's discriminant and Bayes' classifier, expectation-maximisation, unsupervised clustering methods, many image processing filters such as the Gaussian filter, and many others are examples of methods that are mainly based on statistical principles.

Most of the numerical, in contrast to biological, techniques that are recruited to validate and evaluate the generated datasets as well as the results of bioinformatic analysis by supervised classification, unsupervised clustering, optimisation or other methods, are mainly statistical. This includes analysis of variance (ANOVA), analysis of covariance (ANCOVA), classification cross-validation accuracy rates, clustering validation indices, mean-square error (MSE) for optimisation, p-values for non-randomness-significance of observations and results, FDR for the level of positive reliability on the findings, and many others. Therefore, it is important for bioinformatricians to have sufficient understanding of statistics in order to enhance the capabilities of computational methods in high-throughput biological data analysis.

2.7   Software Tools and Technologies

Algorithms, of any class, are complete descriptions of the steps that lead to their desired targets. Though, such descriptions need to be *implemented* and then *executed* in order to actualise the algorithms. Implementation is generally carried out by *coding* the described steps using one of the existing programming languages. Commonly used programming languages and programming environments include MATLAB, *R*, C, C++, Python, Java, C# and Perl. MATLAB and *R* languages are fourth-generation programming languages which represent sophisticated platforms for data analysis, especially when matrices, data visualisation and statistical analysis are heavily involved. C, C++, Python, Java, C# and Perl are all high-level general-purpose third-generation programming languages. Although they are not specialised in data analysis like MATLAB and *R*, they have been widely used in bioinformatics and various bioinformatic packages, and libraries have been produced for them (e.g. the Bioinformatics toolbox in MATLAB and the Bioconductor packages in *R*).

In addition to coding the steps of algorithms, other IT technologies have been used in bioinformatics. For example, database technologies (e.g. SQL, Oracle, MySQL, etc.), spreadsheet applications (e.g. Microsoft Excel and Open Office Calc), MATLAB "MAT" files, markup languages (e.g. XML and HTML), and text files, are amongst the technologies used for data storage, structuring and organisation.

Although different bioinformaticians tend to have different preferences in terms of the technologies they rely upon for different tasks, sufficient familiarity with more technologies facilitates a wider range of tools, and consequently a wider range of possible applications. Moreover, structure standardisation of data and results and the employment of software patterns with loosely-coupled modules ease communication between tools and software packages implemented using different languages. One way for designing loosely-coupled software modules is by allowing them to read their input data and parameters from well-defined and structured text or XML files, and to write their output results to similarly well-defined and structured files. In this case, a module from one language or tool may produce results' files that can be read and analysed by a module from another language or tool. This is important for the reusability of the various efficiently implemented methods and algorithms that have been publically provided for the research community.

2.8   Discussion and Summary

The diversity in the types of high-throughput biological datasets and the diversity in the research questions tackled by bioinformatics have led to corresponding diversity in the classes of computational methods that have been recruited in this growing field of research. For instance, supervised machine learning methods learn the patterns governing a representative set of labelled data objects and generalise that to new unseen data objects. In contrast, unsupervised machine learning methods learn the patterns that are hidden in a set of unlabelled data objects. Data mining, which may recruit machine learning methods, aims at extracting findings and discoveries from large sets of data. Optimisation methods search a massive space of possible solutions to a research question regarding some high-throughput biological data in order to identify the globally optimum solution; in most of the cases, brute force search, that is evaluating all solutions in the space, is infeasible, which necessitates smart heuristic search techniques instead.

Many types of biological high-throughput datasets are generated in the form of massive amounts of images or can be structured in the form of networks. Consequently, methods for image processing and network analysis have been employed in bioinformatics, with the former rising into an emerging sub-area in bioinformatics known as bioimage informatics.

Another class of techniques which is involved in almost all of the aforementioned classes of methods is statistics. Statistical analysis is used in order to quantify the significance, the reliability and the validity of the generated raw datasets as well as the results produced throughout data analysis. A good understanding of the principles of statistics is key for the bioinformaticians who design computational methods as well as for the computational biologists who apply those methods, reason their results and infer new findings from them.

The diversity is not only in the classes of methods used in bioinformatics; it is also in software technologies, that is, tools and programming languages that are used in order to implement those methods. Although most of the methods can be implemented by using any of the available methods, efficient implementations of some of those methods have already been publically provided to the research community. Therefore, the flexibility of the researcher in utilising different technologies and the flexibility of the publically provided implementations in being coupled with modules implemented using other tools and languages enhance the reusability of the growing literature of implemented bioinformatic methods.

This chapter presents an in-breadth review of the computational methods used in bioinformatics. This contextualises clustering analysis, which is the main focus of the rest of this book.

References

Abu-Jamous, B., Fa, R., Roberts, D.J. and Nandi, A.K. (2013). Paradigm of tunable clustering using binarization of consensus partition matrices (Bi-CoPaM) for gene discovery. *Plos One*, **8**(2), e56432.

Altman, N.S. (1992). An introduction to kernel and nearest-neighbour nonparametric regression. *The American Statistician*, **46**(3), pp. 175–185.

Avogadri, R. and Valentini, G. (2008). Ensemble clustering with a fuzzy approach, in *Supervised and Unsupervised Ensemble Methods and their Applications Studies in Computational Intelligence* (ed O. Okun), Springer-Verlag, Berlin.

Bailey, T.L. and Elkan, C. (1994). *Fitting a Mixture Model by Expectation Maximization to Discover Motifs in Biopolymers*. Proceedings of the Second International Conference on Intelligent Systems for Molecular Biology, Stanford, CA, 14–17 August 1994. AAAI Press, Menlo Park, CA, pp. 28–36.

Bolstad, B., Irizarry, R., Astrand, M. and Speed, T. (2003). A comparison of normalization methods for high density oligonucleotide array data based on variance and bias. *Bioinformatics*, **19**, pp. 185–193.

Breiman, L. (1996). Bagging predictors. *Machine Learning*, **24**(2), pp. 123–140.

Chen, B., Fan, W., Liu, J. and Wu Fang-Xiang (2014). Identifying protein complexes and functional modules - from static PPI networks to dynamic PPI networks. *Briefings in Bioinformatics*, **15**(2), pp. 177–194.

Cheng, Y. and Church, G.M. (2000). Biclustering of expression data. Proceedings of the Eighth International Conference on Intelligent Systems for Molecular Biology, Boston, MA, 16–23 August 2000. AAAI Press, Menlo Park, CA, pp. 93–103.

Churchill, G. (2002). Fundamentals of experimental design for cDNA microarrays. *Nature Genetics*, **32**, pp. 490–495.

Cortes, C. and Vapnik, V. (1995). Support-vector networks. *Machine Learning*, **20**(3), pp. 273–297.

Costanzo, M., Baryshnikova, A., Bellay, J. *et al.* (2010). The genetic landscape of a cell. *Science*, **327**(5964), pp. 425–431.

Devroye, L., Györfi, L. and Lugosi, G. (1997). *A Probabilistic Theory of Pattern Recognition*. 2nd ed. Springer, New York.

Dimitriadou, E., Weingessel, A. and Hornik, K. (2002). A combination scheme for fuzzy clustering. *International Journal of Pattern Recognition and Artificial Intelligence*, **16**, pp. 901–912.

Duda, R.O., Hart, P.E. and Stork, D.G. (2000). *Pattern Classification*. 2nd ed. John Wiley & Sons, Inc, New York.

Eisen, M. (1999). *Cluster and TreeView*, http://rana.lbl.gov/manuals/ClusterTreeView.pdf (accessed 26 September 2014).

Fa, R., Roberts, D.J. and Nandi, A.K. (2014). SMART: Unique Splitting-While-Merging Framework for Gene Clustering. *PLoS ONE*, **9**(4), e94141.

Glover, F. (1989). Tabu Search - Part 1. *ORSA Journal on Computing*, **1**(2), pp. 190–206.

Glover, F. (1990). Tabu Search - Part 2. *ORSA Journal on Computing*, **2**(1), pp. 4–32.

Haykin, S. (1999). *Neural Networks – A Comprehensive Foundation*, 3rd Edition, Pearson, Prentice Hall, Singapore.

Horton, P. (2014). Next-generation bioinformatics: connecting bases to genes, networks and disease. *Briefings in Bioinformatics*, **15**(2), p. 137.

Képès, F., ed. (2007). *Biological Networks*, World Scientific Publishing, Singapore.

Kirkpatrick, S., Gelatt, C.D., Jr and Vecchi, M.P. (1983). Optimization by simulated annealing. *Science*, **220**(4598), pp. 671–680.

Li, F., Yin, Z., Jin, G. *et al.* (2013). Chapter 17: Bioimage informatics for systems pharmacology. *Plos Computational Biology*, **9**(4), e1003043.

Liu, K.-H. and Xu, C.-G. (2009). A genetic programming-based approach to the classification of multiclass microarray datasets. *Bioinformatics*, **25**(3), pp. 331–337.

Maetschke, S.R., Madhamshettiwar, P.B., Davis, M.J. and Ragan, M.A. (2014). Supervised, semi-supervised and unsupervised inference of gene regulatory networks. *Briefings in Bioinformatics*, **15**(2), pp. 195–211.

Mitchell, M. (1998). *An Introduction to Genetic Algorithms*. MIT Press, Cambridge.

Naseem, I., Togneri, R. and Bennamoun, M. (2010). Linear regression for face recognition. *IEEE Transactions on Pattern Analysis and Machine Intelligence*, **32**(11), pp. 2106–2112.

Nemhauser, G.L., Rinnooy Kan, A.H.G. and Todd, M.J. (eds) (1989). *Optimization*. Elsevier, New York.

Ng, A.Y., Jordan, M.I. and Weiss, Y. (2001). On Spectral Clustering: Analysis and an algorithm, in *Advances in Neural Information Processing Systems* (eds T.G. Dietterich, S. Becker and Z. Ghahramani), MIT Press, Cambridge, MA, **14**, pp. 849–856.

Nocedal, J. and Wright, S.J. (2006). *Numerical Optimization*. 2nd ed., Springer, New York.

Pena, J.M., Lozano, J.A. and Larranaga, P. (1999). An empirical comparison of four initialization methods for the K-Means algorithm. *Pattern Recognition Letters*, **20**(10), pp. 1027–1040.

Peng, H. (2008). Bioimage informatics: a new area of engineering biology. *Bioinformatics*, **24**(17), pp. 1827–1836.

Pepperkok, R. and Ellenberg, J. (2006). High-throughput fluorescence microscopy for systems biology. *Nature Reviews*, **7**, pp. 690–696.

Quackenbush, J. (2002). Microarray data normalization and transformation. *Nature Genetics*, **32**, pp. 496–501.

Quinlan, J.R. (1986). Induction of decision trees. *Machine Learning*, **1**(1), pp. 81–106.

Salem, S.A., Jack, L.B. and Nandi, A.K. (2008). Investigation of self-organizing oscillator networks for use in clustering microarray data. *IEEE Transactions on Nanobioscience*, **7**(1), pp. 65–79.

Shawe-Taylor, J. and Cristianini, N. (2004). *Kernel Methods for Pattern Analysis*, Cambridge University Press, Cambridge.

Snyman, J.A. (2005). *Practical Mathematical Optimization: An Introduction to Basic Optimization Theory and Classical and New Gradient-Based Algorithms*, Springer, New York.

Tong, A.H., Lesage, G., Bader, G.D. *et al.* (2004). Global mapping of the yeast genetic interaction network. *Science*, **303** (5659), pp. 808–813.

Vega-Pons, S. and Ruiz-Shulcloper, J. (2011). A survey of clustering ensemble algorithms. *International Journal of Pattern Recognition and Artificial Intelligence*, **25**(3), pp. 337–372.

Wu, Z. (2009). A review of statistical methods for preprocessing oligonucleotide microarrays. *Statistical Methods in Medical Research*, **18**, pp. 533–541.

Wu, Z. and Irizarry, R.A. (2007). A statistical framework for the analysis of microarray probe-level data. *The Annals of Applied Statistics*, **1**, pp. 333–357.

Xiao, X., Dow, E.R., Eberhart, R. et al. (2003). Gene Clustering Using Self-organizing Maps and Particle Swarm Optimization. Proceedings of the International Parallel and Distributed Processing Symposium, Nice, France, 22–26 April 2003. IEEE, pp. 154–163.

Part Two

Introduction to Molecular Biology

3

The Living Cell

3.1 Introduction

The *cell* is the main building block of living creatures. It is a busy compartment of thousands of different types of molecules which work cooperatively to meet certain goals such as cell growth, maintenance and reproduction. Although various aspects of cells have been discovered throughout the history of human research, numerous aspects are either vaguely known or completely unknown. Such poorly understood areas act as subjects for researchers to investigate, while prioritising those aspects which, when understood, lead to more important consequences such as better resistance to serious diseases. As understanding the basics of what is known about cells is crucial for bioinformaticians to have fruitful and successful research, we provide such a basic level of knowledge in this chapter, which might be of a greater use to those whose background is in information engineering and computational sciences.

3.2 Prokaryotes and Eukaryotes

There are two main types of cells, *prokaryotes* and *eukaryotes*. In Greek, *karyon* means "kernel", or "nucleus", *pro* means "before", and *eu* means "truly" (Alberts *et al.*, 2008). Therefore, and in simple terms, eukaryotic cells are those which are "truly nucleated", or those cells which have a *nucleus*, while prokaryotic cells are those that do not (Figure 3.1). The nucleus is a relatively large compartment within the cell that has a nuclear membrane. The nuclear membrane has large pores which allow small molecules to pass through. The nucleus in the eukaryotes encapsulates the *genetic material* and decouples RNA synthesis (*transcription*) from protein synthesis (*translation*); the genetic material, transcription and translation are covered in more detail in the following chapter.

Integrative Cluster Analysis in Bioinformatics, First Edition. Basel Abu-Jamous, Rui Fa and Asoke K. Nandi.
© 2015 John Wiley & Sons, Ltd. Published 2015 by John Wiley & Sons, Ltd.

(a)   (b)

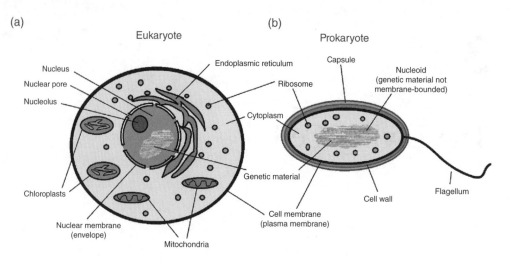

Figure 3.1   Demonstration of (a) the eukaryotic cell and (b) the prokaryotic cell

Prokaryotes are simpler cells which are smaller in size, more basic in functionality and uni-cellular. On the other hand, eukaryotes have larger cells, have more sophisticated functionality and structure and range from unicellular to large multicellular organisms with specialised cells. For example, the prokaryotic cells lack membrane-bound *organelles* which are usually found in the eukaryotic cells such as *mitochondria* and *chloroplasts*. Species belonging to the domains bacteria and archaea are prokaryotes, while protists, fungi, plants and animals are all eukaryotes.

The smallest known free-living eukaryote is the unicellular genus of green alga *Ostreococcus* with an average size of about $0.8\,\mu m^3$ (Courties *et al.*, 1994; Leis *et al.*, 2009). As for prokaryotes, a study by Lancaster and Adams concluded that a size of 172 nm is the smallest hypothetical size for a prokaryote cell with biomedical requirements for growth, metabolism, and reproduction (Lancaster and Adams, 2011). On the other hand, large sulphur bacteria are the largest discovered bacteria with diameters reaching up to 750 μm, representing unicellular prokaryotes that are visible by the naked eye (Salman, Bailey and Teske, 2013).

3.3 Multicellularity

3.3.1 Unicellular and Multicellular Organisms

Organisms can be classified into unicellular and multicellular. As inferred from the name, a unicellular organism is that which consists of a single cell only, while a multicellular organism consists of many cells, which are usually specialised. Although the majority of unicellular organisms are microscopic, reaching volumes of less than $1\,\mu m^3$ (e.g. cyanobacteria), a few of them are macroscopic and can be seen by the naked eye, such as the xenophyophores which are unusually huge unicellular protists that can reach 25 cm in diameter (Kamenskaya, Melnik and Gooday, 2013).

3.3.2   Stem Cells and Cell Differentiation

Multicellular organisms usually have cells with different specialisations. In animals, skin cells, bone cells, blood cells, neurons and retinal cells, are examples of specialised cells that form tissues in a whole multicellular organism. A multicellular organism starts with a single cell known as the *zygote,* which results from the fusion of the male *sperm* with the female *ovum* (Figure 3.2a). This process is known as *fertilisation.* The zygote multiplies by cellular division until an early mass of cells, known as the *blastula,* is formed, which has an outer layer of cells, known as the *blastoderm,* and an inner fluid medium, known as the *blastocoel.* An inner mass of cells is then formed within the blastocoel and is known as the *inner cell mass (ICM)* or the *embryoblast.* Stem cells emerge from the ICM, and then enter a series of divisions to produce the whole organism. While dividing, different stem cells *differentiate* into different types of more specialised cells, which maturate gradually to reach their final mature forms (Figure 3.2a and b).

In order to maintain the population of undifferentiated stem cells, they undergo *mitosis,* which is cell division into two genetically identical daughter cells. This is known as *self-renewal,* and is represented by the loop arrow in Figure 3.2b. They also involve asymmetric division while differentiating such that one of the two daughter cells is identical to the mother cell, that is an undifferentiated stem cell, while the other one is differentiated.

When a stem cell differentiates, it results in a *progenitor cell,* which although it is more specific than the stem cell, it is not a final mature cell and has the potency to differentiate into more specific and mature cell types. Progenitor cells may undergo limited self-renewal divisions before differentiation into mature cells (Figure 3.2b).

Stem cells are classified based on their *potency,* which is their ability to differentiate into many different cell types. A *totipotent* cell is a stem cell that can differentiate into all types of cells in the organism, and the *zygote* is an example of that (Mitalipov and Wolf, 2009; Malaver-Ortega *et al.,* 2012). ICM cells are examples of *pluripotent* stem cells (Figure 3.2a), which are those

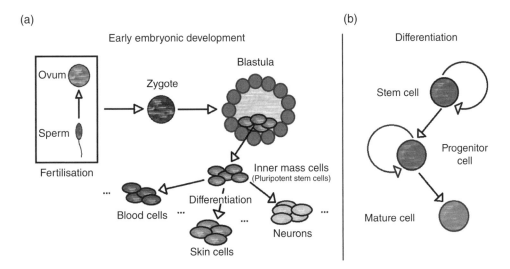

Figure 3.2   (a) Early embryonic development, from fertilisation to differentiation. (b) Stem cell differentiation

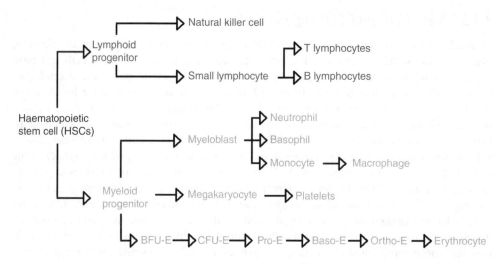

Figure 3.3   Tree of haematopoietic stem, progenitor, precursor and mature cells in mammals

cells that can differentiate into cells of any of the three *germ layers,* namely *endoderm* (interior stomach lining, gastrointestinal tract and the lungs), *mesoderm* (muscles, bones, blood and urogenital), and *ectoderm* (epidermal tissues and the nervous system) (Malaver-Ortega *et al.*, 2012).

After more differentiation, *multipotent* stem/progenitor cells are produced, which, although are capable of differentiating into multiple types of cells, are limited to specific ones (Seaberg and Kooy, 2003). For example, haematopoietic stem cells (HSCs), can differentiate into any type of blood cells, such as red blood cells (RBCs), lymphocytes and platelets, but cannot differentiate into nervous, bone, muscle or germ cells; see Figure 3.3. More differentiated progenitor cells would have less potency as they are pushed to *commit* to a certain branch of cells. If the progenitor cell can only differentiate into a single type of cells, it is known as a *unipotent* cell. Examples of unipotent cells are the burst-forming unit-erythroids (BFU-E) which are committed to the erythroid branch that results in producing erythrocytes (RBCs). These committed unipotent cells which have not reached their final mature form yet are known as *precursor cells* or *blasts.* Figure 3.3 shows the tree of progenitors (precursors) and mature cells that can be derived from HSCs in mammals while providing more details in the erythroid branch and fewer details in other branches.

There is large potential for stem cells in medical therapy as they can be used to produce different types of healthy tissues *in vitro,* that is in the laboratory. The most developed use of stem cells is in bone-marrow transplantation (Gahrton and Björkstrand, 2000; Yen and Sharpe, 2008). They also have therapeutic potentials in many other applications (Singec *et al.*, 2007; Strauer, Schannwell and Brehm, 2009).

3.4   Cell Components

Figure 3.4 shows the main subcellular components of a typical eukaryotic cell from unicellular and multicellular fungi to plants and animals. However, plants additionally have a tough external *cellular wall* formed of cellulose and green components known as *chloroplasts*, which are

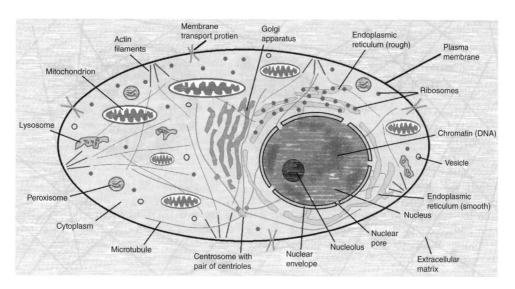

Figure 3.4 Main subcellular components of a typical eukaryotic cell (*See insert for color representation of the figure*)

only shown in Figure 3.1a. The subsequent subsections discuss those components in more detail.

3.4.1 Plasma Membrane and Transport Proteins

The plasma membrane is a selective barrier that separates the internal contents of the cell from the external media. Being selective means that it allows molecule exchange in both directions in a controlled manner, which is a crucial property if the cell is to grow and reproduce as it needs to import raw materials and export waste. This exchange is done by specialised *membrane-transport proteins* which are embedded in the membrane itself. Different transport proteins are specialised for different types of molecules such as sugars, amino acids, peptides, amines, ions and others (Alberts *et al.*, 2008).

The plasma membrane is formed of a *phospholipid bilayer,* which has the interesting property of being *amphiphilic.* An amphiphilic molecule is that which has one *hydrophilic* side (water-soluble) and one *hydrophobic* side (water-insoluble) (Figure 3.5a). When such molecules are dropped in water, the hydrophobic side of each molecule tends to settle at a position with the least possible level of contact with water. This happens by the spontaneous formation of two layers of the molecules such that the hydrophobic sides of one layer face and contact with the hydrophobic sides of the other layer, whereas the hydrophilic sides of both layers face the water (Figure 3.5b). Under the correct conditions, dropping this material in water spontaneously forms vesicles of such a bilayer membrane which separate their content of water from the surrounding medium (Figure 3.5c) (Alberts *et al.*, 2008).

3.4.2 Cytoplasm

The cytoplasm is a dense gel-like fluid contained within the cell membrane and consists of the *cytosol* and the cytoplasmic *organelles* (Alberts *et al.*, 2008). Water comprises most of

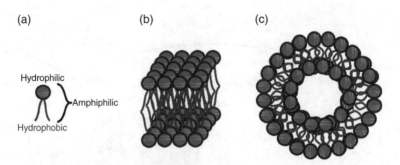

Figure 3.5 Plasma membrane formation of amphiphilic phospholipid bilayers. (a) A single phospholipid molecule with a hydrophilic side and a hydrophobic side. (b) A phospholipid bilayer with the hydrophobic sides facing each other between the two layers of hydrophilic sides; the hydrophilic sides face the surrounding medium, which is a fluid (e.g. water). (c) A vesicle which is formed spontaneously with a phospholipid bilayer plasma membrane when these molecules are dropped in water

the cytoplasm's material, while large amounts of soluble and insoluble macro- and micro-molecules float within this fluid. Many studies have shown that the concentration of molecules within the cytoplasm is high and therefore significantly affects molecular diffusion and the rate of the cellular interactions (Goodsell, 1991; Luby-Phelps, 2000). The *cytosol* is the part of the cytoplasm which is left after excluding the fraction contained within the cellular organelles (Lodish and Matsudaira, 2000).

3.4.3   Extracellular Matrix

The extracellular matrix is the extracellular network of polysaccharides and proteins in a mul-ticellular organism, which differs in nature for different types of species. The extracellular matrix, of any nature, provides the platform on which cells adhere, and provides means for cell communication. In animals, the matrix includes, amongst other things, the basement membrane (aka basal lamina), which separates epithelial sheets and other types of tissues from connective tissues (Alberts *et al.*, 2008). Plants have cell walls of cellulose in their extracellular matrix. The molecules forming the extracellular matrix of a tissue are synthesised and secreted by the cells within that tissue itself.

3.4.4   Centrosome and Microtubules

The centrosome is a central organelle in the cells of animals, and includes a pair of *centrioles*. Microtubules, which are long, hollow, and relatively thick filaments, are attached, from one end, to the centrosomes. In mitosis, which represents the cell division into two genetically iden-tical daughter cells, the centrosome acts as the spindle pole and the microtubules act as spindles. This system represents the main regulator of cell division by increasing its efficiency and fidelity.

3.4.5   Actin Filaments and the Cytoskeleton

The cytoskeleton is a skeleton, or a scaffold, constituted of thread-like proteins within the cytoplasm of the cell, which gives the cell its shape, and supports directed transport of materials. The actin filaments, aka *microfilaments,* are the thinnest of the three types of filaments of the cytoskeleton. Intermediate filaments are thicker than actin filaments, more stable and more strongly bound. The third and thickest type of cytoskeletal filaments is the microtubules, which are, as discussed previously, important for efficient cell division and intracellular transport.

3.4.6   Nucleus

The nucleus is a central membrane-bounded cellular component which occupies about 10% of the volume of the eukaryotic cell. It is bounded by the *nuclear envelope,* which consists of two layers of bilayer lipid membranes that contain many relatively large *nuclear pores* allowing for material exchange with the surrounding cytosol (Alberts *et al.*, 2008).

The main purpose of the nucleus is to include, and to provide integrity for, the *genetic material* formed of *chromatin,* which is the combination of the *deoxyribonucleic acid (DNA)* molecules and the proteins packaging it. The genetic material encodes the required information for building all types of proteins needed by the cell in terms of their molecular structure, time of synthesis, and abundance. In other words, the genetic material in the nucleus controls the cellular biological processes while sensing and responding to the signals from the surrounding medium. The next chapter discusses the structure of the genetic material and programmes in more detail as they will be its core subject.

The most obvious sub-nuclear structure is the *nucleolus* (Figure 3.4). It is an area with high-density aggregation of macromolecules rather than a membrane-bounded component, and it is the factory of the ribosomal ribonucleic acid (rRNA) (Alberts *et al.*, 2008), which will be discussed in the next chapter.

3.4.7   Vesicles

The vesicle is a bubble bounded by a bilayer phospholipid membrane. Vesicles are important for importing, exporting and transporting molecules to, from and within the cell. The molecules being carried by a vesicle for such purposes are referred to as the *cargo.* Figure 3.6a, b, and c summarise three important processes undertaken with the aid of vesicles, namely endocytosis, exocytosis and vesicular transport, respectively.

Endocytosis is the process of importing molecules from the extracellular space to the interior of the cell. It is done by internalising a patch of the plasma membrane which forms a vesicle enclosing the molecules coming from the extracellular space, and then the vesicle is detached from the cell membrane to be released to the cytosol (Figure 3.6a). Exocytosis is the opposite process which exports molecules from the cell to the extracellular space by the fusion of a vesicle with the plasma membrane and opening its membrane from the extracellular side. Therefore the vesicle disappears and its membrane becomes part of the cellular plasma membrane (Figure 3.6b).

Vesicular transport allows for transporting molecules from one cellular compartment to another. This is done by membrane budding from the donor compartment to form a vesicle

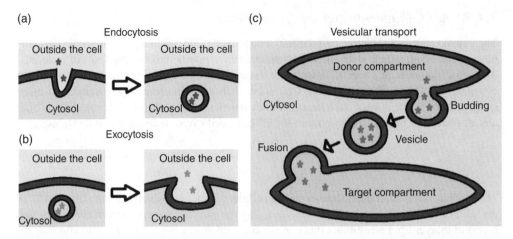

Figure 3.6 (a) Endocytosis: importing contents from the extracellular space. (b) Exocytosis: releasing content to the extracellular space. (c) Vesicular transport: transporting contents from one cellular compartment to another

carrying the molecules, which is released to the cytosol. Then the vesicle, with its cargo, is fused with the membrane of the target compartment to release its cargo to it (Figure 3.6c).

The vesicle formed by endocytosis might be transported itself towards a specific subcellular compartment to release its cargo to it. In this case, this cargo is effectively imported from the extracellular space through endocytosis and then vesicularly transported to a specific target compartment.

3.4.8   Ribosomes

The ribosomes, discovered in the 1950s (Roberts, 1958), are large complexes consisting of both RNAs and proteins. The ribosomes are the protein factories in the cell, as they synthesise proteins by *translating* the instructions carried by the *messenger RNA (mRNA)* molecules from the *genome* residing in the nucleus. More details regarding this process are provided in the next chapter.

Ribosomes can be freely moving within the cytosol or be bound to a membrane, mainly the *endoplasmic reticulum (ER)*. The proteins synthesised by the free ribosomes are secreted directly to the cytosol, while the proteins synthesised by the ribosomes bound to the ER are inserted to ER itself and then may be transported to be used at another location or to be secreted outside the cell.

3.4.9   Endoplasmic Reticulum

The ER is found in all eukaryotic cells, and it is a membrane which extends continuously from the nuclear membrane to form a network of branching tubules and flattened sacs. The area of the ER constitutes more than half of the total membrane area of the cell, and the single common internal space enclosed by the ER and the nuclear membrane occupies about 10% of the volume of the cell (Alberts *et al.*, 2008).

The functions of the ER in the cell are diverse. One notable function is related to the process of manufacturing proteins, that is *translation,* which will be discussed in detail in the next chapter. While synthesising many types of proteins, the protein-manufacturing machines, the *ribosomes,* bind to the ER. These bound ribosomes give the ER a rough appearance, and therefore it is known as the *rough ER.* In contrast, the *smooth ER* is that which has no ribosomes bound to it. Note that the ribosomes do not permanently bind to the rough ER, they are rather bound and released continuously and dynamically.

When a ribosome finishes synthesising a protein and is released from the ER, one end of the nascent protein is kept bound to the ER to be further processed. An important function of the ER thereafter is to translocate this protein to its intended destination. This is done either by direct contact of the destination organelle, for example, the mitochondria, with the ER, or via transport *vesicles* which actively take the protein from the ER to its destination. For example, many newly synthesised proteins and lipids are carried in this way from the ER to the neighbouring *Golgi apparatus,* although their membranes are not physically connected.

The smooth ER has various functions which vary when different types of cells are considered. Synthesising the lipid components of the *lipoprotein* particles in the main liver cell type, *hepatocyte,* depends on the enzymes located in the membrane of the smooth ER. These lipoprotein particles are responsible for carrying lipids in the bloodstream to their destinations. This justifies the fact that the smooth ER is significantly more abundant in this specialised type of cells compared with other types. The smooth ER has many other tasks regarding Ca^{2+} storage, lipid synthesis, transmembrane and water-soluble proteins synthesis, specialised functions to the muscular cells and others.

3.4.10   Golgi Apparatus

The *Golgi apparatus,* named after the Italian physician Camillo Golgi, is a complex stack of flattened membrane-bounded spaces located on the exit route from the ER. The Golgi serves as a buffer which stores, modifies and dispatches the products released from the ER. It is also the main cellular site for carbohydrate synthesis, including *pectin* and *hemicellulose* of the cell wall in plants, and *glycosaminoglycan* of the extracellular matrix in animals.

3.4.11   Mitochondrion and the Energy of the Cell

Mitochondria are the power plants of the cells as they produce most of the energy carrier molecules *adenosine triphosphate (ATP).* ATP is the molecular currency of energy in the cell; it is produced by converting the most basic sugar, glucose, into other molecules with lower *free energy,* where the released energy is stored in the chemical bonds of the ATP molecule. ATP molecules are then consumed by the processes that require this energy.

Anaerobic *glycolysis* is a process that converts one molecule of glucose to the lower-energy molecule of pyruvate while storing the released free energy in two molecules of ATP. This process is carried out in the cytosol and can be found in all prokaryotic and eukaryotic species. Despite that, the pyruvate molecule still has a significant amount of free energy to be released. In eukaryotes, the resulting pyruvate is transported to the mitochondria to be further degraded aerobically, that is, with the existence of oxygen molecules O_2, to produce the basic molecules of carbon dioxide CO_2, water vapour H_2O and released energy. This process of pyruvate oxidation is known as the *citric acid cycle*, or the *Krebs cycle* in reference to Hans Adolf Krebs, winner of the 1953 Nobel Prize of Physiology, who finally identified this cycle in 1937 while at the University of Sheffield.

Glycolysis followed by the citric acid cycle is known as *aerobic respiration*. The theoretical upper limit of ATP production per glucose molecule through aerobic respiration is 38, but owing to many mechanistic and structural complexities, the realistic yield was found to be about 30, which is 15 times of what is produced by anaerobic respiration through glycolysis only (Rich, 2003). Aerobic respiration also occurs in prokaryotes but within the cytosol as they lack mitochondria.

Mitochondria play an important role in efficient glucose oxidation in eukaryotes. A single eukaryotic cell typically includes many mitochondria, which in liver cells for example would reach as many as 1000-2000 (Alberts *et al.*, 2008). While mitochondria normally move within the cytoplasm of the cell, some mitochondria might be fixed or highly concentrated in specific areas within the cell where energy is unusually highly required. Examples are the areas of high ATP consumption between the myofibrils in cardiac muscle cells and around the flagellum in a sperm (Alberts *et al.*, 2008).

Other roles of mitochondria include heat production (Mozo *et al.*, 2005), calcium ions' storage (Miller, 1998), apoptosis (programmed cell death) (Green, 1998), regulation of metabolism (McBride, Neuspiel and Wasiak, 2006), haem synthesis for haemoglobin production in RBCs (Merryweather-Clarke *et al.*, 2011), and others.

In terms of structure, the mitochondrion has two membranes, an inner membrane and an outer membrane. The space between the two membranes is known as the *intermembrane space*, and the space within the inner membrane is known as the *matrix*. Different interactions within the processes carried out in the mitochondria occur in different parts of it. For example, most of the citric acid cycle occurs inside the matrix. Indeed, the transportation of substrates and products across both membranes is an essential part of any of those processes, and, in many cases, partial understanding is what we have about the detailed mechanisms of such transportation (Schultz *et al.*, 2010).

3.4.12   Lysosome

Lysosomes are the membrane-bounded organelles in which intracellular digestion of macromolecules occurs. For most efficient digestion, they maintain an acidic interior with pH of about 4.5–5.0. About 40 different types of digestive enzymes are included within lysosomes. By being encapsulated within the membrane of the lysosome, the cytosol of the cell is protected from the digestive damage that those enzymes would do. Another layer of protection is the fact that, even if those enzymes leak from the lysosome to the cytosol, they will not cause much damage at the cytosolic pH of about 7.2 (Alberts *et al.*, 2008).

The sources of the macromolecules digested by the lysosomes can be proteins, lipids, an excess of and worn-out organelles, viruses and bacteria (Castro-Obregon, 2010). It is interesting that when a eukaryotic cell faces starvation, it would digest its own components through a process called *autophagy* where the lysosomes recycle such components and provide the cell with their basic building blocks (Castro-Obregon, 2010).

3.4.13   Peroxisome

Peroxisomes are membrane-bounded organelles which perform *oxidation* of various types of molecules. Oxidation in this context is the removal of hydrogen atoms from the substrate molecules. This occurs while oxygen O_2 is present, and produces *hydrogen peroxide (H_2O_2)* by binding the hydrogen released from the substrate molecule to the presenting oxygen. The name of this organelle was derived from the name of that molecule, H_2O_2 (Alberts *et al.*, 2008).

By such oxidation, the peroxisomes degrade very long chains of fatty acids, and detoxify many toxic materials such as the alcohol ethanol. Therefore it plays an important role in the liver to purify the materials entering the bloodstream (Alberts *et al.*, 2008).

3.5   Discussion and Summary

Cells are the building blocks of organisms. They are either eukaryotic, with membrane-bound nuclei and subcellular organelles, like protists, fungi, plants and animals, or prokaryotic, with non-membrane-bound nucleoid regions and no subcellular organelles, like bacteria and archaea. Organisms are either unicellular, where the whole organism is constituted of a single cell, like bacteria, archaea and some fungi, or multicellular, where the organism is formed of many cells of various types and roles, like plants, animals and some fungi.

Multicellular organisms start with a single cell, the zygote, which divides to produce multiple cells. Those early-life cells are unspecialised cells which can differentiate into various types of more specialised cells. By series of controlled division and differentiation, ultimately a complete adult organism is formed with all required types of specialised cells.

The cell is surrounded by a cellular membrane which protects the interior of the cell and allows for controlled exchange of materials with the extracellular space by the membrane-transport proteins embedded in it. Outside the cell there is the extracellular matrix, which is a scaffold on which the cells of a multicellular organism adhere, and inside the cell there is the fluid cytosol and the solid filaments of the cytoskeleton. Those filaments give the cell its shape and strength, and play roles in component transportation.

The complete set of information needed for growth, maintenance and duplication of a cell is stored in the genetic material, which is protected by the nuclear double-membrane in eukaryotic cells. Information from and to the nucleus is transmitted through molecules transported via the nuclear pores that are scattered over the nuclear membrane. The cell includes many other types of subcellular components. For instance, vesicles are involved in importing, exporting and transporting molecules to, from and within the cell. Ribosomes with the assistance of the rough ER produce proteins, the smooth ER produces lipids, Golgi apparatus produces carbohydrates and mitochondria produce energy. Lysosomes are the cellular locations for digestion and substrate recycling, while peroxisome degrades very long chains of fatty acids via oxidation. Moreover, the ER and the Golgi apparatus play roles in storing and transporting various types of materials.

Not all cells have all of those types of components. For instance, prokaryotic cells lack membrane-bound organelles like the nucleus, the ER, the Golgi apparatus, mitochondria and lysosomes. Additionally, plant cells and many prokaryotic cells have a cell wall, which is a tough wall of a polysaccharide like cellulose. Plant cells also include a membrane-bound organelle known as the chloroplast, which intakes carbon dioxide, water and sunlight energy to produce oxygen and sugars. By this process, plants synthesise the sugars needed for energy production.

Understanding the physiology of the cell is an important aspect, but a similarly important aspect for bioinformaticians is to understand the genetic programmes which control the cellular processes, and this will be the topic of the next chapter.

References

Alberts, B., Johnson, A., Lewis, J. *et al.* (2008). *Molecular Biology of the Cell*, 5th ed., Garland Science, New York.

Castro-Obregon, S. (2010). The discovery of lysosomes and autophagy. *Nature Education*, **3**(9), p. 49.

Courties, C., Vaquer, A., Troussellier, M. *et al.* (1994). Smallest eukaryotic organism. *Nature*, **370**, p. 255.

Gahrton, G. and Björkstrand, B. (2000). Progress in haematopoietic stem cell transplantation for multiple myeloma. *Journal of Internal Medicine*, **248**(3), pp. 185–201.

Goodsell, D.S. (1991). Inside a living cell. *Trends in Biochemical Sciences*, **16**(6), pp. 203–206.

Green, D.R. (1998). Apoptotic pathways: the roads to ruin. *Cell*, **94**(6), pp. 695–698.

Kamenskaya, O.E., Melnik, V.F. and Gooday, A.J. (2013). Giant protists (xenophyophores and komokiaceans) from the Clarion-Clipperton ferromanganese nodule field (eastern Pacific). *Biology Bulletin Reviews*, **3**(5), pp. 388–398.

Lancaster, W.A. and Adams, M.W.W. (2011) The influence of environment and metabolic capacity on the size of a microorganism, in *The Minimal Cell: The Biophysics of Cell Compartment and the Origin of Cell Functionality* (eds P.L. Luisi and P. Stano), Springer, the Netherlands, pp. 93–103.

Leis, A., Rockel, B., Andrees, L. and Baumeister, W. (2009). Visualizing cells at the nanoscale. *Trends in Biochemical Sciences*, **34**(2), pp. 60–70.

Lodish, H. and Matsudaira, P. (2000). *Molecular Cell Biology*, 4th ed., Freeman, New York.

Luby-Phelps, K. (2000). Cytoarchitecture and physical properties of cytoplasm: volume, viscosity, diffusion, intracellular surface area. *International Review of Cytology*, **192**, pp. 189–221.

Malaver-Ortega, L.F., Sumer, H., Liu, J. and Verma, P.J. (2012). The state of the art for pluripotent stem cells derivation in domestic ungulates. *Theriogenology*, **78**(8), pp. 1749–1762.

McBride, H.M., Neuspiel, M. and Wasiak, S. (2006). Mitochondria: more than just a powerhouse. *Current Biology*, **16**(14), pp. R551–R560.

Merryweather-Clarke, A.T., Atzberger, A., Soneji, S. *et al.* (2011). Global gene expression analysis of human erythroid progenitors. *Blood*, **117**(13), e96–e108.

Miller, R.J. (1998). Mitochondria – the kraken wakes!. *Trends in Neurosciences*, **21**(3), pp. 95–97.

Mitalipov, S. and Wolf, D. (2009) Totipotency, pluripotency and nuclear reprogramming. In: U. Martin, ed. *Advances in Biochemical Engineering/Biotechnology: Engineering of Stem Cells*. Springer, Heidelberg, pp. 185–199.

Mozo, J., Emre, Y., Bouillaud, F. *et al.* (2005). Thermoregulation: what role for UCPs in mammals and birds? *Bioscience Reports*, **25**(3–4), pp. 227–249.

Rich, P. (2003). The molecular machinery of Keilin's respiratory chain. *Biochemical Society Transactions*, **31**(6), pp. 1095–1105.

Roberts, R.B. (1958) Introduction, In: R.B. Roberts, ed. *Microsomal Particles and Protein Synthesis*. Pergamon Press, New York, pp. vii–viii.

Salman, V., Bailey, J.V. and Teske, A. (2013). Phylogenetic and morphologic complexity of giant sulphur bacteria. *Antonie van Leeuwenhoek, International Journal of General and Molecular Microbiology*, **104**(2), pp. 169–186.

Schultz, I.J., Chen, C., Paw, B.H. and Hamza, I. (2010). Iron and porphyrin trafficking in heme biogenesis. *Journal of Biological Chemistry*, **285**(35), pp. 26753–26759.

Seaberg, R.M. and Kooy, D.v.d. (2003). Stem and progenitor cells: the premature desertion of rigorous definitions. *Trends in Neurosciences*, **26**(3), pp. 125–131.

Singec, I., Jandial, R., Crain, A. *et al.* (2007). The leading edge of stem cell therapeutics. *Annual Review of Medicine*, **58**, pp. 313–328.

Strauer, B.E., Schannwell, C.M. and Brehm, M. (2009). Therapeutic potentials of stem cells in cardiac diseases. *Minerva Cardioangiologica*, **57**(2), pp. 249–267.

Yen, A.H.-H. and Sharpe, P.T. (20080. Clinical applications of blood-derived and marrow-derived stem cells for non malignant diseases. *Cell and Tissue Research*, **331**(1), pp. 359–372.

4

Central Dogma of Molecular Biology

4.1 Introduction

Cells are autonomous self-replicating systems which are controlled by extremely sophisticated and complex genetic programmes. The *central dogma of molecular biology* is the explanation of the way in which the information of such programmes is encoded, decoded, maintained, copied, and transmitted within the cell. This chapter discusses those different aspects of the central dogma due to its direct relevance to most of the commonly generated high-throughput datasets, which are, in their turn, the main subjects of analysis in bioinformatics. Microarray datasets, next-generation sequencing (NGS) datasets, proteomic datasets, and chromatin immunoprecipitation-on-chip (ChIP-on-chip) datasets are examples of such datasets. After understanding the central dogma of molecular biology explained in this chapter, the reader will have the required base to move smoothly to the next chapter which explains the aforementioned types of datasets.

4.2 Central Dogma of Molecular Biology Overview

Figure 4.1 shows an abstract overview of the flow of information in cells as described by the central dogma of molecular biology. The first cell of the organism, the *zygote* (see Chapter 3) includes the complete set of genetic information stored in the form of a deoxyribonucleic acid molecule (DNA). When this cell divides, its entire DNA is replicated such that each resulting cell, whether it was specialised or not, includes a complete copy of that set of genetic information.

Integrative Cluster Analysis in Bioinformatics, First Edition. Basel Abu-Jamous, Rui Fa and Asoke K. Nandi.
© 2015 John Wiley & Sons, Ltd. Published 2015 by John Wiley & Sons, Ltd.

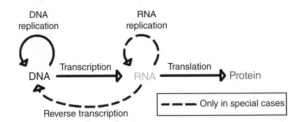

Figure 4.1   Overview of the central dogma of molecular biology: information flow in cells

Pieces of information are copied from the DNA molecule to be encoded in the form of ribonucleic acid (RNA) molecules. This process is known as *transcription* (Figure 4.1). Most of the patches of information copied from the DNA to RNA molecules are in reality codes for protein synthesis, which are synthesised in a process known as *translation*, and then transported to their places of functioning. As proteins are the molecules which perform most of the biological functions in the cell, transcription followed by translation is the mean which transfers information from their storage place in the DNA molecule to their field of application in the cell. Although most of the RNA molecules are *protein coding*, some other types of RNA molecules are involved directly, whilst being in their RNA form, in various molecular functions in the cell.

In special cases, RNA molecules are replicated, that is new RNA molecules are synthesised by directly copying other RNA molecules. Moreover, *reverse transcription*, which is the synthesis of a DNA molecule based on the information stored in an RNA molecule, occurs in special cases, such as in some viruses known as *retroviruses* like HIV.

4.3   Proteins

We will introduce the three main types of macromolecules involved in the central dogma of molecular biology in a different order to that of the natural flow of information shown in Figure 4.1. We will start with proteins, as the main functioning and most dominant molecules of the cell. Then, we will discuss the two nucleic acids, DNA and RNA. Proteins, nucleic acids (DNA and RNA), and carbohydrates are the three main macromolecules constituting all known types of cells. For their relatively less relevance to the central dogma of molecular biology, we shall not cover carbohydrates in this chapter.

Proteins are long linear *polymers* of *amino acids* which are joined with *peptide bonds*. Polymers are large molecules, that is macromolecules, which consist of a large number of repeatedly joined units called *monomers*. In the case of proteins, amino acids joined with peptide bonds represent those monomers (Figure 4.2). Thus, proteins belong to the family of molecules known as *polypeptides*. The protein is that biologically meaningful polypeptide molecule which has a stable conformation in contrast to any arbitrary chain of amino acids.

Twenty different amino acids have been found in the living cells, and are listed in Table 4.1. Each of the 20 amino acids is referred to with its name, one-letter symbol or a three-letter symbol. Moreover, different amino acids are different in terms of their charge and polarity. Five amino acids are charged, and therefore they are water-soluble; three of them are positively charged (basic) and two are negatively charged (acidic). Five amino acids are polar but uncharged, and the ten remaining amino acids are nonpolar.

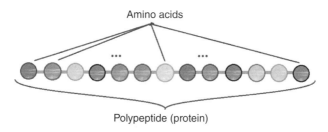

Figure 4.2   The protein is a polymer of repeatedly joined amino acids

Table 4.1   The 20 amino acids

Symbol	Name	Charge and polarity	Symbol	Name	Charge
K (Lys)	Lysine	Basic	A (Ala)	Alanine	Nonpolar
R (Arg)	Arginine	Basic	V (Val)	Valine	Nonpolar
H (His)	Histidine	Basic	L (Leu)	Leucine	Nonpolar
D (Asp)	Aspartic acid	Acidic	I (Ile)	Isoleucine	Nonpolar
E (Glu)	Glutamic acid	Acidic	P (Pro)	Proline	Nonpolar
N (Asn)	Asparagine	Uncharged polar	F (Phe)	Phenylalanine	Nonpolar
Q (Gln)	Glutamine	Uncharged polar	M (Met)	Methionine	Nonpolar
S (Ser)	Serine	Uncharged polar	W (Trp)	Tryptophan	Nonpolar
T (Thr)	Threonine	Uncharged polar	G (Gly)	Glycine	Nonpolar
Y (Tyr)	Tyrosine	Uncharged polar	C (Cys)	Cysteine	Nonpolar

Lengths of proteins vary widely. Neidigh and colleagues were able to design a stable 20-amino acid-length protein-like polypeptide (Neidigh, Fesinmeyer and Andersen, 2002). On the other hand, the largest known protein, with more than 38 000 amino acids, is the giant protein titin, which functions as a molecular spring contributing to the elasticity of muscles in humans (Bang *et al.*, 2001; Opitz *et al.*, 2003). The average length of proteins in the eukaryote *Saccharomyces cerevisiae*, that is baker's yeast, is about 400–450 amino acids (Harrison *et al.*, 2003; Brocchieri and Karlin, 2005). In contrast, bacteria and archaea have average protein lengths of about 250-300 amino acids (Brocchieri and Karlin, 2005).

As for the number of different proteins in species, it is between 20 000 and 25 000 in humans (Collins *et al.*, 2004), ~6300 in budding yeast, ~26 000 in the plant *Arabidopsis thaliana* (thale cress), ~4300 in the *Escherichia coli* bacteria, and less than 500 in the *Mycoplasma genitalium* bacteria (Alberts *et al.*, 2008). These numbers show the large variation between species in terms of the number of proteins as well as their average length.

Although proteins are linear chains of amino acids, they take specific stable 3-D structures when put in water. This is due to the attractive and repulsive forces between the amino acids constituting the protein, especially the charged and polarised ones (Figure 4.3a). This is also affected by the sizes of the amino acids, and their structural occupancy of the 3-D space. Accordingly, the chains of amino acids would form different types of helices, linear sheets and other structures known as protein *secondary structures* (Figure 4.3b). The complete protein might contain many of these secondary structures formed at different patches of its sequence, which in total form the complete 3-D structure of that protein (Figure 4.3b).

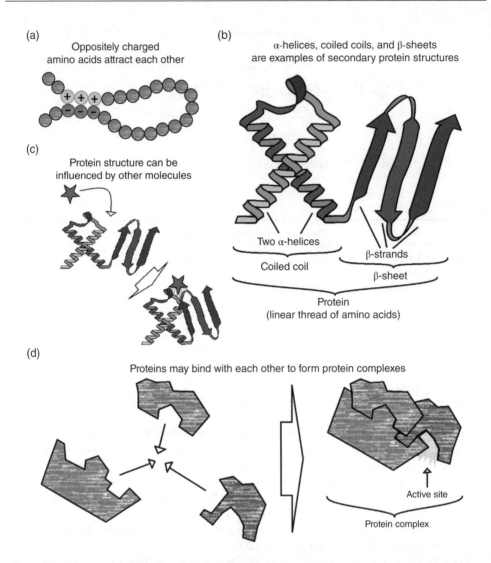

Figure 4.3   Protein structure. (a) Oppositely charged (or polarised) amino acids attract each other, which greatly determines the 3-D structure of the protein. (b) Many secondary structures have been found repeatedly in many proteins, such as helices, coils and sheets. (c) Presence of other molecules, which might be other proteins, affects the 3-D structure of proteins. (d) Many proteins can bind with each other to form protein complexes which can have new chemical and physical properties different from those of their individual components (*See insert for color representation of the figure*)

Moreover, the structure of the protein can be affected by the presence of other molecules whether they were proteins or other types of molecules (Figure 4.3c). These changes in structure might lead to changes in the biochemical properties of the protein, which might be necessary to switch the protein into its active or inactive mode. Furthermore, and as demonstrated in Figure 4.3d, many protein molecules may bind with each other to form a *protein complex*, which gains new physical and chemical properties that allow it to achieve unique tasks which could not be achieved by the individual protein molecules.

4.4 DNA

The DNA molecule that is a polymer of *deoxyribonucleotide units* (or simply *nucleotide units*) encodes and stores the genetic programmes that control the various aspects of the cell. The DNA encodes such information in the form of a linear series of symbols from a four-letter language, namely G, A, C and T, which refer to the nitrogen-containing molecules guanine (G), adenine (A), cytosine (C) and thymine (T), respectively (Figure 4.4b). Any of these four molecules, known as *nucleobases*, or simply, *bases*, can be bound to the five-carbon sugar *deoxyribose* to form a *nucleoside* (Figure 4.5a and Figure 4.4a, c). When a phosphate group

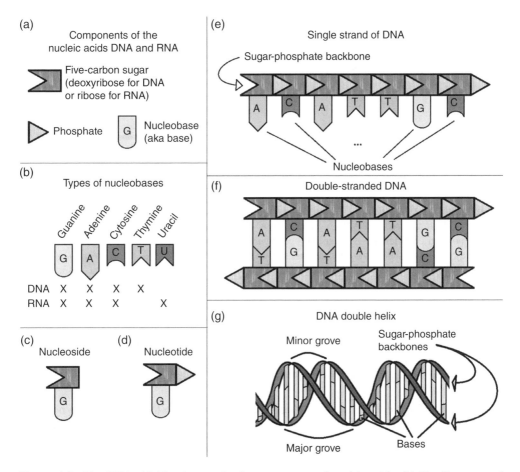

Figure 4.4 The DNA. (a) The three molecular components of nucleic acids. (b) The five types of nucleobases. (c) Nucleoside. (d) Nucleotide. (e) Single strand of DNA. (f) Double-stranded DNA. (g) DNA double helix

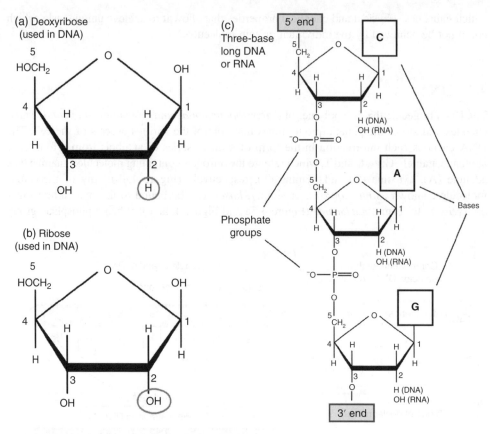

Figure 4.5 The two types of five-carbon sugar used in nucleic acids, namely (a) deoxyribose and (b) ribose which are used in the DNA and the RNA, respectively. The five carbon atoms are numbered and the position of difference between the two sugar rings is highlighted with circles. (c) A three-base sequence of DNA or RNA showing the bases bound to carbon (1) and the phosphate groups joining carbon (5) from one sugar with carbon (3) from the next sugar. The two terminals of the sequences are consequently referred to as the 5′ and the 3′ ends, which are read as five-prime and three-prime ends, respectively

is added to the nucleoside, it becomes a *nucleotide* (Figure 4.4d), which represents the complete monomer (single unit) of the DNA polymer (Figure 4.4d). Nucleosides are joined with each other by phosphate groups to form a sugar-phosphate backbone, to each unit of which a base is bound (Figure 4.4d, e and Figure 4.5c).

As shown in Figure 4.4e, the sugar-phosphate backbone of the DNA molecule is homogeneous, that is its units are identical. On the other hand, the bases which protrude from that backbone differ from one unit to another, and their sequence represents the encoded genetic information.

Owing to their differences in charge polarisation and size, A and T tend to bind to each other with the weak hydrogen bond. Similarly do G and C. This renders the single strand of

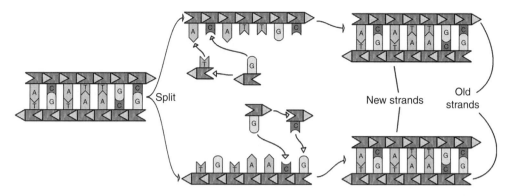

Figure 4.6   DNA replication

DNA shown in Figure 4.4e not in its most stable form. Nucleotides containing matching bases bind to the bases of the single strand by weak hydrogen bonds, and bind with each via their phosphate groups by the strong *phosphodiester bonds*. This process results in the synthesis of a second DNA strand, which perfectly complements the original DNA strand. The two strands are bound with each other by a large number of weak hydrogen bonds, which in total provide the double-stranded molecule high stability (Figure 4.4f). Furthermore, the two bound strands twist to form a double helix, the form at which the DNA molecule is at its highest stability (Figure 4.4g).

The fact that the DNA molecule has two complementary strands serves more important objectives than providing high chemical stability. Since the two strands have a one-to-one matching relationship, either strand can serve as a template to rebuild the other strand, which is exactly how the DNA molecule is faithfully replicated. As shown in Figure 4.6, to replicate the DNA molecule, its two strands are split by breaking the hydrogen bonds between their bases. Then each of the two strands is used as a template which stimulates a new strand to be synthesised to complement it. The result is two new double-stranded DNA molecules identical to the original one. By this accurate replication, all of the cells within a single organism include identical genetic information as they were all produced by successive cell division starting from a single cell, known as the *zygote*. The enzymes which perform the process of polymerising these new strands of DNA that complement the old existing ones, belong to the family of proteins called *DNA polymerases*. Double-stranded DNA molecules also provide means for DNA repair, which is a process that runs actively as parts of the DNA are continuously damaged due to many factors, such as radiations.

4.5   RNA

The RNA molecule is a polymer of *ribonucleotide units*. The structure of the RNA molecule is identical to the single-stranded DNA shown in Figure 4.4e with two core differences. The first difference is that the five-carbon sugar in RNA is the *ribose* in contrast to the *deoxyribose* in the DNA (Figure 4.5). The second difference is that the nucleobase *uracil* (U) is used in RNA instead of the nucleobase T (Figure 4.4b). Similar to T, U tends to form hydrogen bonds with

A. RNA molecules are generated by copying patches of the DNA in the *transcription* process (Figure 4.1 and Section 4.7). The resulting RNA molecule contains the same information that is stored in the copied DNA patch as it adopts a four-letter language similar to the DNA, indeed while replacing T with U.

RNA molecules are single-stranded. Therefore, and due to the aforementioned tendency of certain pairs of nucleobases to form hydrogen bonds, the single strand of RNA folds upon itself in the 3-D space. This is homologous to what has been discussed about protein folding in Section 4.3.

There are many types of RNA molecules in terms of their task in the cell. The most dominant type is the *messenger RNA* (*mRNA*), which serves as a carrier of the genetic information copied from the DNA required to synthesise a protein molecule in a process known as *translation* (Figure 4.1 and Section 4.8).

In contrast, the other types of RNA molecules are functional rather than mere carriers of information as they function in various cellular processes. *Ribosomal RNA* (*rRNA*) molecules, together with many proteins, form the *ribosomes*, which are the protein factories in the cell. In their turn, the ribosomes utilise *transfer RNA* (*tRNA*) molecules to carry out the protein-synthesis process, that is to *translate* the information carried by an mRNA molecule in order to produce a new protein molecule. More about those types of RNA is covered while discussing translation in Section 4.8.

The *small nuclear RNA* (*snRNA*) and the *small nucleolar RNA* (*snoRNA*) carry out *post-transcriptional* processes. These are modifications that are applied to a nascent RNA molecule before it can proceed into its actual biological tasks. *MicroRNA* (*miRNA*), *small interfering RNA* (*siRNA*) and others are involved in *gene regulation*, which will be partially covered in Section 4.7.

4.6 Genes

A *gene* is a fragment of the DNA which is *transcribed*, that is copied, into a single protein-coding or functional RNA molecule. Genes are considered as the molecular units of heredity. In some less common cases, a single gene is transcribed into an intermediate RNA molecule that is spliced post-transcriptionally into more than one protein-coding or functional RNA molecule.

Some genes have *alleles*, which are different versions of the same gene. Different alleles generally result in different traits. For example, the colour of the eye differs from one person to another because they possess different alleles for the genes whose products (coded proteins) are responsible for the colour of the eye. Although this aspect of genes' functions is common amongst non-specialists, it is not the most dominant in reality. Most of the genes encode proteins which are responsible for various commonly essential and favourable biological processes amongst all of the species' individuals such as cell growth, maintenance, reproduction, energy production, waste disposal, signalling and others. In other words, genes are not merely those units which define the traits that distinguish one individual from another.

A gene's sequence includes a *five-prime untranslated region (5′ UTR)*, *three-prime untranslated region (3′ UTR)*, *initiation (start) codon*, *termination (stop) codon*, *introns* and *exons*. When the gene is transcribed, and an RNA molecule is produced, those fragments of the RNA sequence which correspond to the aforementioned parts of the gene are referred to with the same names. For example, the part of the RNA's sequence which corresponds to an intron in the gene's sequence is also called an intron (Figure 4.8a, b).

4.7   Transcription and Post-transcriptional Processes

Transcription is the process of synthesising RNA molecules based on the codes stored in the DNA. The protein complexes which perform this process are referred to as *RNA polymerases*. Three types of RNA polymerases exist in eukaryotes, namely I, II and III, which allow for the synthesis of different types of RNA molecules. Figure 4.7 summarises the process of the most common type of transcription, which depends on RNA polymerase II. This RNA polymerase transcribes all protein-coding genes as well as the genes producing snoRNAs, miRNAs, siRNAs and most snRNAs.

The recruitment of the RNA polymerase protein complex to the start of transcription site is controlled by proteins known as *transcription factors* (*TFs*). TFs are proteins which have the ability to bind to specific DNA sequences in order to promote or block the recruitment of RNA polymerases to transcribe the neighbouring gene(s). The TFs which promote transcription are known as *activators*, while the ones which block transcription are known as *repressors*.

One family of TFs is referred to as *general transcription factors* (*GTFs*), because its member proteins bind to specific DNA sequences that are found next to all types of genes, and their presence is always necessary for the requirement of the RNA polymerases. However, most of the TFs bind to DNA sequences that specifically exist next to a single gene or a few genes in order to activate or repress them. The latter TFs control the differential transcription of different genes at different instances of time and in different types of cells.

The GTFs which are required to recruit RNA polymerase II were named as TFIIB, TFIID and so on. TFIID recognises and binds to a specific DNA sequence known as *TATA box*, which contains repeated alternations between the bases A and T, hence its name. TATA box exists at about 25 bases *upstream* of the start of transcription sites of the genes transcribed by RNA polymerase II (Figure 4.7a). As TFIID binds to the TATA box, it recruits other GTFs including TFIIB, as well as the RNA polymerase II protein complex (Figure 4.7b).

The GTFs and the RNA polymerase II complex unwind the DNA double helix at the start of the transcription site while breaking the hydrogen bonds which join its two strands. RNA polymerase II encompasses one of the two split strands in a groove while keeping the other strand away. This encompassed strand is the template based on which the new RNA molecule is to be synthesised (Figure 4.7c). The process of TFs' binding next to the gene, recruiting the RNA polymerase II complex, stabilising it, and starting transcribing the first few bases, is known as *transcription initiation* (Figure 4.7a, b, c). More than 100 individual proteins are involved in this process, including the GTFs and the sub-units of the RNA polymerase II complex.

As RNA polymerase II has been stabilised in its place and few DNA bases have been transcribed, the GTFs leave the site to be available for transcription of other genes. Before leaving the site, TFIIH *phosphorylates*, that is adds phosphate groups to, two locations on the *C-terminal domain* (*CTD*) of RNA polymerase II (Figure 4.7d). CTD extends like a tail outside the main body of RNA polymerase II, and its phosphorylation changes its structure, which provides the CTD itself with the ability to recruit other proteins that are needed to process the emerging RNA molecule *co-transcriptionally* and *post-transcriptionally*.

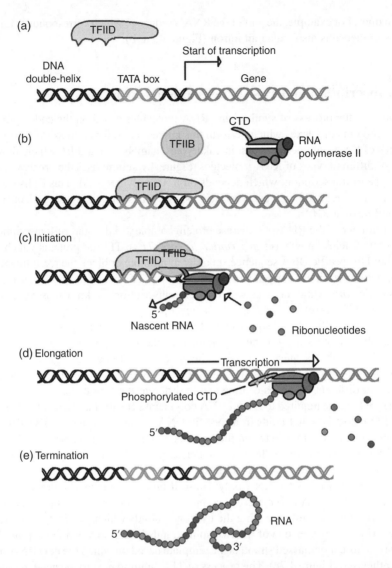

Figure 4.7   The transcription process. (a) The DNA double-helix before the initiation of transcription; the TATA box motif, the gene, and the start of transcription location are marked. (b) TFIID binds to the TATA motif. (c) Transcription initiation; TFIIB and the RNA polymerase II complex are recruited to the start of transcription site, and then transcription of a few codons takes place. (d) Transcription elongation; the CTD of the RNA polymerase II complex is phosphorylated to allow for the polymerase to slide over the gene's sequence, transcribing its codons and thus elongating the nascent RNA. (e) Transcription termination; the RNA polymerase II complex is released from the DNA and the gene is completely transcribed (*See insert for color representation of the figure*)

After initiation, transcription enters the *elongation phase*, in which the RNA polymerase II complex slides through the DNA like a zipper while reading the template strand and synthesising the RNA molecule by adding a single ribonucleotide at a time (Figure 4.7d). As transcription reaches a transcription terminator, that is, a section of nucleic acid sequence that marks the end of the gene, transcription terminates (Figure 4.7e).

Two classes of transcription terminators, namely Rho-independent and Rho-dependent, have been discovered in prokaryotic genomes, while transcription termination in eukaryotes is less understood.

4.7.1 Post-transcriptional Processes

The mRNA molecule delivered by transcription needs to be processed by a number of processes in order to be ready for translation. Three main *post-transcriptional processes* are discussed in this section, namely 5′ capping, RNA splicing, and 3′ polyadenylation (Figure 4.8).

The 5′ capping process starts before transcription terminates as the phosphorylated CTD domain of the RNA polymerase II complex recruits the *capping enzyme (CE)* protein complex to process the 5′ end of the nascent RNA as it emerges from the RNA polymerase (Figure 4.7c, d). The 5′ *cap* is an altered 5′ end of the RNA molecule which stabilises it, protects it from *degradation*, and promotes its consequent translation into a protein (Figure 4.8). Uncapped mRNA molecules are not considered by ribosomes for translation in eukaryotes.

The other end of the RNA, namely the 3′ end, is *polyadenylated*. Polyadenylation is the addition of a long sequence of nucleotides with A bases to the 3′ end of the nascent RNA. This long tail of adenine bases (*poly(A) tail*) is important for the stability of the mRNA molecule, and supports its export and transportation from the nucleus to the cytoplasm for translation (Figure 4.8).

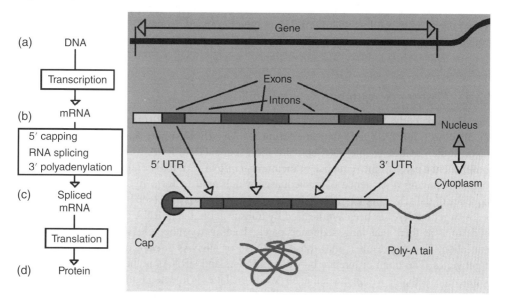

Figure 4.8 Post-transcriptional processes. (a) Transcription. (b) Three main post-transcriptional processes; 5′ capping, RNA splicing, and 3′ polyadenylation. (c) Translation

The other important post-transcriptional process is RNA *splicing*. The sequence of the mRNA includes regions which are translated into the final protein product, and they are called exons. On the other hand, there are *untranslated regions (UTR)* at both ends of the mRNA known as the 5′ UTR and the 3′ UTR, and between the exons known as introns (Figure 4.8b). The RNA splicing process splices out the introns while joining the exons together such that the mature mRNA molecule includes the 5′ UTR, exons and the 3′ UTR, in this respective order (Figure 4.8c). The terms introns, exons, 5′ UTR and 3′ UTR refer to those sequences in the mRNA as well as in the corresponding gene in the DNA.

4.7.2 Gene-specific TFs

A large number of proteins can bind to specific sequences of the DNA which are found next to a specific gene or a few genes. Those proteins may either stabilise the affinity of the RNA polymerase to the gene or interfere with it, which results in either promoting the transcription of the corresponding gene or blocking it, respectively. Because those proteins influence the transcription of genes, they are called *TFs*, and because they bind to specific gene(s), they are called *gene-specific*, in contrast to the GTFs discussed earlier in this section.

The DNA sequence to which a transcription factor binds is referred to as its *binding site*. The *promoter* of a gene is a sequence in its *upstream* sequence, that is before its start of transcription site, which is a binding site of an activator TF. Without the required TF binding to this promoter, the general TFs and the RNA polymerase would neither be able to bind to the gene's sequence nor to transcribe it. On the other hand, an *enhancer* is that DNA sequence which might be upstream, downstream, or even within the gene itself, and might even be many thousands of bases far distant, to which a transcription factor binds and enhances the level of transcription of the gene. The TFs which bind to the enhancers tend to interact with those binding to the promoters as well as with the RNA polymerase itself. Some TFs are formed of single proteins, while many others are protein complexes.

4.7.3 Post-transcriptional Regulation

One of the recently emerging aspects of gene regulation is *post-transcriptional regulation*. MicroRNAs (miRNAs) and small interfering RNAs (siRNAs) play an important role at this stage. At their mature usable state, both are short sequences of RNA which perfectly or mostly complement a patch of the sequence of their *target mRNA*. Complementing the sequence in this context is having a sequence of nucleobases which tend to form hydrogen bonds with the sequence of the target, for example A binds to U, and C binds to G (see Section 4.5). Therefore, as the target mRNA has been synthesised by transcription, the miRNA or the siRNA molecule binds to it at their matching sequence patch, leading to either cleaving it or blocking its translation. The net result is that those miRNAs or siRNAs *silence* or *down-regulate* their target genes. The main difference between miRNAs and siRNAs is that the first is produced *endogenously*, that is within the cell itself, while the latter is injected into the cell from an *exogenous* source such as a virus. Cells use miRNAs, as well as other ways of post-transcriptional regulation, to control the amounts of specific mRNA molecules, and therefore their protein products.

4.8 Translation and Post-translational Processes

Translation is the process of synthesising a protein molecule based on the information stored in an mRNA molecule. Various types of RNA molecules and proteins cooperate to translate the information in the mRNA into a protein accurately and efficiently.

4.8.1 The Genetic Code

The mapping between the four-letter language used in the mRNA (A, U, G, C) and the twenty-letter language used in the proteins (Table 4.1), which is known as the *genetic code*, was deciphered in the early 1960s. Three nucleobases are needed, and are what is actually used, to encode for a single amino acid. This is because each base has four options, and therefore three bases can have any of $4^3 = 64$ combinations, which exceed the needed 20 different types of amino acids. If only two bases were to be used, they can provide up to $4^2 = 16$ combinations, which are not sufficient. A triplet of mRNA bases which are translated into an amino acid is known as a *codon*. As illustrated, there are 64 codons to encode for 20 amino acids, and thus many amino acids have more than one codon encoding for them. The complete genetic code is shown in Table 4.2.

It can be seen in Table 4.2 that the 64 different codons encode for 20 different amino acids in addition to two punctuation marks, namely start and stop. In an overview, translation scans the mRNA from its beginning until the *start codon* is allocated. Then, triplets of bases are read one after another while adding amino acids to an elongating protein accordingly. Once a *stop codon* is met, translation terminates. One can notice that the start codon, AUG, encodes for the amino

Table 4.2 The genetic code: mapping the three-base mRNA codons to their corresponding amino acids

First base (5' end)	Second base				Third base (3' end)
	U	C	A	G	
U	F	S	Y	C	U
	F	S	Y	C	C
	L	S	STOP	STOP	A
	L	S	STOP	W	G
C	L	P	H	R	U
	L	P	H	R	C
	L	P	Q	R	A
	L	P	Q	R	G
A	I	T	N	S	U
	I	T	N	S	C
	I	T	K	R	A
	M + START	T	K	R	G
G	V	A	D	G	U
	V	A	D	G	C
	V	A	E	G	A
	V	A	E	G	G

acid methionine (M) as well. The fact is that its first occurrence in the mRNA serves as the start of translation site, while its consequent occurrences merely encode for M.

4.8.2 tRNA and Ribosomes

tRNA molecules, which are transcribed by RNA polymerase III, represent one important type of functional RNA molecules that participate in translation. Tens to hundreds of different types of tRNA exist in species from bacteria to humans. They are synthesised by RNA polymerase III followed by many post-transcriptional modifications such as trimming, chemical modifications, sequence editing and others. A mature tRNA molecule is about 80 nucleobases long with the affinity to a specific codon, that is three mRNA nucleobases, at one of its ends, and to their corresponding amino acid at the other end. An iconic example of a tRNA molecule decoding for the amino acid tryptophan (W) is shown in Figure 4.9.

Ribosomes are the protein factories in cells as translation takes place in them. They are formed of two sub-units, the large ribosomal sub-unit and the small ribosomal sub-unit (Figure 4.10a). In eukaryotes, the large sub-unit is formed of three rRNA molecules and about 50 proteins while the small sub-unit is formed of one rRNA molecule and about 33 proteins.

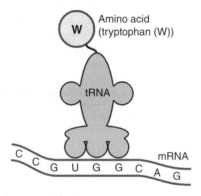

Figure 4.9 A tRNA molecule with the specific affinity to the amino acid W from one end, and to the mRNA sequence UGG from the other end

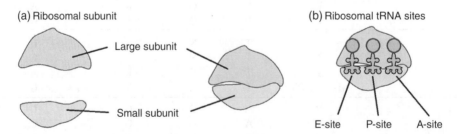

Figure 4.10 (a) Ribosomal large and small sub-units. (b) The three ribosomal sites of tRNA binding, the E, P and A

The numbers of proteins in these sub-units in prokaryotes are lower than that. In terms of weight, the rRNA molecules contribute to two thirds of the weight while the proteins contribute to one third. A typical eukaryotic cell includes millions of ribosomes, which are assembled in the nucleolus after the required ribosomal proteins have been synthesised in the cytoplasm and imported into the nucleus.

Both the eukaryotic and the prokaryotic ribosomes have similar structure and function although they differ in size. The small ribosomal sub-unit serves as a framework for tRNA accurate matching with the mRNA sequence being translated while the large ribosomal sub-unit catalyses binding of the amino acids together with peptide bonds in order to form the polypeptide, that is the protein. Three sites for tRNA binding are available in the ribosome, the E-site, the P-site and the A-site (Figure 4.10b).

4.8.3   The Steps of Translation

The translation process is detailed in Figure 4.11 from its initiation to its termination. First, a tRNA-M complex, that is a tRNA molecule with an M amino acid bound to it, is loaded at the P-site of a small ribosomal sub-unit, and then the small sub-unit with its load is loaded next to the 5′ end of the target mRNA, marked with the existence of the 5′ cap [Figure 4.11 (step 1)]. If the mRNA was not capped, the small ribosomal sub-unit would not recognise it, and therefore the translation process would not initiate. No tRNA-amino-acid complex other than the tRNA-M can bind with stability to a small ribosomal unit without being combined with the large sub-unit. In fact, additional proteins, known as the *eukaryotic initiation factors* (*eIFs*), are loaded with the small sub-unit and the tRNA-M to the 5′ end of the mRNA molecule to support the initiation of translation.

Next, the small ribosomal sub-unit with its load slide on the mRNA sequence until the tRNA-M complex meets its matching codon, namely the start (initiation) codon AUG [Figure 4.11 (step 2)]. At this stage, the large ribosomal sub-unit is loaded [Figure 4.11 (step 3)]. The codon following the start codon resides now at the base of the A-site of the assembled ribosome, and in the example demonstrated in Figure 4.11, it is UGG coding for the amino acid W. Consequently, a matching tRNA-W complex is loaded to the ribosome's A-site and binds to the mRNA's UGG codon [Figure 4.11 (step 4)]. As mentioned earlier, the small sub-unit serves as a framework for such stable binding. In its turn, the large sub-unit catalyses the formation of a peptide bond between the amino acid (M) at the P-site and the amino acid (W) at the A-site while releasing the amino acid (M) from its associated tRNA molecule [Figure 4.11 (step 5)].

Various conformational modifications happen to the two ribosomal sub-units which let them slide over the mRNA molecule for three nucleobases (one codon). After this sophisticated movement, the two tRNA molecules included within the ribosome become residing at the E-site and the P-site instead of the P-site and the A-site, respectively [Figure 4.11 (step 6)]. Consequently, a codon enters the base of the free A-site, which is in this example the UUU codon encoding for the amino acid phenylalanine (F). A matching tRNA-F complex then enters the A-site of the ribosome and binds to the underlining mRNA's codon while the tRNA in the E-site exits the ribosome [Figure 4.11 (step 7)].

A new peptide bond is formed between the most recently added amino acid (W) and the following amino acid (F), which results in a short polypeptide chain (protein) emerging from the ribosome [Figure 4.11 (step 8)]. Note that the most recently added amino acid is always

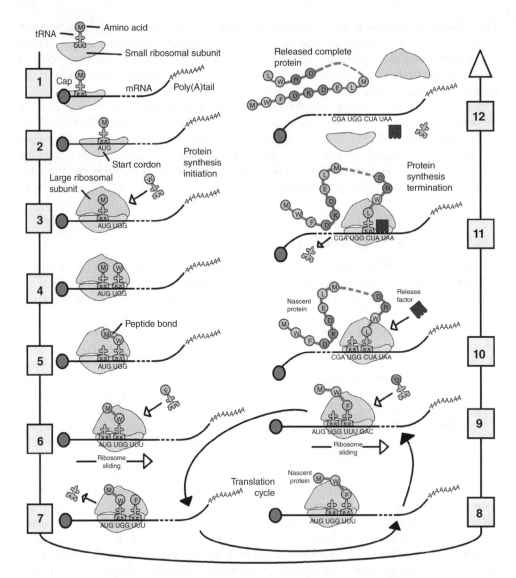

Figure 4.11 The translation process from initiation to termination (*See insert for color representation of the figure*)

kept bound to its corresponding tRNA molecule. Again, and through complex series of conformational modifications, the ribosome slides for one more codon towards the 3' end of the mRNA (towards the poly(A) tail) [Figure 4.11 (step 9)].

The steps 7–9, which are similar to the steps 3–6 except that there was no tRNA molecule initially at the E-site, are referred to as the *translation cycle*. Every time they are repeated, the ribosome slides for one more codon over the mRNA molecule, and a new amino acid is added to the nascent polypeptide (protein) chain. The rate of translation varies from about two to 20 amino acids per second for some eukaryotic species and bacteria, respectively.

The translation cycle terminates when it faces a *stop codon*. As shown in Table 4.2, there are three stop codons in the genetic code, namely UAA, UAG and UGA. Step 10 in Figure 4.11 shows a long nascent protein emerging from the ribosome after many repetitions of the translation cycle where the next codon, that is the one at the base of the A-site, is the stop codon UAA. A special protein, known as *release factor*, recognises the stop codon in the mRNA sequence, and binds to it once it is at the A-site of the ribosome [Figure 4.11 (step 11)]. This causes the protein to be unbound from the tRNA molecule residing at the P-site, and leads to the disassembly of the two ribosomal sub-units, the tRNA molecules, the release factor, the mRNA, and the newly synthesised protein [Figure 4.11 (step 12)].

4.8.4   Polyribosomes (Polysomes)

For higher efficiency and throughput, as soon as the ribosome slides away from the start codon for a sufficient number of codons, another ribosome assembles at the 5′ end of the same mRNA molecule. By this mechanism, multiple ribosomes pursue translation for the same mRNA molecule simultaneously, each residing at a different patch of the mRNA sequence, and therefore at a different stage of translation. These multiple ribosomes are called a *polyribosome*, or a *polysome* (Figure 4.12). Moreover, and to increase the efficiency even more, the 3′ poly(A) tail is joined with the 5′ cap of the mRNA by a number of binding proteins to form a loop of translating ribosomes. In this manner, the ribosome which is released from the 3′ end of the mRNA is at a very appropriate location to reassemble at the 5′ end, and consequently to synthesise another copy of the target protein (Figure 4.12).

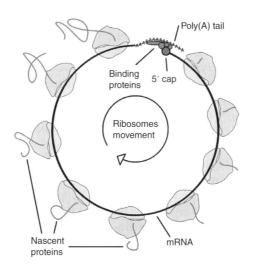

Figure 4.12   Polyribosomes (polysomes): multiple ribosomes translate the same mRNA molecule in a pipeline

4.8.5   Post-translational Processes

After translation, the protein undergoes many modifications before becoming a mature functional protein. One important process is *protein folding*, in which the protein's linear chain of amino acids folds upon itself to settle at a stable 3-D conformation. This was discussed in details in Section 4.3; see Figure 4.3.

Other modifications include binding of functional chemical groups to the protein chain, such as phosphate, lipids, carbohydrates and others. Such modifications might lead to conformational changes, as illustrated in Figure 4.3c. Moreover, as all nascent proteins start with the amino acid M because its codon represents the start codon (Figure 4.11, step 2), this starting amino acid is removed in post-translational modification from many proteins.

If the ribosomes were bound to the surface of the rough endoplasmic reticulum, as illustrated in the previous chapter, the nascent protein emerges from the large ribosomal sub-unit into the endoplasmic reticulum immediately. Then, it is transported, through the Golgi apparatus, vesicles or other means, to its destination.

4.9   Discussion and Summary

Most of the functions in living cells are carried out by the most dominant macromolecule type, proteins. Proteins are linear chains of amino acids bound to each other with peptide bonds. There are 20 different types of amino acids (Table 4.1), which represent the alphabet from which the different types of proteins are composed. Attractive and repulsive forces between different amino acids within the protein chain cause this linear chain to fold upon itself to form a specific stable structure in the 3-D space. Differences in physical structures and chemical properties of different types of proteins lead to the differences in their functions within the cell.

The complete set of information needed to determine when, where and how to synthesise all types of proteins within an organism, as well as to perform the other cellular functions, is stored in the DNA molecule, which resides in the nucleus of the eukaryotic cell or the nucleoid region of the prokaryotic cell. The DNA is a double-stranded molecule where each strand is a linear sequence of nucleobases. Four types of nucleobases are found in DNA, and therefore the DNA stores information using a four-letter language. The two strands of the DNA are complementary, such that each one of them can be used as a template to reform the other one accurately. By unbinding the two strands from each other and then building new complementary strands for each, the cell replicates its genetic material, that is the DNA, when it divides into two daughter cells.

The gene is that patch of the DNA sequence which includes complete information on synthesising a single type of protein or functional RNA molecules. However, the gene in the DNA is not used directly for protein synthesis; it is rather copied into an RNA molecule, which is exported from the nucleus of eukaryotes and then used for protein synthesis. The RNA molecule is a single-stranded nucleic acid composed of nucleobases of four different types. It is very similar to the structure of a single strand of DNA except in a subtle difference in its chemical form (in its five-carbon sugar part), and in that it uses the nucleobase U instead of T. The process of copying that DNA patch, that is the gene, into an RNA molecule is known as transcription.

After transcription, the nascent RNA molecule is processed with post-transcriptional processes such as splicing, adding a cap at its 5′ end (the first end), and adding a polyadenine tail to its 3′ end (the other end). These processes stabilise the RNA molecule, help in its transportation within the cell to the place in which it is to be used, and assist its usage in synthesising proteins.

There are different types of RNA molecules, the most common one of which is the mRNA, which functions as a carrier of the genetic information in the DNA molecule required to synthesise a protein. Other types of RNAs are known as functional RNA molecules, where they themselves perform cellular functions without being translated into proteins. Examples of these include ribosomal RNAs (rRNAs), which constitute about two thirds of the protein factories known as ribosomes, transfer RNAs (tRNAs), which cooperate with the ribosomes in translating mRNAs into proteins, and siRNAs and miRNAs, which both participate in post-transcriptional regulation. The latter perform post-transcriptional regulation by interfering with the newly synthesised mRNA molecules and blocking their translation into proteins; this helps in regulating protein synthesis within the cell when the amounts of some proteins need to be reduced at certain stages. Other types include snRNA, snoRNA and others and these participate in various functions including processing and modifying other types of RNA molecules.

The mature mRNA molecule is merely a message from the DNA with the required information for synthesising a protein molecule. Ribosomes identify these mRNA molecules, bind to them at their 5′ end (the first end), slide over them while translating their codes into nascent proteins, and finally release them. Every three consecutive nucleobases in the mRNA molecule are called a codon, and are translated into a single amino acid in the generated protein. With three nucleobases, each of which has one out of four possible symbols, there are 64 different possible codons, encoding for 20 different types of amino acids, indeed with redundancy (Table 4.2). Specific codons are preserved as start and stop codons, which mark starting and ending points of translation within the mRNA molecule. After translation, the protein is folded into its most chemically and physically stable 3-D shape, and is transported into its destination where it shall perform its required biological functions.

To summarise the central dogma of molecular biology (Figure 4.1), the main source of genetic information is the DNA molecule. This information is transmitted through generations of cells and individuals through DNA replication, where two daughter DNA molecules are produced identical to the mother DNA. Within a cell, the information in the DNA is transmitted to RNA molecules via transcription. In some cases those RNA molecules are functional and will perform biological tasks themselves, but in most of the cases the RNA molecules are messages of information copied from the DNA and translated into protein molecules. The proteins then perform most of the biological functions required by the cell.

References

Alberts, B., Johnson, A., Lewis, J. *et al.* (2008). *Molecular Biology of the Cell*, 5th edn, Garland Science, New York.

Bang, M.-L., Centner, T., Fornoff, F. *et al.* (2001). The complete gene sequence of titin, expression of an unusual approximately 700- kDa titin isoform, and its interaction with obscurin identify a novel Z-line to I-band linking system. *Circulation Research*, **89**(11), pp. 1065–1072.

Brocchieri, L. and Karlin, S. (2005). Protein length in eukaryotic and prokaryotic proteomes. *Nucleic Acids Research*, **33**(10), pp. 3390–3400.

Collins, F., Lander, E., Rogers, J. and Waterson, R. (2004). Finishing the euchromatic sequence of the human genome. *Nature*, **431**(7011), pp. 931–945.

Harrison, P.M., Milburn, D., Zhang, Z. *et al.* (2003). Identification of pseudogenes in the Drosophila melano-
    gaster genome. *Nucleic Acids Research*, **31**(3), pp. 1033–1037.
Neidigh, J.W., Fesinmeyer, R.M. and Andersen, N.H. (2002). Designing a 20-residue protein. *Nature Structural
    Biology*, **9**, pp. 425–430.
Opitz, C.A., Kulke, M., Leake, M.C. *et al.* (2003). Damped elastic recoil of the titin spring in myofibrils of
    human myocardium. *Proceedings of the National Academy of Sciences*, **100**(22), pp. 12688–12693.

Part Three

Data Acquisition and Pre-processing

5

High-throughput Technologies

5.1 Introduction

Decades ago, experimental biology was based on technologies which measure or edit different biological parameters at a low-throughput level, that is, while considering single or few objects and parameters at any single technique's application. Examples of those techniques include mass spectrometry, which dates back to the beginning of the twentieth century (Aston, 1919). Southern blotting, which was proposed by the English molecular biologist Sir Edwin Southern (1975), detects a specific given DNA sequence in DNA samples. Following that, many blotting methods were proposed such as *northern blotting* (Alwine, Kemp and Stark, 1977) and *western blotting* (Burnette, 1981) which detect a specific RNA or a specific protein in a sample, respectively. Chromatin immunoprecipitation (ChIP) is another type of those techniques, which detects specific DNA-protein binding (Gilmour and Lis, 1985).

In the last couple of decades, new technologies have emerged extending some of the aforementioned low-throughput technologies, allowing for the measurement of different biological parameters at a high-throughput level, that is, while considering large numbers of objects or parameters at any single-technique application. For example, DNA microarrays extend Southern blots by detecting the relative abundance of thousands of different specific DNA sequences in a DNA sample. ChIP-chip extends ChIP by combining it with the microarray technology to identify the genome-wide DNA sequences to which some proteins of interest bind.

As understanding the mechanisms adopted by high-throughput technologies is important for successful and reliable analysis of their data, we cover a number of the most common high-throughput technologies in this chapter, including DNA and protein microarrays, RNA next-generation sequencing (NGS), ChIP-chip and others.

Integrative Cluster Analysis in Bioinformatics, First Edition. Basel Abu-Jamous, Rui Fa and Asoke K. Nandi.
© 2015 John Wiley & Sons, Ltd. Published 2015 by John Wiley & Sons, Ltd.

5.2   Microarrays

Microarrays are high-density chips of probes, where each probe has high affinity to be bound by a specific target molecule and low affinity to be bound by the rest of the molecules possibly existing in the provided sample (Figure 5.1). The sample of molecules to be processed is dyed with fluorescent labels, and then exposed to the microarray (Figure 5.1a, b). The probes reflect the relative amounts of their target molecules by radiating respective levels of intensity caused by different levels of dyed target binding (Figure 5.1c). After that, the microarray chip is scanned and image-processed to quantify the intensities of those probes (Figure 5.1d), which ideally will be linearly proportional to the amounts of their target molecules in the sample (Figure 5.1e).

The history of microarrays dates back to 1983 when Tse-Wen Chang proposed coating antibodies with different specificities to different cell surface antigens in a matrix-like manner on a small glass chip in order to measure the relative proportions of cells with different surface antigens in a mixed population of cells (Chang, 1983). However, many other types of microarrays emerged thereafter, like DNA microarrays, which are the most widely spread type of microarrays, protein microarrays, carbohydrate microarrays and others. Moreover, many modern high-throughput technologies involve microarrays in their pipelines, such as ChIP-chip, which combines ChIP with the DNA microarray "chip" technology.

5.2.1   DNA Microarrays

DNA microarrays, representing the most common type of microarrays, are generally used to measure the expression of a large number of genes in a population of cells. The *expression* of a gene refers to the amount of transcripts (RNA) transcribed from it. The population of RNA molecules are *reverse transcribed* to form *complementary DNA (cDNA)* molecules, which are dyed with fluorescent labels and exposed to the microarray.

Different types of microarrays use different approaches of probe organisation. Spotted cDNA microarrays contain probes whose sequences represent complete complementary sequences of the target mRNA molecules. On the other hand, oligonucleotide microarrays contain probe-sets of 10–25 probes in each. Each probe-set corresponds to one mRNA/gene sequence where each of its probes has a short oligonucleotide sequence complementary to a specific fragment of the target mRNA molecule. Each of these short sequences usually contains around 25 bases.

Many microarray platforms were designed, and the most commonly used are Affymetrix, Agilent and Illumina (Calza and Pawitan, 2010). Affymetrix microarrays are oligonucleotide arrays with probe sets containing 11–20 probe-pairs each of which is 25 bases long. Each probe-pair has two probes, which are called *perfect match (PM)* and *mismatch (MM)* respectively. The PM probe has the exact target sequence while the MM has that same exact sequence but with a single altered base at the middle of the probe's sequence, that is, at the base number 13 (out of 25). The reason for this setup is explained in Chapter 7 (Wu, 2009).

Another variation in the design of DNA microarrays is the number of channels – either one or two channels; also called one- or two-colour microarrays. In the case of two-channel arrays, two different mRNA samples are used, which are labelled with different fluorescent labels that radiate different orthogonal colours (usually Cy3 for green and Cy5 for red). After

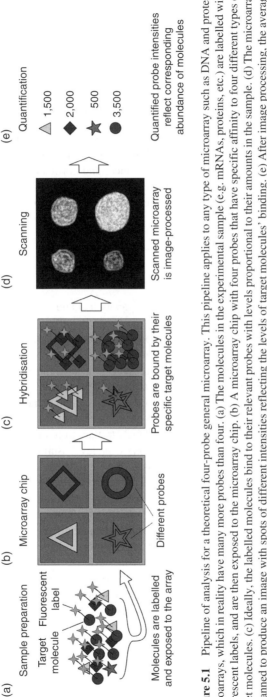

Figure 5.1 Pipeline of analysis for a theoretical four-probe general microarray. This pipeline applies to any type of microarray such as DNA and protein microarrays, which in reality have many more probes than four. (a) The molecules in the experimental sample (e.g. mRNAs, proteins, etc.) are labelled with fluorescent labels, and are then exposed to the microarray chip. (b) A microarray chip with four probes that have specific affinity to four different types of target molecules. (c) Ideally, the labelled molecules bind to their relevant probes with levels proportional to their amounts in the sample. (d) The microarray is scanned to produce an image with spots of different intensities reflecting the levels of target molecules' binding. (e) After image processing, the average intensities of probes are recorded as quantified measurements for the relative abundance of target molecules in the sample

hybridisation and scanning, each probe's colour is quantified into two intensities for the green and the red components respectively. The ratio between these two intensities reflects the ratio of the abundance of the corresponding gene in the two provided samples. These samples are usually from different biological phenotypes (e.g. cancer cells versus normal cells), or one of them can be an invariant control sample which serves as a reference for the other one.

The variabilities in the many intermediate steps between the real amount of mRNA in the sample and the final quantified intensity stimulate questioning the reliability of these raw gene expression values. Such variabilities can be caused by the preparation of the biological sample, fluorescent labelling, specific hybridisation, non-specific hybridisation, scanning, image processing and others (Calza and Pawitan, 2010). Moreover, in most of the microarray datasets, many mRNA samples are taken and measured by multiple microarray chips/slides; these samples can be from different types of tissues (e.g. cancer and normal tissues), at different chronological stages or time points within a biological process, or from different samples belonging to different biological conditions. Such aspects are targeted by *normalisation* methods, which are thoroughly discussed in Chapter 7.

5.2.2   Protein Microarrays

There are two main types of protein microarrays, analytical and functional. *Analytical protein microarrays* are arrays of antibody probes that have specific affinity to different types of proteins. By exposing proteins dyed with fluorescent labels to such arrays, the expression (abundance) levels of proteins in a population of cells are measured. In contrast, the probes of *functional protein microarrays* are actual protein molecules rather than antibodies to proteins. As in the case of DNA microarrays, protein microarrays can be one- or two-colour arrays (Haab, Dunham and Brown, 2001). Various objectives can be achieved by using functional protein microarrays by considering different types of target molecules. For example, exposing a sample of lipid molecules with fluorescent labels to that array of protein probes reveals the levels of protein–lipid interactions at the *proteome-wide* level. Similarly, if the exposed sample of labelled molecules included other proteins, DNA fragments, drugs, small molecules or peptides, the functional array would measure protein–protein, protein–DNA, protein–small molecules or protein–peptide interactions, respectively (Zhu *et al.*, 2001; Chen and Zhu, 2006; Pratsch, Wellhausen and Seitz, 2014).

One of the main challenges regarding analytical protein microarrays, which are extensions of antibody microarrays, is the design of protein antibodies that have high sensitivity and specificity, that is, any single antibody in the array should have significantly high affinity to its target protein and significantly low affinity to the rest of the proteins. Generally, there is no similar problem in DNA microarrays because the best probe which detects a DNA sequence is its complementary DNA sequence, while proteins do not possess such a global complementary rule. Until today, there is no analytical protein microarray which covers the entire proteome of a model species like *E. coli* bacteria, *S. cerevisiae* yeast or humans. Nevertheless, arrays with several hundreds of protein antibodies have already been prepared for commercial use, like the arrays KAM-850, Panorama@ Antibody Array – XPRESS Profiler725, and others. However, the number of proteins from different species whose specific antibodies have been identified is increasing.

5.2.3   Carbohydrate Microarrays (Glycoarrays)

The *glycome* is the complete set of *glycans* (carbohydrates) in an organism. A glycan is a chain of monosaccharides (monosugars) linked glycosidically. It was estimated that the human glycome includes more than 7000 different glycan-recognition determinants (Song *et al.*, 2014). Glycan microarrays (glycoarrays) are functional, that is, they are arrays of glycan molecules immobilised on a solid surface, which are exposed to different types of molecules, mainly proteins, to identify their levels of glyco-binding (Laurent, Voglmeir and Flitsch, 2008).

Glycoarrays have been used to serve many objectives. The most notable one is the investigation of glycan–protein interactions through the identification of carbohydrate-binding proteins (CBPs) and their glycan ligands (Liu *et al.*, 2007; Blixt and Westerlind, 2014; Palma *et al.*, 2014). *Glycoproteins* are proteins which are glycosylated, that is, one or more polysaccharide chains (glycans) are bound to its amino-acid sequence. If the protein core is heavily glycosylated such that the glyco- content of the resulting complex was more than the protein content, it is called a *proteoglycan*. Glycoproteins and proteoglycans participate in many processes in the cell including cell-surface structure and adhesion and immunity (Laurent, Voglmeir and Flitsch, 2008; Arthur, Cummings and Stowell, 2014).

Glycoarrays are also involved in cell-adhesion studies, in which whole cells are exposed to the glycoarrays to mimic and thereafter analyse cell–cell interface and interaction (Laurent, Voglmeir and Flitsch, 2008). Other objectives of glycoarrays include the identification of antibody specificity to glycans (Liang *et al.*, 2011; Galban-Horcajo *et al.*, 2014), glycosyltransferase specificities (Laurent, Voglmeir and Flitsch, 2008), diagnosis (Geissner, Anish and Seeberger, 2014), and others.

The current version (version 5.1) of the glycan microarray generated by the Consortium for Functional Glycomics (www.functionalglycomics.org) represents just over 600 glycans, comprising less than 10% of the complete glycome. Therefore, glycan microarrays will be greatly enhanced by defining larger portions of the glycomes of humans and other model species, as well as identifying the procedures needed for their synthesis or gathering. This has been responded to by using the Shotgun Glycan Microarray (Song *et al.*, 2011; Arthur, Cummings and Stowell, 2014). Additional glycoarrays challenges include glycan-immobilisation strategies (Laurent, Voglmeir and Flitsch, 2008; Chevolot *et al.*, 2014), and the proper glycan presentation and orientation on the microarrays (Song *et al.*, 2014).

5.2.4   Other Types of Microarrays

The idea of an array of micro-biochemical probes on a solid platform for various analytical and functional studies is not limited to macromolecules such as nucleic acids, proteins and carbohydrates (glycans); rather, it has been applied to other types of biochemical molecules. Small-molecule microarrays (SMM) represent an example of a successful microarray platform for small-molecule–protein interaction investigation (Hong *et al.*, 2014). Peptides, which are short chains of amino acids, were also arrayed (Voskuhl, Brinkmann and Jonkheijm, 2014). Peptides and proteins are conventionally distinguished based on their length, where proteins generally include hundreds of amino acids while peptides can be as short as including only two. On the other side of the spectrum, complexes as large as whole cells were arrayed to detect antigen-specific cells and their responsiveness (Chen and Davis, 2006).

5.3   Next-generation Sequencing (NGS)

5.3.1   DNA Sequencing

is the process of identifying the ordered linear sequence of nucleobases in a given DNA strand. Automated Sanger sequencing has been the most commonly used method for DNA sequencing over the last couple of decades (Metzker, 2010), and it was used to produce the first and only *finished-grade* human genome sequence (Collins *et al.*, 2004). However, that generation of sequencing methods has many limitations, mainly high cost and long running time. A class of NGS methods has emerged in the last decade with the most appealing ability of producing enormous amounts of sequencing data with low costs (Mardis, 2008; Metzker, 2010; Dijk, Jaszczyszyn and Thermes, 2014).

NGS involves (i) template preparation, (ii) sequencing, and (iii) imaging. Numerous NGS methods were designed by adopting variants of the techniques developed for each of these three steps. In DNA sequencing, templates are prepared by fragmenting the provided DNA sample into short vectors, which are joined with known adapter sequences that can be bound by universal primers. Most of the NGS protocols amplify the template sequences by producing thousands to millions of copies (clones) from each template to provide the following NGS steps with sufficient quantities of DNA material (Metzker, 2010). Variants of the *polymerase chain reaction (PCR)* are normally adopted to achieve this objective. However, PCR has been challenged for selective biases in sequence amplification, and was either modified or replaced with other techniques like multiple displacement amplification (MDA) (Dijk, Jaszczyszyn and Thermes, 2014). Some other NGS protocols were designed to not require amplification in order to overcome the biases and mutations caused by the amplification step, but these protocols require, on the other hand, relatively large quantities of input materials, which might not be always available (Dijk, Jaszczyszyn and Thermes, 2014).

Cyclic reversible termination (CRT) is one of the techniques used for sequencing (Metzker, 2010) (Figure 5.2). After fragmentation, the DNA fragments are ligated with known predefined adapter sequences which support the immobilisation of the templates on a solid platform, normally made of glass (Figure 5.2c, d). Modified nucleotides are added with DNA polymerases to the platform (Figure 5.2e). Those nucleotides possess two key features; they are labelled with fluorescent labels which have four different colours corresponding to the four different nucleobases A, T, G, and C and they have reversible polymerisation terminators which terminate polymerisation upon their addition by blocking the addition of more nucleotides. The platform is then washed to remove all of the extra unbound nucleotides. At this stage, each template would have a single fluorescent nucleotide corresponding to the complement of the first nucleotide in the target sequence (Figure 5.2f). Imaging is performed to store a snapshot of this first nucleotide in the sequence of each of the templates (Figure 5.2f).

After imaging, the fluorescent labels are cleaved and the terminators are reversed to result in templates with non-fluorescent complement strands which are unblocked and are ready to accept the next nucleotide in the sequence (Figure 5.2 g). The step in Figure 5.2e is repeated for a second cycle of modified nucleotides' addition (Figure 5.2 h). The nucleotides complementing the second nucleotides in the template sequences bind to the templates and, similar to the first cycle, they terminate polymerisation, that is, they block the template from accepting more nucleotides (Figure 5.2i). Again, the platform is washed and an image is taken (Figure 5.2i). As the steps in Figure 5.2 g-i are repeated, series of snapshots are produced capturing the sequence of all of the hundreds of millions of templates immobilised on the platform in parallel (Metzker, 2010).

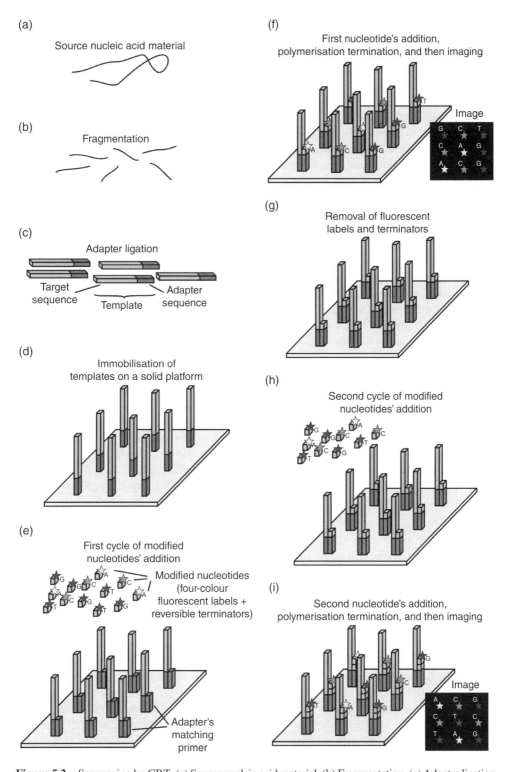

Figure 5.2 Sequencing by CRT. (a) Source nucleic acid material. (b) Fragmentation. (c) Adapter ligation. (d) Immobilisation of templates on a solid platform. (e) First cycle of modified nucleotides' addition; primers matching the adapters are bound. (f) First nucleotide's addition followed by polymerisation termination and then imaging; the image's probes' colours indicate the type of nucleotide bound to each of the templates in this cycle. (g) Removal of fluorescent labels and terminators. (h) Second cycle of modified nucleotides' addition. (i) Second nucleotide's addition followed by polymerisation termination and then imaging. The steps (h–i) are repeated iteratively until the entire template is sequenced (*See insert for color representation of the figure*)

Some NGS platforms, such as Illumina/Solexa, consider four-colour labelling as shown in Figure 5.2. Other platforms, such as Helicos BioSciences, consider one-colour labelling (not shown in the figure). In the latter case, nucleotides of a specific type of base [e.g. adenine (A)] are added in the first cycle. The added nucleotides will hybridise only with the templates whose next sequence base is complemented by it, while the other templates will not be polymerised in this cycle. After washing and imaging, the hybridised templates will show bright spots in the image, while the other templates will show dark spots. Therefore, this snapshot provides binary information regarding which templates have the adenine (A) base's complement next in their sequences. By providing different nucleotide types in consecutive cycles, the final series of snapshots will include the complete information regarding the sequence of all of the templates (Metzker, 2010).

Other sequencing methods are considered in NGS protocols, other than the CRT, such as sequencing by ligation, single-nucleotide addition (pyrosequencing) and real-time sequencing. The latter does not consider nucleotides with polymerisation terminators; rather it measures the fluorescence pulse in real time with sufficiently high sampling rate while series of nucleotides are added to the template. This technique, used by Pacific Biosciences, can read longer consecutive sequences but has higher error rates unless consensus reads are performed (Metzker, 2010).

With NGS technologies, the cost of sequencing a whole genome has dropped significantly. The estimated cost of the first, and only one until now, finished-grade human genome sequenced in 2004 was US$300 million. Other whole human genomes have been sequenced since then, but none at the finished-grade quality level. By now, the cost of sequencing a whole human genome has reached around US$5000 (Hayden, 2014). The US National Human Genome Research Institute (NHGRI) launched a huge grant scheme, which has the greatest credit in the huge drop of sequencing costs. The future aim is to reach the $1000 whole-genome limit, which opens the horizon for a much wider range of applications through feasible personal-genome sequencing (Pareek, Smoczynski and Tretyn, 2013).

5.3.2 RNA Sequencing (Transcripome Analysis)

RNA sequencing can be performed by reverse transcribing the RNA molecules, before or after fragmentation, to produce cDNA molecules, which are sequenced as explained above. Nevertheless, some platforms sequence RNA templates directly without reverse transcription. One major application of RNA sequencing is gene expression analysis as an alternative approach to DNA microarrays (Mardis, 2008; Metzker, 2010; Dijk, Jaszczyszyn and Thermes, 2014). First, mRNA molecules are selectively selected, or the rRNA molecules are selectively depleted. This is because most of the RNA content belongs to the rRNA class while the protein-coding mRNAs are the main target of such analysis (Dijk, Jaszczyszyn and Thermes, 2014). Then, the sequenced mRNA templates are aligned with the sequences of the target genome's genes in order to quantify the absolute or the relative number of transcripts per gene, that is, the level of genetic expression.

NGS techniques surpass microarrays in their larger dynamic range, lower background noise, higher sensitivity to low expression levels, and the ability to discover the expression of rare RNA transcripts as well as splicing alternatives and variants even if their sequences have not been previously discovered and stored (Mardis, 2008; Metzker, 2010; Mori et al.,

2013). This latter capability facilitates transcriptomic analysis for non-model organisms and the organisms that lack complete reference genomes, such as many microorganisms (Fang, Martin and Wang, 2012).

Another important application of RNA sequencing is the sequencing of non-coding RNAs (ncRNAs), which although have many key roles in the regulation of cellular processes, have not been well understood yet (Mardis, 2008).

5.3.3 Metagenomics

Many environments, such as oceans, soil, buildings, the human body and others, include large numbers of members from different species. Metagenomic analysis is the analysis of the genetic material derived from an environment at its genomic scale. By sequencing such genetic material and mapping it to libraries of known genomes, questions regarding as to which species these samples belong to can be answered (Mardis, 2008; Metzker, 2010; Dijk, Jaszczyszyn and Thermes, 2014). The analysis of the *microbiome* residing in any of those environments is indeed targeted by metagenomics. The microbiome is the complete set of microorganisms that reside in a specific environment (Kembel *et al.*, 2014).

5.3.4 Other Applications of Sequencing

The relatively cheap high-throughput DNA sequencing offered by the NGS technologies has stimulated many applications in genomics. Examples include studying the genome-wide epigenetic modifications, such as methylation (MeDip-seq) (Ku *et al.*, 2011; Dijk, Jaszczyszyn and Thermes, 2014), exome sequencing, the analysis of DNase sensitivity (DNase-seq), the analysis of DNA-associated proteins (ChIP-seq; discussed in Section 5.4) (Liang *et al.*, 2013), single-nucleotide polymorphisms (SNPs) and structural variants (SV) analysis (Metzker, 2010), re-sequencing the human genome (Metzker, 2010), and sequencing personal genomes (Metzker, 2010). The latter would have many benefits including the investigation of the personal variants in the genome which are related to personal phenotypes in disease or health.

5.4 ChIP on Microarrays and Sequencing

The chromatin is the combination of the DNA molecule and the proteins packaging it in the nuclei of eukaryotic cells. Many proteins, such as chromatin-modifying proteins and transcription factors, meet their roles in the cell through binding to the chromatin, whether that was by direct binding to the DNA molecule or to the proteins packaging it. Such chromatin-binding proteins can be selective in terms of the DNA sequence or the chromatin marks to which they bind. For example, all of the non-generic transcription factors selectively bind to specific short sequences of DNA (motifs) which reside close to those transcription factors' target genes. The identification of the proteins selectively binding to the proximities of specific genes greatly enhances our understanding of gene regulation and the relations between the regulators and their targets.

ChIP is a technique by which the association of proteins with the chromatin is measured. When combined with microarray chips (ChIP-chip) or DNA sequencing (ChIP-seq), high-throughput data quantifying the levels and locations of a specific protein association with the DNA are generated. The ChIP process starts by chemically cross-linking the DNA and its associated proteins. Then, DNA, with its cross-linked proteins, is extracted from the nuclei and fragmented. Antibodies to the specific protein(s) under investigation are used to *immuno-precipitate* the fragments of DNA associated with this (those) specific protein(s) while filtering out the rest of the DNA fragments. After that, the chemical cross-links are released to keep bare DNA fragments, which are either exposed to a microarray chip (ChIP-chip) or sequenced (ChIP-seq) in order to identify the DNA sequences (and nearby genes) that are associated with that (those) protein(s) (Mardis, 2008; Liang *et al.*, 2013; Pareek, Smoczynski and Tretyn, 2013).

5.5 Discussion and Summary

Technological advancements in the last two decades have greatly boosted the rate of data generation. The consequence is the emergence of huge amounts of high-throughput datasets which report various biological variables at very large scales.

One of the most widely adopted high-throughput techniques is the microarray. This is a small chip on which a large number (tens to hundreds of thousands) of immobilised specific probes are implanted. A sample of molecules dyed with fluorescent labels is exposed to this chip. Depending on their chemical and physical properties, some of those molecules would bind to some of the probes. When the chip is scanned, probes bound by more fluorescent molecules show light signals with higher intensities. Finally, the intensities are reported as values quantifying the levels of specific association between the sample of molecules and the specific corresponding probes. DNA microarrays and protein analytical microarrays (antibody microarrays) are used to measure the levels of gene expression and protein expression (abundance), respectively. Functional protein microarrays, carbohydrate microarrays, microRNA microarrays, SMM and many other types of microarrays are tools for high-throughput screening of the functions of various types of biological molecules in the cell.

Sequencing methods have been used for more than two decades to identify the nucleobase-by-nucleobase sequences of whole genomes or fragments of DNA or RNA molecules. However, it had been very expensive before the introduction of the modern NGS techniques. Thanks to these NGS techniques, the cost of human genome sequencing dropped (in US dollar terms) from hundreds of millions to a few thousands in a decade, and the aim is to reach the $1000 line, which allows for numerous important applications that had been infeasible ten years ago.

NGS techniques compete with microarrays in some areas like gene expression analysis and high-throughput ChIP. The latter is a technique which investigates the locations and levels of protein association with the DNA. ChIP analysis, whether by using microarray chips (ChIP-chip) or sequencing techniques (ChIP-seq), produces high-throughput data that enhance our understanding of key processes in gene regulation, epigenetics and others.

Taken together, microarrays, NGS and other techniques of high-throughput data generation are being heavily and increasingly used to produce massive amounts of raw datasets. Additionally, developments are still taking place which aim at providing more accurate, cheaper and more productive techniques in the future. This imposes challenges on the developers of the computational methods which are increasingly required in order to boost the pace of data analysis and discovery inference to cope with that pace of data generation.

References

Alwine, J.C., Kemp, D.J. and Stark, G.R. (1977). Method for detection of specific RNAs in agarose gels by transfer to diazobenzyloxymethyl-paper and hybridization with DNA probes. *Proceedings of the National Academy of Sciences*, **74**(12), pp. 5350–5354.

Arthur, C.M., Cummings, R.D. and Stowell, S.R. (2014). Using glycan microarrays to understand immunity. *Current Opinion in Chemical Biology*, **18**, pp. 55–61.

Aston, F.W. (1919). A positive ray spectrograph. *Philosophical Magazine Series 6*, **38**(228), pp. 707–714.

Blixt, O. and Westerlind, U. (2014). Arraying the post-translational glycoproteome (PTG). *Current Opinion in Chemical Biology*, **18**, pp. 62–69.

Burnette, W.N. (1981). "Western Blotting": electrophoretic transfer of proteins from sodium dodecyl sulfate-polyacrylamide gels to unmodified nitrocellulose and radiographic detection with antibody and radioiodinated protein A. *Analytical Biochemistry*, **112**(2), pp. 195–203.

Calza, S. and Pawitan, Y. (2010). Normalization of gene-expression microarray data. *Methods in Molecular Biology*, **673**, pp. 37–52.

Chang, T.-W. (1983). Binding of cells to matrixes of distinct antibodies coated on solid surface. *Journal of Immunological Methods*, **65**(1–2), pp. 217–223.

Chen, D.S. and Davis, M.M. (2006). Molecular and functional analysis using live cell microarrays. *Current Opinion in Chemical Biology*, **10**(1), pp. 28–34.

Chen, C.-S. and Zhu, H. (2006). Protein microarrays. *BioTechniques*, **40**(4), pp. 423–429.

Chevolot, Y., Laurenceau, E., Phaner-Goutorbe, M. *et al.* (2014). DNA directed immobilization glycocluster array: applications and perspectives. *Current Opinion in Chemical Biology*, **18**, pp. 46–54.

Collins, F., Lander, E., Rogers, J. and Waterson, R. (2004). Finishing the euchromatic sequence of the human genome. *Nature*, **431**(7011), pp. 931–945.

Dijk, E.L.v., Jaszczyszyn, Y. and Thermes, C. (2014). Library preparation methods for next-generation sequencing: tone down the bias. *Experimental Cell Research*, **322**(1), pp. 12–20.

Fang, Z., Martin, J. and Wang, Z. (2012). Statistical methods for identifying differentially expressed genes in RNA-Seq experiments. *Cell & Bioscience*, **2**, p. 26.

Galban-Horcajo, F., Halstead, S.K., McGonigal, R. and Willison, H.J. (2014). The application of glycosphingolipid arrays to autoantibody detection in neuroimmunological disorders. *Current Opinion in Chemical Biology*, **18**, pp. 78–86.

Geissner, A., Anish, C. and Seeberger, P.H. (2014). Glycan arrays as tools for infectious disease research. *Current Opinion in Chemical Biology*, **18**, pp. 38–45.

Gilmour, D.S. and Lis, J.T. (1985). In vivo interactions of RNA polymerase II with genes of Drosophila melanogaster. *Molecular and Cellular Biology*, **5**(8), pp. 2009–2018.

Haab, B.B., Dunham, M.J. and Brown, P.O. (2001). Protein microarrays for highly parallel detection and quantitation of specific proteins and antibodies in complex solutions. *Genome Biology*, **2**(2), pp. research0004.1–research0004.13.

Hayden, E.C. (2014). The $1,000 genome. *Nature*, **507**, pp. 294–295.

Hong, J.A., Neel, D.V., Wassar, D. *et al.* (2014). Recent discoveries and applications involving small-molecule microarrays. *Current Opinion in Chemical Biology*, **18**, pp. 21–28.

Kembel, S.W., Meadow, J.F., O'Connor, T.K. *et al.* (2014). Architectural design drives the biogeography of indoor bacterial communities. *Plos One*, **9**(1), e87093.

Ku, C.S., Naidoo, N., Wu, M. and Soong, R. (2011). Studying the epigenome using next generation sequencing. *Journal of Medical Genetics*, **48**(11), pp. 721–730.

Laurent, N., Voglmeir, J. and Flitsch, S.L. (2008). Glycoarrays—tools for determining protein–carbohydrate interactions and glycoenzyme specificity. *Chemical Communications*, **7**(37), pp. 4400–4412.

Liang, C.-H., Wang, S.K., Lin, C.W. *et al.* (2011). Effects of neighboring glycans on antibody–carbohydrate interaction. *Angewandte Chemie International Edition*, **50**(7), pp. 1608–1612.

Liang, F., Xu, K., Gong, Z.J. *et al.* (2013). ChIP-seq: a new technique for genome-wide profiling of protein-DNA interaction. *Progress in Biochemistry and Biophysics*, **40**(3), pp. 216–227.

Liu, Y., Feizi, T., Campanero-Rhodes, M.A. *et al.* (2007). Neoglycolipid probes prepared via oxime ligation for microarray analysis of oligosaccharide-protein interactions. *Chemistry and Biology*, **14**(7), pp. 847–859.

Mardis, E.R. (2008). Next-generation DNA sequencing methods. *Annual Review of Genomics and Human Genetics*, **9**, pp. 387–402.

Metzker, M.L. (2010). Sequencing technologies — the next generation. *Nature Reviews Genetics*, **11**(1), pp. 31–46.

Mori, A., Deola, S., Xumerle, L. *et al.* (2013). Next generation sequencing: new tools in immunology and hematology. *Blood Research*, **48**(4), pp. 242–249.

Palma, A.S., Feizi, T., Childs, R.A. *et al.* (2014). The neoglycolipid (NGL)-based oligosaccharide microarray system poised to decipher the meta-glycome. *Current Opinion in Chemical Biology*, **18**, pp. 87–94.

Pareek, C.S., Smoczynski, R. and Tretyn, A. (2013). Sequencing technologies and genome sequencing. *Journal of Applied Genetics*, **52**(4), pp. 413–435.

Pratsch, K., Wellhausen, R. and Seitz, H. (2014). Advances in the quantification of protein microarrays. *Current Opinion in Chemical Biology*, **18**, pp. 16–20.

Song, X., Lasanajak, Y., Xia, B. *et al.* (2011). Shotgun glycomics: a microarray strategy for functional glycomics. *Nature Methods*, **8**(1), pp. 85–90.

Song, X., Heimburg-Molinaro, J., Cummings, R.D. and Smith, D.F. (2014). Chemistry of natural glycan microarrays. *Current Opinion in Chemical Biology*, **18**, pp. 70–77.

Southern, E.M. (1975). Detection of specific sequences among DNA fragments separated by gel electrophoresis. *Journal of Molecular Biology*, **98**(3), pp. 503–517.

Voskuhl, J., Brinkmann, J. and Jonkheijm, P. (2014). Advances in contact printing technologies of carbohydrate, peptide and protein arrays. *Current Opinion in Chemical Biology*, **18**, pp. 1–7.

Wu, Z. (2009). A review of statistical methods for preprocessing oligonucleotide microarrays. *Statistical Methods in Medical Research*, **18**, pp. 533–541.

Zhu, H., Bilgin, M., Bangham, R. *et al.* (2001). Global analysis of protein activities using proteome chips. *Science*, **293**(5537), pp. 2101–2105.

6

Databases, Standards and Annotation

6.1 Introduction

One of the aspects that are tackled by bioinformatics is the storage, organisation, standardisation and annotation of high-throughput datasets in a way that enhances data retrieval and consolidates efficient knowledge transfer between the members of the research community. Therefore, parts of the funds given to researchers in order to establish the recent revolution in molecular biological research have been spent on building massive centralised and standardised high-throughput data repositories. Although managed by centralised consortiums or committees, the data content of such repositories is generally provided by the entirety of the bioinformatics research community from all over the world.

Data submission to centralised repositories has become the norm, and the community is increasingly adhering to those centralised standards and protocols of annotation. Furthermore, many research journals do not publish studies that include analysis of newly generated datasets without in tandem publishing of the datasets, and the encouragement is towards publishing those datasets through public centralised repositories in contrast to private websites.

The databases provided by the National Centre for Biotechnology Informatics (NCBI) and the European Bioinformatics Institute (EBI) represent major resources in this field. Thus, we start by introducing them. Then, we introduce some databases that are specialised in one species or a number of related species. While introducing the databases and repositories, we describe the standard labels and identifiers that are used to annotate the relevant biological variables.

6.2 NCBI Databases

The NCBI was founded in 1988 as a division of the National Library of Medicine (NLM), which is a department at the U.S. National Institute of Health (NIH). The major objective

Integrative Cluster Analysis in Bioinformatics, First Edition. Basel Abu-Jamous, Rui Fa and Asoke K. Nandi.
© 2015 John Wiley & Sons, Ltd. Published 2015 by John Wiley & Sons, Ltd.

Figure 6.1   NCBI's entrez search bar

of this centre is to establish systems which organise storing and analysing data and information about molecular biology and genetics through databases, software tools and standards for data deposition, exchange and biological nomenclature.

The NCBI hosts large number of databases which are specialised in different aspects related to molecular biology and genetics. Excellently designed, the entries of any of those databases are well linked to their related entries in the other databases, which altogether represent a comprehensively integrated data and knowledge resource for the research community.

The contents of the databases are available through both the NCBI HTTP website (http://www.ncbi.nlm.nih.gov) and the FTP website (ftp://ftp.ncbi.nlm.nih.gov). *Entrez* is the NCBI's universal search engine with which users can search for specific keywords or identifiers in all of the databases or in specific ones of them. Entrez search is carried out by using the search bar at the top of the NCBI's HTTP home page (Figure 6.1). The drop-down list provided by this search bar is a list of all of the searchable NCBI databases, and selecting one of them from this list indicates that it is the target database over which search will be performed. An option "All Databases" is available which indicates that all of the databases will be searched for the given keywords.

In this section we will describe some of the NCBI databases which are most relevant to the scope of this book, and we will describe how to navigate through them and through the links between their entries. While we briefly describe some other useful NCBI databases, we refer the reader to the webpage at http://www.ncbi.nlm.nih.gov/guide/all/#databases for the complete list of databases.

6.2.1   Literature Databases (PubMed, PMC, the Bookshelf and MeSH)

PubMed is a citation repository which accesses the *MEDLINE* database of life sciences literature and other databases (Acland *et al.*, 2014). More than 20 million citations have been stored in MEDLINE by 2014 with their abstracts and, in many cases, links to full-texts. PubMed search is performed by selecting "PubMed" from the Entrez search drop-down list (Figure 6.1). Information such as PubMed unique identifier (PMID), title, date, source, authors, volume, pages, digital object identifiers (doi), links to related entries in other NCBI databases, and others are provided for each applicable article accessed by PubMed. The PMID of a PubMed entry is a number that uniquely identifies it, and is used to refer to it from the rest of the NCBI databases. Moreover, PubMed citations can be downloaded as files that are importable by some of the most popular citation managers such as EndNote and Reference Manager (Acland *et al.*, 2014).

PubMed Central (*PMC*) is a repository of full-texts of journal papers that is fed by collaborations with journals in life sciences as well as direct submissions by contributors (Acland *et al.*, 2014). By 2014, about 1500 journals have fully participated in PMC enrichment, about 2500 journals have selectively participated, and about three million full-text articles have been stored in the PMC database. Articles in PMC are identified with unique identifiers starting with the prefix 'PMCID'. If a PubMed citation article has a PMC full-text, its PMCID and a link to the PMC full-text are provided by the PubMed record.

In addition to PubMed and the PMC, the NCBI provides the *Bookshelf* database, which indexes books, reports, and documents in life sciences, brief information about them and their full-texts (Hoeppner, 2013; Acland *et al.*, 2014). Each book in this database is identified by a unique identifier that starts with the letters 'NBK' followed by numerical digits. Furthermore, the NCBI provides access to the U.S. NLM catalogue of journals which stores information about journals in life sciences with all PubMed journal articles linked to their corresponding NLM journal entries. By 2014, more than 220 000 books, reports and documents have been indexed by the Bookshelf (Acland *et al.*, 2014).

The medical subject headings (MeSH) database, hosted by the NLM and accessed through NCBI Entrez search, organises controlled terms that represent headings of medical subjects in tree-form (Acland *et al.*, 2014). Any node in the tree represents a subheading within its parent node, and a parent-heading for its descendant nodes. For example, the heading "gene expression" is a subheading of the term "genetic processes", which is a subheading of the term "genetic phenomena", which is a subheading of the term "phenomena and processes category", which is the root of one of the main trees of heading terms in MeSH. This tree, "phenomena and processes category", is uniquely labelled with the tree letter 'G', and its child term "genetic phenomena" is accordingly labelled as G05. The unique tree identifier for the heading "gene expression" in MeSH is G05.355.310, which corresponds to the sequence of headings starting from the root tree "G", through the terms "genetic phenomena" (G05) and "genetic processes" (G05.355), and ending at the desired node "gene expression" (G05.355.310). By 2014, more than 240 000 terms have been recorded by the MeSH database (Acland *et al.*, 2014).

6.2.2   GenBank (Nucleotide Database)

The *GenBank* database (Benson *et al.*, 2014), assumed by the NCBI in 1992, stores DNA sequences submitted by individual laboratories as well as through regular data exchange with databases like the European Nucleotide Archive (ENA) of the European Molecular Biology Laboratory (EMBL) and the DNA Database of Japan (DDBJ). By 2014, the GenBank has stored more than half a trillion nucleotide base pairs for more than 280 000 formally described species. The source of most of this data is whole-genome shotgun sequencing, which is still growing rapidly (Benson *et al.*, 2014).

Searching the GenBank database can be performed by selecting "Nucleotide" as the Entrez search target database. Moreover, the *basic local alignment search tool (BLAST)*, provided by the NCBI at the website http://blast.ncbi.nlm.nih.gov, facilitates very fast sequence similarity searches over the GenBank database, as well as other sequence databases (Benson *et al.*, 2014).

6.2.3   Reference Sequences (RefSeq) Database

RefSeq is a collection of sequences for nucleotides (DNA and RNA) and proteins (Pruitt *et al.*, 2012). RefSeq unique accession identifiers are characterised with a prefix of two letters followed by an underscore then a number. Genomic (DNA) sequences have the prefixes AC, NC, NG, NT, NW, NS and NZ; RNA sequences have the prefixes NM, NR, XM and XR; and protein sequences have the prefixes AP, NP, YP, XP and ZP (Pruitt *et al.*, 2002). The contents of RefSeq can be accessed through the relevant entries in the NCBI GenBank (Nucleotide) and Protein databases. Reporting an increase of more than 60-fold over a decade, more than 300 billion nucleotide bases and 11 billion protein amino acid residues have been collected and stored by RefSeq by July 2013 (Acland *et al.*, 2014).

6.2.4   Gene Database

The Entrez *Gene* database is a very comprehensive database of integrated information about genes from species with completely sequenced genomes, as well as the genomes which have active research communities, with more than 14 million gene records having been stored by 2014 (Maglott *et al.*, 2011; Acland *et al.*, 2014). Each gene is provided a unique, stable and tracked integer identifier known as the *Entrez Gene ID*, which is used to refer to this gene from the relevant entries in all of the other NCBI databases.

A gene's profile in this database is comprehensive as it includes information regarding the gene's official name and symbol, type (e.g. protein coding), organism, other names (synonyms), a summary paragraph, sequence and locus information, genomic regions, transcripts, products, single-nucleotide polymorphisms (SNPs), list of relevant literature that cite this gene, gene's references to functions, phenotypes, variations, pathways, interactions with other genes and their products, homology, gene ontology (GO), protein information, reference sequences, related sequences, epigenomics and others. Most of those pieces of information are stored in their corresponding specialised NCBI databases (e.g. sequences in GenBank (Benson *et al.*, 2014), protein information in Protein database, SNPs in the dbSNP database (Sherry *et al.*, 2001; Bhagwat, 2010), bibliography in PubMed etc.), where their relevant entries are linked to that gene through its gene ID. In other words, a gene's profile is its basic information followed by an integrated compilation of those other databases' relevant pieces of information. In a symmetric manner, this gene's profile links out to the corresponding entries in those source databases.

6.2.5   Protein Database

The protein database includes summaries about protein sequences that are translated from the nucleic acid sequences in the GenBank, RefSeq, and Third Party Annotation (TPA) databases and others. More than 90 million protein sequences have been stored in this database by 2014 (Acland *et al.*, 2014). Each protein's sequence has an accession number which is a unique identifier. Those accessions are indexed by the RefSeq database which includes nucleotide sequences (DNA and RNA) as well as protein sequences. Protein accession identifiers always include the letter 'P' as the second letter of their prefixes; for example, the accession identifier for the sequence of the human protein encoded by the CPOX gene is NP_000088.

6.2.6   Gene Expression Omnibus

The Gene Expression Omnibus (GEO) is a repository of microarray and next-generation sequencing (NGS) datasets (Barrett *et al.*, 2013). The study records are organised as series of samples, where each series is stored alongside its description and related information such as the organism of the study, the microarray or NGS platform, the submitter's contact, affiliation and country, and the list of samples belonging to the series. GEO database stores structured records for microarray and NGS platforms as well as for the series' samples. Series, platforms and samples are identified with unique accession identifiers starting with the respective prefixes 'GSE', 'GPL' and 'GSM', followed by unique numbers.

Series, samples and platforms are cross-linked through their accession numbers. Each series record links out to the record of the platform it was generated with as well as the records of its samples, each sample's record links out to its containing series and the base platform, and each platform's record lists hyperlinked accession identifiers for all of the series and samples generated using it. Moreover, each of those types of GEO records links out to the Taxonomy database record which represents the species to which this GEO record belongs, and the BioProject database's record which represents the project (study) under which this sample or series was generated.

Importantly, complete raw or preprocessed data files for those series, samples and platforms are downloadable from the NCBI HTTP webpages presenting those records as well as through the corresponding FTP web folder. They can be retrieved by the MATLAB function '*getgeodata*' by providing the GEO accession number as the sole argument to the function.

When a series is read through the HTTP site, the FTP site or a research environment like MATLAB, the retrieved data will be a matrix of rows representing microarray probes or NGS sequence identifiers, and columns representing samples. Indeed if a sample is read, it will include a single column of data in addition to some meta-data. Despite that, there will be no detailed meta-data describing the microarray probes or the NGS sequences. This detailed information, such as the genes represented by specific probes or sequences, is included in the platform's record, which can be retrieved in the same manner as retrieving series' and samples' records.

6.2.7   Taxonomy and HomoloGene Databases

The NCBI *Taxonomy* database is a repository of all of the organisms for which there are genetic or protein sequences stored by the International Nucleotide Sequence Database Collaboration (INSDC). This collaboration comprises the NCBI GenBank database, the ENA-EMBL, and the DDBJ (Federhen, 2012). The taxonomies, of which there are more than one million, are labelled with unique integer identifiers. Whenever the other NCBI databases report the organism of a gene, a protein, a data sample, a gene expression series, a gene expression platform or the like, the reported organism is linked to its corresponding record in the Taxonomy database.

The NCBI also provides the *HomoloGene* database, which links homologous genes from the same species (paralogues) or different species (orthologues) through HomoloGene groups. The relation between the HomoloGene database and the Gene database is a one-to-many relation where a single HomoloGene group, which is identified by a unique integer identifier, is associated with a group of homologous genes through their Gene IDs. Entries in this database consider genes from 21 completely sequenced eukaryotic species with more than 44 000 recorded HomoloGene groups by 2014 (Acland *et al.*, 2014).

6.2.8   Sequence Read Archive

The Sequence Read Archive (SRA) database, founded in 2009 as part of the INSDC, stores records from NGS studies (Kodama, Shumway and Leinonen, 2012). SRA experiments, studies, runs and analysis records are labelled with unique identifiers that have the prefixes 'SRX', 'SRP', 'SRR' and 'SRZ', respectively. By April 2014, more than 2400 Terabases have been stored by the SRA database, where more than 1100 Terabases of them are available as open-access sequence reads.

6.2.9   Genomic and Epigenomic Variations

Genomics short variations, including SNPs, and large variations, including insertions, deletions, translocations and inversions, are archived in the NCBI databases dbSNP (Sherry *et al.*, 2001; Saccone *et al.*, 2011) and dbVar, respectively (Church *et al.*, 2010). The records in both databases have exceeded 300 million and 3.5 million records, respectively, by 2014 (Acland *et al.*, 2014).

Epigenetic variations, the variations that occur to the genome other than sequence variations, are also archived by the NCBI in their specific database, *Epigenomics* (Fingerman *et al.*, 2013). Examples of epigenetic variations include post-translational histone protein modification, DNA methylation, chromatin modification and non-coding RNA expression (Acland *et al.*, 2014).

6.2.10   Other NCBI Databases

The NCBI provides many databases other than the ones described in the previous sub-sections. The *BioProject* is a repository of projects (studies) that involve high-throughput data generation and analysis, their attributes and their associated NCBI datasets (e.g. GEO series) (Barrett *et al.*, 2012). The *BioSample* database is a repository of biological samples that belong to datasets submitted to the NCBI such as sequencing data, gene expression data, epigenomics and others (Barrett *et al.*, 2012). The *BioSystems* database archives biological systems and their related biological molecules. A record in this repository includes a list of genes, proteins and small molecules, indeed through their NCBI cross-linked identifiers, and includes their interactions within the considered system which might be a metabolic pathway, a signalling pathway, a disease profile or any other type of biological system (Geer *et al.*, 2010).

The *PubChem* databases, namely *PubChem Substance, PubChem Compound* and *PubChem BioAssay,* are databases of records related to the chemical compounds in biology and their properties (Wang *et al.*, 2009, 2012). PubChem Substance is a repository of chemical substances and their attributes as submitted by the researchers in the research community. PubChem Compound, which links to the PubChem Substance database, is a repository of validated chemical structures. Finally, the PubChem BioAssay database stores data and descriptions of the bioactivity assays that screen the chemical substances and compounds in the two former PubChem databases.

The NCBI database of Genotypes and Phenotypes (*dbGaP*) stores records that associate genotypes (genomic characteristics) with phenotypes (observed characteristics). Genotypes include genomic and RNA sequence data, genomic variations (SNP and large variations), epigenomic variations and gene copy number variations. The records of dbGaP include studies, sets of analysis, datasets and documents (Tryka *et al.*, 2014). The Entrez's Molecular Modelling Database (*MMDB*) is a repository of proteins' 3D structures (Chen *et al.*, 2003), the *Genome* database archives whole genomes, whether they are complete or in progress (Acland *et al.*, 2014), and the *UniGene* database stores sets of transcript sequences which appear to come from the same transcriptional locus (Acland *et al.*, 2014).

Indeed, each record in any of those databases has a unique identifier, which in most of the cases starts with a prefix that indicates the hosting database (e.g. 'BSID' for BioSystems records' identifiers). Moreover, those databases are heavily cross-linked. For example, references to genes or proteins in any of the databases' records consider those genes' or proteins' identifiers in the NCBI Gene or Protein databases, references to the literature articles that relate to any record in any of the databases that consider the PMIDs of such articles, and so on.

6.3 The EBI Databases

The EBI or EMBL-EBI, which is a part of the EMBL, provides a number of databases and tools that are publically accessible by the bioinformatics research community (Brooksbank *et al.*, 2014). Together with the U.S. NCBI databases and the DDBJ, they constitute the major species-generic publically accessible high-throughput data repositories for molecular biology. The EBI databases adopt user-centred design (UCD), which concentrates on the easiness and intuitiveness in use by target users (Brooksbank *et al.*, 2014). The homepage of the EMBL-EBI is at www.ebi.ac.uk.

The databases of the EBI are organised in a manner that corresponds to the layers of the central dogma in molecular biology; those layers, in order, are genomics (DNA), transcriptomics (RNA), proteomics, structures, chemical biology and systems. As supplementary to that, there are databases for the literature, ontologies, and cross-domain tools and resources (Brooksbank *et al.*, 2014).

DNA and RNA databases include Ensembl, Ensembl Genomes, the European Genome-phenome Archive (EGA), the Metagenomics portal, the ENA, the Genome-Wide Association Studies (GWAS) archive and others (Brooksbank *et al.*, 2014). *Ensembl,* which is a joint project between the EMBL-EBI and the Wellcome Trust Sanger Institute, is a repository of genomic data for model vertebrate species. Each gene in Ensembl is assigned a unique identifier with the prefix of 'ENSG' (Flicek *et al.*, 2014). By 2014, 73 vertebrate species have been supported by the Ensembl database, in addition to three non-vertebrate model species – the *Caenorhabditis elegans* worm, the *Drosophila melanogaster* fruit fly, and the *Saccharomyces cerevisiae* baker's yeast (Flicek *et al.*, 2014). Corresponding gene records in the Ensembl and NCBI Gene databases are mutually cross-linked, facilitating easy cross-navigation between those databases. *Ensembl Genomes* database extends the Ensembl's taxonomic coverage of reference genomes by including thousands of genomes from non-vertebrate animals, plants and bacteria (Kersey *et al.*, 2012). The *EGA* links genetic characteristics (genotypes) with their consequent observed traits (phenotypes) (Brooksbank *et al.*, 2014). The *Metagenomics* portal archives metagenomic studies, which are studies of collective genomic information from an environment containing various species (Hunter *et al.*, 2014). The ENA stores nucleotide sequence data ranging from raw reads to assemblies and functional annotation (Pakseresht *et al.*, 2014). Finally, the GWAS Catalogue, which is a collaborative database between EMBL-EBI and the US National Human Genome Research Institute (NHGRI), represents a resource for SNP information (Welter *et al.*, 2014).

Genetic and protein-expression EMBL-EBI databases include the ArrayExpress Archive, Expression Atlas, Proteomics Identifications (PRIDE) database and MetaboLights (Brooksbank *et al.*, 2014). ArrayExpress Archive complements and overlaps with the NCBI GEO database and the DDBJ Omics Archive for gene expression data archiving from both microarray- and sequencing-based experiments (Rustici *et al.*, 2013). The Expression Atlas database adds another layer of gene expression analysis to the datasets stored in the ArrayExpress Archive by facilitating queries about genetic differential expression across biological conditions such as tissues and cell types. The underlying data sources for this Atlas include the datasets in the ArrayExpress Archive, the NCBI GEO datasets and the ENA (Adamusiak *et al.*, 2012). The PRIDE database is a repository of mass spectrometry - (MS)-based proteomics data, such as protein-expression data (Vizcaíno *et al.*, 2013). Lastly, MetaboLights is a public repository for metabolomics data, including raw experimental data and associated meta-data (Haug *et al.*, 2013).

EMBL-EBI's Protein information databases include UniProt, InterPro and others (Brooksbank *et al.*, 2014). UniProt, which is managed by a collaboration between the EMBL-EBI, the Swiss Institute of Bioinformatics (SIB), the University of Georgetown and the University of Delaware, is a unified database of information regarding protein sequences and functions (The UniProt Consortium, 2014). The UniProt database is well integrated with the Ensembl and Ensembl Genomes database, and is complemented by the InterPro database. The latter is an archive of protein families, domains, functional sites and motifs (Hunter *et al.*, 2009).

The Electron Microscopy Data Bank (EMDB) and the Protein Data Bank in Europe (PDBe) are amongst the EMBL-EBI's databases of molecular and cellular structures (Brooksbank *et al.*, 2014; Gutmanas *et al.*, 2014). ChEMBL and ChEBI are two databases of chemical biology which archive drugs, biochemical activities, reference chemical structures, nomenclature and ontological classification, and natural compound classification (Willighagen *et al.*, 2013; Brooksbank *et al.*, 2014).

Systems databases by the EMBL-EBI include BioModels, IntAct, the Reactome and the Enzyme Portal (Brooksbank *et al.*, 2014). BioModels database archives computational models of biological processes (Chelliah, Laibe and Novère, 2013), IntAct is a repository of molecular interactions (Kerrien *et al.*, 2012; Orchard *et al.*, 2014), the Reactome is a comprehensive database of curated human pathways including molecular interactions as well as structural and expression data (Croft *et al.*, 2014) and the Enzyme Portal is an integrated database of enzymes and their functions, resolved structures, reactions, pathways, substrates, products, and related diseases and literature (Alcántara *et al.*, 2013).

As for literature databases, the EMBL-EBI manages the Europe PubMed Central (Europe PMC) in collaboration with the University of Manchester and the British Library. The Europe PMC is part of the international PMC run by the NCBI (McEntyre *et al.*, 2011; Brooksbank *et al.*, 2014).

The EMBL-EBI researchers play key roles in feeding the GO Consortium initiative, which involves many research groups and institutes aiming at unifying the representation of the attributes of genes and their products. The GO repository includes associations of genes to their or their products' attributes such as their biological processes, molecular functions and cellular components. The EMBL-EBI has provided added-value GO databases such as the Experimental Factor Ontology (EFO) database. This database aims at annotating gene expression datasets and GWAS, as well as integrating genomic and disease data (Malone *et al.*, 2010; Brooksbank *et al.*, 2014).

The EBI BioSamples database is a cross-domain resource of biological samples. It resembles the BioSample databases of the US NCBI and the DDBJ in terms of types of stored information and function. The three bodies agreed on a common accessioning scheme, and consequently data exchange takes place between them (Barrett *et al.*, 2012; Faulconbridge *et al.*, 2014).

6.4   Species-specific Databases

The majority of the databases provided by the US NCBI and the EBI are general with regards to the species they consider. More specialised databases exist for more focused investigation of a specific model species or a group of related species. We review some of the commonly used species-specific data sources in this section.

6.4.1   Animals

The Human Metabolome Database (*HMDB*), whose homepage is at www.hmdb.ca, was developed by the Wishart Research Group at the University of Alberta in Canada as part of the Human Metabolome Project. This database archives chemical, clinical and biochemical information about the metabolites (small molecules) found in human bodies (Wishart *et al.*, 2013). Version 3.5 contains more than 40 000 metabolite entries and more than 5600 linked protein sequences. The included metabolites are from a wide range of abundance level from rare to very abundant, and with different properties such as being water-soluble or lipid-soluble.

A model mammal organism of research interest is the house mouse *Mus musculus*. The Mouse Genome Informatics (*MGI*) databases, with a homepage address at www.informatics.jax.org, constitute a comprehensive resource for mouse genomic data and analysis. MGI is run by the non-profit molecular biology research institute, the Jackson Laboratory, and funded mainly by the US NHGRI, which is part of the US NIH. Different databases are included within the MGI, such as the Mouse Genome Database (MGD) (Blake *et al.*, 2014), the mouse Gene Expression Database (GXD) (Smith *et al.*, 2014), the Mouse Tumour Biology database (MTB) (Begley *et al.*, 2012), the Mouse Phenome Database (MPD) (Grubb, Bult and Bogue, 2014) and others.

Another useful resource for mouse molecular biology data is the Edinburgh Mouse Atlas Project (*EMAP*), which has its homepage at www.emouseatlas.org. EMAP includes the E-Mouse Anatomy Atlas (EMA) as well as the Mouse Gene Expression Spatial Database (EMAGE) (Hayamizu *et al.*, 2013; Richardson *et al.*, 2014).

Another model mammalian organism, the Norway rat *Rattus norvegicus*, is the focus of the Rat Genome Database (*RGD*) hosted at rgd.mcw.edu. This database is run by the group of Dr Howard J. Jacob at the Medical College of Wisconsin, and funded by the National Heart Lung and Blood Institute (NHLBI) of the US NIH. This database provides data regarding rat's genome, genes, phenotypes, strains, diseases, physiology and nutrition (Cruz *et al.*, 2005; Laulederkind *et al.*, 2013).

Beyond mammals, the databases *Xenbase*, hosted at www.xenbase.org (James-Zorn *et al.*, 2013), Zebrafish Information Network (*ZFIN*), hosted at www.zfin.org (Howe *et al.*, 2013), *WormBase*, hosted at www.wormbase.org (Harris *et al.*, 2014), and *FlyBase*, hosted at www.flybase.org (St Pierre *et al.*, 2014), are data resources that focus on selected model organisms belonging to amphibians, fish, worms and insects, respectively. Those databases provide different ranges of integrated data about genomics, gene expression, sequences, anatomy and others.

6.4.2 Plants

The *GreenPhyl* database (Rouard *et al.*, 2011) and the Munich Information Centre for Protein Sequences (MIPS) *PlantDB* (Nussbaumer *et al.*, 2013) databases are focused on integrative and comparative analysis of plant genomes and building their phylogenetic trees. The homepage of the former is at www.greenphyl.org while the homepage of the latter is at mips.helmholtz-muenchen.de/plant/genomes.jsp.

In contrast, some databases are focused on narrower subsets of botanical species. For instance, The Arabidopsis Information Resource (*TAIR*) (Lamesch *et al.*, 2012), hosted at www.arabidopsis.org, and the Maize Genetics and Genomics Database (*MaizeGDB*) (Schaeffer *et al.*, 2011), hosted at www.maizegdb.org, store integrative data regarding the genes, proteins, metabolic pathways and others for the model plant *Arabidopsis thaliana* (thale cress) and the model crop *Zea mays* (maize), respectively.

6.4.3 Fungi

The Stanford University's Saccharomyces Genome Database (*SGD*) is the most popular database that is specialised in baker's yeast (*Saccharomyces cerevisiae*) and relative species molecular biology data. The SGD database is manually curated from the studies in the literature in a well organised and easily accessible manner (Cherry *et al.*, 2012). Within the SGD website www.yeastgenome.org, every yeast gene has a single webpage that has tabs focusing on the locus history, relevant literature articles, GO, phenotypes, interactions, expression profiles, genetic regulation, gene's product (protein) and the corresponding Wikipedia page. A summary tab is also available which summarises the key aspects from all of the other tabs, and this is the default tab that is shown when the gene's page is requested. SGD assigns each gene a unique identifier (SGDID) which starts with the prefix 'S', and provides links to the relevant gene or gene product records in many other databases such as the NCBI Entrez Gene, the EMBL-EBI UniProt and the MIPS CYGD databases.

The literature in the SGD database is thoroughly manually curated where each article is given a unique SGD identifier and the genes for which an article is considered as a primary or secondary source are identified. Indeed, links to those articles' records in their publishing journals as well as the PubMed and the PMC databases are provided. SGD provides various tools for sequence and GO analysis such as BLAST, GO Slim Mapper and GO Term Finder. It has recently included gene regulation information as well (Costanzo *et al.*, 2014).

By adopting the same design and scheme as the SGD database, a group at Stanford University provides the *Candida* Genome Database (*CGD*) at the web address www.candidagenome.org

(Binkley *et al.*, 2014). *Candida* is a genus of budding yeasts that belongs to the same family of the genus *Saccharomyces*.

The US Department of Energy Joint Genome Institute has developed a fungal genomics portal, *MycoCosm*, which includes integrated fungal genomic data, analysis and tools (Grigoriev *et al.*, 2014). MycoCosm, with a homepage at genome.jgi.doe.gov/programs/fungi, also promotes for the 1000 fungal genomes project, hosted at 1000.fungalgenomes.org, aiming at filling more gaps in the huge phylogenetic tree of fungi.

The *MIPS* Comprehensive Yeast Genome Database (CYGD), which is hosted at mips. helmholtz-muenchen.de/genre/proj/yeast, archives up-to-date data about *Saccharomyces cerevisiae* (baker's yeast) molecular structures and functional networks (Güldener *et al.*, 2005). Additionally, the Yeast Metabolome Database (*YMDB*), which is hosted at www.ymdb.ca, is a manually curated database of baker's yeast small molecules (metabolites), and their properties (Jewison *et al.*, 2012), and *YeastNet*, which is hosted at www.inetbio.org/yeastnet, is a resource for integrated functional gene networks for *Saccharomyces cerevisiae* (Kim *et al.*, 2014).

6.4.4   Archaea and Bacteria

The Microbial Genome Database (*MBGD*), funded by the Japan Society for the Promotion of Science (JSPS), is a resource for comparative genome analysis of the growing numbers of completely sequenced microbial genomes (Uchiyama *et al.*, 2013). This database, with a homepage at mbgd.genome.ad.jp, mainly includes species from archaea and bacteria, in addition to a few eukaryotic species.

The Pathosystems Resource Integration Centre (*PATRIC*), hosted at patricbrc.org, is a bacterial bioinformatics resource centre that provides the research community with integrated genomic, transcriptomic, proteomic, structural, sequencing and other types of bacterial data, metadata and tools. More than 10 000 genomes have been annotated within PATRIC in a consistent way by employing the Rapid Annotations using Subsystems Technology (*RAST*) (Overbeek *et al.*, 2014; Wattam *et al.*, 2014).

The Integrated Microbial Genome (*IMG*), whose homepage is at img.jgi.doe.gov, is a resource for integrated annotation, analysis and distribution of microbial genomic and metagenomic datasets (Markowitz *et al.*, 2014a, b). IMG and IMG Metagenomics (IMG/M) provide analysis tools as well as support for teaching relevant courses. Full access to IMG and IMG/M requires registration and logging in to their system.

The Leibniz Institute DSMZ of the German Collection of Microorganisms and Cell Cultures runs the Bacterial Diversity Metadatabase (*BacDive*), hosted at bacdive.dsmz.de. This database contains taxonomic, morphologic, physiologic, environmental and molecular-biological information about more than 23 000 bacterial and archaeal strains (as per September 2013); it also facilitates data download as well as easy but detailed search throughout its contents (Söhngen *et al.*, 2014).

The Collection of Microarrays for Bacterial Organisms (COLOMBOS) database, which is hosted at www.colombos.net, is a resource for exploring and analysing cross-platform microarray datasets for organism-specific bacteria (Meysman *et al.*, 2014). The raw microarray and NGS datasets considered by COLOMBOS are mainly from the NCBI GEO and the EMBL-EBI ArrayExpress databases. Information from cross-platform datasets is combined and analysed in COLOMBOS to provide the user with high-level information and interactive search for specific genes, pathways, transcription-regulation mechanisms, high-level condition ontology and others. An application programming interface (API) is also provided as a web service to allow researchers to access COLOMBOS through user-specific software tools. Rcolombos is an R language package that utilises that API, and it is downloadable from the webpage cran. r-project.org/web/packages/Rcolombos.

With a narrower but deeper scope, the *PortEco* database represents a comprehensive resource for knowledge and data about the model bacterial organism *Escherichia coli*, mainly the laboratorial strain *K-12* (Hu *et al.*, 2014). *E coli* is amongst the most understood organisms due to its relatively small size and ease of manipulate, and the knowledge driven from its investigation feeds our understanding regarding various aspects of molecular and cellular biology. PortEco, with a homepage at www.porteco.org, is run by a US national consortium including laboratorial biologists and computational biologists, and is funded by the US NIH.

6.4.5 *Viruses*

The ViralZone database, developed and funded by the SIB and hosted at viralzone.expasy.org, is a knowledge and data resource for viral bioinformatics. The database includes information about DNA viruses, RNA viruses and retro-viruses regarding their structures, replication cycles, interactions with hosts and other molecular-biological types of data (Masson *et al.*, 2013).

The US NCBI provides the virus-specific database Virus Variation Resource at the webpage www.ncbi.nlm.nih.gov/genomes/VirusVariation. This database archives viral gene and protein sequence annotations and relevant meta-data, as well as means to access such data resource through an HTTP and an FTP websites (Brister *et al.*, 2014).

6.5 Discussion and Summary

Large numbers of laboratories and institutes around the world participate in the generation and analysis of high-throughput biological datasets. Centralised repositories and knowledge bases which comprehensively collect such amounts of data and provide the research community with practical means of data access and retrieval are needed for efficient progression in the global understanding of molecular biology and related sciences. Indeed, centralised repositories necessitate centralised protocols and standards for data annotation and labelling. A number of such databases and repositories are listed in Table 6.1.

The US NCBI, the EBI, the SIB, and the DDBJ are the most recognisable bodies that provide centralised resorts for data depository and retrieval. The NCBI and the EBI databases, thoroughly discussed in this chapter, cover all aspects of molecular biology including DNA, RNA, and protein sequencing, genomics, genetics, proteomics, gene and protein expression, taxonomy classification, homology, genetic and epigenetic variations, chemical substances and bioactivity assays, relevant literature citations and full-texts and others. They label all of the records in those databases with unique identifiers which facilitate easy referral to them from the other databases within the same repositories or from other repositories. With this, those databases are heavily integrated through cross-linking; for example, a record summarising a gene's information has links to the identifiers of the gene's product (protein), homology group, DNA sequence, known SNPs or other genetic or epigenetic variations, relevant literature articles, GO terms and external links to this gene's records in other independent databases.

In addition to those generic data resources, there are many databases that focus on a specific species or group of related species. Many of those databases selectively concentrate on model organisms such as human, mouse and rat from mammals, thale cress and maize from plants, baker's yeast from fungi and *E. coli* from bacteria. In contrast, many other species-specific databases stretch over a wider range of species for comparison objectives, mainly regarding microorganisms such as some fungi, archaea, bacteria and viruses.

The ways in which those data repositories can be exploited are diverse. In addition to the more intuitive ways which each dataset suggests, cross-aspect analysis can be carried out by clever manipulation of the data resources. For example, cross-species analysis can be performed by considering the homology groups provided by the NCBI HomoloGene database as criteria which map different species' genes to each other. Studies considering genomic-proteomic comprehensive analysis would observe the cross-links between the records in a database of genes and the records of their products in a database of proteins. This can be extended to any sets of analysis which collectively consider different aspects of molecular biology. Moreover, bioinformaticians can exploit the FTP websites and the APIs provided by those data resources in order to access massive amounts of data in an automatic and structured manner, which is an excellent utility for meta-analysis and the next-generation intensively collective and comprehensive computational methods in bioinformatics.

Table 6.1 High-throughput data resources

Database	Provider	Description
PubMed	NCBI	Literature citations
PMC	NCBI	Full-text journal papers
Bookshelf	NCBI	Books, reports, and documents
MeSH	NCBI/US NLM	Medical subject headings
GenBank	NCBI	DNA sequences
RefSeq	NCBI	DNA, RNA, and protein sequences (accessed by GenBank (Nucleotide) and Protein databases)
Entrez Gene	NCBI	Genes
Protein	NCBI	Proteins
GEO Series	NCBI	Microarray and NGS expression datasets
GEO Platforms	NCBI	Microarray and NGS platforms
GEO Samples	NCBI	Microarray and NGS expression samples
Taxonomy	NCBI	Organisms
HomoloGene	NCBI	Genes' homologues
SRA	NCBI	NGS experiments and results
dbSNP	NCBI	Single nucleotide polymorphisms and short sequence variations
dbVar	NCBI	Large sequence variations (insertions, deletions, translocations and inversions)
BioProject	NCBI	Biological projects (studies)
BioSample	NCBI	Biological samples (e.g. expression data and epigenomics)

(*continued overleaf*)

Table 6.1 (*continued*)

Database	Provider	Description
BioSystems	NCBI	Biological systems (e.g. metabolic and signalling pathways)
PubChem Substance	NCBI	Chemical substances submitted by researchers
PubChem Compound	NCBI	Validated chemical structures – links to PubChem Substance
PubChem BioAssay	NCBI	Bioactivity assays which screen chemical substances
dbGaP	NCBI	Associations of genotypes and phenotypes
MMDB	NCBI	Protein 3D structures
Genome	NCBI	Whole genomes
UniGene	NCBI	Transcript sequences that appear to come from the same transcriptional locus
Ensembl	EMBL-EBI & Wellcome Trust Sanger Institute	Genomic data for model vertebrate species and few model non-vertebrate species (e.g. *C. elegans*, *D. melanogaster* and *S. cerevisiae*)
Ensembl Genomes	EMBL-EBI	Extends Ensembl database by including thousands of genomes
EGA	EMBL-EBI	Associations of genotypes and phenotypes
ENA	EMBL-EBI	Nucleotide (DNA and RNA) sequences
GWAS	EMBL-EBI & US NHGRI	Single-nucleotide polymorphisms
ArrayExpress Archive	EMBL-EBI	Microarray and NGS expression data
Expression Atlas	EMBL-EBI	Additional layer of analysis to the data in the ArrayExpress Archive
PRIDE	EMBL-EBI	Proteomic data (e.g. protein-expression data)
MetaboLights	EMBL-EBI	Metabolomic data
Omics Archive	DDBJ	Microarray and NGS expression data
UniProt	EMBL-EBI and others	Protein sequences and information
InterPro	EMBL-EBI	Protein families, domains, functional sites and motifs
EMDB	EMBL-EBI	Electron microscopy data
PDBe	EMBL-EBI	Protein molecular structures and cellular structures
ChEMBL	EMBL-EBI	Chemical biology data
ChEBL	EMBL-EBI	Chemical biology data
BioModels	EMBL-EBI	Computational models of biological processes
IntAct	EMBL-EBI	Molecular interactions
Reactome	EMBL-EBI	Curated human pathways
Enzyme portal	EMBL-EBI	Enzymes functions, structures, reactions, pathways, substrates, and so on
Europe PMC	EMBL-EBI and others	Full-text literature; part of the NCBI international PMC
GO	GO Consortium	Gene ontologies (biological processes, molecular functions, and cellular components)
EFO	EMBL-EBI	An added-value GO databases annotating gene expression, genome-wide associations, and integrating genomic and disease data
EBI BioSamples	EMBL-EBI	Biological samples (gene expression, epigenomics and others)

Table 6.1 (*continued*)

Database	Provider	Description
HMDB	University of Alberta	Chemical, clinical, and biochemical information about human metabolites
MGI/MGD	Jackson Laboratory	Mouse genomes
MGI/GXD	Jackson Laboratory	Mouse gene expression data
MGI/MTB	Jackson Laboratory	Mouse tumour data
MGI/MPD	Jackson Laboratory	Mouse phenome data
EMAP/EMA	UK MRC, Jackson Lab, and Heriot-Watt University	Mouse anatomy
EMAP/EMAGE	UK MRC, Jackson Lab, and Heriot-Watt University	Mouse gene expression spatial data
RGD	Medical College of Wisconsin	Rat genome, genes, phenotypes, strains, diseases, physiology and nutrition
Xenbase	International community	Frog genus *Xenopus* genomic, expression and functional data
ZFIN	University of Oregon	Zebrafish genetic, genomic and developmental data
WormBase	International community	Worm *C. elegans* and related roundworms genetic and genomic data
FlyBase	International community	Fruit fly genus *Drosophila* genetic and genomic data
GreenPhyl	Biodiversity International and CIRAD	Plant genomes integrative and comparative analysis
PlantDB	Munich MIPS	Plant genomes integrative and comparative analysis
TAIR	Phoenix Bioinformatics Corporation	Thale cress (*A. thaliana*) genetic, proteomic, metabolic and other data
MaizeDB	International community	Maize crop (*Z. mays*) genetic, proteomic, metabolic and other data
SGD	Stanford University	Baker's budding yeast (*Saccharomyces* genus) genetic, genomic, proteomic, structural, literature and other data
CGD	Stanford University	Budding yeast (*Candida* genus) genetic, genomic, proteomic, structural, literature and other data
MycoCosm	US Department of Energy Joint Genome Institute	Fungal genome portal for integrated fungal genomic data and promotes for the 1000 genomes project
CYGD	Munich MIPS	Baker's yeast (*S. cerevisiae*) molecular structures and functional networks
YMDB	University of Alberta, Canada	Baker's yeast (*S. cerevisiae*) metabolome database
YeastNet	Yonsei University, Korea	Baker's yeast (*S. cerevisiae*) integrated functional gene networks
MBGD	JSPS	Comparative genome analysis of completely sequenced microbial genomes (bacteria, archaea and a few eukaryotes)
PATRIC	Virginia Bioinformatics Institute	Bacterial integrated genomic, transcriptomic, proteomic, structural and sequencing data

(*continued overleaf*)

Table 6.1 (*continued*)

Database	Provider	Description
IMG and IMG/M	University of California	Integrated annotation, analysis and distribution of microbial genomic and metagenomic datasets
BacDive	Leibniz Institute DSMZ	Bacterial and archaeal taxonomic, morphologic, physiologic, environmental and molecular-biological information
COLOMBOS	International community	Cross-platform microarray datasets for organism-specific bacteria
PortEco	A consortium of US laboratories	Model bacteria *E. coli* comprehensive resource
ViralZone	Swiss SIB	Viral bioinformatics resource
Virus Variation Resource	NCBI	Viral gene and protein sequence annotations and relevant meta-data

References

Acland, A., Agarwala, R., Barrett, T. *et al.* (2014). Database resources of the National Center for Biotechnology Information. *Nucleic Acids Research*, **42**(Database), pp. D7–D17.

Adamusiak, M.K.T., Kapushesky, M., Burdett, T. *et al.* (2012). Gene expression atlas update—a value-added database of microarray and sequencing-based functional genomics experiments. *Nucleic Acids Research*, **40**(Database), pp.D1077–D1081.

Alcántara, R., Onwubiko, J., Cao, H. *et al.* (2013). The EBI enzyme portal. *Nucleic Acids Research*, **41**(Database), pp.D773–D780.

Barrett, T., Clark, K., Gevorgyan, R. *et al.* (2012). BioProject and BioSample databases at NCBI: facilitating capture and organization of metadata. *Nucleic Acids Research*, **40**(Database), pp. D57–D63.

Barrett, T., Wilhite, S.E., Ledoux, P. *et al.* (2013). NCBI GEO: archive for functional genomics data sets—update. *Nucleic Acids Research*, **41**(D1), pp. D991–D995.

Begley, D.A., Krupke, D.M., Neuhauser, S.B. *et al.* (2012). The Mouse Tumor Biology Database (MTB): a central electronic resource for locating and integrating mouse tumor pathology data. *Veterinary Pathology*, **49**(1), pp. 218–223.

Benson, D.A., Clark, K., Karsch-Mizrachi, I. *et al.* (2014). GenBank. *Nucleic Acids Research*, **42**(Database), pp. D32–D37.

Bhagwat, M. (2010). Searching NCBI's dbSNP database. *Current Protocols in Bioinformatics*, Issue SUPP. 32, Unit 1.19.

Binkley, J., Arnaud, M.B., Inglis, D.O. *et al.* (2014). The Candida Genome Database: the new homology information page highlights protein similarity and phylogeny. *Nucleic Acids Research*, **42**(Database), pp. D711–D716.

Blake, J.A., Bult, C.J., Eppig, J.T. *et al.* (2014). The Mouse Genome Database: integration of and access to knowledge about the laboratory mouse. *Nucleic Acids Research*, **42**(Database), pp. D810–D817.

Brister, J.R., Bao, Y., Zhdanov, S.A. *et al.* (2014). Virus variation resource—recent updates and future directions. *Nucleic Acids Research*, **42**(Database), pp. D660–D665.

Brooksbank, C., Bergman, M.T., Apweiler, R. *et al.* (2014). The European Bioinformatics Institute's data resources 2014. *Nucleic Acids Research*, **42**(Database), pp. D18–D25.

Chelliah, V., Laibe, C. and Novère, N.L. (2013). BioModels Database: a repository of mathematical models of biological processes. *Methods in Molecular Biology*, **1021**, pp. 189–199.

Chen, J., Anderson, J.B., DeWeese-Scott, C. *et al.* (2003). MMDB: entrez's 3D-structure database. *Nucleic Acids Research*, **31**(1), pp. 474–477.

Cherry, J.M., Hong, E.L., Amundsen, C. *et al.* (2012). Saccharomyces Genome Database: the genomics resource of budding yeast. *Nucleic Acids Research*, **40**(Database), pp. D700–D705.

Church, D.M., Lappalainen, I., Sneddon, T.P. *et al.* (2010). Public data archives for genomic structural variation. *Nature Genetics*, **42**(10), pp. 813–814.

Costanzo, M.C., Engel, S.R., Wong, E.D. *et al.* (2014). Saccharomyces genome database provides new regulation data. *Nucleic Acids Research*, **42**(Database), pp. D717–D725.

Croft, D., Mundo, A.F., Haw, R. *et al.* (2014). The reactome pathway knowledgebase. *Nucleic Acids Research*, **42**(Database), pp. D472–D477.

Cruz, N.d.l., Bromberg, S., Pasko, D. *et al.* (2005). The Rat Genome Database (RGD): developments towards a phenome database. *Nucleic Acids Research*, **33**(Database), pp. D485–D491.

Faulconbridge, A., Burdett, T., Brandizi, M. *et al.* (2014). Updates to BioSamples database at European Bioinformatics Institute. *Nucleic Acids Research*, **42**(Database), pp. D50–D52.

Federhen, S. (2012). The NCBI Taxonomy database. *Nucleic Acids Research*, **40**(Database), pp. D136–D143.

Fingerman, I.M., Zhang, X., Ratzat, W. *et al.* (2013). NCBI epigenomics: what's new for 2013. *Nucleic Acids Research*, **41**(Database), pp. D221–D225.

Flicek, P., Amode, M.R., Barrell, D. *et al.* (2014). Ensembl 2014. *Nucleic Acids Research*, **42**(Database), pp. D749–D755.

Geer, L.Y., Marchler-Bauer, A., Geer, R.C. *et al.* (2010). The NCBI BioSystems database. *Nucleic Acids Research*, **38**(Database), pp. D492–D496.

Grigoriev, I.V., Nikitin, R., Haridas, S. *et al.* (2014). MycoCosm portal: gearing up for 1000 fungal genomes. *Nucleic Acids Research*, **42**(Database), pp. D699–D704.

Grubb, S.C., Bult, C.J. and Bogue, M.A. (2014). Mouse Phenome Database. *Nucleic Acids Research*, **42**(Database), pp.D825–D834.

Güldener, U., Münsterkötter, M., Kastenmüller, G. *et al.* (2005). CYGD: the Comprehensive Yeast Genome Database. *Nucleic Acids Research*, **33**(Database), pp. D364–D368.

Gutmanas, A., Alhroub, Y., Battle, G.M. *et al.* (2014). PDBe: Protein Data Bank in Europe. *Nucleic Acids Research*, **42**(Database), pp. D285–D291.

Harris, T.W., Baran, J., Bieri, T. *et al.* (2014). WormBase 2014: new views of curated biology. *Nucleic Acids Research*, **42**(Database), pp. D789–D793.

Haug, K., Salek, R.M., Conesa, P. *et al.* (2013). MetaboLights—an open-access general-purpose repository for metabolomics studies and associated meta-data. *Nucleic Acids Research*, **41**(Database), pp. D781–D786.

Hayamizu, T.F., Wicks, M.N., Davidson, D.R. *et al.* (2013). EMAP/EMAPA ontology of mouse developmental anatomy: 2013 update. *Journal of Biomedical Semantics*, **4**(1), 15.

Hoeppner, M.A. (2013). NCBI Bookshelf: books and documents in life sciences and health care. *Nucleic Acids Research*, **41**(Database), pp. D1251–D1260.

Howe, D.G., Bradford, Y.M., Conlin, T. *et al.* (2013). ZFIN, the Zebrafish Model Organism Database: increased support for mutants and transgenics. *Nucleic Acids Research*, **41**(Database), pp. D854–D860.

Hu, J.C., Sherlock, G., Siegele, D.A. *et al.* (2014). PortEco: a resource for exploring bacterial biology through high-throughput data and analysis tools. *Nucleic Acids Research*, **42**(Database), pp. D677–D684.

Hunter, S., Apweiler, R., Attwood, T.K. *et al.* (2009). InterPro: the integrative protein signature database. *Nucleic Acids Research*, **37**(Database), pp. D211–D215.

Hunter, S., Corbett, M., Denise, H. *et al.* (2014). EBI metagenomics—a new resource for the analysis and archiving of metagenomic data. *Nucleic Acids Research*, **42**(Database), pp. D600–D606.

James-Zorn, C., Ponferrada, V.G., Jarabek, C.J. *et al.* (2013). Xenbase: expansion and updates of the Xenopus model organism database. *Nucleic Acids Research*, **41**(Database), pp. D865–D870.

Jewison, T., Knox, C., Neveu, V. *et al.* (2012). YMDB: the Yeast Metabolome Database. *Nucleic Acids Research*, **40**(Database), pp. D815–D820.

Kerrien, S., Aranda, B., Breuza, L. *et al.* (2012). The IntAct molecular interaction database in 2012. *Nucleic Acids Research*, **40**(Database), pp. D841–D846.

Kersey, P.J., Staines, D.M., Lawson, D. *et al.* (2012). Ensembl genomes: an integrative resource for genome-scale data from non-vertebrate species. *Nucleic Acids Research*, **40**(Database), pp. D91–D97.

Kim, H., Shin, J., Kim, E. *et al.* (2014). YeastNet v3: a public database of data-specific and integrated functional gene networks for saccharomyces cerevisiae. *Nucleic Acids Research*, **42**(Database), pp. D731–D736.

Kodama, Y., Shumway, M. and Leinonen, R. (2012). The sequence read archive: explosive growth of sequencing data. *Nucleic Acids Research*, **40**(Database), pp. D54–D56.

Lamesch, P., Berardini, T.Z., Li, D. *et al.* (2012). The Arabidopsis Information Resource (TAIR): improved gene annotation and new tools. *Nucleic Acids Research*, **40**(Database), pp. D1202–D1210.

Laulederkind, S.J.F., Hayman, G.T., Wang, S.J. *et al.* (2013). The Rat Genome Database 2013-data, tools and users. *Briefings in Bioinformatics*, **14**(4), pp. 520–526.

Maglott, D., Ostell, J., Pruitt, K.D. and Tatusova, T. (2011). Entrez Gene: gene-centered information at NCBI. *Nucleic Acids Research*, **39**(Database), pp. D52–D57.

Malone, J., Holloway, E., Adamusiak, T. *et al.* (2010). Modeling sample variables with an Experimental Factor Ontology. *Bioinformatics*, **26**(8), pp. 1112–1118.

Markowitz, V.M., Chen, I.-M.A., Chu, K. *et al.* (2014a). IMG/M 4 version of the integrated metagenome comparative analysis system. *Nucleic Acids Research*, **42**(Database), pp. D568–D573.

Markowitz, V.M. Chen, I.-M.A., Palaniappan, K. *et al.* (2014b). IMG 4 version of the integrated microbial genomes comparative analysis system. *Nucleic Acids Research*, **42**(Database), pp. D560–D567.

Masson, P., Hulo, C., De Castro, E. *et al.* (2013). ViralZone: recent updates to the virus knowledge resource. *Nucleic Acids Research*, **41**(Database), pp. D579–D583.

McEntyre, J.R., Ananiadou, S., Andrews, S. *et al.* (2011). UKPMC: a full text article resource for the life sciences. *Nucleic Acids Research*, **39**(Database), pp. D58–D65.

Meysman, P., Sonego, P., Bianco, L. *et al.* (2014). COLOMBOS v2.0: an ever expanding collection of bacterial expression compendia. *Nucleic Acids Research*, **42**(Database), pp. D649–D653.

Nussbaumer, T., Martis, M.M., Roessner, S.K. *et al.* (2013). MIPS PlantsDB: a database framework for comparative plant genome research. *Nucleic Acids Research*, **41**(Database), pp. D1144–D1151.

Orchard, S., Ammari, M., Aranda, B. *et al.* (2014). The MIntAct project—IntAct as a common curation platform for 11 molecular interaction databases. *Nucleic Acids Research*, **42**(Database), pp. D358–D363.

Overbeek, R., Olson, R., Pusch, G.D. *et al.* (2014). The SEED and the Rapid Annotation of microbial genomes using Subsystems Technology (RAST). *Nucleic Acids Research*, **42**(Database), pp. D206–D214.

Pakseresht, N., Alako, B., Amid, C. *et al.* (2014). Assembly information services in the European Nucleotide Archive. *Nucleic Acids Research*, **42**(Database), pp. D38–D43.

Pruitt, K., Brown, G., Tatusova, T. and Maglott, D. (2002). Chapter 18 of the Reference Sequence (RefSeq) Database. In: J. McEntyre and J. Ostell, eds. *The NCBI Handbook [Internet]*. Bethesda, MD: National Center for Biotechnology Information (US).

Pruitt, K.D., Tatusova, T., Brown, G.R. and Maglott, D.R. (2012). NCBI Reference Sequences (RefSeq): current status, new features and genome annotation policy. *Nucleic Acids Research*, **40**(Database), pp. D130–D135.

Richardson, L., Venkataraman, S., Stevenson, P. *et al.* (2014). EMAGE mouse embryo spatial gene expression database: 2014 update. *Nucleic Acids Research*, **42**(Database), pp. D835–D844.

Rouard, M., Guignon, V., Aluome, C. *et al.* (2011). GreenPhylDB v2.0: comparative and functional genomics in plants. *Nucleic Acids Research*, **39**(Database), pp. D1095–D1102.

Rustici, G., Kolesnikov, N., Brandizi, M. *et al.* (2013). ArrayExpress update—trends in database growth and links to data analysis tools. *Nucleic Acids Research*, **41**(Database), pp. D987–D990.

Saccone, S.F., Quan, J., Mehta, G. *et al.* (2011). New tools and methods for direct programmatic access to the dbSNP relational database. *Nucleic Acids Research*, **39**(Database), pp. D901–D907.

Schaeffer, M.L., Harper, L.C., Gardiner, J.M. *et al.* (2011). MaizeGDB: curation and outreach go hand-in-hand. *Database*, **2011**, p. bar022.

Sherry, S.T., Ward, M.H., Kholodov, M. *et al.* (2001). DbSNP: the NCBI database of genetic variation. *Nucleic Acids Research*, **29**(1), pp. 308–311.

Smith, C.M., Finger, J.H., Hayamizu, T.F. *et al.* (2014). The mouse Gene Expression Database (GXD): 2014 update. *Nucleic Acids Research*, **42**(Database), pp. D818–D824.

Söhngen, C., Bunk, B., Podstawka, A. *et al.* (2014). BacDive—the Bacterial Diversity Metadatabase. *Nucleic Acids Research*, **42**(Database), pp. D592–D599.

St Pierre, S.E., Ponting, L., Stefancsik, R. *et al.* (2014). FlyBase 102—advanced approaches to interrogating FlyBase. *Nucleic Acids Research*, **42**(Database), pp. D780–D788.

The UniProt Consortium (2014). Activities at the Universal Protein Resource (UniProt). *Nucleic Acids Research*, **42**(Database), pp. D191–D198.

Tryka, K.A., Hao, L., Sturcke, A. *et al.* (2014). NCBI's Database of Genotypes and Phenotypes: dbGaP. *Nucleic Acids Research*, **42**(Database), pp. D975–D979.

Uchiyama, I., Mihara, M., Nishide, H. and Chiba, H. (2013). MBGD update 2013: the microbial genome database for exploring the diversity of microbial world. *Nucleic Acids Research*, **41**(Database), pp. D631–D635.

Vizcaíno, J.A., Côté, R.G., Csordas, A. *et al.* (2013). The Proteomics Identifications (PRIDE) database and associated tools: status in 2013. *Nucleic Acids Research*, **41**(Database), pp. D1063–D1069.

Wang, Y., Xiao, J., Suzek, T.O. *et al.* (2009). PubChem: a public information system for analyzing bioactivities of small molecules. *Nucleic Acids Research*, **37**(Web Server), pp. W623–W633.

Wang, Y., Xiao, J., Suzek, T.O. *et al.* (2012). PubChem's BioAssay Database. *Nucleic Acids Research*, **40**(Database), pp. D400–D412.

Wattam, A.R., Abraham, D., Dalay, O. *et al.* (2014). PATRIC, the bacterial bioinformatics database and analysis resource. *Nucleic Acids Research*, **42**(Database), pp. D581–D591.

Welter, D., MacArthur, J., Morales, J. *et al.* (2014). The NHGRI GWAS Catalog, a curated resource of SNP-trait associations. *Nucleic Acids Research*, **42**(Database), pp. D1001–D1006.

Willighagen, E.L., Waagmeester, A., Spjuth, O. *et al.* (2013). The ChEMBL database as linked open data. *Journal of Cheminformatics*, **5**(1), p. 23.

Wishart, D.S., Jewison, T., Guo, A.C. *et al.* (2013). HMDB 3.0—the Human Metabolome Database in 2013. *Nucleic Acids Research*, **41**(Database), pp. D801–D807.

7

Normalisation

7.1 Introduction

Reliable quantification of gene expression is that which faithfully reflects the true mRNA levels in a given sample. However, much variability exists in the available technologies (e.g. microarrays) which perturb the measurements so that they are no longer reliable in their raw form. Such variability can be caused by the preparation of the biological sample, fluorescent labelling, specific hybridisation, non-specific hybridisation, scanning, image processing and others (Calza and Pawitan, 2010). Moreover, in most of the microarray datasets, many mRNA samples are taken and measured by multiple microarray chips/slides; these samples can be from different types of tissues (e.g. cancer and normal tissues), at different chronological stages or time points within a biological process, or from different samples contained in different biological conditions. Thus, not only the comparability of intensities of different genes within one slide is questioned, but also the comparability of intensities of a single gene amongst different slides (samples) is questioned.

Normalisation aims at eliminating these technical variations within a single slide or between multiple slides. This is so that remaining variations of intensities reliably represent actual biological variations, which are what such experiments desire to measure. The necessity of the normalisation step was reported as one of the six important issues listed by the Minimum Information about a Microarray Experiment (MIAME) protocol (Brazma et al., 2001).

In this chapter, we discuss the issues which exist in microarrays and are tackled by normalisation, and then we present some of the most commonly used normalisation methods.

Integrative Cluster Analysis in Bioinformatics, First Edition. Basel Abu-Jamous, Rui Fa and Asoke K. Nandi.
© 2015 John Wiley & Sons, Ltd. Published 2015 by John Wiley & Sons, Ltd.

7.2   Issues Tackled by Normalisation

Before delving into some examples of normalisation methods, it is appropriate to introduce some of the core issues which have been tackled by normalisation. Different normalisation methods tend to lean on different hard presumptions about the nature of gene expression or about the way in which different sources of variation bias the measured intensity from the original mRNA concentration (Wu, 2009). This section reviews some of these issues, each in a separate sub-section.

7.2.1   Within-slide and Between-slides Normalisation

As mentioned above, variations cause the comparability of expression intensities between genes within one slide or between different slides to be questionable. Therefore, different methods were proposed to perform within-slide and/or between-slides normalisation, where in many cases both are necessary (Yang *et al.*, 2002; Sievertzon, Nilsson and Lundeberg, 2006; Verducci *et al.*, 2006).

In two-channel microarrays, within-slide normalisation also tackles the uneven distribution of the Cy3 and Cy5 dyes' intensities within the same slide (Yang *et al.*, 2002; Sievertzon, Nilsson and Lundeberg, 2006). Moreover, in spotted cDNA arrays, the microarray slide usually has a grid of probes' regions such that each region is printed by a different print-tip (pen). Within-slide normalisation methods can be applied locally on each of these regions; in this case, it is called within-print-tip normalisation. Then, between-slides normalisation methods can be applied between different print-tips' regions and is called between-print-tips normalisation (Yang *et al.*, 2002).

7.2.2   Normalisation Based on Non-differentially Expressed Genes

Many normalisation methods are based on the presumption that the distribution of gene expression values in different slides is the same and that the distribution of up-regulated genes is even with that of down-regulated ones; of course this includes the assumption that the genome as a whole is mostly non-differentially expressed between different slides. Within a single slide, it is presumed that the distribution of up-regulated and down-regulated genes, when spanned over all of the orders of intensities, is even.

Although most methods use the entire genome as a reference for normalisation based on the assumption that the entire genome collectively is fairly invariant across slides, this assumption has been questioned by many researchers (Calza and Pawitan, 2010). The suggestion by many has been to use a set of non-differentially expressed genes, for example housekeeping genes which are involved in essential cell metabolic processes, as controls to be used for distribution estimation; this led many microarray manufacturers to include such controls within their arrays (Tseng *et al.*, 2001; Reilly, Wang and Rutherford, 2003; Abruzzo *et al.*, 2005; Sievertzon, Nilsson and Lundeberg, 2006; Wu, 2009; Calza and Pawitan, 2010).

However, even the assumption itself that these specific genes are always invariant across different slides may not be as accurate as required (Sievertzon, Nilsson and Lundeberg, 2006; Wu, 2009; Calza and Pawitan, 2010). If that was the case, the uncounted-for variabilities in the small subset of genes expected to be non-differentially expressed would result in worse affects than if such variabilities occurred at the full set level. Therefore, using the full set of genes would be preferred in many cases (Sievertzon, Nilsson and Lundeberg, 2006; Calza and Pawitan, 2010).

This alternation in the literature between using the entire genome and a subset of non-differentially expressed genes seems to be one of the most varying subjects in microarray data normalisation; the general trend is towards using the entire genome though.

Examples of the methods which select a subset of invariant genes include the least variant set (LVS) method (Calza, Valentini and Pawitan, 2008; Calza and Pawitan, 2010) and a data-driven model-based procedure (Li and Wong, 2001b).

7.2.3   Background Correction

It is well known that optical noise and non-specific binding (NSB) are of the reasons for the technical variations in microarrays (Wu *et al.*, 2004). NSB is when sequences other than the intended target bind to the probe and add to its measured intensity value. This additional undesired value is considered as an additive background noise which needs to be eliminated (Yang, Buckley and Speed, 2001; Konishi, 2004; Wu, 2009). Different normalisation methods address this issue by following different approaches.

As a trial to overcome this issue at the level of microarray chip design, and as mentioned in Chapter 5, Affymetrix microarrays include for each target sequence a *perfect match* (*PM*) probe, which perfectly matches the target's sequence, paired with a *mismatch* (*MM*) probe, which differs from the PM by only the middle nucleotide. According to the manufacturer, the MM is supposed to measure the background bias due to optical noise and NSB for the corresponding PM. Accordingly, the intensity difference (*PM* − *MM*) was considered as the corrected intensity (Wu, 2009).

However, it was shown by many researchers that MMs do not represent merely the background (Naef and Magnasco, 2003; Ahmed, 2006); for example, in many cases the MM's intensity was consistently higher than the PM's (Naef *et al.*, 2002; Roberts, 2008; Wu, 2009). Other studies also showed that background adjustment merely by subtracting the MM values results in inflated variance of the gene expression estimates (Irizarry *et al.*, 2003a; Wu, 2009). So, many studies used PM only (Wu, 2009). In this case, the gain in precision overweights the loss in accuracy when using (*PM* − *MM*) (Cope *et al.*, 2004; Wu, 2009). For example, the popular method MAS5 uses the intensity difference (*PM* − *MM*) while the popular method RMA only uses the PM value.

Wu and Irizarry summarised the background noise and the other sources of variation which affect the measured intensity in a common additive-background-multiplicative-measurement (ABME) error model (Wu and Irizarry, 2007b; Wu, 2009). The model is represented as shown in Equation (7.1),

$$Y_{ij} = B_{ij} + S_{ij} = \left(B_j + \delta_{ij}\right) + e^{a_j + \gamma_i + \mu_{ij} + \varepsilon_{ij}} \qquad (7.1)$$

where Y_{ij} is the intensity of the jth probe on the array i and $B_{ij} = \left(B_j + \delta_{ij}\right)$ is the background noise for the jth probe on the array i; B_j is the mean background on the jth probe and δ_{ij} is an additive error term, which is sometimes absorbed by the background term. The term S_{ij} represents the specific binding intensity which is composed of many terms in a multiplicative manner, a_j is the efficiency of the probe, γ_i is the array effect for all probes, μ_{ij} is the log concentration of the target, and ε_{ij} is the multiplicative measurement error.

By taking the array effect γ_i outside the exponential and naming it f_i, the model becomes as shown in Equation (7.2):

$$Y_{ij} = \left(B_j + \delta_{ij}\right) + f_i e^{a_j + \mu_{ij} + \varepsilon_{ij}} \qquad (7.2)$$

If a probe-set g was used to represent one gene, the model becomes as shown in Equation (7.3),

$$Y_{ij} = \left(B_{gj} + \delta_{gij}\right) + f_i e^{a_{gj} + \mu_{gi} + \varepsilon_{gij}} \qquad (7.3)$$

where μ_{gi} is the log concentration of the target gene. Although these models were mentioned mainly in the context of Affymetrix matrices, it appears that they apply to many others including Illumina (Lin et al., 2008; Wu, 2009).

Many older studies had suggested that the background noise is multiplicative (Kerr, Martin and Churchill, 2000; Dudoit et al., 2002), but Wu and colleagues did empirically demonstrate that it is additive with non-zero mean, and they supported the aforementioned ABME model (Wu et al., 2004).

7.2.4   Logarithmic Transformation

Either for one-channel microarrays intensities or for two-channel microarrays intensity ratios, logarithmic transformation is a very common step included within most normalisation methods (Eisen, 1999; Kerr, Martin and Churchill, 2000; Roberts, 2008). Wu stated that this transformation is usually found to remove or reduce the dependency between the mean and the variance (Wu, 2009); this is because the relation between the variance and the mean was found to be heteroscedastic (Kepler, Crosby and Morgan, 2002; Motakis et al., 2006).

Many researchers have mentioned that the variations in microarrays follow a multiplicative model, and that the intensities follow a lognormal distribution; thus, logarithmic transformation converts the variations to be additive and the distribution to be fairly normal (Kerr, Martin and Churchill, 2000; Tseng et al., 2001; Konishi, 2002, 2004; Olshen and Jain, 2002; Kreil and Russell, 2005).

However, as this transformation is usually performed after background elimination, it might be less effective at low intensities. This is because negative values might appear by background subtraction. So, it has been proposed by many studies to add an offset before taking the logarithm (Irizarry, Wu and Jaffee, 2006; Roberts, 2008). Moreover, some studies showed that the intensities distribution sometimes does not perfectly fit a lognormal distribution model (Kerr, Martin and Churchill, 2000; Hoyle et al., 2002; Konishi, 2004).

Sapir and Churchill compared logarithmic transformation with some other options such as square roots and reciprocals; logarithmic transformation was superior in such comparisons (Sapir and Churchill, 2000). Konishi investigated the transcriptome based on the laws of thermodynamics and proposed using a model which justifies why the transcriptome tends to follow a lognormal distribution (Konishi, 2005).

In the two-channel microarrays case, the log-ratio ($\log_2(intensity1/intensity2)$) makes fold changes in both directions symmetrical. For example, $\log_2\left(\dfrac{5}{1}\right) = 2.32$ and $\log_2\left(\dfrac{1}{5}\right) = -2.32$, which is more intuitive to work with than the non-log-ratios (Quackenbush, 2002; Sievertzon, Nilsson and Lundeberg, 2006; Lee and Saeed, 2007).

7.2.5 Intensity-dependent Bias – (MA) Plots

It was observed in many studies that the bias in the measured intensities due to technical variation is intensity-dependent; that is, the bias is different at different orders of intensity values. This applies to both one-channel and two-channel microarrays.

To visualise this fact in two-channel microarrays, MA plots were proposed (Dudoit *et al.*, 2002) and used in many instances of research studies (Calza and Pawitan, 2010). The MA plot is a plot of the log-ratio $M = \log_2(Cy5/Cy3)$ versus the abundance $A = \log_2\sqrt{Cy5 \times Cy3}$ (Dudoit *et al.*, 2002). Based on the assumption that the log-ratios represented by the M-axis are evenly distributed around the zero axis, the expected MA plot for an unbiased dataset would have that property met at all orders of intensities represented by the A-axis.

An example of an MA plot for un-normalised data is shown in Figure 7.1. It is very clear that the distribution of the data points is not symmetrical around the $M = 0$ axis.

Such plots show that there is a bias between the Cy5-dye and the Cy3-dye, that is the presence of intensity-dependent effects (Irizarry *et al.*, 2003a; Sievertzon, Nilsson and Lundeberg, 2006). Another version of MA plots is called the ratio-intensity (RI) plot, which plots the log-ratio $R = \log_2(Cy5/Cy3)$ versus the intensity $I = \log_{10}(Cy5 \times Cy3)$ (Quackenbush, 2002). Many normalisation methods, such as the very popular lowess method, tackle the problem of normalisation based on this MA plot (Quackenbush, 2002; Yang *et al.*, 2002; Xie *et al.*, 2004).

7.2.6 Replicates and Summarisation

To increase the reliability of the measurements, replicates of the samples are considered. Replicates can be biological or technical. Biological replicates are actually multiple biological samples taken for the same condition; they provide information about the biological variation as well as the stochastic variation in the pre-processing of the samples. Technical replicates consider measuring the expression of the same biological sample by multiple arrays (chips).

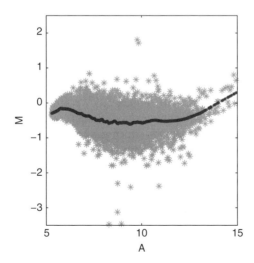

Figure 7.1 Sample MA plot for an un-normalised yeast genome data sample with the NCBI accession number GSM81075

Such replicates can provide information about the stochastic technical variations in the arrays' probes or the scanner (Churchill, 2002; Quackenbush, 2002).

In Affymetrix GeneChips there are 11–20 probes per gene with an average of four probes per exon. In Illumina there is one probe per gene but there are many replicates for it (about 30) (Wu, 2009). Summarisation is to obtain one measurement which represents the collective summary of the various readings for the same gene (Wu, 2009).

In the case of two-channel arrays, this can be done by taking the geometric mean of the intensities, which is equivalent to taking the arithmetic mean of the log-ratios (Quackenbush, 2002). Some normalisation methods, such as RMA and GCRMA (defined in Section 7.3.2), perform summarisation as a step within a wider framework of normalisation.

7.3   Normalisation Methods

This section describes a number of the most commonly used methods for normalisation, background adjustment and/or expression index estimation.

7.3.1   Microarray Suite 5 (MAS 5.0)

Initially, MAS 4.0 was introduced by Affymetrix, where background elimination was simply done by subtracting each MM probe value from its corresponding PM probe value. It was then shown that many MM values actually exceed the values of PM probes because MM probes do have a level of specific binding in addition to NSB. Thus, Affymetrix introduced MAS 5.0 which adopts the background correction procedure ($PM - MM^*$) such that the MM^* value is a tweaked MM, which guarantees the corrected intensity not to be negative. This is provided by Affymetrix as their default package for normalisation (Hubbell, Liu and Mei, 2002; Affymetrix, 2002a, b; Irizarry et al., 2003a; Wu et al., 2004; Roberts, 2008).

As discussed earlier in Section 7.2.3, the usage of MM values in background correction has been criticised by many researchers, which led to the introduction of many normalisation methods that depend merely on PM values (Naef et al., 2001; Calza and Pawitan, 2010). For example, Irizarry and colleagues showed that the variance of $\log(PM - MM^*)$ is generally larger than that of $\log(PM)$ values (Irizarry et al., 2003b). They also noticed that the relation between the variance of $\log(PM - MM^*)$ and the average intensity of the corresponding probe set is reciprocal; that is, it is significantly higher at lower intensities (Irizarry et al., 2003b).

Even though, MAS 5.0 is still a popular method (Cahan et al., 2007). In some studies, it was shown to outperform many other methods which used to be known as better than it such as RMA. The very recent study by Furlotte and colleagues used their proposed literature-based procedure to evaluate different microarray-normalisation methods. The cases included in their study showed an outperformance of the MAS5 method over positional-dependent nearest neighbour (PDNN) and RMA (Furlotte et al., 2011).

7.3.2   Robust Multi-array Average (RMA)

In order to overcome the problematic issues related to the methods that use MM values for background correction in Affymetrix microarrays, Irizarry and colleagues proposed a method which depends only on PM probes and tackles the background-correction problem based on the analysis of the empirical distribution of the intensities; this method is called *RMA* (Bolstad *et al.*, 2003; Irizarry *et al.*, 2003a).

RMA is not only a method for background correction, but also a framework of background correction, expression normalisation and summarisation. By the time this chapter is being written, more than 4500 citations have been recorded in Scopus for that Irizarry and colleagues' publication which proposed the RMA method (Irizarry *et al.*, 2003a). This supports what many researchers have mentioned in that this has become a very popular method in Affymetrix microarrays normalisation (Cahan *et al.*, 2007; Vendettuoli, Doyle and Hofmann, 2011).

RMA assumes a common mean background using only PM data. The background is subtracted, then the intensities are adjusted for identical distributions. This includes performing logarithmic transformation followed by quantile normalisation, cyclic lowess or contrast normalisation; quantile normalisation is usually the most rapid of them. Then RMA performs summarisation (Irizarry *et al.*, 2003a; Vendettuoli, Doyle and Hofmann, 2011).

The summarisation part is performed by fitting the linear model presented by Equation (7.4):

$$Y_{ijn} = \mu_{in} + \alpha_{jn} + \epsilon_{ijn} \tag{7.4}$$

The value Y_{ijn} represents the background adjusted, normalised and log transformed PM intensity value for the jth probe pair within the nth gene's probe-set in the ith array. The average log-scale expression for this n^{th} gene in the ith array is μ_{in}, and this is the parameter to be estimated as the gene expression value. The value α_{jn} is the probe affinity and is assumed to meet the constraint $\sum_j \alpha_j = 0$ for all probe-sets. The last term, ϵ_{ijn}, represents an independent identically distributed (i.i.d.) error term with zero-mean (Irizarry *et al.*, 2003a).

7.3.2.1 GCRMA

Wu and colleagues investigated the way in which the RMA method tackles Affymetrix microarrays background adjustment, pointed out the loss of accuracy caused by RMA's sacrifice in favour of gains in precision, and proposed a different background-adjustment approach (Wu *et al.*, 2004). They proposed plugging this background-adjustment approach with the original RMA's normalisation and summarisation steps. Thus, this is considered as a variant of the RMA method, and it is named *GCRMA*, where the 'GC' part is taken from the two symbols 'G' and 'C' representing the two nucleobases – Guanine and Cytosine, respectively (Wu *et al.*, 2004).

The movement from RMA to GCRMA was motivated by Naef and Magnasco's observation that probes' affinity is sequence-dependent. This means that both specific and NSB are significantly affected by the sequence of the probe (Naef and Magnasco, 2003). Each of the Affymetrix probes consists of 25 nucleobases, and it was found that the contribution of each of these bases to the overall affinity of the probe depends on both the type of the nucleobase (A, T, G or C) and its position. Thus, the affinity of the probe (α) is modelled as the summation of the contributions of its 25 nucleobases [Equation (7.5)],

$$\alpha = \sum_{k=1}^{25} \mu_{b_k, k} \tag{7.5}$$

where $\mu_{b_k,k}$ is the contribution of the nucleobase ($b_k \in \{A,T,G,C\}$) found at the kth position. The values of $\mu_{b_k,k}$ were estimated empirically from large amounts of microarray data by fitting a polynomial of degree three (Naef and Magnasco, 2003), or by fitting a spline with five degrees of freedom (Wu et al., 2004). The results of the latter fitting clearly showed the significant differences of contributions to the affinity for different types of nucleobases and for different positions within the probe's sequence.

The measurements of the PM and MM values are modelled as shown in Equations (7.6) and (7.7) (Wu et al., 2004),

$$PM = O_{PM} + N_{PM} + S \tag{7.6}$$

$$MM = O_{MM} + N_{MM} + \phi S \tag{7.7}$$

where O_{PM} and O_{MM} represent the optical noise of the PM and the MM probes respectively, N_{PM} and N_{MM} represent the NSB contribution to the intensity for the PM and the MM probes respectively, S is a quantity proportional to the mRNA concentration (quantity of interest), and $\phi \in [0,1]$ is a factor which accounts for the fact that MM probes might have specific binding which is larger than zero but less than that of PM probes.

The optical noise is assumed to be lognormally distributed, but because its variance is ignorable compared with that of NSB, it is assumed as an array-dependent constant. The estimation of the optical noise ($\hat{O}$) is considered to be just smaller than the minimum probe intensity in that array ($\hat{O} = \min\left\{ \min_j PM_j, \min_j MM_j \right\} - 1$). The value of the factor ϕ is set to zero as the empirical results showed that this would not affect the performance of the method (Wu et al., 2004).

The NSB log-values $\log(N_{PM})$ and $\log(N_{MM})$ are assumed to have a bivariate-normal distribution with the respective means of m_{PM} and m_{MM}, the common variance σ^2, and the common correlation constant across the probes ρ. These means are $m_{PM} \equiv h(\alpha_{PM})$ and $m_{MM} \equiv h(\alpha_{MM})$, such that $h(.)$ is a smooth function (almost linear), and the α values are computed as in Equation (7.5).

The value of the correlation constant ρ can be estimated from the data, and because it should not change from array to array, it was estimated to be 0.7 (Wu et al., 2004). If the parameters m_{PM}, m_{MM} and σ^2 were estimated from the data and h was known, the statistical problem of the GCRMA method can then be formalised as the problem of estimating the value of interest S.

Wu and colleagues proposed three main approaches to carry out this estimation – maximum-likelihood estimate (MLE), mean squared error (MSE) and an empirical Bayes estimate (Wu et al., 2004). The results of their most emphasised approach, the empirical Bayes estimate, showed that ignoring the MM probes in estimation does not significantly affect the performance. They even showed that, for highly expressed genes, the only-PM-based estimations show better results (Wu et al., 2004).

This normalisation method has been very popular in Affymetrix microarray data analysis (Cahan et al., 2007). It is implemented in MATLAB as the function 'gcrma' within the bioinformatics toolbox. It is also implemented as an R language package for the Bioconductor project (Wu and Irizarry, 2007a).

7.3.2.2   Other RMA Variants

Frozen RMA (fRMA) was proposed as a variation of the RMA method in McCall, Bolstad and Irizarry (2010) and was used in McCall, Jaffee and Irizarry (2012). Thawing fRMA is a variation of the fRMA method proposed in McCall and Irizarry (2011). Another modification over the RMA method was proposed in Ritchie *et al.* (2007) and recommended for being used in Calza and Pawitan (2010). It is worth mentioning that Rafael Irizarry, the professor at the Johns Hopkins Bloomberg School of Public Health in Baltimore, Maryland, is an author in the papers which propose RMA, GCRMA, fRMA and thawing fRMA. In 2010, he was ranked as the second-most cited mathematical scientist in the world by Essential Science Indicators. This observation might indicate that the developments of these variants of the RMA method have been carried out in a series of enhancements by the same research group, or directed by the same guide.

7.3.3   Quantile Normalisation

This method, which was proposed by Bolstad and colleagues (2003), has become the most popular method for normalising one-channel microarray datasets (Cahan *et al.*, 2007; Roberts, 2008). This method is based on the assumption that all of the arrays have a similar signal distribution, which is typical for most of the microarray datasets (Bolstad *et al.*, 2003; Roberts, 2008; Calza and Pawitan, 2010). However, for the cases in which different samples are taken from very different tissue types, quantile normalisation should be avoided as the underlying assumption would not be valid any more (Roberts, 2008; Calza and Pawitan, 2010; Wang *et al.*, 2012).

The steps of quantile normalisation are summarised as follows:

1. Given M arrays of length (number of elements) N, form X of dimension $N \times M$ where each column represents an array.
2. Sort each column to get X_{sorted}.
3. Take the means across rows of X_{sorted} and assign this mean to each element of that row to get X'_{sorted}.
4. Rearrange the elements in the columns of X'_{sorted} to have the same order as in X. This results in the normalised array $X_{normalised}$.

Bolstad and collaborators discussed then that this forces the quantiles to be equal in all of the given arrays, which might not be very accurate at very high intensities, although Bolstad and collaborators followed up by mentioning that, because probe-set expression values were calculated by considering multiple probes, this problem did not seem to be a major problem any more (Bolstad *et al.*, 2003).

Irizarry and colleagues used some controlled datasets to show the importance of normalisation for oligonucleotide microarrays (Irizarry *et al.*, 2003a). They used a 'dilution dataset' in which a range of six known proportions of the cRNA taken from human liver tissues were considered; five replicates were taken per each of these six samples and were scanned by five different scanners. Another dataset which they used was a spike-in data in which all genes are expected to be non-differentially expressed except for 20 genes from which fragments at known concentrations were added. These datasets are real (not simulated) datasets but with controls that provide the ground truth information. Their analysis of the distribution of intensities and log-ratios from these datasets showed the necessity of normalisation and showed that quantile normalisation meets the requirements as needed (Irizarry *et al.*, 2003a).

Calza and Pawitan included quantile normalisation in their recent review as one of the most commonly used techniques for normalising one-channel arrays (Calza and Pawitan, 2010). They mentioned that it can deal with non-linear intensity distributions, is simple to understand and implement, and is fast to run. They also mentioned that it is usually performed over the entire set of probes before summarisation so as to exploit as much information as possible (Calza and Pawitan, 2010). However, this method did not perform well in some studies; namely, when compared with some less popular methods over DNA methylation microarray datasets (Adriaens *et al.*, 2012).

7.3.4   Locally Weighted Scatter-plot Smoothing (Lowess) Normalisation

This was proposed by Yang and colleagues (2002) based on the statistical regression model proposed in Cleveland (1979). The excellent review of normalisation methods in *Nature Genetics* by Quackenbush also presented this method in a very clear way and showed its strength in normalisation (Quackenbush, 2002). It takes non-linearity into consideration and it is the most commonly used method in the case of within-slide two-channel normalisation (Smyth and Speed, 2003; Sievertzon, Nilsson and Lundeberg, 2006; Roberts, 2008; Calza and Pawitan, 2010).

The method was motivated by the obvious bias between the Cy5-dye and the Cy3-dye in two-channel microarrays, that is, the intensity-dependent effects (Irizarry *et al.*, 2003a; Sievertzon, Nilsson and Lundeberg, 2006). MA plots, as introduced in Section 7.2.5 (see Figures 7.1 and 7.2a), show this bias clearly. The lowess normalisation method aims at correcting this bias (Quackenbush, 2002; Yang *et al.*, 2002; Xie *et al.*, 2004).

7.3.4.1   Mathematical Formulation

Assume that $x_i = \log_2 \sqrt{R_i G_i}$ and $y_i = \log_2 (R_i/G_i)$, where R_i and G_i are the red and the green intensities of the ith probe/gene. An MA plot would plot y versus x. A robust lowess smoother

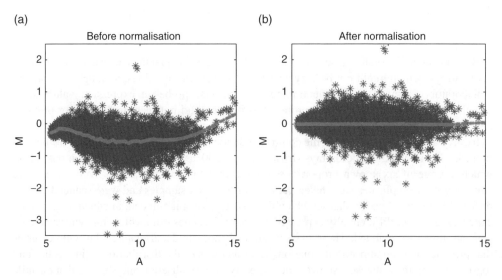

Figure 7.2   MA plots for the yeast genome sample with the NCBI accession number GSM81075 (a) before and (b) after normalisation

is used for regression in order to estimate $y(x_k)$ which represents the best-fit average based on the experimentally observed values (Quackenbush, 2002; Sievertzon, Nilsson and Lundeberg, 2006). While estimating the y value for the point x, a fraction of points closest to the point x can be considered instead of the entire sample set; this fraction of points is called the *span*. If the span is too small, it leads to over-fitting, while if it is too large it leads to inefficient normalisation (Sievertzon, Nilsson and Lundeberg, 2006). Spans of about 0.3 (30%) are usually used (Quackenbush, 2002; Yang *et al.*, 2002; Sievertzon, Nilsson and Lundeberg, 2006).

Then, log-ratio correction is applied in a point-by-point manner by subtracting the best-fit estimate from the original log-ratio. This is represented by Equation (7.8)

$$\log_2\left(T_i'\right) = \log_2(T_i) - y(x_i) = \log_2(T_i) - \log_2\left(2^{y(x_i)}\right) \tag{7.8}$$

or by Equation (7.9).

$$\log_2\left(T_i'\right) = \log_2\left(T_i \times \frac{1}{2^{y(x_i)}}\right) = \log_2\left(\frac{R_i}{G_i} \times \frac{1}{2^{y(x_i)}}\right) \tag{7.9}$$

In terms of intensity correction, this is equivalent to Equation (7.10).

$$G_i' = G_i \times 2^{y(x_i)} \quad \text{and} \quad R_i' = R_i \tag{7.10}$$

An example of lowess normalised data is shown in Figure 7.2b. Lowess normalisation was used successfully by many other studies (Quackenbush, 2002; Xie *et al.*, 2004; Önskog *et al.*, 2011).

7.3.4.2   Global and Local Normalisation

Lowess as well as many other normalisation algorithms can be applied either locally or globally. For example, it might be that different regions of the microarray chip were printed by different pins, or it might be any other reason for which different regions of the microarray have different local properties. In this case, the condition for such local normalisation to be valid is that it must, on its own, satisfy the conditions for the global normalisation. For example, this local region should not have a bias such as containing mostly differentially expressed genes, otherwise lowess normalisation over this region would be invalid (Quackenbush, 2002; Yang *et al.*, 2002; Xie *et al.*, 2004; Sievertzon, Nilsson and Lundeberg, 2006).

Even though Önskog and colleagues noted that global lowess normalisation showed marginally better results over local print-tip lowess in their comprehensive comparative study, their main conclusion was that normalisation is significantly better than no normalisation for microarray data analysis (Önskog *et al.*, 2011).

7.3.4.3   Variance Regularisation

In local lowess normalisation (Huber *et al.*, 2002; Quackenbush, 2002; Yang *et al.*, 2002), after making the mean of the log-ratios within each region (sub-grid) equal to zero, the variance of the ratios among the sub-grids might vary. To solve this issue, the log-ratios are scaled to have similar variance values across sub-grids. Because the mean has become zero, the variance within the nth sub-grid will be as given in Equation (7.11),

$$\sigma_n^2 = \sum_{j=1}^{N} \left[\log_2(T_j) \right]^2 \tag{7.11}$$

where N is the number of elements within this sub-grid. Let a_k be the scaling factor for the kth sub-grid. It is calculated by dividing the variance within this sub-grid by the geometric mean of the variance values of all of the M_{grids} sub-grids. It is formulated as shown in Equation (7.12).

$$a_k = \frac{\sigma_k^2}{\left[\prod_{n=1}^{M_{grids}} \sigma_n^2 \right]^{\frac{1}{M_{grids}}}} \tag{7.12}$$

Accordingly, the corrected log-ratio values are as shown in Equation (7.13)

$$\log_2(T_i) = \frac{\log_2(T_i)}{a_k} \tag{7.13}$$

and the corrected intensities are given in Equation (7.14).

$$G_i' = [G_i]^{1/a_k} \quad \text{and} \quad R_i' = [R_i]^{1/a_k} \tag{7.14}$$

7.3.4.4   Cyclic Lowess

Cyclic lowess is a one-channel probe arrays-normalisation technique based on the two-channel lowess normalisation technique (Bolstad et al., 2003). In this technique, the correction happens in a pairwise manner amongst the different arrays rather than between the different channels. Based on the kth array, the pairwise lowess adjustments for all of the other arrays paired with this array are calculated and applied. Empirically, after running some iterations of adjustments, the changes become small. However, this method might be time consuming (Bolstad et al., 2003; Calza and Pawitan, 2010).

7.3.4.5   Confounding Studies

It was shown by Schmidt and colleagues that, in some cases, normalisation methods would affect the results of microarray data analysis significantly. In their case of study, differentially expressed genes selected after performing lowess normalisation were incoherent and had many contradictory observations (Schmidt et al., 2011). When applied over two-channel ChIP-chip datasets, it showed a worse performance than some other unpopular methods such as T-quantile and Tukey's bi-weight scaling (Adriaens et al., 2012).

7.3.5   Scaling Methods

7.3.5.1   One-channel Arrays

This method, which can be used for one-channel arrays, is described in Bolstad et al. (2003). First of all, a baseline array is chosen which has the median of the medians within all of the arrays. Let x_{base} be the baseline array, x_i be the ith array, $\widetilde{x}_{base}$ and $\widetilde{x}_i$ be the trimmed mean intensities for the baseline array and the ith array respectively, and x_i' be the normalised i^{th} array. The trimmed mean intensities can be calculated after excluding, for example, the highest and the lowest 2% intensity values. Normalisation is then achieved by using Equations (7.15) and (7.16).

$$\beta_1 = \frac{\widetilde{x}_{base}}{\widetilde{x}_i} \tag{7.15}$$

$$x_i' = \beta_i x_i \tag{7.16}$$

This method is equivalent to linear fitting; a non-linear method was proposed in Bolstad *et al.* (2003).

7.3.5.2 Two-channel Arrays

Yang and colleagues proposed a scaling method to be performed over log-ratios of two-channel arrays after within-slide normalisation had been carried out (Yang *et al.*, 2002). In this case, the log-ratios within each slide (array) would be centred about zero but might have varying variances.

Assume that the variance of the ith slide is represented by $\sigma_i^2 = a_i^2 \sigma^2$, where the term σ^2 is the true log-ratios variance and the term a_i^2 is the scaling factor for this slide. Normalisation is done by estimating the values of a_i then eliminating them. Based on the assumption that log-ratios are normally distributed, the forced constraint is $\sum_{i=1}^{I} a_i^2$, where I is the number of slides. So, the MLE for a_i will be as given in Equation (7.17),

$$\hat{a}_i^2 = \frac{\sum_{j=1}^{n_i} M_{ij}^2}{\sqrt[I]{\prod_{k=1}^{I} \sum_{j=1}^{n_k} M_{kj}^2}} \tag{7.17}$$

where M_{ij}^2 is the log-ratio for the jth gene/probe in the ith slide and n_i is the number of genes/probes in the ith slide. A robust alternative estimation is given by Equation (7.18):

$$\hat{a}_i = \frac{MAD_i}{\sqrt[I]{\prod_{i=1}^{I} MAD_j}} \tag{7.18}$$

where $MAD_i = \underset{j}{\text{median}} \left\{ \left| M_{ij} - \underset{j}{\text{median}} (M_{ij}) \right| \right\}$.

7.3.6 Model-based Expression Index (MBEI)

Li and Wong statistically analysed the oligonucleotide microarrays expression at the probe level (Li and Wong, 2001a). They suggested a multiplicative model for the intensities of the *MM* and the *PM* probes as in Equations (7.19) and (7.20), respectively. The probe-pair difference (*PM* − *MM*) is shown in Equation (7.21).

$$MM_{ij} = v_j + \theta_i \alpha_j + \varepsilon \tag{7.19}$$

$$PM_{ij} = v_j + \theta_i \alpha_j + \theta_i \phi_j + \varepsilon \tag{7.20}$$

$$y_{ij} = PM_{ij} - MM_{ij} = \theta_i \phi_j + \varepsilon_{ij} \tag{7.21}$$

The values MM_{ij} and PM_{ij}, respectively, represent the MM and PM average intensity levels of the jth probe-pair in the ith array for the gene under consideration. v_j is due to non-specific hybridisation, θ_i is the mRNA concentration of the target, α_j is the rate of linear increase of the intensity of the MM_{ij} probe with respect to the increase of the mRNA concentration, ϕ_j is the additional rate of increase of the intensity of the PM_{ij} probe with respect to the increase of the mRNA concentration, and ε is a generic random error term. The rates α_j and ϕ_j are non-negative values. The reasoning behind this model is that the PM and the MM probes have a background caused by non-specific hybridisation, over which the intensities increase linearly with respect to the concentration of the mRNA. The rate of increase in the PM probe is expected to be higher than that in the MM probe ($\alpha_j + \phi_j$ vs. α_j) (Li and Wong, 2001a). They empirically fitted this model to some real datasets and compared it with an additive model. The multiplicative model was shown to fit the data much better than did the additive one (Li and Wong, 2001a).

This analysis resulted in identifying the outlier arrays and probes, and then excluding them. The method is based on estimating the variables in the aforementioned model, then identifying the probes with high standard error values compared with the fitted model as outliers (Li and Wong, 2001a).

The same authors published another work soon after that, in which they generalise the MBEI method to be used to PM-only microarrays as well (Li and Wong, 2001b). They also point out that they have made the *DNA-Chip Analyser* software available publically at http://www.dchip.org.

7.3.7   Other Normalisation Methods

7.3.7.1   Contrast Normalisation

This was proposed with full details by Åstrand (2003) and was used with a brief description in Bolstad *et al.* (2003). This is for single-channel oligonucleotide arrays and is a non-linear method for normalising feature intensities (Åstrand, 2003). It involves changing the basis of the data through a transformation with an orthonormal transformation matrix, then using local regression (loess) non-linear fitting curves to normalise all of the arrays versus a baseline array.

7.3.7.2   Median/Mean Normalisation

This is a global normalisation method performed by simply subtracting a constant from the intensity log-ratios within a slide. This constant is usually the log-ratios' median or mean (Dudoit *et al.*, 2002; Schmidt *et al.*, 2011). In some studies, this has been shown to perform better than some of the more sophisticated methods such as lowess (Schmidt *et al.*, 2011).

7.3.7.3   Three-parameter Normalisation

This was proposed by Konishi and is meant mainly to normalise different one-channel microarray datasets so they become comparable (Konishi, 2004). This method assumes that the distribution within any single microarray is stable. The background is estimated and considered uniform for any single dataset; this background is estimated as the constant which, when subtracted from the entire dataset, allows the revised dataset to become the best fit to the given model. The following text is the description of the method as detailed in Konishi (2004):

The method assumes that the original intensity data, (r_i) for $i = 1, 2 \ldots n$, obey a lognormal distribution. The probability density function (pdf) of the intensity data used data used is given in Equation (7.22),

$$f(r_i) = \left(\frac{1}{\sigma\sqrt{2\pi}}\right) e^{-\log(r_i - \gamma) + \frac{\mu^2}{2\sigma^2}} \ for \ r_i > \gamma \tag{7.22}$$

where σ, μ and γ are the shape, scale and threshold parameters, respectively.

The parameter σ is found through trial and improvement calculation processes. In the trial, the distribution of $\log(r_i - \gamma)$ is checked by normal probability plotting (NIST, 2010), and the value that gives the best fit to the model is selected for γ. The fitness is evaluated by the sum of absolute differences between the model and $\log(r_i - \gamma)$, within the interquartile range of data. The parameter μ is the median of $\log(r_i - \gamma)$, and the parameter σ is found from the interquartile range of $\log(r_i - \gamma)$; these are known as robust alternatives for the arithmetic mean and standard deviation, respectively. Parameters μ and σ are found for each data grid, a group of data for DNA spots that were printed by an identical pin in order to avoid divergences caused by pin-based differences (Schuchhardt et al., 2000). Z-normalisation is carried out for each datum as shown in Equation (7.23).

$$Z_{ri} = \frac{\log(r_i - \gamma) - \mu}{\sigma} \tag{7.23}$$

Intensity data (r_i) less than γ are treated as 'data not detected', since such data might contain negative noise larger than the signal.

This model allows for finding the lowest value unaffected by additive noise and the highest value unaffected by saturation. This is done by noticing that, beyond these two values, the values deviate from the expected model distribution (Konishi, 2004; Konishi et al., 2008).

7.3.7.4   Total Intensity Normalisation

For two-channel arrays, this is based on many assumptions as discussed in Quackenbush (2002). Assume that the red and green channels' intensities are R and G, respectively. The ratio T would be as given in Equation (7.24).

$$T = \frac{R}{G} \tag{7.24}$$

If the number of elements (genes) in the array is N_{array}, N_{total} will be the *normalisation factor*, and is expressed as shown in Equation (7.25).

$$N_{total} = \frac{\sum_{i=1}^{N_{array}} R_i}{\sum_{i=1}^{N_{array}} G_i} \tag{7.25}$$

One or both intensities are scaled accordingly, for example [Equation (7.26)]

$$R' = R, \ G' = N_{total}G \tag{7.26}$$

Then the adjusted ratio will be as given in Equation (7.27).

$$T' = \frac{1}{N_{total}} \cdot \frac{R}{G} \tag{7.27}$$

This makes the mean ratio be equal to unity. In the log-ratio case, this is equivalent to Equation (7.28).

$$\log_2(T') = \log_2(T) - \log_2(N_{total}) \tag{7.28}$$

7.3.7.5 Average Difference (AD or AvDiff)

This is the one-channel normalisation procedure implemented in the Affymetrix software GeneChip (Åstrand, 2003). It is based on the average difference (AD) between the PM intensity and the MM intensity. Many variations were proposed regarding how to remove the outliers before averaging (Irizarry *et al.*, 2003a). The mathematical expression of the AvDiff is shown in Equation (7.29),

$$AvDiff = \frac{1}{|A|} \sum_{j \in A} (PM_j - MM_j) \tag{7.29}$$

where A is the set of probe-pairs within the same probe-set after eliminating the outliers (Åstrand, 2003), and $|A|$ is the cardinality of A.

7.3.7.6 Other Methods

Many ratio-based scheme-normalisation methods have been proposed in the literature (Chen, Dougherty and Bittner, 1997; Tseng *et al.*, 2001; Durbin *et al.*, 2002; Huber *et al.*, 2002), but they were criticised for either showing intensity-dependent fluctuations or having questionable stability (Konishi, 2004).

Other methods include analysis of variance-based (ANOVA-based) normalisation (Kerr, Martin and Churchill, 2000), variance stabilisation normalisation (VSN) (Motakis *et al.*, 2006), self-consistency and local regression (Kepler, Crosby and Morgan, 2002). Sievertzon and colleagues discussed the latter one and noticed that it is a similar approach to lowess normalisation but with a different local regression method (Sievertzon, Nilsson and Lundeberg, 2006). Another method is the probe logarithmic intensity error (PLIER), which is a newer algorithm developed by Affymetrix that uses an improved $(PM - MM)$ correction and quantile normalisation (Affymetrix, 2005; Roberts, 2008).

7.4   Discussion and Summary

Microarrays measure the expression values of thousands of genes simultaneously and quantify those measurements in the form of colour intensities. One-channel microarrays provide a single expression measurement for any single gene at a particular condition, while two-channel micro-arrays provide two values, one for the target condition and one for the control; the ratio between the experimental and the control values represents the value considered for further analysis. Various issues affect the accuracy of both types of microarrays' measurements and are tackled by normalisation methods. One issue is the validity of comparing the values of different genes within one microarray sample (aka chip or slide), and their values across multiple samples (chips or slides). Those two issues are tackled by within-slide methods and between-slides methods, respectively (Yang *et al.*, 2002; Sievertzon, Nilsson and Lundeberg, 2006; Verducci *et al.*, 2006). Quantile normalisation is commonly used to correct the within-slide differences in one-channel microarrays (Bolstad *et al.*, 2003; Roberts, 2008) while the lowess normalisation is commonly used in two-channel microarrays (Quackenbush, 2002; Yang *et al.*, 2002).

Another issue is the background noise, its causes, modelling, and correction. Different approaches were proposed to tackle this issue, such as the addition of a MM probe paired with each PM probe. This approach assumed that the MM probe would measure the background noise mainly caused by NSB, and the PM probe would measure the original signal added to that noise (Wu, 2009). However, many following studies showed that in various cases the value measured by the MM probe was larger than that of the PM, which does not conform to the original assumption (Naef and Magnasco, 2003; Ahmed, 2006). Consequently, the Microarray Suite 5 (MAS 5.0) provided by Affymetrix considered a modified version of the MM probes' measurements that never exceed that of the PM probes (Affymetrix, 2002a; Roberts, 2008). On the other hand, many normalisation methods were designed to ignore MM probes and use PM probes only, such as the RMA method (Bolstad *et al.*, 2003; Irizarry *et al.*, 2003a).

As for the statistical distribution of the expression values, it was found to be lognormal in most of the cases. Therefore logarithmic transformation of the expression values has been commonly adopted by microarray analysis studies (Kreil and Russell, 2005).

Other types of major issues include intensity-based bias, which is the fact that the sensitivity of the microarray probes changes non-linearly with their intensity, that is with the amount of exposed mRNA. MA scatter plots, which plot log-ratios (M) versus abundance values (A) of two-channel microarray expression values, are used to visualise this bias, and then to correct it by using the lowess normalisation method (Dudoit *et al.*, 2002).

To increase the reliability of the microarray measurements, several replicates are usually considered for the same biological condition. Consequently, multiple values are obtained for the same condition and need to be summarised (Churchill, 2002; Quackenbush, 2002). Summarisation can be simply by taking the median or the mean of the values. If the number of replicates was small, for example three, the median would be a better choice because it is more robust than the mean at such low numbers of replicates.

Taken together, datasets vary in their types and statistical properties such as the number of channels, the statistical distribution, the types of probes, the number of replicates and others. Given that many variations of normalisation methods exist while being differentially custo-mised to different types of datasets, understanding such variables is crucial for the correct choice of normalisation method (Table 7.1).

Table 7.1 Normalisation software packages

Method	Platform/Package	Function
Background correction (general)	R/Bioconductor – affy	expresso
		bgcorrect
Contrast normalisation	R/Bioconductor – affy	normalize.contrasts
		normalize.AffyBatch.contrasts
Cyclic loess normalisation	R/Bioconductor – limma	normalizeCyclicLoess
Filter low absolute value genes	MATLAB	genelowvalfilter
Filter low entropy genes	MATLAB	geneentropyfilter
Filter small range genes	MATLAB	generangefilter
Filter small variance genes	MATLAB	genevarfilter
GC-RMA	MATLAB	affygcrma
		gcrma
		gcrmabackadj
Invariant set normalisation	MATLAB	affyinvarsetnorm
		mainvarsetnorm
	R/Bioconductor –	normalize.invariantset
	preprocessCore	normalize.AffyBatch.invariantset
Loess/lowess normalisation	MATLAB	malowess
	R/Bioconductor –	normalize.loess
	preprocessCore	normalize.AffyBatch.loess
MAS 5.0	R/Bioconductor – affy	mas5
MBEI	R/Bioconductor – affy	expresso (with specific set of
		parameters)
Normalisation (general)	R/Bioconductor – affy	normalize
Quantile normalisation	MATLAB	quantilenorm
	R/Bioconductor –	normalize.quantiles
	preprocessCore	normalize.quantiles.robust
		normalize.AffyBatch.quantiles
RMA	MATLAB	affyrma
		rmabackadj
		rmasummary
	R/Bioconductor – affy	rma
Scaling and centring	MATLAB	manorm
	R/Bioconductor –	normalize.constant
	preprocessCore	
Summarisation (general)	R/Bioconductor – affy	summary

References

Abruzzo, L.V., Lee, K.Y., Fuller, A. *et al.* (2005). Validation of oligonucleotide microarray data using micro-fluidic low-density arrays: a new statistical method to normalize real-time RT - PCR data. *BioTechniques*, **38**, pp. 785–792.

Adriaens, M.E., Jaillard, M., Eijssen, L.M. *et al.* (2012). An evaluation of two-channel ChIP -on-chip and DNA methylation microarray normalization strategies. *BMC Genomics*, **13**, e42.

Affymetrix (2002a). *Affymetrix Microarray Suite User Guide 5.0*, Affymetrix, Santa Clara, CA.

Affymetrix (2002b). *Statistical algorithms description document*, Affymetrix, Santa Clara, CA.

Affymetrix (2005). *Guide to Probe Logarithmic Intensity Error (PLIER) Estimation*, www.affymetrix.com/support/technical/technotes/plier_technote.pdf (accessed 21 July 2014).

Ahmed, F.E. (2006). Microarray RNA transcriptional profiling: part I. Platforms, experimental design and standardization. *Expert Review of Molecular Diagnostics*, **6**, pp. 535–550.

Åstrand, M. (2003). Contrast normalization of oligonucleotide arrays. *Journal of Computational Biology*, **10**, pp. 95–102.

Bolstad, B., Irizarry, R., Astrand, M. and Speed, T. (2003). A comparison of normalization methods for high density oligonucleotide array data based on variance and bias. *Bioinformatics*, **19**, pp. 185–193.

Brazma, A., Hingamp, P., Quackenbush, J. *et al.* (2001). Minimum information about a microarray experiment (MIAME)-toward standards for microarray data. *Nature Genetics*, **29**, pp. 365–371.

Cahan, P., Rovegno, F., Mooney, D. *et al.* (2007). Meta-analysis of microarray results: challenges, opportunities, and recommendations for standardization. *Gene*, **401**, pp. 12–18.

Calza, S. and Pawitan, Y. (2010). Normalization of gene-expression microarray data. *Methods in Molecular Biology*, **673**, pp. 37–52.

Calza, S., Valentini, D. and Pawitan, Y. (2008). Normalization of oligonucleotide arrays based on the least-variant set of genes. *BMC Bioinformatics*, **9**, p. 140.

Chen, Y., Dougherty, E.R. and Bittner, M.L. (1997). Ratio-based decisions and the quantitative analysis of cDNA microarray images. *Journal of Biomedical Optics*, **2**, pp. 364–374.

Churchill, G. (2002). Fundamentals of experimental design for cDNA microarrays. *Nature Genetics*, **32**, pp. 490–495.

Cleveland, W.S. (1979). Robust locally weighted regression and smoothing scatter-plots. *Journal of the American Statistical Association*, **74**, pp. 829–836.

Cope, L.M., Irizarry, R.A., Jaffee, H.A. *et al.* (2004). A benchmark for Affymetrix GeneChip expression measures. *Bioinformatics*, **20**, pp. 323–331.

Dudoit, S., Yang, Y.H., Callow, M.J. and Speed, T.P. (2002). Statistical methods for identifying differentially expressed genes in replicated cDNA microarray experiments. *Statistica Sinica*, **12**, pp. 111–139.

Durbin, B.P., Hardin, J.S., Hawkins, D.M. and Rocke, D.M. (2002). A variance-stabilizing transformation for gene-expression microarray data. *Bioinformatics*, **18**, pp. S105–110.

Eisen, M. (1999). *Cluster and TreeView*, http://rana.lbl.gov/manuals/ClusterTreeView.pdf (accessed 21 July 2014).

Furlotte, N.A., Xu, L., Williams, R.W. and Homayouni, R. (2011). *Literature-based evaluation of microarray normalization procedures*. IEEE International Conference on Bioinformatics and Biomedicine (BIBM), Atlanta, GA, November 2011, pp. 608–612.

Hoyle, D.C., Rattray, M., Jupp, R. and Brass, A. (2002). Making sense of microarray data distributions. *Bioinformatics*, **12**, pp. 576–584.

Hubbell, E., Liu, W. and Mei, R. (2002). Robust estimators for expression analysis. *Bioinformatics*, **18**, pp. 1585–1592.

Huber, W., von Heydebreck, A., Sültmann, H. *et al.* (2002). Variance stabilization applied to microarray data calibration and to the quantification of differential expression. *Bioinformatics*, **18**, pp. S96–104.

Irizarry, R.A., Bolstad, B.M., Collin, F. *et al.* (2003a). Summaries of Affymetrix GeneChip probe level data. *Nucleic Acids Research*, **31**, e15.

Irizarry, R.A., Hobbs, B., Collin, F. *et al.* (2003b). Exploration, normalization, and summaries of high density oligonucleotide array probe level data. *Biostatistics*, **4**, pp. 249–264.

Irizarry, R.A., Wu, Z. and Jaffee, H.A. (2006). Comparison of Affymetrix GeneChip expression measures. *Bioinformatics*, **22**, pp. 789–794.

Kepler, T.B., Crosby, L. and Morgan, K.T. (2002). Normalization and analysis of DNA microarray data by self-consistency and local regression. *Genome Biology*, **28**, RESEARCH0037.

Kerr, M.K., Martin, M. and Churchill, G.A. (2000). Analysis of variance for gene expression microarray data. *Journal of Computational Biology*, **7**, pp. 819–837.

Konishi, T. (2002). Parametric treatment of cDNA microarray data. *Genome Informatics*, **13**, pp. 280–281.

Konishi, T. (2004). Three-parameter lognormal distribution ubiquitously found in cDNA microarray data and its application to parametric data treatment. *BMC Bioinformatics*, **5**, p. 5.

Konishi, T. (2005). A thermodynamic model of transcriptome formation. *Nucleic Acids Research*, **33**, pp. 6587–6592.

Konishi, T., Konishi, F., Takasaki, S. *et al.* (2008). Coincidence between transcriptome analyses on different microarray platforms using a parametric framework. *PLoS ONE*, **3**, e3555.

Kreil, D.P. and Russell, R.R. (2005). There is no silver bullet – a guide to low-level data transforms and normalisation methods for microarray data. *Briefings in Bioinformatics*, **6**, pp. 86–97.

Lee, N.H. and Saeed, A.I. (2007). Microarrays: an overview. (eds E. Hilario and J. Mackay), *Methods in Molecular Biology*. Human Press Inc., Totowa, NJ, pp. 265–300.

Li, C. and Wong, W.H. (2001a). Model-based analysis of oligonucleotide arrays: expression index computation and outlier detection. *Proceedings of the National Academy of Sciences*, **98**, pp. 31–36.

Li, C. and Wong, W.H. (2001b). Model-based analysis of oligonucleotide arrays: model validation, design issues and standard error application. *Genome Biology*, **2**, RESEARCH0032.

Lin, S.M., Du, P., Huber, W. and Kibbe, W.A. (2008). Model-based variance-stabilizing transformation for illumina microarray data. *Nucleic Acids Research*, **36**, e11.

McCall, M.N. and Irizarry, R.A. (2011). Thawing frozen robust multi-array analysis (fRMA). *BMC Bioinformatics*, **12**, e369.

McCall, M.N., Bolstad, B.M. and Irizarry, R.A. (2010). Frozen robust multiarray analysis (fRMA). *Biostatistics*, **11**, pp. 242–253.

McCall, M.N., Jaffee, H.A. and Irizarry, R.A. (2012). fRMA ST: frozen robust multiarray analysis for Affymetrix Exon and Gene ST arrays. *Bioinformatics*, **28**(23), pp. 3153–3154.

Motakis, E.S., Nason, G.P., Fryzlewicz, P. and Rutter, G.A. (2006). Variance stabilization and normalization for one-color microarray data using a data-driven multiscale approach. *Bioinformatics*, **22**, pp. 2547–2553.

Naef, F. and Magnasco, M.O. (2003). Solving the riddle of the bright mismatches: labeling and effective binding in oligonucleotide arrays. *Physical Review E*, **68**(1 Pt 1), p. 011906.

Naef, F., Lim, D.A., Patil, N. and Magnasco, M.O. (2001). From features to expression: high density oligonucleotide array analysis revisited. *arXiv:physics/0102010* [*physics.bio-ph*], **1**, pp. 1–9.

Naef, F., Lim, D.A., Patil, N. and Magnasco, M. (2002). DNA hybridization to mismatched templates: a chip study. *Physical Review*, **65**(4 Pt 1), p. 040902.

NIST (2010). *NIST/SEMATECH e-Handbook of Statistical Methods*, National Institute of Standards and Technology (NIST).

Olshen, A.B. and Jain, A.N. (2002). Deriving quantitative conclusions from microarray expression data. *Bioinformatics*, **18**, pp. 961–970.

Önskog, J., Freyhult, E., Landfors, M. *et al.* (2011). Classification of microarrays; synergistic effects between normalization, gene selection and machine learning. *BMC Bioinformatics*, **12**, e390.

Quackenbush, J. (2002). Microarray data normalization and transformation. *Nature Genetics*, **32**, pp. 496–501.

Reilly, C., Wang, C. and Rutherford, M. (2003). A method for normalizing microarrays using genes that are not differentially expressed. *Journal of the American Statistical Association*, **98**, pp. 868–879.

Ritchie, M.E., Silver, J., Oshlack, A. *et al.* (2007). A comparison of background correction methods for two-colour microarrays. *Bioinformatics*, **23**, pp. 2700–2707.

Roberts, P.C. (2008). Gene expression microarray data analysis demystified. *Biotechnology Annual Review*, **14**, pp. 29–61.

Sapir, M. and Churchill, G.A. (2000). *Estimating the Posterior Probability of Differential Gene Expression from Microarray Data*, http://sapir.us/DataAnalysis/SapirChurchillPoster.pdf (accessed 22 July 2014).

Schmidt, M.T., Handschuh, L., Zyprych, J. *et al.* (2011). Impact of DNA microarray data transformation on gene expression analysis — comparison of two normalization methods. *Acta Biochimica Polonica*, **58**, pp. 573–580.

Schuchhardt, J., Beule, D., Malik, A. *et al.* (2000). Normalization strategies for cDNA microarrays. *Nucleic Acids Research*, **28**(10), E47.

Sievertzon, M., Nilsson, P. and Lundeberg, J. (2006). Improving reliability and performance of DNA microarrays. *Expert Review of Molecular Diagnostics*, **6**, pp. 481–492.

Smyth, G.K. and Speed, T. (2003). Normalization of cDNA microarray data. *Methods*, **31**, pp. 265–273.

Tseng, G.C., Oh, M.K., Rohlin, L. *et al.* (2001). Issues in cDNA microarray analysis: quality filtering, channel normalization, models of variations and assessment of gene effects. *Nucleic Acids Research*, **29**, pp. 2549–2557.

Vendettuoli, M., Doyle, E. and Hofmann, H. (2011). Clustering microarray data to determine normalization method. *Advances in Experimental Medicine and Biology*, **696**, pp. 145–153.

Verducci, J.S., Melfi, V.F., Lin, S. *et al.* (2006). Microarray analysis of gene expression: considerations in data mining and statistical treatment. *Physiological Genomics*, **25**, pp. 355–363.

Wang, D., Cheng, L., Zhang, Y. *et al.* (2012). Extensive up-regulation of gene expression in cancer: the normalised use of microarray data. *Molecular BioSystems*, **8**, pp. 818–827.

Wu, Z. (2009). A review of statistical methods for preprocessing oligonucleotide microarrays. *Statistical Methods in Medical Research*, **18**, pp. 533–541.

Wu, Z. and Irizarry, R. (2007a) Description of Gcrma Package, http://www.bioconductor.org/packages/2.0/bioc/vignettes/gcrma/inst/doc/gcrma2.0.pdf (accessed 22 July 2014).

Wu, Z. and Irizarry, R.A. (2007b). A statistical framework for the analysis of microarray probe-level data. *Annals of Applied Statistics*, **1**, pp. 333–357.

Wu, Z., Irizarry, R.A., Gentleman, R. *et al.* (2004). A model-based background adjustment for oligonucleotide expression arrays. *Journal of the American Statistical Association*, **99**, pp. 909–917.

Xie, Y., Jeong, K.S., Pan, W. *et al.* (2004). A case study on choosing normalization methods and test statistics for two-channel microarray data. *Comparative and Functional Genomics*, **5**, pp. 432–444.

Yang, Y.H., Buckley, M.J. and Speed, T.P. (2001). Analysis of cDNA microarray images. *Briefings in Bioinformatics*, **2**, pp. 341–349.

Yang, Y.H., Dudoit, S., Luu, P. *et al.* (2002). Normalization for cDNA microarray data: a robust composite method addressing single and multiple slide systematic variation. *Nucleic Acids Research*, **30**(4), e15.

8

Feature Selection

8.1 Introduction

In gene expression datasets [e.g. microarrays and next-generation sequencing (NGS)], the number of genes N tends to be much larger than the number of samples or biological conditions M ($N \gg M$) (Tarca, Romero and Draghici, 2006). However, most of the genes in such a dataset would be irrelevant to the problem in hand. Depending on the type of downstream analysis and research objectives, a subset of the genes would be selected before progression to the following steps of analysis. The same issue applies to the other types of expression datasets such as proteomics, glycomics and metabolomics (Haynes *et al.*, 2013).

Feature selection (*FS*) techniques are employed to select the subset of features which is expected to result in better separability of the given data objects. For example, *supervised classification* techniques can be used to classify a set of samples in a gene expression dataset based on their expression values over genes. Genes are considered in this case as features for those samples, and FS techniques would be utilised to select the subset of genes expected to lead to building better supervised classifiers; this is known as *gene selection.* A *classifier* in the context of gene expression data, as discussed in Chapter 2, is a model which is trained by using the expression profiles of a number of samples with known classes, in order to be able to predict the class of a previously unseen sample. For example, a classifier would be trained to distinguish between samples from healthy individuals and individuals with breast cancer based on their gene expression profiles (Tarca, Romero and Draghici, 2006; Hassanien, Al-Shammari and Ghali, 2013).

It is a major issue in designing classifiers that the number of genes (features) is usually much larger than the number of samples ($N \gg M$). This is known as the *curse of dimensionality* and is tackled by gene selection in order to reduce the dimensionality, that is, the number of genes (Fang, Martin and Wang, 2012). Gene selection aims at identifying the non-redundant subset

Integrative Cluster Analysis in Bioinformatics, First Edition. Basel Abu-Jamous, Rui Fa and Asoke K. Nandi.
© 2015 John Wiley & Sons, Ltd. Published 2015 by John Wiley & Sons, Ltd.

of genes which are expressed differentially between the different classes or conditions between which the classifier should differentiate. Two conditions are required in the selected subset of genes – first, they need to be informative in that they show significant difference in their expression levels across classes or conditions; second, they need to be non-redundant, that is, if two or more genes provide the same useful piece of information, including one of them is sufficient while including all of them leads to redundancy.

Moreover, machine-learning techniques, especially clustering, would be applied to genes while considering the expression samples as features. In this case, FS represents sample selection rather than gene selection. However, it is more common to employ *feature generation* (FG) methods instead of FS methods in such cases, which produce a new set of features, usually fewer than the original ones, that represent linear or non-linear combinations of the original features.

In contrast, many studies hold the objective of identifying all of the genes that are differentially expressed between two or more biological conditions because this differential expression itself is biologically meaningful. Having said that, dimensionality reduction and the improvement of a classifier's feasibility and accuracy are not within the scope of such studies, and the issue of redundancy is irrelevant to them (Tarca, Romero and Draghici, 2006).

This chapter covers the aspect of FS and GS, while the following chapter covers the aspect of differential expression. Although this chapter mainly considers gene expression datasets, the same concepts apply to the other expression datasets such as proteomic and metabolomic datasets. However, the presentation will be clearer and smoother by providing complete discussion about the more mature gene expression datasets instead of naming the possibilities: gene expression, protein expression, metabolic expression, and so on repetitively throughout the chapter.

8.2   FS and FG – Problem Definition

A genetic expression dataset with N genes and M samples can be formulated as given in Equation (8.1).

$$X_{N \times M} = \begin{bmatrix} x_{11} & \cdots & x_{1M} \\ \vdots & \ddots & \vdots \\ x_{N1} & \cdots & x_{NM} \end{bmatrix} \tag{8.1}$$

FS is the process of selecting the subset of non-redundant features that is most informative regarding downstream analysis. It can be performed over either dimension of the matrix depending on the type of following analysis, that is, it can be by considering genes as features and selecting a subset of genes accordingly by FS, and can be by considering samples as features and selecting a subset of samples accordingly.

In order to generalise the problem, we consider that the data is composed of N objects, each of which is represented by a row vector containing values of M features. This dataset can be formulated as shown in Equation (8.2).

$$X = \begin{bmatrix} \vec{x}_1 \\ \vec{x}_2 \\ \vdots \\ \vec{x}_N \end{bmatrix} = \begin{bmatrix} x_{11} & \cdots & x_{1M} \\ \vdots & \ddots & \vdots \\ x_{N1} & \cdots & x_{NM} \end{bmatrix}, \ \vec{x}_i = \begin{bmatrix} x_{i1} & x_{i2} & \ldots & x_{iM} \end{bmatrix} \tag{8.2}$$

Let the original set of features be $F = \{f_1, f_2, \ldots, f_M\}$; FS selects a subset of m features from F, labelled as S, where $S \subset F$.

In contrast, FG methods generate a new set of m features, S, which are linear or non-linear combinations of the original features (Mu, Nandi and Rangayyan, 2007). FS and FG are expressed in the following two Equations, (8.3) and (8.4), respectively,

$$F = \{f_1 \ldots f_M\} \xrightarrow{FS} S = \{f_{s_1} \ldots f_{s_m}\} \subset F, \ \{s_1 \ldots s_m\} \subset \{1 \ldots M\} \tag{8.3}$$

$$F = \{f_1 \ldots f_M\} \xrightarrow{FG} S = \{g_i | 1 \leq i \leq m, \ g_i = h_i(F)\} \tag{8.4}$$

where $h_i(F)$ calculates the value of the ith generated feature as a function of the original features F.

FS and FG methods are classified into two main classes: filter methods (open-loop) and wrapper methods (closed-loop). In the context of supervised classification analysis, filter methods select features without considering the classifier that will be used, while wrapper methods, as closed-loop methods, consider iterations of feedback from the classifier (Day and Nandi, 2011). Such methods aim at maximising the classification accuracy with respect to the selected subset of genes.

As open-loop methods are more relevant to the context of this book, that is, unsupervised clustering, the following sections present some open-loop FS and FG methods.

8.3   Consecutive Ranking

Consecutive ranking is a recursive process which depends on the information content in the subset of selected features, denoted as $I(S)$. Two main classes of consecutive ranking exist, namely *forward search* (most informative first admitted) and *backward elimination* (least useful first eliminated) (Kung and Mak, 2008), and they are detailed below.

8.3.1   Forward Search (Most Informative First Admitted)

Forward search starts with an empty list of selected features S, and then admits the feature which contains the largest amount of information. In each round, the feature which adds the largest amount of information content to the set S is admitted. The mathematical annotation for the iteration (t) is shown in Equation (8.5);

$$S^{t+1} \xleftarrow{admit} \underset{f_i \notin S(t)}{\mathrm{argmax}} \ I(S^t \cup f_i) \tag{8.5}$$

when the size of S reaches the predefined number of selected features m, the process terminates.

8.3.2   Backward Elimination (Least Useful First Eliminated)

Backward elimination is a form of *over-select-then-prune* strategy. It starts with the complete list of M features contained in S, and then eliminates features iteratively until S has only m features. In each round, the feature which if eliminated causes the least loss of information content is the one which is eliminated. The tth iteration of backward elimination is represented as shown in Equation (8.6).

$$S^{t+1} \xleftarrow{eliminate} \operatorname*{argmax}_{f_i \in S(t)} I(S^t - f_i) \tag{8.6}$$

Although the termination criterion can be that the size of S has reached m, other criteria can also be considered. Pre-setting the number of the selected features m is not an easy task and cannot be chosen arbitrarily. In many cases the number of selected features is kept flexible while the quality of the set S is considered as the termination criterion. For example, the process can be terminated when $I(S)$ reaches about 90% of the information contained in the complete set $I(complete\ set)$.

8.4   Individual Ranking

Individual ranking methods rank the features in a descending order, and then select the top m of them. Different criteria have been employed here, and some of them are detailed in the following subsections.

8.4.1   Information Content

This criterion is used in unsupervised learning and is defined in different ways. The most common information content metric is *Shannon's entropy* (Kadota *et al.*, 2006; Kung and Mak, 2008), which is defined as given in Equations (8.7) and (8.8),

$$H(j) = -\sum_{i=1}^{N} p_{ij} \log_2 \left[p_{ij} \right] \tag{8.7}$$

$$p_{ij} = \frac{x_{ij}}{\sum_{i^*=1}^{N} x_{i^*j}} \tag{8.8}$$

where $H(j)$ is the entropy, that is, the information content, of the jth feature, N is the number of objects, p_{ij} is the ratio between the jth feature of the ith object and the total feature values of that object, and x_{ij}/x_{i^*j} is the value of the jth feature of the ith/i^*th object.

The value of the entropy for each feature ranges from 0 to $\log_2 N$. The entropy is equal to $\log_2 N$ when the feature is equally valued across all of the objects, while it becomes equal to zero when the feature has a high value for a single object and low values for the others. This means that: the smaller the value of the entropy, the more object-specific the feature.

8.4.2   SNR Criteria

This class of criteria is applied to datasets in which the objects belong to a number of classes (groups). The argument on which all the *Signal-to-noise Ratio* (*SNR*) criteria rely is that better (more influencing, more informative) features distribute the data objects such that the values of the objects which belong to the same class are close to each other while being far from the values of the objects in the other classes, that is, small intra-class variance and large inter-class variance. Higher mean distances and lower intra-class variance values lead to better separability for classes. Figure 8.1 shows an illustrative example for a two-class problem where different couples of features are selected and lead to different separability characteristics.

The general *SNR* formula is shown in Equation (8.9).

$$SNR = \frac{Signal}{Noise} = \frac{Distance}{Intra-Class\ Variance} \qquad (8.9)$$

Below are some of the most common SNR-based gene-ranking criteria for two-class problems (except for the F-ratio) where the two classes are denoted as C^+ and C^- respectively, and $\mu_j^+, \mu_j^-, \sigma_j^+, \sigma_j^-$ are the class-conditional means and standard deviations for both classes,

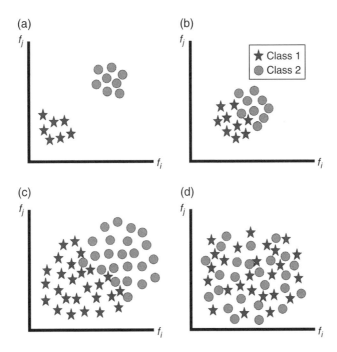

Figure 8.1   Two-class separability problem; (a) high inter-class distance and low intra-class variance – the best separability; (b) low inter-class distance and low intra-class variance; (c) high inter-class distance and high intra-class variance; (d) low inter-class distance and high intra-class variance – the worst separability

respectively. Each of the following formulas (8.10)–(8.12) calculates the *SNR* for the *j*th feature.

1. Signed Fisher's discriminant ratio (signed-FDR):

$$Signed-FDR(j) = \frac{\mu_j^+ - \mu_j^-}{\sigma_j^+ + \sigma_j^-} \tag{8.10}$$

2. Second-order FDR:

$$FDR(j) = \frac{\left(\mu_j^+ - \mu_j^-\right)^2}{\left(\sigma_j^+\right)^2 + \left(\sigma_j^-\right)^2} \tag{8.11}$$

3. T-statistic:

$$z_j = \frac{\mu_j^+ - \mu_j^-}{\sqrt{\frac{\left(s_j^+\right)^2}{N^+} + \frac{\left(s_j^-\right)^2}{N^-}}} \tag{8.12}$$

where N^+ and N^- are the numbers of objects in the classes C^+ and C^- respectively, $\left(s_j^+\right)^2$ and $\left(s_j^-\right)^2$ are defined in Equation (8.13)

$$\left(s_j^+\right)^2 = \frac{\sum_{k \in C^+}\left(x_{jk} - \mu_j^+\right)^2}{N^+ - 1} \text{ and } \left(s_j^-\right)^2 = \frac{\sum_{k \in C^-}\left(x_{jk} - \mu_j^-\right)^2}{N^- - 1} \tag{8.13}$$

4. Symmetric divergence (SD) [Equation (8.14)]

$$D(j) = \frac{1}{2}\left(\frac{\left(\sigma_j^+\right)^2}{\left(\sigma_j^-\right)^2} + \frac{\left(\sigma_j^-\right)^2}{\left(\sigma_j^+\right)^2}\right) - 1 + \frac{1}{2}\left(\frac{\left(\mu_j^+ - \mu_j^-\right)^2}{\left(\sigma_j^+\right)^2 + \left(\sigma_j^-\right)^2}\right) \tag{8.14}$$

5. F-ratio: This is a generalisation of the t-statistic to problems with (K) classes $\{C^1, C^2 \dots C^K\}$, Equation (8.15),

$$F_j = \frac{\frac{\sum_{k=1}^{K} N^k\left(\mu_j^k - \mu_j\right)^2}{K-1}}{\frac{\sum_{k=1}^{K}\left(\sigma_j^k\right)^2 (N^k - 1)^2}{N - K}} \tag{8.15}$$

where N^k is the number of objects in class C^k, N is the total number of objects, μ_j^k and σ_j^k are the class-conditional mean and standard deviation for objects in class C^k considering the feature (j), and μ_j is the global mean for all objects considering the feature (j).

8.5 Principal Component Analysis

Principal component analysis (PCA) is an FG technique which orthogonally transforms the data points (objects) in a dataset into a new space of orthogonal, linearly uncorrelated features. The number of features in the new space is less than or equal to the original number of features.

PCA can be applied by performing singular value decomposition (SVD) over a zero-centred and normalised data matrix (Alter, Brown and Botstein, 2000) or eigenvalue decomposition over a data covariance or correlation matrix (Trefethen and Bau, 1997). Other techniques are also used to perform PCA, such as alternating least squares (ALS).

The result of PCA analysis is a list of orthogonally independent principal components (PCs) (generated features) that are combinations of the original features and have the same number as them. The PCs are ordered from the most contributing to the variance of the data to the least, with an associating weighting value, known as the *eigenvalue,* with each.

The most influential PCs with the highest eigenvalues can be considered as the new features for the data while discarding the rest of the PCs. The overall result is therefore transforming of the data from a high-dimensional real-feature space into a relatively low-dimensional orthogonally independent generated-feature space.

8.6 Genetic Algorithms and Genetic Programming

The genetic algorithm is a popular optimisation method in which a population of proposed solutions for the problem is driven through different modifying operators to move closer to the global optimum solution. In FS, the solution is modelled as a list of selected features, or as a binary string with a length equal to the total number of features where each feature has a corresponding binary bit filled with 'one' to indicate the inclusion of the feature or with 'zero' to indicate its exclusion. The genetic algorithm for FS has been applied to many bioinformatic applications (Nandi *et al.*, 2006) as well as other applications (Jack and Nandi, 2000, 2002).

Genetic programming (GP) assumes a population of solutions and progresses in a similar way to the genetic algorithms (GA) with a main difference; an individual genetic programming solution, in the context of FG, is a mathematical function (programme) that generates new features by transforming the original features. Genetic programming has been utilised in a wide range of applications related to FG, such as in breast cancer diagnosis (Guo and Nandi, 2006), mechanical systems (Zhang *et al.*, 2005), and audio-signal processing (Day and Nandi, 2007).

8.7 Discussion and Summary

High dimensionality is an attribute of biological high-throughput data such as microarrays and NGS datasets. Applying machine-learning methods, such as supervised classification and unsupervised clustering, to those datasets at their original dimensions can be infeasible or inefficient. Thus, feature selection (FS), that is, selecting a subset of the originally provided features, during pre-processing can reduce the complexity of analysis and enhance feasibility and efficiency.

In a gene expression dataset, the expression values of thousands of genes are measured over multiple samples taken from similar or different biological conditions. If classification or clustering is to be applied to those samples while considering each gene as a feature, FS techniques aim at selecting the subset of irredundant genes which are expected to provide the best separability between the samples. On the other hand, if machine-learning techniques are to be applied to the genes (e.g. gene clustering) while considering the samples as features, FS techniques aim at selecting the subset of samples which provides best separability between genes. The same concepts apply to other high-throughput expression datasets from proteomics, metabolomics, glycomics and others.

Two key attributes are observed in features for them to be selected by FS, to be informative and not redundant. Informative features are those which contribute to better separability between the objects that belong to different classes. However, two or more features can be informative but redundant when they provide the same useful information regarding separability. In such cases, successful FS would not select all of those redundant features.

FS methods belong to two main classes, filter (open-loop) and wrapper (closed-loop) methods. Wrapper methods are useful when supervised classification is to be applied to the dataset following FS. This is because they involve the prospective classifier itself within FS by considering a feedback loop. However, the scope of this book is focused on unsupervised clustering rather than supervised classification; therefore, filter (open-loop) methods, such as consecutive ranking and individual ranking, are more relevant here.

With an overlapping but different view of the features, feature generation (FG) methods, like PCA, generate a new set of features which are linear or non-linear combinations of the original features. The aim here is to transform the dataset into a new feature space with lower dimensionality, in which the objects from different groups have better separability. Table 8.1 lists some resources for feature selection and feature-generation methods.

Table 8.1 Resources for feature selection and FG methods

Method	Platform/Package	Function
Feature ranking	MATLAB	Rank features
	R/F Selector	cutoff.k
		cutoff.k.percent
		cutoff.biggest.diff
Random feature selection	MATLAB	Rand features
PCA	MATLAB	pca
	R/stats	princomp
Genetic algorithm	MATLAB	ga
	R/genalg	rbga
		rbga.bin

References

Alter, O., Brown, P.O. and Botstein, D. (2000). Singular value decomposition for genome-wide expression data processing and modeling. *Proceedings of the National Academy of Sciences*, **97**(18), pp. 10101–10106.

Day, P. and Nandi, A.K. (2007). Robust text-independent speaker verification using genetic programming. *IEEE Transactions on Audio, Speech, and Language Processing*, **15**(1), pp. 285–295.

Day, P. and Nandi, A.K. (2011). Evolution of super features through genetic programming. *Expert Systems*, **28**(2), pp. 167–184.

Fang, Z., Martin, J. and Wang, Z. (2012). Statistical methods for identifying differentially expressed genes in RNA-Seq experiments. *Cell & Bioscience*, **2**(1), article 26.

Guo, H. and Nandi, A.J. (2006). Breast cancer diagnosis using genetic programming generated feature. *Pattern Recognition*, **39**(5), pp. 980–987.

Hassanien, A.E., Al-Shammari, E.T. and Ghali, N.I. (2013). Computational intelligence techniques in bioinformatics. *Computational Biology and Chemistry*, **47**, pp. 37–47.

Haynes, W.A., Higdon, R., Stanberry, L. *et al.* (2013). Differential expression analysis for pathways. *PLoS Computational Biology*, **9**(3), e1002967.

Jack, L.B. and Nandi, A.K. (2000). Genetic algorithms for feature selection in machine condition monitoring with vibration signals. *IEE Proceedings – Vision, Image and Signal Processing*, **147**(3), pp. 205–212.

Jack, L.B. and Nandi, A.K. (2002). Fault detection using support vector machines and artificial neural networks, augmented by genetic algorithms. *Mechanical Systems and Signal Processing*, **16**(2–3), pp. 373–390.

Kadota, K., Ye, J., Nakai, Y. *et al.* (2006). ROKU: A novel method for identification of tissue-specific genes. *BMC Bioinformatics*, **7**, p. 294.

Kung, S.-Y. and Mak, M.-W. (2008). Feature selection for genomic and proteomic data mining, in *Machine Learning in Bioinformatics*, (eds Y. Zhang and J.C. Rajapakse), John Wiley & Sons, Inc., Hoboken, pp. 1–46.

Mu, T., Nandi, A.K. and Rangayyan, R.M. (2007). Classification of breast masses via nonlinear transformation of features based on a kernel matrix. *Medical & Biological Engineering & Computing*, **45**(8), pp. 769–780.

Nandi, R.J., Nandi, A.K., Rangayyan, R.M. and Scutt, D. (2006). Classification of breast masses in mammograms using genetic programming and feature selection. *Medical and Biological Engineering and Computing*, **44**(8), pp. 683–694.

Tarca, A.L., Romero, R. and Draghici, S. (2006). Analysis of microarray experiments of gene expression profiling. *American Journal of Obstetrics and Gynecology*, **195**(2), pp. 373–388.

Trefethen, L.N. and Bau, D., III (1997). *Numerical Linear Algebra*, Society for Industrial and Applied Mathematics (SIAM), Philadelphia.

Zhang, L., Jack, L.B. and Nandi, A.K. (2005). Fault detection using genetic programming. *Mechanical Systems and Signal Processing*, **19**(2), pp. 271–289.

References

9

Differential Expression

9.1 Introduction

Many genes, proteins, or other biological molecules, are required to be differentially expressed between different biological conditions, that is, their expression levels are purposefully and consistently different between such conditions. For example, when cells are moved from growth conditions to stress conditions, the genes that produce the machineries which synthesise proteins in large amounts needed for growth are *down-regulated,* that is, their levels of expression and activity are severely decreased. On the other hand, those genes which are required to strengthen a cell's resistance and respond to stress are *up-regulated,* in that their expression levels are highly increased (Gasch *et al.*, 2000).

The identification of differentially expressed genes is one of the key steps in the analysis of gene, protein or other expression datasets that are composed of expression measurements over multiple time-points or samples from different biological conditions. This step needs proper understanding of the statistical distribution of the underlying data, and appropriate assumptions therein. Therefore, different types of datasets, such as microarrays and next-generation sequencing (NGS), have been analysed while considering different assumptions.

In order to obtain sufficient statistical significance, such datasets usually include multiple *replicates* for the same condition. Replicates are of two types, biological and technical. Biological replicates are different samples taken from different individuals or cultures with the same biological condition (e.g. cells under growth conditions), while technical replicates measure the expression of the same biological sample but by using multiple microarray or NGS chips. Biological replicates compensate for the biological variation caused by other than the particular condition under consideration, while technical replicates compensate for the technical variation and noise that are embodied in the used technology.

Integrative Cluster Analysis in Bioinformatics, First Edition. Basel Abu-Jamous, Rui Fa and Asoke K. Nandi.
© 2015 John Wiley & Sons, Ltd. Published 2015 by John Wiley & Sons, Ltd.

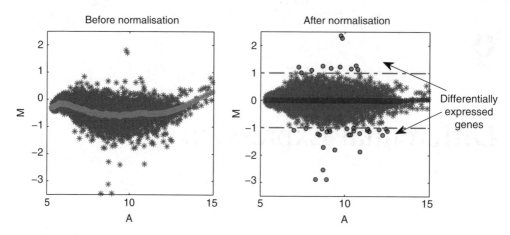

Figure 9.1 Two-fold change differentially expressed genes based on MA plots and lowess normalisation. The left panel shows the plot before normalisation while the right panel shows it after normalisation. The data are from the two-colour sample with the GEO accession GSM81075

In this chapter, we discuss different methods for the identification of differentially expressed genes or proteins based on microarrays or the NGS technology.

9.2 Fold Change

The most intuitive and the first proposed method to identify differentially expressed genes between two conditions is *fold change*. According to this method, genes which show K-fold change, usually two-fold change, or more between conditions are selected as being differentially expressed (Tarca, Romero and Draghici, 2006; Colak *et al.*, 2012).

Figure 9.1 (left) shows an MA plot, which is a scatter plot of log-ratio values versus abundance (Dudoit *et al.*, 2002), for the expression values belonging to the two channels of the microarray sample with the GEO accession GSM81075. If the expression of the ith gene is x_i in the first condition (e.g. microarray channel) and y_i in the other channel, the MA plot scatters the genes on a 2D plain where the horizontal axis represents the abundance values $A_i = \log_2 \sqrt{x_i y_i}$ and the vertical axis represents the log-ratio values $M = \log_2 (x_i/y_i)$. The right panel of Figure 9.1 shows the MA plot for same genes after performing lowess normalisation (explained Chapter 7) (Yang *et al.*, 2002). Those genes that cross the two log-ratio dashed lines in the Figure ($M = 1.0$ and $M = -1.0$) are the ones which have at least two-fold changes; they are marked with solid green circles, and they are identified as differentially expressed.

Similarly, fold changes based on MA plots have been considered in proteomics to identify proteins with differential expression (abundance) (Ting *et al.*, 2009).

9.3 Statistical Hypothesis Testing – Overview

Researchers have argued against the use of fold changes to identify differentially expressed genes because some important genes, like transcription factors, might be differentially expressed in a biologically significant manner with fewer than two folds (Tarca, Romero and Draghici, 2006). Therefore, the trend in this field has moved towards setting null and research hypotheses regarding differential expression of the genes after being statistically modelled. The null hypothesis is usually that the gene is not differentially expressed, that is, it belongs to the mainstream expression distribution of the data, while the research hypothesis is that it is differentially expressed.

Microarray data, as discussed in Chapter 7, are generally believed to belong to a lognormal distribution; thus, it is log-transformed prior to further analysis (Kreil and Russell, 2005). Afterwards, differentially expressed genes are identified by a statistical test, most commonly, variants of *t*-statistic (Tarca, Romero and Draghici, 2006). In contrast, NGS expression data do not follow a lognormal distribution; they have rather been modelled by Poisson or relative distributions (Fang, Martin and Wang, 2012). The difference is that microarrays measure continuous analogue expression levels of transcripts while NGS technologies measure discrete integer counts of transcripts. The rationale behind Poisson modelling of NGS data is that the measured expression of a transcript is the total number of short sequences aligned to it, and is therefore the total of many random events. Each single random event is the alignment of a single short sequence to the target transcript, and is assumed to follow a Bernoulli distribution with a success probability that is equal to the actual relative expression of that transcript. By assuming independence between alignment events of different short sequences, the total number of successful alignments, which is the measured expression value of the transcript, can be estimated by a Poisson distribution (Fang, Martin and Wang, 2012).

The rest of this section describes the concept of *p-values,* the visualisation tool *volcano plots,* and the required adjustments of p-values. The next section describes many statistical tests that can be used to identify differentially expressed genes from microarrays or the NGS technology, as well as for protein differential expression (differential abundance). The suitability of each method to those different technologies will be clearly stated.

9.3.1   p-Values and Volcano Plots

Statistical methods employed for differential expression analysis generally measure the statistical significance of differential expression by calculating the p-value. The p-value is the probability of obtaining the observed difference in expression for the gene (or protein) being examined, or obtaining more extreme differences. The null hypothesis is rejected and the corresponding gene (or protein) is considered differentially expressed if the p-value is very low (e.g. < 0.01), that is, it is very unlikely that such difference in expression was obtained merely by chance.

To combine the concepts of statistically derived p-values and fold changes in differential expression decision making, volcano plots were proposed (Cui and Churchill, 2003). A volcano plot is a scatter plot on which genes (or proteins, metabolites, glycans, etc.) are scattered, where the horizontal axis represents the base-two logarithm of the fold change and the vertical axis represents the negative base-ten logarithm of the p-value. Thresholds regarding both measures, fold changes and p-values, can be imposed on the figure to identify differentially expressed genes (Figure 9.2).

9.3.2   The Multiple-hypothesis Testing Problem

The statistical tests examining genes (or proteins, metabolites or other variables) for differential expression are designed to test multiple hypotheses in parallel. In other words, the test is applied

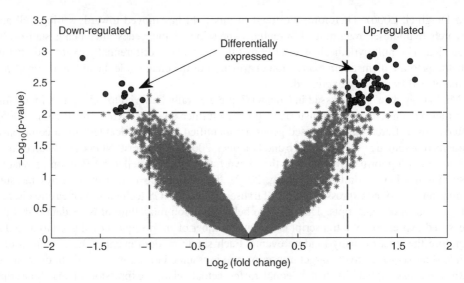

Figure 9.2 Volcano plot for differential expression analysis. The analysed data are composed of the samples 1, 2, 3, 10, 11 and 12 from the microarray dataset with the GEO accession GSE22552 where the first three samples of which are replicates for one condition and the last three are replicates for another condition. The dataset has 54 675 probes (representatives of genes). The horizontal axis of the plot is the base-two logarithm of the fold change between intra-condition mean expression values, and the vertical axis is the negative of the base-ten logarithm of the p-value. In this example, genes with at least two fold changes (either direction) and having p-values less than or equal to 0.01 are labelled as differentially expressed. The two vertical dashed lines mark two fold changes and the horizontal dashed line marks the p-value of 0.01. P-values here were calculated using the moderated *t*-statistic (explained below)

to thousands of genes in parallel and calculates a p-value for each of them, which is used to decide which genes are differentially expressed and which genes are not. As each of those multiple hypotheses can be modelled as a random event, it is statistically expected that a fraction of them will pass the threshold of success merely by chance, leading to many false positives. If the probability of this by-chance success of hypotheses was 0.01, about 200 out of approximately 20 000 human genes, or 60 genes out of approximately 6000 yeast genes, would be identified as differentially expressed by chance. This is a well-known issue when testing a large number of hypotheses in parallel; it is known as the *multiple-hypothesis testing problem,* and has been tackled by different adjustments or corrections for the ordinary p-value (Storey, 2002).

One of the approaches to overcome this problem is to calculate the *q-values* based on the p-values. As proposed by Storey, the q-value can be calculated by following four main steps (Storey, 2002):

1. The p-values are calculated for all of the N genes based on the adopted statistical test.
2. The p-values are ordered such that $p_1 \leq p_2 \leq \cdots \leq p_N$.
3. The q-value for the gene with the largest p-value is set equal to its estimated value of the positive false discovery rate (pFDR) [Equation (9.1)]:

$$q(p_N) = \mathrm{pFDR}_\lambda(p_N) \tag{9.1}$$

The pFDR value for the jth gene is estimated by the formula given in Equation (9.2):

$$\text{pFDR}_\lambda(p_j) = \frac{\pi_0 \times p_j \times N}{R(p_j)} \tag{9.2}$$

where p_j is the p-value, $\lambda \in (0,1)$ is a tuning parameter, π_0 is the probability that the null-hypothesis is true, and $R(p_j)$ is the ascending rank of p_j amongst the p-values of all of the genes. The rank $R(p_i)$ is equal to the number of genes with p-values equal to or less than p_i. The value of π_0 is estimated by using Equation (9.3),

$$\pi_0 = \frac{W(\lambda)}{(1-\lambda)N} \tag{9.3}$$

where $W(\lambda)$ is the number of genes with p-values greater than λ. Different methods have been proposed to estimate the best λ-value such as bootstrapping and polynomial fitting. In both cases, many λ-values covering the range $(0,1)$ are tested in order to choose the best one of them. In bootstrapping for example, the π_0-values corresponding to each tested λ-value are calculated based on the original complete set of p-values and based on each bootstrap resampling set; then, the λ-value which leads to the minimum mean square error (MSE) relative to the minimum π_0 from the original set is chosen as the best λ-value (Storey, 2002).

4. The following genes' q-values are calculated in a respective order for the p-values p_{N-1} to p_1 based on the formula shown in Equation (9.4).

$$q(p_j) = \min\{\text{pFDR}_\lambda(p_j), q(p_{j+1})\} \tag{9.4}$$

The MATLAB function *mafdr* as well as the *R* language package (*qvalue*) of the Bioconductor project calculate the q-values for a set of p-values or an expression dataset; they also estimate the internal parameters automatically and calculate related statistics such as pFDR.

Other popular p-value adjustment methods for the multiple-hypothesis testing problem include Bonferroni's and Holm's adjusted p-values (Holm, 1979; Fang, Martin and Wang, 2012).

9.4   Statistical Hypothesis Testing – Methods

9.4.1   t-Statistic, Modified t-Statistics and the Analysis of Variance (ANOVA)

Given n_1 and n_2 replicates for the two biological conditions c_1 and c_2 between which differential expression is to be examined, the t-statistic for the jth gene (or protein) is calculated as shown in Equation (9.5),

$$t_j = \frac{m_{j1} - m_{j2}}{\sqrt{\dfrac{s_{j1}^2}{n_1} - \dfrac{s_{j2}^2}{n_2}}} \tag{9.5}$$

where m_{j1} and m_{j2} are the mean values for the expression of the jth gene in the c_1 and the c_2 replicates, respectively, and s_{j1} and s_{j2} are similarly and respectively the standard deviations.

This statistic measures how dissimilar the two groups of measurements are, and therefore indicates the level of differential expression. This test follows a Student's t-distribution with v degrees of freedom, calculated by using Equation (9.6).

$$v = \frac{\left(\frac{s_1^2}{n_1} + \frac{s_2^2}{n_2}\right)^2}{\frac{\left(\frac{s_1^2}{n_1}\right)^2}{(n_1-1)} + \frac{\left(\frac{s_2^2}{n_2}\right)^2}{(n_2-1)}} \tag{9.6}$$

The t-test assumes that the tested variables are normally distributed, which is the case for log-transformed normalised gene microarray data (Tarca, Romero and Draghici, 2006) as well as proteomic microarray data (Ting $et\ al.$, 2009), but not the NGS data (Fang, Martin and Wang, 2012).

This ordinary t-test has fundamental problems when expression datasets are considered, due to the very small numbers of replicates that are usually available. These small n-values render the standard deviation estimations unreliable as there is a considerable probability that small standard deviations occur by chance and lead to large false discovery rates (FDR). In order to overcome this, S-statistic (Tusher, Tibshirani and Chu, 2001) and moderated t-statistic (Baldi and Long, 2001) were proposed.

S-statistic penalises the denominator of the t-statistic in order to compensate for low variances (Tusher, Tibshirani and Chu, 2001), as shown in Equation (9.7),

$$S = \frac{m_{j1} - m_{j2}}{a + \sqrt{\frac{s_{j1}^2}{n_1} - \frac{s_{j2}^2}{n_2}}} \tag{9.7}$$

where a is the penalty parameter. Efron and colleagues suggested using the 90th percentile of the standard errors ($s^2/\sqrt{n}$) of all genes as an estimation for the value of a (Efron $et\ al.$, 2000).

The moderated t-statistic, proposed by Baldi and Long, is a t-statistic which considers a moderated standard deviation that adopts an empirically derived penalty for low variances (Baldi and Long, 2001). The estimated moderated standard deviation $\tilde{s}$ is calculated based on the formula given in Equation (9.8),

$$\tilde{s} = \sqrt{\frac{v_0 \sigma_0^2 + (n-1)s^2}{v_0 + n - 2}} \tag{9.8}$$

where s is the well known standard deviation, v_0 is the number of the additional $pseudo\ observations$ with which the empirical variance is modulated, and σ_0 is the background standard deviation associated with the pseudo observations. The standard deviation of the entire set of observations can be an estimation for the background standard deviation σ_0. As for the number of pseudo observations v_0, it can be set by following the rule of thumb suggested by Baldi and Long that $v_0 + n$ would always be equal to a number K, where K is a sufficient number of observations for standard deviation estimation. For example, K can be set to 10 and

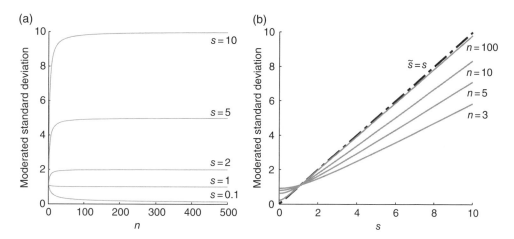

Figure 9.3 Comparison between the moderated standard deviation $\tilde{s}$ and the ordinary standard deviation s. Calculations are based on Equation (9.8) with $v_0 = 5$ and $\sigma_0 = 1$; (a) as the number of replicates (n) increases from one to 500, $\tilde{s}$ approaches s; this is true for all of the five different considered s-values 0.1, 1, 2, 5 and 10; (b) moderation mainly boosts the very small s-values in a nonlinear manner while linearly decreasing the large s-values. For larger n-values, the moderation is less influential at both high and low s-values as the transformation curves become closer to the identity curve $\tilde{s} = s$; this is because the ordinary standard deviation (s) is expected to be more representative and reliable when estimated by using a large number of observations (n)

consequently $v_0 = 10 - n$ (Baldi and Long, 2001). The webserver *Cyber-T*, which is an online tool that calculates moderated t-statistic, sets v_0 to 5 unless otherwise specified by the user (Kayala and Baldi, 2012). The degrees of freedom in this case are equal to $(v_0 + n)$.

If the number of samples n is big, the moderated standard deviation $\tilde{s}$ approaches the ordinary standard deviation s because the moderation would not be needed. Moreover, the moderated standard deviation lifts the very low ordinary standard deviations s while maintaining a linear relation with large s-values. Those two facts are demonstrated in Figure 9.3a and b, respectively.

After finding the t-statistic, the two-tailed p-value is calculated as the probability of obtaining such a t-statistic or larger in absolute value by chance, and is equal to twice the value of the t-cumulative distribution function (CDF) at the negative of the absolute value of the measured t-statistic.

The t-statistic can only be applied to compare two groups of observations. *ANOVA* statistics generalise the t-test by allowing for comparisons between several groups. Similar to the t-test, the null-hypothesis in ANOVA is that the mean values of the observations in the several groups are equal, while the research (alternative) hypothesis is that they are differentially expressed.

9.4.2 B-Statistic

Another way to overcome the drawback of the ordinary t-statistic at low variances is Bayes log posterior odds B-statistic proposed by Lönnstedt and Speed (2002). Given the log ratio M_g for the gene g, the B-statistic (B_g) is equal to the logarithm of the ratio between the probability that the gene's log-ratio belongs to a normal distribution with a non-zero mean (differentially expressed) and the probability that the gene's log-ratio belongs to a normal distribution with a zero mean (not differentially expressed). This is formulated as shown in Equation (9.9),

$$B_g = \log \frac{\Pr\left(I_g = 1 | M_g\right)}{\Pr\left(I_g = 0 | M_g\right)} \tag{9.9}$$

where $M_g \sim N\left(\mu_g, \sigma_g^2\right)$ and I_g is a binary value which is equal to unity when $\mu_g \neq 0$ and is equal to zero when $\mu_g = 0$. After derivation, the value of B_g can be calculated by using Equation (9.10),

$$B_g = \log \frac{p}{1-p} \cdot \frac{1}{\sqrt{1+nc}} \left[\frac{a + s_g^2 + M_g^2}{a + s_g^2 + \dfrac{M_g^2}{1+nc}} \right]^{v + \frac{n}{2}} \tag{9.10}$$

where n is the number of replicates/observations, s_g^2 is the variance of the observations of the gene g, p is the proportion of differentially expressed genes in the experiment, a is a parameter which compensates for small variances, v is the degrees of freedom for an inverse gamma prior distribution for the variances, and c is a hyperparameter in the normal prior of the non-zero means $\{\mu_g | I_g = 1\}$. Estimation of those parameters was discussed by Lönnstedt and Speed (2002); p can be feasibly fixed to a sensible value like 0.01 or 0.001, v and a can be estimated based on the inverse gamma distribution of the variances such that $\dfrac{na}{2\sigma_i^2} \sim \Gamma(v,1)$ for all i, and c can be estimated based on comparisons between the observed normal density of the averages $\{M_g\}$ for the top p proportion of genes, and the observed normal density of the averages $\{M_g\}$ for all of the genes (Lönnstedt and Speed, 2002).

The R language package *limma* of the *R Bioconductor* project estimates those parameters by the function *eBayes* and finds the *B*-statistic consequently by the function *topTable*.

9.4.3   Fisher's Exact Test

This test was introduced by the English statistician Sir Ronald Aylmer Fisher in the 1920s (Fisher, 1922). The test suites two-class NGS data with one sample for each class, as it is based on the assumption that the data are discrete with Poisson sampling. Let $x_j^{(1)}$ and $x_j^{(2)}$ be the expression values (counts of aligned short sequences) of the jth gene in the samples 1 and 2 which represent conditions 1 and 2 respectively, and let the sequencing depth of the kth sample $(k \in \{1, 2\})$, which is that sample's total number of short sequences, be $L^{(k)} = \sum_{\forall i} x_j^{(k)}$. A 2×2 contingency matrix can be drawn from this as shown in Table 9.1.

Table 9.1   Contingency matrix for Fisher's exact test over the jth gene

	Gene j	Not gene j
Class/condition 1	$x_j^{(1)}$	$L^{(1)} - x_j^{(1)}$
Class/condition 2	$x_j^{(2)}$	$L^{(2)} - x_j^{(2)}$

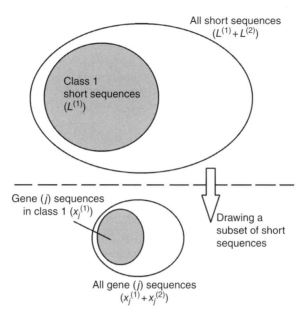

Figure 9.4   Venn diagram demonstration for the hypergeometric modelling of the genetic expression (short sequences' count) of the gene (j)

The counts in this two-class problem are assumed to follow a hypergeometric distribution. Here, the complete set of sequences from all genes in both classes contains ($L^{(1)} + L^{(2)}$) short sequences, where $L^{(1)}$ of which belong to class 1. The short sequences that are aligned to the gene (j) represent a subset of that complete set with the size of ($x_j^{(1)} + x_j^{(2)}$). Given that, and as graphically demonstrated in Figure 9.4, the hypergeometric probability distribution function (pdf), formulated in Equation (9.11), calculates the chance (probability) of obtaining $x_j^{(1)}$ short sequences that belong to class 1 in a subset of ($x_j^{(1)} + x_j^{(2)}$) short sequences that were uniformly randomly drawn from the entire pool.

$$Pr_{hypergeometric}\left(X = x_j^{(1)}\right) = \frac{\binom{L^{(1)}}{x_j^{(1)}}\binom{L^{(2)}}{x_j^{(2)}}}{\binom{L^{(1)} + L^{(2)}}{x_j^{(1)} + x_j^{(2)}}} \tag{9.11}$$

The hypergeometric p-value is therefore the probability of observing, by chance, such an expression value or a value which is more favourable by the research (alternative) hypothesis of differential expression. If the alternative hypothesis is that this gene is differentially expressed such that it has significantly higher expression in class/condition 1 compared with class/condition 2, the p-value will be as shown in Equation (9.12).

$$p_{greater} = Pr_{hypergeometric}\left(X \geq x_j^{(1)}\right) = \sum_{x=x_j^{(1)}}^{min(L^{(1)},\, x_j^{(1)} + x_j^{(2)})} \frac{\binom{L^{(1)}}{x}\binom{L^{(2)}}{x_j^{(1)} + x_j^{(2)} - x}}{\binom{L^{(1)} + L^{(2)}}{x_j^{(1)} + x_j^{(2)}}} \tag{9.12}$$

On the other hand, if the alternative hypothesis is that it has significantly higher expression in class/condition 2, the p-value will be as shown in Equation (9.13).

$$p_{less} = Pr_{hypergeometric}\left(X \le x_j^{(1)}\right) = \sum_{x=0}^{x_j^{(1)}} \frac{\binom{L^{(1)}}{x}\binom{L^{(2)}}{x_j^{(1)}+x_j^{(2)}-x}}{\binom{L^{(1)}+L^{(2)}}{x_j^{(1)}+x_j^{(2)}}} \tag{9.13}$$

Calculating the hypergeometric p-value avoids the biases that might be caused by direct fold change calculations if the sequencing depth was different for different samples. For example, if $L^{(1)} = 1000$, $L^{(2)} = 5000$, $x_j^{(1)} = 10$ and $x_j^{(2)} = 50$, the p-value testing the alternative hypothesis that $x_j^{(1)}$ is significantly lower than $x_j^{(2)}$ is $p_{less} = 0.58$, which is a very high value that definitely leads to the rejection of the alternative hypothesis and to considering the gene as being not-differentially expressed. In contrast, a plain fold change calculation shows that there is a five-fold change in expression, and this would normally lead to considering the gene as being differentially expressed. It is clear that this gene has lower expression in condition 1 (10 vs. 50) because of the fact that the total number of sequences in condition 1's samples is much smaller (1000 vs. 5000). This bias is perfectly considered by the hypergeometric p-value. Another example with a genuine five-fold change would be when $L^{(1)} = 1000$, $L^{(2)} = 1000$, $x_j^{(1)} = 10$ and $x_j^{(2)} = 50$; the p-value at this case is $p_{less} = 5.4 \times 10^{-8}$, which is extremely small and indicates that this gene is significantly differentially expressed.

Although this test has many positives, it fails if the assumption that the alignment of short sequences to genes follows Poisson distribution fails, which is the case when biological replicates exist. This is because those replicates have high dependency, which violates the Poisson model. Consequently, other tests should be used for such cases. If only technical replicates exist, the counts from the replicates in each class can be added to each other to obtain total count values to be used in the contingency table and the following analysis (Fang, Martin and Wang, 2012).

9.4.4   Likelihood Ratio Test

In this test, which is applied to NGS data, the genetic expression (short sequences' count) for the jth gene in the ith sample of the kth class/condition ($m_{ji}^{(k)}$) is modelled by using a Poisson distribution $Poi\left(\mu_j^{(k)} = L_i^{(k)} v_j^{(k)}\right)$, where $\mu_j^{(k)}$ is the mean of the distribution, $L_i^{(k)}$ is the sequencing depth (total number of short sequences) in the ith sample of the kth class, and $v_j^{(k)}$ is the proportion of short sequences in $L_i^{(k)}$ that are aligned to this gene. Accordingly, in a two-class problem ($k \in \{1, 2\}$), the likelihood ratio test statistic (Λ_j) regarding the jth gene for the two-sided alternative hypothesis ($v_j^{(1)} > v_j^{(2)}$ OR $v_j^{(1)} < v_j^{(2)}$) can be formulated as given in Equation (9.14).

$$\Lambda_j = -2\sum_{k=1}^{2}\left(\left(\sum_i m_{ji}^{(k)}\right) \times \log\left(\frac{\sum_{k'}\sum_i m_{ji}^{(k')}}{\sum_i m_{ji}^{(k)}} \cdot \frac{\sum_i L_i^{(k)}}{\sum_{k'}\sum_i L_i^{(k')}}\right)\right) \tag{9.14}$$

The two-sided alternative hypothesis p-value is the right-tailed cumulative Chi-squared (χ^2) probability of the statistic shown in Equation (9.15),

$$pvalue_{two-sided} = 1 - \frac{1}{\Gamma\left(\frac{1}{2}\right)}\gamma\left(\frac{1}{2}, \frac{\Lambda_j}{2}\right) \tag{9.15}$$

where $\Gamma(.)$ is the gamma function and $\gamma(., .)$ is the incomplete gamma function. The p-value for the one-sided alternative hypothesis ($v_j^{(1)} > v_j^{(2)}$) is given by Equation (9.16).

$$pvalue_{one-sided} = \begin{cases} \frac{1}{2}pvalue_{two-sided,} & \dfrac{\sum_i m_{ji}^{(1)}}{\sum_i L_i^{(1)}} > \dfrac{\sum_i m_{ji}^{(2)}}{\sum_i L_i^{(2)}} \\ 0.5 & \text{, otherwise} \end{cases} \tag{9.16}$$

The likelihood ratio test remains valid for both technical and biological replicates as long as the Poisson's assumption holds true, that is, the mean of counts is similar to their standard deviation. However, if over-dispersion occurs, that is, when the observed variance is greater than the assumed variance, this test becomes invalid (Fang, Martin and Wang, 2012).

9.4.5   Methods for Over-dispersed Poisson Distribution

Many models were designed, by extensions of the Poisson distribution, to analyse differential expression in NGS datasets with over-dispersion, that is, the datasets in which the observed variance is larger than the expected variance based on the ordinary Poisson distribution (expected variance = mean). This is usually the realistic case when biological replicates from the same biological condition/class exist due to their unavoidable dependency (Fang, Martin and Wang, 2012; Soneson and Delorenzi, 2013).

Methods targeting such data include a two-stage Poisson model, which first classifies the genes into genes with or without over-dispersion, and secondly calculates the statistics for the two groups of genes by using two different models which would be appropriate accordingly (Auer and Doerge, 2011). Moreover, other methods targeting this problem include the *R* language packages *edgeR* (Robinson, McCarthy and Smyth, 2010), *DESeq* (Anders and Huber, 2010), and *baySeq* (Hardcastle and Kelly, 2010), which consider an underlying negative binomial distribution of the data. Additionally, the *BBSeq R* language package implements a method based on beta-binomial distribution (Zhou, Xia and Wright, 2011).

9.5   Discussion and Summary

Those genes that show statistically significantly different levels of expression across certain different biological conditions are known as *differentially expressed* under such conditions. The same applies to proteins, glycans (carbohydrates), metabolites and any other biological molecules or entities whose expression or abundance can be measured. It is known that only a relatively small subset of biological molecules (e.g. genes) is involved in any single biological process under investigation. Therefore, identifying differentially expressed genes (or proteins, glycans, etc.) can be considered as means for the identification of a list of candidate participants in such biological processes and conditions.

The most intuitive method to identify differentially expressed genes is fold change, in which a gene which shows at least K-fold change (e.g. two-fold change) is considered differentially expressed. However, most of the modern techniques consider statistical hypothesis testing, where the null-hypothesis is that the gene belongs to the mainstream statistical distribution of the data, that is, it is not differentially expressed, while the alternative hypothesis is that the gene is differentially expressed.

Gene and protein microarray datasets contain continuous (analogue) expression values, and tend to show lognormal distributions, and therefore tests based on the moderated t-statistic,

Table 9.2 Differential expression analysis software packages

Method	Platform/Package	Function
Adjusted p-value	R	p.adjust
	R/Bioconductor – limma	topTable
ANOVA	MATLAB	anova1
		anova2
		kruskalwallis
		multcompare
ANOVA	R	aov
		anova
baySeq	R/baySeq	
BBSeq	R/BBSeq	
B-statistic	R/Bioconductor – limma	eBayes
		topTable
DESeq	R/DESeq	
edgeR	R/edgeR	
FDR for multiple hypotheses	MATLAB	mafdr
Fisher's exact test	MATLAB	By the hypergeometric CDF: hygecdf
Fisher's exact test	R	fisher.test
Fold change	R/Bioconductor – limma	topTabe
Fold change	MATLAB	mavolcanoplot
Likelihood ratio test	MATLAB	lratiotest
Likelihood ratio test	R/lmtest	lrtest
MA scatter plot	MATLAB	mairplot
MA scatter plot	R/Bioconductor – limma	plotMA
Moderated *t*-test	R/Bioconductor – limma	topTable
q-value	MATLAB	mafdr
q-value	MATLAB	topTable
Student's *t*-test	MATLAB	ttest
		ttest2
		mattest
Student's *t*-test	R	t.test
Volcano plot	MATLAB	mavolcanoplot
Volcano plot	R/Bioconductor – limma	volcanoplot

S-statistic and B-statistic can be applied to them. In contrast, NGS data are digital because they represent integer counts of the short sequences found in a sample and aligned to the target gene (or protein). Tests like Fisher's exact test, and the maximum likelihood ratio test, can be applied to such datasets while assuming that they belong to Poisson distribution. However, when over-dispersion occurs, that is, when the observed variance of the data is larger than the expected variance, assuming an ordinary Poisson distribution becomes inaccurate. Other methods were proposed for such cases while assuming other distributions such as the negative binomial distribution or the beta-binomial distribution.

Statistical tests usually provide a p-value for each object (e.g. gene), which represents the probability that this object's observed profile, or a profile more favourable by the alternative hypothesis, is due to mere chance. For example, if the p-value of a gene's differential expression is equal to 0.01, there is a 1% chance that the observed expression profile, or any more differentially expressed profile, would occur by chance. Because multiple hypotheses are tested simultaneously, for example the differential expression of each of thousands of genes in a single dataset, the rate of false-positive discoveries would be high. Many methods were proposed to adjust raw p-values to compensate for this fact, such as q-values and Bonferroni's and Holm's adjusted p-values. Some software packages for differential expression analysis are provided in Table 9.2.

References

Anders, S. and Huber, W. (2010) Differential expression analysis for sequence count data. *Genome Biology*, **11**(10), R106.

Auer, P.L. and Doerge, R.W.(2011). A two-stage poisson model for testing RNA-seq data. *Statistical Applications in Genetics and Molecular Biology*, **10**(1), pp. 1–26.

Baldi, P. and Long, A.D. (2001). A Bayesian framework for the analysis of microarray expression data: regularized t-test and statistical inferences of gene changes. *Bioinformatics*, **17**(6), pp. 509–519.

Colak, T., Cine, N., Bamac, B. *et al.* (2012). Microarray-based gene expression analysis of an animal model for closed head injury. *Injury*, **43**(8), pp. 1264–1270.

Cui, X. and Churchill, G.A. (2003). Statistical tests for differential expression in cDNA microarray experiments. *Genome Biology*, **4**(4), 210. doi:10.1186/gb-2003-4-4-210. PMC 154570

Dudoit, S., Yang, Y.H., Callow, M.J. and Speed, T.P. (2002). Statistical methods for identifying differentially expressed genes in replicated cDNA microarray experiments. *Statistica Sinica*, **12**, pp. 111–139.

Efron, B., Tibshirani, R., Goss, V. and Chu, G. (2000). *Microarrays and their use in a comparative experiment*, Stanford University, Stanford, CA.

Fang, Z., Martin, J. and Wang, Z. (2012). Statistical methods for identifying differentially expressed genes in RNA-Seq experiments. *Cell & Bioscience*, **2**(1), p. 26.

Fisher, R.A. (1922). On the interpretation of χ^2 from contingency tables, and the calculation of P. *Journal of the Royal Statistical Society*, **85**(1), pp. 87–94.

Gasch, A.P., Spellman, P.T., Kao, C.M. *et al.* (2000). Genomic expression programs in the response of yeast cells to environmental changes. *Molecular Biology of the Cell*, **11**, pp. 4241–4257.

Hardcastle, T.J. and Kelly, K.A. (2010). BaySeq: empirical bayesian methods for identifying differential expression in sequence count data. *BMC Bioinformatics*, **11**, p. 422.

Holm, S. (1979). A simple sequentially rejective multiple test procedure. *Scandinavian Journal of Statistics*, **6**(2), pp. 65–70.

Kayala, M.A. and Baldi, P. (2012). Cyber-T web server: differential analysis of high-throughput data. *Nucleic Acids Research*, **40**(Web Server), pp. W553–W559.

Kreil, D.P. and Russell, R.R. (2005). There is no silver bullet – a guide to low-level data transforms and normalisation methods for microarray data. *Briefings in Bioinformatics*, **6**, pp. 86–97.

Lönnstedt, I. and Speed, T. (2002). Replicated microarray data. *Statistica Sinica*, **12**, pp. 31–46.

Robinson, M.D., McCarthy, D.J. and Smyth, G.K. (2010). edgeR: a bioconductor package for differential expression analysis of digital gene expression data. *Bioinformatics,* **26**(1), pp. 139–140.

Soneson, C. and Delorenzi, M. (2013). A comparison of methods for differential expression analysis of RNA - seq data. *BMC Bioinformatics,* **14**, p. 91.

Storey, J.D. (2002). A direct approach to false discovery rates. *Journal of the Royal Statistical Society: Series B,* **64**(3), pp. 479–498.

Tarca, A.L., Romero, R. and Draghici, S. (2006). Analysis of microarray experiments of gene expression profiling. *American Journal of Obstetrics and Gynecology,* **195**(2), pp. 373–388.

Ting, L., Cowley, M.J., Hoon, S.L. *et al.* (2009). Normalization and statistical analysis of quantitative proteomics data generated by metabolic labeling. *Molecular & Cellular Proteomics,* **8**(10), pp. 2227–2242.

Tusher, V.G., Tibshirani, R. and Chu, G. (2001). Significance analysis of microarrays applied to the ionizing radiation response. *Proceedings of the National Academy of Sciences,* **98**(9), pp. 5116–5121.

Yang, Y.H., Dudoit, S., Luu, P. *et al.* (2002). Normalization for cDNA microarray data: a robust composite method addressing single and multiple slide systematic variation. *Nucleic Acids Research,* **30**, e15.

Zhou, Y.-H., Xia, K. and Wright, F.A. (2011). A powerful and flexible approach to the analysis of RNA sequence count data. *Bioinformatics,* **27**(19), pp. 2672–2678.

Part Four

Clustering Methods

10

Clustering Forms

10.1 Introduction

In Part Three, we introduced biological data acquisition and pre-processing. In our pipeline of processing and analysing these biological data, we reach the stage of clustering. The task of clustering is to group data objects into a set of disjoint classes, called clusters, so that objects within a class have high similarity to each other, while objects in separate classes are more dissimilar. Before we dive deeply into clustering algorithms, in this chapter, we would like to overview the world of clustering methods and briefly explain the common rationale behind all clustering methods.

There is a very rich literature about clustering algorithms, which has been developed to solve different clustering problems in specific fields for more than five decades. With the advent of high-throughput data-collecting techniques in the biology and biomedicine fields, people have to invoke the help of clustering algorithms against the dramatically increasing biological data and to squeeze knowledge from a large amount of data.

Suppose that the biological data, whether gene expression data or protein-expression data, DNA sequence data or protein sequence data, or others, is in a form of data matrix $X = \{x_{nm} | n = 1, \ldots, N; m = 1, \ldots, M\}$, which is called a pattern matrix, with N rows representing N objects (genes, proteins, metabolites, glycans, etc.) and M columns representing M features (samples from different time points, tissues or species). Having this matrix, we can cluster it in three ways: first, we can cluster it row-wise to group genes or proteins into clusters based on how similar the patterns of their features are; second, we can cluster it column-wise to group features into clusters based on how many genes or proteins behave similarly in these features; and last, we can cluster it in both row-wise and column-wise manners to group those genes that behave similarly in many, if not all, features together.

Integrative Cluster Analysis in Bioinformatics, First Edition. Basel Abu-Jamous, Rui Fa and Asoke K. Nandi.
© 2015 John Wiley & Sons, Ltd. Published 2015 by John Wiley & Sons, Ltd.

Alternatively, the biological data can also be organised into a square matrix, $S = \{s_{ij} | i, j = 1, \ldots, N\}$, which is called a similarity matrix or proximity matrix, where both its rows and columns represent data objects and its entry, s_{ij}, represents the similarity or dissimilarity between the ith object and the jth object. Apparently, this matrix is a symmetric one. For the rest of this chapter, we start with proximity measures, which are fundamental elements of clustering, then move to an overview of the world of clustering algorithms.

10.2   Proximity Measures

There are many different terms for proximity measurement; for example, similarity, dissimilarity, distance and correlation. Conceptually, they are the same thing in that higher similarity or higher correlation means that two patterns are more similar (with less dissimilarity) or geometrically closer (with less distance). So the next questions, naturally, are what kind of standards need to be used for measuring the distance or similarity and how to measure them between two objects. Sometimes we also need to measure them between one object and one cluster or between two clusters.

There are a number of distance metrics in the literature. They can be employed in different applications based on their different features. Since the data can be quantitative or qualitative, continuous or binary, nominal or ordinal, the distance metrics can be roughly classified into two categories: (1) measuring distances of discrete feature objects, and (2) measuring distances of continuous feature objects. Nevertheless, these distance metrics share some common properties (Xu and Wunsch, 2005):

1. Symmetry: $D(x_i, x_j) = D(x_j, x_i)$ or $S(x_i, x_j) = S(x_j, x_i)$ for all x_i, x_j;
2. Positivity: $D(x_i, x_j) \geq 0$ or $1 \geq S(x_i, x_j) \geq 0$ for all x_i, x_j;
3. Triangle   inequality: $D(x_i, x_j) \leq D(x_i, x_k) + D(x_k, x_j)$ or $S(x_i, x_j) S(x_j, x_k) \leq [S(x_i, x_j) + S(x_j, x_k)] S(x_i, x_k)$ for all x_i, x_j and x_k;
4. Reflectivity: $D(x_i, x_j) = 0$ or $S(x_i, x_j) = 1$, if $x_i = x_j$

where $D(x_i, x_j)$ and $S(x_i, x_j)$ denote the dissimilarity and the similarity between x_i and x_j, respectively.

10.2.1   Distance Metrics for Discrete Feature Objects

Let us suppose that the discrete value objects $x_n = \{x_{nm} \in \mathcal{A}\}$ and $\mathcal{A}$ is an alphabet $\{a_i | i = 1, \cdots, b\}$ which can be binary $\{0, 1\}$ ($\{$false, true$\}$), or any finite number letters alphabet, say four-letter genetic code alphabet, and 20-letter proteinic code alphabet and so on.

10.2.1.1   Hamming Distance

Hamming distance between two discrete value objects with equal length is the number of positions at which the corresponding discrete values are different.

Example 9.1: The hamming distance between binary data vectors '10110110101' and '11001011001' is 7 and the hamming distance between discrete data vectors '324321432331' and '424221431331' is 3. If we imagine that the latter example is to calculate the hamming distance between two DNA sequences and $\{1, 2, 3, 4\}$ is the labelling of four nucleotides $\{A, T, G, C\}$, those two sequences are actually 'GTCGTACGTGGA' and 'CTCTTACGAGGA', respectively, and the hamming distance between them is 3.

10.2.1.2   Matching Coefficient

Let us define a matching vector that indicates how many positions of two data vectors with equal length are the same. The entry of the matching vector is 1 if two elements at the same position of two data vectors are same; otherwise, the entry is 0. Thus we may obtain the number of 1's, m_1 and the number of 0's, m_0 in the matching vector. The matching coefficient is defined as $m_1/(m_0 + m_1)$. Let us consider the cases in Example 9.1, the matching coefficient of the two binary data vectors is $4/(4 + 7) = 0.364$; the matching coefficient of the two DNA sequences is $9/(9 + 3) = 0.75$.

10.2.2   Distance Metrics for Continuous Feature Objects

We summarise the metrics for continuous feature objects in Table 10.1.

In MATLAB, there is a subroutine called **pdist** that can calculate all the metrics shown in the table except jackknife correlation. In bioinformatics applications, Euclidean distance and Pearson correlation are the two metrics with the greatest popularity (Scherf *et al.*, 2000; Stein *et al.*, 2004; D'haeseleer, 2005). Mahalanobis distance is also widely used, especially in the family of the model-based clustering, which we will detail in a later chapter.

10.3   Clustering Families

Different starting criteria may lead to different taxonomies of clustering algorithms. We consider a widely agreed classification to group all important clustering algorithms within bioinformatic applications into eight families, namely *partitional clustering, hierarchical clustering, fuzzy clustering, neural network-based clustering, mixture model clustering, graph-based clustering, consensus clustering* and *biclustering*.

10.3.1   Partitional Clustering

Partitional clustering, literally, attempts to directly decompose the dataset into a set of disjoint partitions based on a pre-specified optimisation criterion. This is a definition in a broad sense, and in this definition, partitional clustering could cover many clustering families, such as self-organising clustering, mixture model clustering and so on. But throughout this book, we restrict it within the algorithms based on squared error minimisation criterion, also known as squared error-based clustering (Xu and Wunsch, 2010). The best examples of this family are *k*-means

Table 10.1 Summary of the dissimilarity and the similarity measures of continuous feature objects

Measures	Formula	Comments
Minkowski distance	$D_{ij} = \left(\sum_{m=1}^{M} \left| x_{im} - x_{jm} \right|^p \right)^{\frac{1}{p}}$	Minkowski distance is a metric on Euclidean space, and is also considered as a generalisation of Euclidean distance, Manhattan distance and Chebyshev distance
Euclidean distance	$D_{ij} = \left(\sum_{m=1}^{M} \left| x_{im} - x_{jm} \right|^2 \right)^{\frac{1}{2}}$	Euclidean distance is a special case of Minkowski distance at $p = 2$
Manhattan distance	$D_{ij} = \sum_{m=1}^{M} \left| x_{im} - x_{jm} \right|$	Manhattan distance is a special case of Minkowski distance at $p = 1$
Chebyshev distance	$D_{ij} = \max_{1 \leq m \leq M} \left| x_{im} - x_{jm} \right|$	Chebyshev distance is a special case of Minkowski distance at $p \rightarrow +\infty$
Mahalanobis distance	$D_{ij} = \sqrt{ \left(x_i - x_j \right) \Sigma^{-1} \left(x_i - x_j \right) }$	Σ is the within-cluster covariance matrix
Pearson correlation	$S_{ij} = \left(1 + R_{ij} \right)/2$	Pearson correlation coefficient R_{ij} does not have the positivity property, since it ranges in $[-1, 1]$. Nevertheless, we can transform it to a range of $[0, 1]$ by $S_{ij} = \left(1 + R_{ij} \right)/2$, and the dissimilarity can be $D_{ij} = 1 - S_{ij}$
Jackknife correlation?	$J_{ij} = \min \left\{ R_{ij}^{(1)}, R_{ij}^{(2)}, \ldots, R_{ij}^{(M)} \right\}$ $S_{ij} = \left(1 + J_{ij} \right)/2$	Jackknife correlation, after the well known jack-knifing statistics, is robust to the single outlier. $R_{ij}^{(l)}$ denotes the correlation between the ith and jth data object with lth feature deleted (Heyer, Kruglyak and Yooseph, 999)
Spearman's rank correlation	Same form as Pearson correlation, but with ranking orders	It does not require the Gaussian distribution assumption
Cosine similarity	$S_{ij} = \dfrac{x_i \cdot x_j}{\|x_i\| \|x_j\|}$	Cosine similarity is widely used in text mining and information-retrieval applications

and k-medoids, which is also known as partitioning around medoids (PAM). We will detail partitional clustering and its applications in bioinformatics in Chapter 11.

10.3.2   Hierarchical Clustering

Different from partitional clustering, hierarchical clustering clusters the dataset with a hierarchical set of nested clusters which can be graphically presented as a tree structure. The tree structure is called a *dendrogram* and the ultimate clustering results are obtained by cutting this dendrogram. Hierarchical clustering can be further classified into divisive (top-down) approaches and agglomerative (bottom-up) approaches. Furthermore, in each of these two classes, there are many different algorithms, based on different divisive methods or linkage methods. We will detail hierarchical clustering in Chapter 12.

10.3.3   Fuzzy Clustering

Fuzzy clustering is also one of the most popular clustering families in bioinformatics applications. The rationale behind this is that it relaxes the restriction of hard clustering by allowing each data object to be associated with all clusters with a degree of membership, which means that each data object can be assigned to multiple clusters rather than to only one. The concept of a membership function derives from *fuzzy logic*; therefore, any clustering algorithm that employs fuzzy membership belongs to this family; for example, fuzzy c-means (FCM, also known as soft-k-means), fuzzy k-medoids, possibilistic c-means (PCM) and so on. We will detail fuzzy clustering in Chapter 13.

10.3.4   Neural Network-based Clustering

Neural network-based clustering starts with a set of nodes (also called neurons) that are all the same except for some parameters initialised randomly which make each node behave slightly differently. Then these nodes learn from the data in a competitive fashion: active nodes reinforce their neighbourhood within certain regions, while suppressing the activities of other nodes. The typical examples are self-organising map (SOM), self-organising oscillator networks (SOON), adaptive resonance theory (ART) and other algorithms in competitive learning. We will detail them in Chapter 14.

10.3.5   Mixture Model Clustering

Mixture model clustering is another important family of clustering, which has attracted more and more attention recently. It is based on formulating a clustering kernel for each individual component in terms of a sampling density $p(X|\theta)$, where θ is an unknown parameter set. Compared with Euclidean distance-based algorithms, mixture model clustering provides meaningful results in many cases where Euclidean distance-based algorithms fail, especially in time series (time series gene expression) and categorical (DNA or protein sequence) datasets. Mixture model clustering can be classified into two large groups, namely finite mixture models (parametric models) and infinite mixture models (nonparametric models). We will detail the algorithms within this family in Chapter 15.

10.3.6   Graph-based Clustering

Graph-based clustering, as a very useful clustering family, not only is able to do the same job that other clusterings can do (e.g. the similarities or dissimilarities are calculated and organised in a graphical form), but also is able to cluster the relational data even though other data are not available, which is difficult to do with other clustering algorithms. Therefore, in studies of molecular and cellular biology, the graph representation has been widely used in the analysis of protein–protein interaction networks, gene regulatory networks and metabolic networks. We will detail the algorithms within this family in Chapter 16.

10.3.7   Consensus Clustering

To combine different clustering results, also known as ensemble clustering, consensus clustering or cluster aggregation has received a lot of attention and is considered as a solution to the problem of inconsistency of stochastic clustering algorithms or clusterings with different parameters. Ensemble clustering formalises the idea that combining different clusterings into a single representative or consensus would emphasise the common organisation in the different clustering results. Ensemble clustering has attracted a lot of interest during the past decade. We will detail the algorithms within this family in Chapter 17.

10.3.8   Biclustering

The concept of simultaneously clustering dataset in both dimensions can be traced back to 1972, and it did not attract much attention until (Cheng and Church, 2000) developed a biclustering algorithm to analyse gene expression data. In Chapter 18 we will discuss the basic concept of biclustering and introduce the types of bicluster.

10.4   Clusters and Partitions

Clusters and partitions are the products of clustering algorithms. In this section, we introduce many formations of the clustering products. The first one is in the form of an index vector, which is defined as $Z = \{z_n | n = 1, \ldots, N\}$ called a partition and $z_n \in [1, \ldots, K]$, where K is the number of clusters. The second form is a partition matrix $U^{N \times K} = \{u_{k,n} | k = 1, \ldots, K; n = 1, \ldots, N\}$. In the crisp partitions, the nth object belongs to the kth cluster if the binary entry $u_{k,n}$ of the partition matrix is 1, otherwise the object does not; while in fuzzy partitions, the entry $u_{k,n} \in [0,1]$ represents the degree to which the nth object belongs to the kth cluster. Lastly, partitions can be represented as a cluster set $C = \{C_1, \ldots, C_K\}$, where C_k denotes the kth cluster and $k = 1, \ldots, K$. The number of memberships in the kth cluster is n_k, and $\sum_{k=1}^{K} n_k = N$.

10.5   Discussion and Summary

In this chapter, we have introduced distance metrics for discrete features and continuous features. Hamming distance and matching coefficient are the most popular discrete data distance metrics. Continuous data distance metrics are listed in Table 10.1. The choice of the distance metric is critical to the final clustering results since different distance metrics may lead to different clustering results. Therefore, the distance metric has to be chosen to represent the characteristics of the data as much as possible. Then, we briefly introduced the eight clustering families, namely partitional clustering, hierarchical clustering, fuzzy clustering, neural network-based clustering, mixture model clustering, graph-based clustering, consensus clustering and biclustering. These clustering algorithms have been extensively used in the analysis of molecular biology, genetics, genomics and proteomics data. We also defined symbols to represent clusters and partitions, which may be frequently used in the forthcoming chapters. Now, we will start our journey to the technical world of clustering.

References

Cheng, Y. and Church, G.M. (2000). Biclustering of expression data. Proceedings of the International Conference on Intelligent Systems for Molecular Biology, San Diego, 19–23 August 2000,. International Society for Computational Biology (ISCB), pp. 93–103.

D'haeseleer, P. (2005). How does gene expression clustering work? *Nature Biotechnology,* **23**(12), pp. 1499–1502.

Heyer, L.J., Kruglyak, S. and Yooseph, S. (1999). Exploring expression data: identification and analysis of coexpressed genes. *Genome Research,* **9**(11), pp. 1106–1115.

Scherf, U., Ross, D.T., Waltham, M. *et al.* (2000). A gene expression database for the molecular pharmacology of cancer. *Nature Genetics,* **24**(3), pp. 236–244.

Stein, T., Morris, J.S., Davies, C.R. *et al.* (2004). Involution of the mouse mammary gland is associated with an immune cascade and an acute-phase response, involving LBP, CD14 and STAT3. *Breast Cancer Research,* **6**(2), pp. R75–R91.

Xu, R. and Wunsch, D. (2005). Survey of clustering algorithms. *IEEE Transactions on Neural Networks,* **16**(3), pp. 645–678.

Xu, R. and Wunsch, D. (2010). Clustering algorithms in biomedical research: a review. *IEEE Reviews in Biomedical Engineering,* **3**, pp. 120–154.

11

Partitional Clustering

11.1 Introduction

The problem of partitional clustering can be stated as follows. Given a dataset with N data objects in an M-dimensional feature space, the task of partitional clustering is to determine a partition of K groups or clusters, such that the data objects in a cluster are more similar to each other than to data objects in other clusters. In theory, there are $(1/K!)\sum_{i=1}^{K}(-1)^{K-i}\binom{K}{i}i^{N}$ possibilities to group N data objects into K clusters (Jain and Dubes, 1988). Therefore, exhaustive enumeration of all possibilities is not computationally feasible. On the other hand, the definition is in its broadest sense. With this definition, partitional clustering could cover many clustering families, such as neural network-based clustering, mixture model clustering and so on, in terms of different clustering criteria. A clustering criterion has to be adopted. In this chapter, we restrict the criterion of partitional clustering to be squared-error.

The best examples of this family are k-means and k-medoids (also known as partition around medoids (PAM)). As one of those oldest clustering algorithms that we can trace back to the fifties of the last century (Steinhaus, 1956; MacQueen, 1967), k-means was designed for similarity grouping and relevant clustering. In communication and signal-processing fields, similar algorithms were developed for data compression, for example vector quantisation (VQ) algorithm and Lloyd algorithm (Linde, Buzo and Gray, 1980; Lloyd, 1982). The k-medoids algorithm was proposed by (Kaufman and Rousseeuw, 1987, 1990).

With the advent of high-throughput data-collecting techniques in the molecular biology fields, for example, microarrays, DNA sequencing and RNA sequencing, k-means and k-medoids have been widely used in bioinformatics applications. Herwig and colleagues developed a clustering procedure based on a modified k-means, called sequential k-means, to

Integrative Cluster Analysis in Bioinformatics, First Edition. Basel Abu-Jamous, Rui Fa and Asoke K. Nandi.
© 2015 John Wiley & Sons, Ltd. Published 2015 by John Wiley & Sons, Ltd.

analyse high-throughput cDNA fingerprinting data (Herwig *et al.*, 1999). Tavazoie and colleagues employed k-means as the clustering tool to analyse microarray gene expression data for *Saccharomyces cerevisiae* (budding yeast) (Tavazoie *et al.*, 1999). In their studies of the chromatin signatures of promoters and enhancers, Heintzman and colleagues used k-means as one of computational methods in the processing pipeline (Heintzman *et al.*, 2007, 2009). Pollard and van de Laan analysed microarray gene expression data using k-medoids because of its better ability against outliers (Pollard and van de Laan, 2002). Harbison and colleagues developed a computational method to discover the transcriptional regulator binding sites, where k-medoids was used to cluster the significant motifs of transcriptional regulators (Harbison *et al.*, 2004). Cheung and colleagues employed k-medoids in the cluster analysis of DNA copy-number data, as a part of their genome-wide profiling of follicular lymphoma (FL), which revealed regional copy-number imbalances and identified prognostic correlates in relation to both survival and transformation risk (Cheung *et al.*, 2009). Recently, both k-means and k-medoids were employed by Elsheikh and colleagues in a breast cancer study to identify the presence of variants in global levels of histone markers in different classes of invasive breast cancer (Elsheikh *et al.*, 2009).

There have also been many extensions and variations based on the original k-means and k-medoids algorithms, for example fuzzy c-means and fuzzy k-medoids (will be detailed in Chapter 13), kernel k-means (Zhang and Rudnicky, 2002; Yu *et al.*, 2012), genetic k-means (Krishna and Murty, 1999; Lu *et al.*, 2004) and genetic k-medoids (Sheng and Liu, 2006), spherical k-means (Dhillon and Modha, 2001; Hornik *et al.*, 2012) and spherical k-medoids (Dikmen, Hoiem and Huang, 2012), and so on. In the rest of this chapter, we will detail the principles of k-means and k-medoids, their variations and their applications.

11.2 k-Means and its Applications

11.2.1 Principles

The k-means algorithm is the most common partitional clustering algorithm. The basic idea is to obtain the partition that minimises the squared-error for a given number of clusters, K. The squared-error is also known as within-cluster variation. Suppose that the given partition is formed in a cluster set $C = \{C_1,...,C_K\}$ for a dataset with N data objects, as we introduced in the last chapter. We define the centroid, c_k, of the kth cluster C_k as the mean vector of the members in C_k, which is written as shown in Equation (11.1).

$$c_k = \frac{1}{n_k} \sum_{x_n \in C_k} x_n \tag{11.1}$$

Thus, the squared-error, or within-cluster variation, of the kth cluster is the sum of the squared Euclidean distances between all members in C_k and the centroid c_k, mathematically written as Equation (11.2),

$$E_k = \sum_{x_n \in C_k} (x_n - c_k)^T (x_n - c_k) \tag{11.2}$$

where $(\cdot)^T$ is the transpose operator. The total squared-error for the whole partition is the sum of within-cluster variations of K clusters, which is given by Equation (11.3).

$$E = \sum_{k=1}^{K} E_k \qquad (11.3)$$

The squared-error clustering criterion is to find the optimal partition by minimising the total squared-error, E. The optimisation problem to find the optimal partition minimising the total squared-error is a non-deterministic polynomial-time hard (NP-hard) problem. According to their experimental results, the random and KA methods work much better than the Forgy and MacQueen methods.

The k-means algorithm, as a heuristic algorithm, is computationally efficient and converges quickly. The basic procedure of the k-means is given in Table 11.1. In Step 1, the random initialisation partitions the dataset at random; deterministic initialisation calculates the representative objects subsequently in a deterministic way. There have been many methods to randomly initialise k-means. The first one, which is also the most usual one, is to generate the partition randomly and update the centroids accordingly; the second one, which was developed by (Forgy, 1965), is to select K data objects randomly as the centroids and assign the rest of data objects to the cluster represented by the nearest centroid; the last one, which was proposed by (MacQueen, 1967), is to select K data objects randomly, assign the rest of the objects to the cluster with the nearest centroid following the data-object order, and update the centroid after each assignment. The Kaufman approach (KA) is a deterministic initialisation method, which is summarised in (Kaufman and Rousseeuw, 1990). Peña, Lozano, and Larrañaga compared the abovementioned four classical initial partition methods based on effectiveness, robustness and convergence speed (Peña, Lozano and Larranaga, 1999) (Table 11.2). According to their experimental results, the random and KA methods work much better than the Forgy and MacQueen methods.

Table 11.1   The basic procedure of k-means

Step 1	Initialise K clusters either randomly or deterministically;
Step 2	Assign each data object to its nearest cluster C_k, where $k = \arg\min_k \|x_n - c_k\|$;
Step 3	Update the centroid of each changed cluster, $c_k = (1/n_k) \sum_{x_n \in C_k} x_n$;
Step 4	Repeat Steps 2 and 3 until there is no change in any cluster.

Table 11.2   The Kaufman approach (KA)

Step 1	Select the most centrally located data object as the first centroid;
Step 2	FOR every unselected data object x_i DO
Step 2.1	FOR every unselected data object x_j DO calculate $\beta_{ji} = \max(D_j - d_{ij}, 0)$, where $d_{ij} = \|x_i - x_j\|$ and $D_j = \min_s d_{sj}$, where s represents one of selected representatives;
Step 2.2	Calculate the gain of selecting x_j by $\sum_j \beta_{ji}$;
Step 3	Select the unselected data object maximising $\sum_j \beta_{ji}$ as the new centroid;
Step 4	IF there are K selected representatives, THEN stop ELSE go to Step 2.

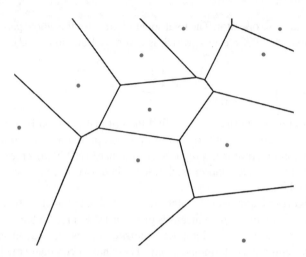

Figure 11.1 Illustration of Voronoi diagram in two-dimensional space. The space is partitioned into 10 Voronoi cells

Mathematically speaking, the second step of the k-means algorithm partitions the observations according to the Voronoi diagram. The input space is divided into Voronoi cells corresponding to a set of prototype vectors, which is illustrated in Figure 11.1. Each data object in a certain Voronoi cell is closer to its centroid than any other centroids. The computational complexity of k-means is of the order of $\mathcal{O}(NMKT)$, where T is the number of iterations. The value of T depends on the initial cluster centres, distribution of patterns and the size of the clustering problem. However, in practice, an upper limit of T can be specified. The k-means method works well for many practical problems, particularly when the resulting clusters are compact and hyper-spherical in shape. Many platforms have k-means implemented, for example, the routine **kmeans** in the Statistics Toolbox of MATLAB (MathWorks, 2013), the routine **kmeans** in Package 'stats' of R (R-Core-Team, 2013), and **SimpleKMeans** class in weka java package (Witten *et al.*, 1999).

Example 11.1: Given a two-dimensional dataset with 20 objects labelled by letters A—U (excluding letter O), the procedure of k-means clustering is illustrated in Figure 11.2. Suppose that the number of clusters is known to be two, the k-means algorithm is initialised by randomly assigning objects into two clusters. Thus the centroids of two clusters can be obtained, as depicted with solid shapes in Figure 11.2a. Subsequently, the objects are re-assigned to the two clusters in terms of the distances between them and two centroids, and the centroids of changed clusters are updated, which is illustrated in Figure 11.2b. The iteration will continue until there is no change in any cluster.

11.2.2   Variations

11.2.2.1   Kernel k-means

Kernel k-means is an extension of standard k-means by nonlinearly mapping data objects to a higher-dimensional feature space. Thus, it has the ability to discover clusters that are not linearly separable in the original space. Usually the extension from k-means to kernel k-means is

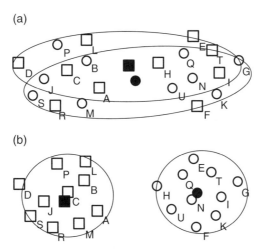

Figure 11.2 Two-dimensional example with 20 objects using k-means. (a) 20 objects are randomly assigned into 2 clusters, where the symbol square represents cluster 1 and the symbol circle represents cluster 2. Solid symbols represent the centroids. (b) Clustering results using k-means

Table 11.3 Examples of popular kernel functions

Gaussian kernel (also referred as Radial basis function)	$\kappa(x_i, x_j) = e^{-\frac{\|x_i - x_j\|^2}{2\sigma^2}}, \sigma > 0$
Polynomial kernel	$\kappa(x_i, x_j) = \left(\alpha x_i^T x_j + 1\right)^d, \alpha \in \mathbb{R}, d \in \mathbb{N}$
Sigmoid kernel	$\kappa(x_i, x_j) = \tanh\left(c x_i^T x_j + \theta\right)$
Exponential kernel	$\kappa(x_i, x_j) = e^{-\frac{\|x_i - x_j\|}{2\sigma^2}}, \sigma > 0$
Laplacian kernel	$\kappa(x_i, x_j) = e^{-\frac{\|x_i - x_j\|}{\sigma}}, \sigma > 0$

simply realised by expressing the distance in the form of kernel function which is expressed as shown in Equation (11.4),

$$\kappa(x_i, x_j) = \Phi(x_i)^T \Phi(x_j) \tag{11.4}$$

where $\Phi(\cdot)$ is the nonlinear mapping function, which sometimes is not infeasible. Some popular kernel functions are listed in Table 11.3. Thus, a straightforward way to transform the calculation of Euclidean distance in the feature space into the kernel version is to use the kernel trick (Girolami, 2002) as in Equation (11.5),

$$D_E^\kappa(x_i, x_j) = \sqrt{\|\Phi(x_i) - \Phi(x_j)\|^2} = \sqrt{\|\Phi(x_i)\|^2 + \|\Phi(x_i)\|^2 - 2\Phi(x_i)^T \Phi(x_j)}$$

$$= \sqrt{\kappa(x_i, x_i) + \kappa(x_j, x_j) - 2\kappa(x_i, x_j)} = \sqrt{2 - 2\kappa(x_i, x_j)} \tag{11.5}$$

and the kernel version of the modified Pearson correlation is given by Equation (11.6).

$$S_E^\kappa\left(x_i,x_j\right) = \frac{\Phi(x_i)^T \Phi(x_j)}{\sqrt{\|\Phi(x_i)\|^2}\sqrt{\|\Phi(x_j)\|^2}} = \kappa\left(x_i,x_j\right) \tag{11.6}$$

The core part of kernel k-means is to calculate the Euclidean distance between the objects and the centroids of the clusters in the feature space, which is written as given in Equation (11.7).

$$D_E^\kappa\left(x_i, c_k^\Phi\right) = \sqrt{\left\|\Phi(x_i) - \frac{1}{N_k}\sum_{x_n \in C_k}\Phi(x_n)\right\|^2}$$

$$= \sqrt{\kappa(x_i,x_i) - \frac{1}{N_k}\sum_{x_n \in C_k}\kappa(x_i,x_n) + \frac{1}{N_k^2}\sum_{x_n \in C_k}\sum_{x_l \in C_k}\kappa(x_l,x_n)} \tag{11.7}$$

Then, the rest of the algorithm is the same process as the standard k-means. The main drawback of this algorithm is the high clustering cost due to the repeated calculations of kernel values, or insufficient memory to store the kernel matrix.

11.2.2.2 Spherical k-means

Spherical k-means was proposed by Dhillon and Modha (Dhillon and Modha, 2001) for clustering of large sparse text data in text-mining applications. The algorithm employs the cosine similarity as the distance measure and produces K disjoint clusters, each with a concept vector that is the centroid of the cluster normalised to have unit Euclidean norm and is mathematically written as shown in Equation (11.8).

$$c_k' = \frac{c_k}{\|c_k\|} \tag{11.8}$$

The concept vector has a very important property: For any unit vector $z \in \mathbb{R}^M$, based on the Cauchy–Schwarz inequality, we obtain Equation (11.9).

$$\sum_{x_n \in C_k} x_n^T c_k' \geq \sum_{x_n \in C_k} x_n^T z \tag{11.9}$$

The concept vector, thus, is the closest vector in cosine similarity to all data objects in the cluster C_k.

The problem turns out to be a maximisation problem of an objective function, and the solution is the optimal partition, which is mathematically given by Equation (11.10).

$$\{C_k\}_{k=1}^K = \arg\max_{\{C_k\}_{k=1}^K} \sum_{k=1}^K \sum_{x_n \in C_k} x_n^T c_k' \tag{11.10}$$

The solution of this optimisation problem is NP-hard. Spherical k-means is an approximation algorithm using an effective and efficient iterative heuristic, which is summarised in

Table 11.4 The basic procedure of spherical k-means

Step 1	Initialise K clusters with arbitrary partitioning of the data objects;
Step 2	Assign each data object to its nearest cluster C_k, where $k = \arg\max_k x_n^T c_k'$;
Step 3	Update the centroid of each changed cluster, $c_k = (1/n_k)\sum_{x_n \in C_k} x_n$ and $c_k' = c_k/\|c_k\|$;
Step 4	Repeat Steps 2 and 3 until there is no change in any cluster.

Table 11.5 Summary of genetic k-means

Step 1	Initialise the algorithm: population size N_p; mutation probability p; maximum number of generation MAX_GEN; the population $\mathbf{P}$; $s^* = \mathbf{P}(0)$;
Step 2	FOR $i = 1, \cdots$, MAX_GEN DO
Step 2.1	Calculate Fitness values of strings in $\mathbf{P}$;
Step 2.2	$\widehat{\mathbf{P}} = \text{Selection}(\mathbf{P})$;
Step 2.3	FOR $n = 1, \cdots, N_p$ DO $\mathbf{P}_n = $ Mutation $(\widehat{\mathbf{P}}_n)$
Step 2.4	FOR $n = 1, \cdots, N_p$ DO K-Means $(\mathbf{P}_n)$
Step 2.5	IF $E_{s^*} > E_s$ THEN $s^* = s$, where $s \in \mathbf{P}$
Step 3	Output s^*

Table 11.4. In package 'skmeans' of R, the routine **_skmeans_** is implemented for the spherical k-means algorithm.

11.2.2.3   Genetic k-means

Genetic algorithm (GA), which was originally proposed in the 1970s (Holland, 1975), has been used in clustering by employing either an expensive crossover operator to generate valid child chromosomes from parent chromosomes or a costly fitness function or both. Krishna and Murty proposed a hybrid clustering algorithm combining GA and the k-means method, called genetic k-means algorithm (GKA) (Krishna and Murty, 1999). If we define a string s with length of N as the chromosome and each element of the chromosome is an allele from $\{1, \ldots, K\}$, the search space of the GKA is all possible chromosomes.

The core of GKA contains a selection operator, a mutation operator and a k-means operator. It starts with a population of chromosomes, say L chromosomes, which are randomly generated. The selection operator selects the chromosomes based on their merits or fitness values to survive in the next generation. The mutation operator changes allele values depending on the distance of the cluster centroids from the corresponding data object. The k-means operator is introduced to accelerate the convergence speed. After a certain number of generations, GKA selects the string s^* minimising the total within-cluster variations E in Equation (11.3) as the output. The GKA is summarised in Table 11.5.

Lu and colleagues developed an incremental genetic k-means algorithm (IGKA) for gene expression data analysis (Lu *et al.*, 2004). Similarly as GKA, IGKA also has a selection operator, a mutation operator and a k-means operator. IGKA, however, calculates the total within-cluster variations and cluster centroids incrementally, differently from GKA. In order to obtain the new centroids and new total within-cluster variations, IGKA maintains the difference values between the old solution and the new solution when the allele changes. However, if the mutation probability is large, too many alleles change their cluster membership, the maintenance of difference values becomes expensive, and IGKA becomes inferior to other GKA algorithms in performance. GKA can be obtained by using the routine **skmeans** in package 'skmeans' of R, with parameter 'method' being set to 'genetic'.

11.2.3 Applications

11.2.3.1 Case Study 1

Tavazoie and colleagues conducted a study to discover distinct expression patterns in the *Saccharomyces cerevisiae* microarray mRNA dataset and then identify upstream DNA sequence patterns specific to each expression cluster (Tavazoie *et al.*, 1999). In this study, k-means was employed as the clustering tool and $K = 30$ clusters were produced. Researchers found that the members of many clusters are significantly enriched for genes with similar functions. They mapped the genes in each cluster to the 199 functional categories in the Martinsried Institute of Protein Sciences (MIPS) functional classification scheme database. To determine the statistical significance for functional category enrichment, the hypergeometric distribution was used to obtain the chance probability of observing similar or higher numbers of genes from a particular MIPS function category within each cluster. The most notable cluster they found was the one in which 64 out of 164 genes encode ribosome proteins and P-value is 10^{-54}.

Next, Tavazoie and colleagues carried out a blind and systematic search for upstream DNA sequence motifs that were common to members of each cluster. They found 18 motifs from 12 different clusters, where 7 had been experimentally identified. For the clusters whose members belong to known regulons, the expected *cis*-regulatory element(s) emerged as the highest scoring motif(s) in every case. Some newly discovered motifs, namely M14a, M14b, M3a and M3b, are very likely to have roles in the global regulation of protein synthesis. It is worth noting that both function category enrichment analysis and upstream motifs discovery are based on the clustering results by k-means.

11.2.3.2 Case Study 2

The human body is composed of diverse cell types with distinct functions and the lineage specification depends on cell-specific gene expression, which is driven by many regulatory elements, namely promoters, enhancers, insulators and other *cis*-regulatory DNA sequences. Heintzman and colleagues carried out a study of the relative roles of these regulatory elements, wherein they discovered that the chromatin signatures at promoters, CTCF (insulator-binding protein, encoded by CTCF gene) occupancy and CTCF enrichment patterns are remarkably similar across all cell types, while they observed that enhancers are marked with highly cell-type-specific histone modification patterns, strongly correlate to cell-type-specific gene expression programs on a global scale, and are functionally active in a cell-type-specific manner (Heintzman *et al.*, 2007, 2009). Over 55 000 potential transcriptional enhancers in the human genome were defined, which significantly expands the current catalogue of human enhancers and highlights the role of these elements in cell-type-specific gene expression. In this study, k-means was used to cluster the chromatin modifications found ±5 kb from 414 promoters, and cluster 1423 non-redundant enhancers predicted on the basis of chromatin signatures.

11.3   *k*-Medoids and its Applications

11.3.1   Principles

The *k*-medoids algorithm, also known as PAM, is based on a similar strategy to *k*-means, that is, searching for *k* representatives in the dataset. Differently from *k*-means, *k*-medoids looks for *k* representative objects, which are called medoids, among all the objects, rather than their mean values. *k*-Medoids is more robust to outliers than is *k*-means, because outliers may affect the centroids but hardly affect medoids. The procedure of *k*-medoids is similar to that of *k*-means shown in Table 11.1, except that in Step 3 it updates the medoid of each changed cluster.

Example 11.2: Considering the same dataset as the one in Example 11.1, we cluster the dataset using *k*-medoids. The *k*-medoids algorithm randomly selects two objects as medoids, which, in this case, are object H and object I, and assigns all other objects to two clusters in terms of the distances between them and the medoids, as depicted in Figure 11.3a. Then the medoids are updated in terms of the members in the same clusters, which turn out to be object C and object N in our example, and all objects are re-assigned subsequently, as illustrated in Figure 11.3b. The iteration will continue until there is no change in any cluster. The main difference between *k*-means and *k*-medoids is that *k*-means uses centroids while *k*-medoids uses medoids.

11.3.2   Variations

11.3.2.1   Spherical *k*-medoids

Dikmen and colleagues proposed spherical *k*-medoids to define a vocabulary of image patches (Dikmen, Hoiem and Huang, 2012). Spherical *k*-medoids employs cosine similarity

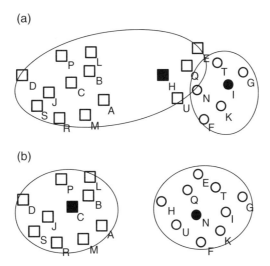

Figure 11.3   Two-dimensional example with 20 objects using *k*-medoids. (a) Two medoids, which are represented by the solid square and circle, were randomly selected, and then two clusters were initialised, where the symbol square represents cluster 1 and the symbol circle represents cluster 2. (b) Clustering results using *k*-medoids

measurement, which is similar to spherical k-means. Differently from spherical k-means, spherical k-medoids updates the medoid in every cluster by finding the object maximising the pairwise similarities in each respective cluster.

11.3.2.2 Genetic k-medoids

Sheng and Liu proposed the genetic k-medoids algorithm, which is a hybrid GA for k-medoids clustering (Sheng and Liu, 2006). Variable-length chromosomes that encode different numbers of clusters were employed for evolution and a modified Davies–Bouldin (MDB) index (see Chapter 20 for the detail of the DB index) was used as a measure of the fitness. In doing so, first of all, the number of clusters in the population ranged from 2 to k_{max} and clustering was initialised by randomly selecting medoids from the dataset. Genetic operators, including selection operators, crossover operators and mutation operators, functioned generation by generation for evolution. However these genetic operators took a long time to converge. Therefore a heuristic operator was designed and integrated with the global search for k-medoids clustering. This operator efficiently improved the fitness of the offspring by updating the medoids encoded in them such that the total dissimilarity or distance within each cluster represented by the medoid is minimised. Interestingly, for the genetic k-medoids method it was claimed that there was no need to specify the exact number of clusters *a priori*. By using variable-length chromosomes and the MDB-based fitness function, the genetic k-medoids can converge to true number of clusters and a good partition.

11.3.3   Applications

11.3.3.1   Case Study 1

Harbison and colleagues conducted a study of yeast's transcriptional regulatory code (Harbison *et al.*, 2004). In this study, the sequence elements that are bound by regulators under various conditions were identified and an initial map of these transcriptional regulatory codes was constructed. In doing so, first of all, the genomic occupancy of 203 DNA-binding transcriptional regulators in rich media conditions was determined by using genome-wide location analysis. To identify the *cis*-regulatory sequences that are likely to serve as recognition sites for transcriptional regulatory, six motifs-discovery algorithms were employed and 68 279 DNA sequence motifs for the 147 regulators that bound more than ten probes were discovered. The resulting specificity predictions were filtered for significance and 4259 motifs were left. Significant motifs were clustered with the use of the k-medoids algorithm. Aligned motifs within each cluster were averaged to produce consensus motifs and filtered according to their conservation. This procedure typically produced several distinct consensus motifs for each regulator.

11.3.3.2   Case Study 2

Approximately 85% of FL cases are associated with a specific balanced translocation t(14;18) (q32;q21), that leads to over-expression of the anti-apoptotic gene *BCL2*. However, this genetic abnormality alone is unlikely to produce clinical FL, and t(14;18)-bearing lymphocytes have been frequently demonstrated in healthy individuals. A number of large studies have been carried out to investigate the chromosomal imbalances in FL by using a combination of techniques including conventional karyotyping, comparative genomic hybridisation (CGH) and single nucleotide polymorphism (SNP) technology. Cheung and colleagues conducted a study using a combination of a high-resolution genomic analysis and a large FL cohort composed exclusively of diagnostic biopsies (Cheung *et al.*, 2009). In this study, whole-genome tiling-path bacterial artificial chromosome (BAC)-array CGH, with a resolution of at least 200 kb for detection of copy-number alterations in clinical specimens and a reported tolerance of up to 70% contamination by non-tumour cells, was applied to a cohort of 106 FL diagnostic specimens with complete clinical information. Seventy-one regional alterations that were recurrent in at least 10% of cases were identified. These altered regions ranged in size from 200 kb to 44 Mb. As a part of the study, 4912 BAC clones were extracted from the 71 regions of alterations and *k*-medoids was applied to the 106 cases to find clusters using the Hamming distance metric.

11.4 Discussion and Summary

The definition of partitional clustering in the broadest sense may include many clustering families, which partition datasets into groups in terms of different clustering criteria. We restricted the criterion of partitional clustering to be squared-error. In this chapter, we discussed the partitional clustering algorithms, essentially, *k*-means and *k*-medoids, as well as their variations. All introduced algorithms and their publically accessible software are summarised in Table 11.6. Furthermore, we discussed some real applications in bioinformatics using these partitional clustering algorithms. Virtually, there are two main issues of *k*-means-type and *k*-medoids-type clustering: the first is that the number of clusters, K, which is an essential parameter to the algorithms, is difficult to set *a priori*; the second is that the squared-error criterion may not match the structure of some datasets, which may result in performance degradation. Nonetheless, as one group of the most popular clustering algorithms, partitional clustering algorithms, including *k*-means, *k*-medoids and their variations, have been widely used in cluster analysis in the bioinformatics field because of their lower computational complexity.

Table 11.6 Summary of all partitional clustering algorithms introduced in this chapter

Algorithm Name	Year	Platform	Package	Function
k-means	1967	*R*/MATLAB/JAVA	Statistics/stats/weka	kmeans/kmeans/SimpleKMeans
k-medoids	1990	*R*	cluster	pam
Genetic *k*-means	1999	*R*	skmeans	skmeans(method = 'genetic')
Spherical *k*-means	2001	*R*	skmeans	skmeans
Kernel *k*-means	2002	*R*	kernlab	kkmeans
Spherical k-medoids	2006			
Genetic k-medoids	2006			

References

Cheung, K.-J.J., Shah, S.P., Steidl, C. *et al.* (2009). Genome-wide profiling of follicular lymphoma by array comparative genomic hybridization reveals prognostically significant DNA copy number imbalances. *Blood*, **113**(1), pp. 137–148.

Dhillon, I.S. and Modha, D.S. (2001). Concept decompositions for large sparse text data using clustering. *Machine Learning*, **42**(1-2), pp. 143–175.

Dikmen, M., Hoiem, D. and Huang, T.S. (2012). A Data Driven Method for Feature Transformation. IEEE Conference on Computer Vision and Pattern Recognition (CVPR), 16–21 June 2012, IEEE, Providence, RI, pp. 3314–3321.

Elsheikh, S.E., Green, A.R., Rakha, E.A. *et al.* (2009). Global histone modifications in breast cancer correlate with tumor phenotypes, prognostic factors, and patient outcome. *Cancer Research*, **69**(9), pp. 3802–3809.

Forgy, E.W. (1965). Cluster analysis of multivariate data: efficiency versus interpretability of classifications. *Biometrics*, **21**, pp. 768–769.

Girolami, M. (2002). Mercer kernel-based clustering in feature space. *IEEE Transactions on Neural Networks*, **13**(3), pp. 780–784.

Harbison, C.T., Gordon, D.B., Lee, T.I. *et al.* (2004). Transcriptional regulatory code of a eukaryotic genome. *Nature*, **431**(7004), pp. 99–104.

Heintzman, N.D., Stuart, R.K., Hon, G.C. *et al.* (2007). Distinct and predictive chromatin signatures of transcriptional promoters and enhancers in the human genome. *Nature Genetics*, **39**(3), pp. 311–318.

Heintzman, N.D., Hon, G.C., Hawkins, R.D. *et al.* (2009). Histone modifications at human enhancers reflect global cell-type-specific gene expression. *Nature*, **459**(7243), pp. 108–112.

Herwig, R., Poustka, A.J., Müller, C. *et al.* (1999). Large-scale clustering of cDNA -fingerprinting data. *Genome Research*, **9**(11), pp. 1093–1105.

Holland, J.H. (1975). *Adaptation in Natural and Artificial Systems: An Introductory Analysis with Applications to Biology, Control, and Artificial Intelligence*, University of Michigan Press, Ann Arbor, MI.

Hornik, K., Feinerer, I., Kober, M. and Buchta, C. (2012). Spherical k-means clustering. *Journal of Statistical Software*, **50**(10), pp. 1–22.

Jain, A.K. and Dubes, R.C. (1988). *Algorithms for Clustering Data*, Prentice-Hall, New York.

Kaufman, L. and Rousseeuw, P.J. (1987). *Clustering by Means of Medoids*, Delft Faculty of Mathematics and Informatics.

Kaufman, L. and Rousseeuw, P.J. (1990). *Finding Groups in Data: an Introduction to Cluster Analysis*, John Wiley & Sons, Inc, New York.

Krishna, K. and Murty, M.N. (1999). Genetic K-means algorithm. *Systems, Man, and Cybernetics, Part B; IEEE Transactions on Cybernetics*, **29**(3), pp. 433–439.

Linde, Y., Buzo, A. and Gray, R. (1980). An algorithm for vector quantizer design. *IEEE Transactions on Communications*, **28**(1), pp. 84–95.

Lloyd, S. (1982). Least squares quantization in PCM. *IEEE Transactions on Information Theory*, **28**(2), pp. 129–137.

Lu, Y., Lu, S., Fotouhi, F. *et al.* (2004). Incremental genetic K-means algorithm and its application in gene expression data analysis. *BMC Bioinformatics*, **5**(1), p. 172.

MacQueen, J. (1967). Some Methods for Classification and Analysis of Multivariate Observations. Proceedings of the Fifth Berkeley Symposium on Mathematical Statistics and Probability, 21 June–18 July 1965 and 27 December 1965–7 January 1966, University of California Press, Berkeley, CA, **14**, pp. 281–297.

MathWorks (2013). *MATLAB Version 8.1.0 (R2013a)*, The MathWorks Inc., Natick, Massachusetts.

Peña, J.M., Lozano, J.A. and Larrañaga, P. (1999). An empirical comparison of four initialization methods for the K-means algorithm. *Pattern Recognition Letters*, **20**(10), pp. 1027–1040.

Pollard, K.S. and van de Laan, M.J. (2002). Statistical inference for simultaneous clustering of gene expression data. *Mathematical Biosciences*, **176**(1), pp. 99–121.

R-Core-Team (2013). *R: A Language and Environment for Statistical Computing*, R Foundation for Statistical Computing, Vienna.

Sheng, W. and Liu, X. (2006). A genetic k-medoids clustering algorithm. *Journal of Heuristics*, **12**(6), pp. 447–466.

Steinhaus, H. (1956). Sur la division des corps matériels en parties. *Bulletin de l'Académie Polonaise des Sciences*, **1**, pp. 801–804.

Tavazoie, S., Hughes, J.D., Campbell, M.J. *et al.* (1999). Systematic determination of genetic network architecture. *Nature Genetics,* **22**(3), pp. 281–285.

Witten, I.H., Frank, E., Trigg, L. *et al.* (1999). *Weka: Practical Machine Learning Tools and Techniques with Java Implementations,* http://citeseerx.ist.psu.edu/viewdoc/download?doi=10.1.1.44.4026&rep=rep1&type=pdf (accessed 21 July 2014).

Yu, S., Tranchevent, L.C., Liu, X. *et al.* (2012). Optimized data fusion for kernel K-means clustering. *IEEE Transactions on Pattern Analysis and Machine Intelligence,* **34**(5), pp. 1031–1039.

Zhang, R. and Rudnicky, A. (2002) A Large Scale Clustering Scheme for Kernel K-Means. Proceedings of the 16th International Conference on Pattern Recognition, 11–15 August 2002, Québec City, Canada, **4**, pp. 289–293.

12

Hierarchical Clustering

12.1 Introduction

Hierarchical clustering is one of most popular clustering methods in the literature. In contrast to partitional clustering, which attempts to directly decompose the dataset into a set of disjoint clusters, a hierarchical clustering method is a procedure for transforming a proximity matrix into a nested partition, which can be graphically represented by a tree, called a ***dendrogram.*** To obtain the number of clusters and the corresponding partition, we have to cut the dendrogram at a certain level. Cutting it at different levels will lead to different clustering results with different levels of resolution.

Hierarchical clustering algorithms are mainly classified into agglomerative methods (bottom-up methods) and divisive methods (top-down methods), based on how the hierarchical dendrogram is formed. Agglomerative methods start with N clusters initially (basically regard each object as a cluster). Gradually, they merge the closest pair of clusters in terms of different linkage methods until all the groups are merged into one cluster and a dendrogram is formed. There are many linkage methods, namely single linkage, complete linkage, average linkage, Ward's linkage and others. Divisive algorithms are initialised with one cluster containing all objects, and they then split the dataset step by step until only singleton clusters remain. If all $\left(2^{N-1} - 1\right)$ possible divisions into two sub-clusters of N objects considered at each step, it is very expensive in computation. Therefore, divisive algorithms are not commonly used in practice. There are two divisive algorithms, which will be discussed later, namely monothetic analysis (MONA) and divisive analysis (DIANA).

Integrative Cluster Analysis in Bioinformatics, First Edition. Basel Abu-Jamous, Rui Fa and Asoke K. Nandi.
© 2015 John Wiley & Sons, Ltd. Published 2015 by John Wiley & Sons, Ltd.

12.2   Principles

In this section, we overview the principles of hierarchical clustering in terms of hierarchy strategies, that is bottom-up or top-down, which correspond to agglomerative methods or divisive methods.

12.2.1   Agglomerative Methods

Agglomerative methods are dominant in the hierarchical clustering family. Without further specification, the term 'hierarchical clustering' was often used to refer to agglomerative hierarchical clustering. The general agglomerative hierarchical clustering is summarised in Table 12.1. It is worth noting that in Step 2, every time when the merging occurs, the two merged clusters are graphically connected by a link. Therefore, the outcome of the agglomeration is a dendrogram. For example, we employ the data reported by Golub and colleagues (Golub *et al.*, 1999), which consist of 38 bone marrow samples obtained from acute leukaemia patients at time of diagnosis. The samples include three groups: 11 acute myeloid leukaemia (AML) samples, 8 T-lineage acute lymphoblastic leukaemia (ALL) samples and 19 B-lineage ALL samples. The cluster dendrogram of the samples is shown in Figure 12.1. This example gives us an intuitive impression of what the dendrogram looks like. In Step 3, there are many different definitions of the distance between clusters, which lead to different clustering algorithms/linkage techniques algorithms, namely single linkage, complete linkage, average linkage and Ward's linkage. In the next subsections, we introduce these linkage methods.

12.2.1.1   Single Linkage

Single linkage clustering is also known as shortest-distance linkage or nearest-neighbour clustering. In Step 3 of Table 12.1, the distance between two clusters is represented by a single pair of objects with the shortest distance. Mathematically, the linkage function – the distance $D(C_i, C_j)$ between clusters C_i and C_j – is described by Equation (12.1),

$$D(C_i, C_j) = \min_{x_p \in C_i, x_q \in C_j} d(x_p, x_q) \tag{12.1}$$

where C_i and C_j are any two sets of objects considered as clusters, and $d(x_p, x_q)$ denotes the distance between the two objects x_p and x_q. Single linkage often produces a skewed hierarchy (called the chaining problem, where clusters formed via single-linkage clustering may be forced together due to single elements being close to each other, even though many of the elements in each cluster may be very distant to each other) and is therefore not very useful for

Table 12.1   Specifications of all agglomerative hierarchical clustering methods

Step 1	Start with N clustering; basically, each object is a cluster; calculate the proximity matrix for N clusters;
Step 2	Find minimum distance in the proximity matrix and merge the two clusters with the minimal distance;
Step 3	Update the proximity matrix using the new distances between new cluster and other clusters;
Step 4	Repeat Steps 2 and 3 until all objects are in one cluster.

Cluster dendrogram

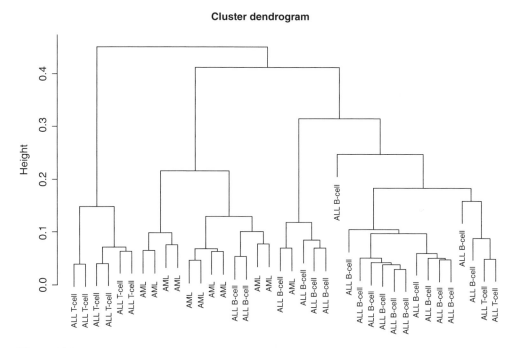

Figure 12.1 Cluster dendrogram of the leukaemia dataset reported by Golub and colleagues (1999), which contains 38 bone marrow samples using average linkage: 11 acute myeloid leukaemia (AML) samples, 8 T-lineage acute lymphoblastic leukaemia (ALL) samples and 19 B-lineage ALL samples

summarising data. However, outlying objects are easily identified by this method, as they will be the last to be merged.

12.2.1.2 Complete Linkage

Complete linkage is also known as farthest-neighbour clustering. In Step 3 of Table 12.1, the distance between two clusters is represented by a single pair of objects with the longest distance. Mathematically, the distance function is given by Equation (12.2).

$$D(C_i, C_j) = \max_{x_p \in C_i, x_q \in C_j} d(x_p, x_q) \tag{12.2}$$

Complete-linkage clustering avoids a drawback of single-linkage clustering, that is, the chaining problem. Complete linkage tends to find compact clusters of approximately equal diameters. However, complete linkage tends to be less desirable when there is a considerable amount of noise present in the data.

12.2.1.3 Average Linkage

There are two ways of defining average linkage. One is called weighted pair group method with arithmetic mean (WPGMA) or McQuitty's method, and the other one is called unweighted pair group method with arithmetic mean (UPGMA) (Sneath and Sokal, 1973). The distance between clusters in WPGMA is calculated as a simple average. For example, suppose we wish to calculate the distance between C_i and C_j, where C_j is composed of C_m and C_n; the distance function of WPGMA, therefore, is written as shown in Equation (12.3).

$$D\left(C_i, C_j\right) = \frac{D(C_i, C_m) + D(C_i, C_n)}{2} \tag{12.3}$$

Though computationally easier, when there are unequal numbers of objects in the clusters the distances in the original proximity matrix do not contribute equally to the intermediate calculations.

UPGMA is a superior method, in which averages are weighted by the number of objects in each cluster at each step such that every object is treated equally. Its distance function is written as shown in Equation (12.4).

$$D\left(C_i, C_j\right) = \frac{D(C_i, C_m)|C_m| + D(C_i, C_n)|C_n|}{|C_m| + |C_n|} \tag{12.4}$$

Note that Equation (12.4) implies an iterative implementation of UPGMA, and is equivalent to the expression given in Equation (12.5).

$$D\left(C_i, C_j\right) = \frac{1}{|C_i| \cdot |C_j|} \sum_{x_p \in C_i} \sum_{x_q \in C_j} d\left(x_p, x_q\right) \tag{12.5}$$

12.2.1.4   Centroid Linkage

In centroid linkage clustering, a centroid is assigned to each cluster, and this centroid is used to compute the distances between the current cluster and all other clusters to update the original proximity matrix. The vector is the average of the vectors of all actual objects contained within the cluster. Thus, when a new cluster is formed by joining together two clusters, the new cluster is assigned a centroid that is the average of all objects it contains, and not the average of the two joined. Mathematically, the centroid is written as Equation (12.6).

$$c_i = \frac{1}{|C_i|} \sum_{x_p \in C_i} x_p \tag{12.6}$$

Therefore, the distance between two centroids is simply given by Equation (12.7).

$$D\left(C_i, C_j\right) = d\left(c_i, c_j\right) \tag{12.7}$$

Centroid linkage is also called unweighted pair group method centroid (UPGMC) (Sneath and Sokal, 1973).

12.2.1.5   Median Linkage

Median linkage is also called weighted pair group method centroid (WPGMC) (Sneath and Sokal, 1973), where a weighted centroid is defined as in Equation (12.8),

$$w_j = \frac{1}{2}(w_m + w_n) \qquad (12.8)$$

where w_m and w_n are weighted centroids of C_m and C_n respectively. Cluster C_j is composed of clusters C_m and C_n. Thus the distance between two clusters is mathematically given by Equation (12.9).

$$D(C_i, C_j) = d(w_i, w_j) \qquad (12.9)$$

12.2.1.6 Ward's Linkage

Ward's linkage is also known as Ward's minimum variance method, originally presented by Ward (Ward, 1963). Ward's linkage minimises the total within-cluster variance. To do so, in Step 2 of Table 12.1, Ward's linkage merges two clusters with minimum between-cluster distance, that is, two clusters that lead to the minimum increase in total within-cluster variance after merging. Therefore, the distance function between two clusters in Ward's linkage is defined as within-cluster variance by considering them as one cluster. The within-cluster variance is also called error sum of squares (ESS), mathematically written as shown in Equation (12.10).

$$\text{ESS} = \sum_{x_n \in C} \|x_n - \bar{x}\|_2 \qquad (12.10)$$

In general, this method is regarded as very efficient; however, it tends to create clusters of small size.

12.2.1.7 Generalised Representation

All the hierarchical methods mentioned above can be implemented recursively by the *Lance–Williams* algorithm (Cormack, 1971). Suppose that clusters (objects) C_m and C_n are agglomerated into cluster $C_j = C_m \cup C_n$. The Lance–Williams algorithm calculates the distance between C_i and C_j recursively by means of Equation (12.11),

$$D(C_i, C_j) = \alpha_m D(C_i, C_m) + \alpha_n D(C_i, C_n) + \beta D(C_m, C_n) + \gamma |D(C_i, C_m) - D(C_i, C_n)| \qquad (12.11)$$

where parameters α_m, α_n, β and γ define the agglomerative criterion, parameter α_n with index n is defined identically to parameter α_m with index m. The values of the parameters for several linkage methods are given in Table 12.2.

Let us take an example of the single linkage method; using $\alpha_m = \alpha_n = 1/2, \beta = 0$ and $\gamma = -1/2$ gives the formula (12.12),

$$D(C_i, C_j) = \frac{1}{2}D(C_i, C_m) + \frac{1}{2}D(C_i, C_n) - \frac{1}{2}|D(C_i, C_m) - D(C_i, C_n)| \qquad (12.12)$$

which can be also rewritten as $D(C_i, C_j) = \min\{D(C_i, C_m), D(C_i, C_n)\}$, equivalent to Equation (12.1).

The 'method' argument of function *linkage* in MATLAB determines the algorithm for computing distance between clusters. The routine of agglomerative methods in MATLAB is

Table 12.2 The general procedure of agglomerative hierarchical clustering

Linkage methods	α_m	β	γ
Single linkage	$\dfrac{1}{2}$	0	$-\dfrac{1}{2}$
Complete linkage	$\dfrac{1}{2}$	0	$\dfrac{1}{2}$
Average linkage (UPGMA)	$\dfrac{|C_m|}{|C_m|+|C_n|}$	0	0
Average linkage (WPGMA)/McQuitty's method	$\dfrac{1}{2}$	0	0
Centroid linkage (UPGMC)	$\dfrac{|C_m|}{|C_m|+|C_n|}$	$-\dfrac{|C_m||C_n|}{(|C_m|+|C_n|)^2}$	0
Median linkage (WPGMC)	$\dfrac{1}{2}$	$-\dfrac{1}{4}$	0
Ward's linkage	$\dfrac{|C_m|+|C_i|}{|C_m|+|C_n|+|C_i|}$	$-\dfrac{|C_i|}{|C_m|+|C_n|+|C_i|}$	0

Table 12.3 Specifications of all agglomerative hierarchical clustering methods

Method	Description
Average	Unweighted average distance (UPGMA)
Centroid	Centroid distance (UPGMC), appropriate for Euclidean distances only
Complete	Complete linkage/Furthest distance
Median	Weighted centre of mass distance (WPGMC), appropriate for Euclidean distances only
Single	Single linkage/Shortest distance
Ward's	Minimum variance algorithm, appropriate for Euclidean distances only
Weighted	Weighted average distance (WPGMA)/McQuitty's methods

function *linkage.* One of the input arguments is 'method', which defines the linkage method, as shown in Table 12.3. The routine of agglomerative methods in R is *hclust* in 'stats' package. The argument 'method', similarly to the function *linkage* in MATLAB, selects the linkage method to be one of 'ward', 'single', 'complete', 'average', 'mcquitty', 'median' or 'centroid'.

12.2.2 Divisive Methods

12.2.2.1 Divisive Analysis

DIANA splits up a cluster into two smaller ones, until finally all clusters contain only a single element. We use an example to illustrate the DIANA algorithm. The data consist of six objects, which are numbered in the form $\{1, 2, 3, 4, 5, 6\}$ and its dissimilarity matrix is given by Matrix 12.1.

$$
\begin{bmatrix}
 & 1 & 2 & 3 & 4 & 5 & 6 \\
1 & 0.0 & 2.0 & 6.0 & 1.0 & 9.0 & 10.0 \\
2 & 2.0 & 0.0 & 10.0 & 2.0 & 8.0 & 9.0 \\
3 & 6.0 & 10.0 & 0.0 & 7.0 & 3.0 & 4.0 \\
4 & 1.0 & 2.0 & 7.0 & 0.0 & 10.0 & 9.0 \\
5 & 9.0 & 8.0 & 3.0 & 10.0 & 0.0 & 3.0 \\
6 & 10.0 & 9.0 & 4.0 & 9.0 & 3.0 & 0.0
\end{bmatrix}
\qquad \text{(Matrix 12.1)}
$$

The DIANA algorithm starts with a single cluster, $\{1, 2, 3, 4, 5, 6\}$. In the first step, the dataset has to be split into two clusters. Rather than considering all possible divisions, which is a computationally demanding process and sometimes infeasible, DIANA finds the farthest object away from other objects and forms a new cluster; then some others join the new cluster until a kind of equilibrium is attained. In our example, calculating the average dissimilarity with other objects, we find the sixth object is the one with the largest dissimilarity $((10.0 + 9.0 + 4.0 + 9.0 + 3.0)/5 = 7)$. Thus the sixth object is chosen to initiate the so-called *splinter group*. Then for each object in the larger cluster we calculate its average dissimilarity with the remaining objects, and compare it with the average dissimilarity with the objects in the splinter group. Therefore, the largest difference indicates the object that is likely to join the new cluster. In our example, the fifth object is the one with the largest difference $((9.0 + 8.0 + 3.0 + 10.0)/4 - 3.0 = 4.5)$. We continue the process to find other objects to join the new cluster until there is no positive difference. We find that the third object is the next one to join the new cluster $((6.0 + 10.0 + 7.0)/3 - (3.0 + 4.0)/2 = 4.17)$, and there are no more objects to join the new cluster. The process stops and we have completed the first divisive step, which splits the data into two clusters, $\{1, 2, 4\}$ and $\{3, 5, 6\}$.

In the next step, we split the cluster with the largest diameter first. The diameter of a cluster is defined as the largest dissimilarity between two of its objects. The dissimilarity matrices of clusters $\{1, 2, 4\}$ and $\{3, 5, 6\}$ are shown respectively as Matrices 12.2 and 12.3.

$$
\begin{array}{c}
\begin{matrix} & 1 & 2 & 4 \end{matrix} \\
\begin{matrix} 1 \\ 2 \\ 4 \end{matrix}
\begin{bmatrix}
0.0 & 2.0 & 1.0 \\
2.0 & 0.0 & 2.0 \\
1.0 & 2.0 & 0.0
\end{bmatrix}
\end{array}
\quad \text{and} \quad
\begin{array}{c}
\begin{matrix} & 3 & 5 & 6 \end{matrix} \\
\begin{matrix} 3 \\ 5 \\ 6 \end{matrix}
\begin{bmatrix}
0.0 & 3.0 & 4.0 \\
3.0 & 0.0 & 3.0 \\
4.0 & 3.0 & 0.0
\end{bmatrix}
\end{array}
\qquad \text{(Matrix 12.2 and Matrix 12.3)}
$$

Obviously, cluster $\{3, 5, 6\}$ has a larger diameter, which is 4.0. DIANA uses the same method in the first step to split the cluster $\{3, 5, 6\}$ and also other clusters step by step. The algorithm stops when each object is assigned to a singleton cluster. The finial dendrogram of our example is shown in Figure 12.2. The routine in *R* for DIANA is **diana** in package 'cluster'.

12.2.2.2   Monothetic Analysis

MONA considers only binary variables, corresponding to the presence or absence of some attributes. In MONA, each separation is carried out using a single variable, which is the reason why it is called *monothetic*. Methods like DIANA, which use all variables simultaneously, are called *polythetic*.

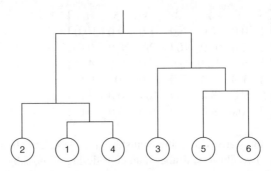

Figure 12.2 The dendrogram of a demonstration example for the DIANA algorithm. This example contains six objects numbered in $\{1, 2, 3, 4, 5, 6\}$

Suppose that there is no missing value in the dataset. First of all, we consider a two-by-two contingency table of two variables k and l as shown in Matrix 12.4.

$$\begin{array}{cc} k\,|\,l & 1 \quad 0 \\ \begin{array}{c} 1 \\ 0 \end{array} & \begin{bmatrix} x & b \\ c & y \end{bmatrix} \end{array} \qquad\qquad \text{(Matrix 12.4)}$$

Thus the association between variables k and l is defined as $a_{kl} = |xy - cb|$. The algorithm constructs a clustering hierarchy, starting with one large cluster. At a separation step, it selects one of the variables and divides the set of objects into two groups, one for which the selected variable equals 0 and the other for which it equals 1. The variable used for splitting a cluster is selected as follows. For each variable k, the association measures with all other variables are added, giving the total association given by Equation (12.13).

$$A_k = \sum_{l \neq k} a_{kl} \qquad\qquad (12.13)$$

The variable used for splitting a cluster is $\mathrm{argmax}_k A_k$. However, if the same maximal value is found for several variables, the variable is chosen as the one appearing first. The process is continued until each cluster consists of objects having identical values for all variables. Such clusters cannot be split any more. A final cluster is then a singleton or an indivisible cluster.

12.3   Discussion and Summary

Hierarchical clustering is one of most widely used clustering algorithms in the bioinformatics field, because of the following three reasons: the first is that hierarchical clustering is easy to use and there are very few parameters to set; the second is its low computational load; and the third is that the results of clustering can be visualised in a dendrogram clearly.

Some pioneering works in gene expression analysis by (Eisen *et al.*, 1998) and (Spellman *et al.*, 1998) employed hierarchical clustering and graphically represented the clustered dataset. Since then, clustering has become one of the main exploratory tools in bioinformatics.

Perou and colleagues conducted a study of molecular portraits of human breast cancer tumours (Perou *et al.*, 2000). Gene expression patterns in 65 surgical specimens of human breast tumours from 42 individuals, employing cDNA macroarrays representing 8102 genes, were clustered using hierarchical clustering. This study concluded that the tumours could be classified into subtypes distinguished by pervasive differences in their gene expression patterns. Soon after, Sørlie and colleagues continued the study of classifying breast carcinomas based on variations in gene expression patterns derived from cDNA microarrays, thereby correlating tumour characteristics with clinical outcome (Sørlie *et al.*, 2001). In this study, hierarchical clustering was also employed to analyse a total of 85 cDNA microarray experiments representing 78 cancerous, three fibroadenomas, and four normal breast tissues. The light shed on subtype classification of breast carcinomas by these two studies made it clear that there are four molecular 'intrinsic' subtypes of breast cancer, namely Luminal A, Luminal B, HER2-enriched and Basal-like. As genomic studies evolve, further sub-classification of breast tumours into new molecular entities is expected to occur and a new breast cancer intrinsic subtype, known as Claudin-low, has been identified in human tumours, again by using hierarchical clustering (Herschkowitz *et al.*, 2007).

Other examples include a series of studies of diffuse large B-cell lymphoma (DLBCL), which is the most common subtype of non-Hodgkin's lymphoma (Alizadeh *et al.*, 2000; Shipp *et al.*, 2002; Monti *et al.*, 2005), where hierarchical clustering was used effectively to discovery subtypes of DLBCL. Furthermore, the applications of hierarchical clustering are not restricted to microarray gene expression analysis. In protein-expression profile clustering analysis, hierarchical clustering was also employed to diagnose ovarian carcinomas and borderline tumours (Alaiya *et al.*, 2002) and classify breast cancer subtypes (Abd *et al.*, 2005). In the Chapter 19, we will discuss more details about a number of applications of hierarchical clustering in the bioinformatics field.

References

Abd, D.M., Ball, G., Pinder, S.E. *et al.* (2005). High-throughput protein expression analysis using tissue microarray technology of a large well-characterised series identifies biologically distinct classes of breast cancer confirming recent cDNA expression analyses. *International Journal of Cancer*, **116**(3), pp. 340–350.

Alaiya, A.A., Franzén, B., Hagman, A. *et al.* (2002). Molecular classification of borderline ovarian tumors using hierarchical cluster analysis of protein expression profiles. *International Journal of Cancer*, **98**(6), pp. 895–899.

Alizadeh, A.A., Eisen, M.B., Davis, R.E. *et al.* (2000). Distinct types of diffuse large B-cell lymphoma identified by gene expression profiling. *Nature*, **403**(6769), pp. 503–511.

Cormack, R.M. (1971). A review of classification. *Journal of the Royal Statistical Society. Series A (General)*, **134**, pp. 321–367.

Eisen, M.B., Spellman, P.T., Brown, P.O. and Botstein, D. (1998). Cluster analysis and display of genome-wide expression patterns. *PNAS*, **95**(25), pp. 14863–14868.

Golub, T.R., Slonim, D.K., Tamayo, P. *et al.* (1999). Molecular classification of cancer: class discovery and class prediction by gene expression monitoring. *Science*, **286**(5439), pp. 531–537.

Herschkowitz, J.I., Simin, K., Weigman, V.J. *et al.* (2007). Identification of conserved gene expression features between murine mammary carcinoma models and human breast tumors. *Genome Biology*, **8**(5), R76.

Monti, S., Savage, K.J., Kutok, J.L. *et al.* (2005). Molecular profiling of diffuse large B-cell lymphoma identifies robust subtypes including one characterized by host inflammatory response. *Blood*, **105**(5), pp. 1851–1861.

Perou, C.M., Sørlie, T., Eisen, M.B. *et al.* (2000). Molecular portraits of human breast tumours. *Nature*, **406** (6797), pp. 747–752.

Shipp, M.A., Ross, K.N., Tamayo, P. *et al.* (2002). Diffuse large B-cell lymphoma outcome prediction by gene-expression profiling and supervised machine learning. *Nature Medicine*, **8**(1), pp. 68–74.

Sneath, P.H. and Sokal, R.R. (1973). *Numerical taxonomy. The principles and practice of numerical classification*, W.H. Freeman & Co Ltd, New York.

Sørlie, T., Perou, C.M., Tibshirani, R. *et al.* (2001). Gene expression patterns of breast carcinomas distinguish tumor subclasses with clinical implications. *Proceedings of the National Academy of Sciences*, **98**(19), pp. 10869–10874.

Spellman, P.T., Sherlock, G., Zhang, M.Q. *et al.* (1998). Comprehensive identification of cell cycle-regulated genes of the yeast saccharomyces cerevisiae by microarray hybridization. *Molecular Biology of the Cell*, **9** (12), pp. 3273–3297.

Ward, J.H. (1963). Hierarchical grouping to optimize an objective function. *Journal of the American Statistical Association*, **58**(301), pp. 236–244.

13

Fuzzy Clustering

13.1 Introduction

We have introduced partitional clustering and hierarchical clustering in the preceding chapters. All those clustering algorithms discussed the assignment of data objects into many non-overlapping clusters and assigned each object to one and only one of these clusters. The clustering algorithms which have this property are called 'hard' or 'crisp' clustering algorithms. As mentioned in Chapter 10, hard clustering produces partition matrices with binary entries, that is, one represents the presence of the object in the cluster; zero represents its absence. However, hard clustering often does not always reflect the description of real data, where boundaries between clusters might be fuzzy, and where a more nuanced description of the object's affinity to the specific cluster is required.

The term 'fuzzy logic' emerged with the development of fuzzy set theory by Zadeh (1965). Since its advent, fuzzy logic has been applied to many fields, from control theory to pattern recognition and artificial intelligence. The concept of using fuzzy sets in clustering was proposed by Ruspini (1969, 1970). The convenience of fuzzy clustering over hard clustering was discussed. Assigning each point a degree of 'belongingness' to each cluster provides a way of characterising overlapping objects. Dunn developed fuzzy k-partition algorithms which minimise certain fuzzy extensions of the k-means least-squared-error criterion function (Dunn, 1973, 1974). The investigations of Bezdek (1974, 1976; Bezdek and Dunn, 1975) showed that fuzzy extensions of k-means are superior to ordinary k-means. A parameter $m(m >1)$, which is called the fuzzifier or weighting exponent, was introduced to generalise fuzzy k-partition algorithms (Bezdek, 1976). Eventually, the generalised algorithm was named fuzzy c-means (FCM) (Bezdek, 1981). FCM and its variants have been widely used in various pattern-recognition applications, particularly in biomedical engineering and bioinformatics (Gath and Geva,

Integrative Cluster Analysis in Bioinformatics, First Edition. Basel Abu-Jamous, Rui Fa and Asoke K. Nandi.
© 2015 John Wiley & Sons, Ltd. Published 2015 by John Wiley & Sons, Ltd.

1989; Ahmed *et al.*, 2002; Dougherty *et al.*, 2002; Dembele and Kastner, 2003; Wang *et al.*, 2003; Zhang and Chen, 2004; Tari, Baral and Kim, 2009).

FCM uses the probabilistic constraint that the memberships of a data object across clusters must sum to 1. This constraint came from generalising a crisp *k*-partition of a dataset, and was used to generate the membership-update equations for an iterative algorithm based on the minimisation of a least-squares type of criterion function. The constraint on memberships used in the FCM algorithm is meant to avoid the trivial solution of all memberships being equal to zero, and it does give meaningful results in applications where it is appropriate to interpret memberships as probabilities or degrees of sharing. However, since the memberships generated by this constraint are relative numbers, they are not suitable for applications in which the memberships are supposed to represent 'typicality', or compatibility with an elastic constraint. Krishnapuram and Keller proposed possibilistic *c*-means (PCM) algorithms to restrain the membership value of an object in a cluster such that it represents the typicality of the object in the cluster, or the possibility of the object belonging to the cluster, rather than a relative number (Krishnapuram and Keller, 1993, 1996). Although the PCM algorithms were reported to provide coincident clusters, very likely because no link exists between clusters, some light in respect of the possibilistic algorithms has been shed on fuzzy clustering, illuminating the fact that there are two main types of membership; that is, a relative type, termed probabilistic, and an absolute or possibilistic type, indicating the strength of the attribution to any cluster independent from the rest. Consequently, a number of fuzzy clustering algorithms have been proposed to improve both FCM and PCM by combining them (Pal, Pal and Bezdek, 1997; Timm *et al.*, 2004; Zhang and Leung, 2004; Pal *et al.*, 2005; Masulli and Rovetta, 2006).

There have also been many other fuzzy clustering algorithms in the literature, which combine fuzzy logic with other clustering families; for example, fuzzy agglomerative hierarchical clustering (FAHC) (Horng *et al.*, 2005), fuzzy self-organising map (FSOM) (Tsao, Bezdek and Pal, 1994), fuzzy *c*-shell (FCS) (Dave, 1990, 1992; Klawonn, Kruse and Timm, 1997), and fuzzy adaptive resonance theory (fuzzy ART) (Carpenter, Grossberg and Rosen, 1991). There have also been many comprehensive review papers and books (Baraldi and Blonda, 1999a, 1999b; de Oliveira and Pedrycz, 2007). In this chapter, we will detail some of the most popular fuzzy clustering algorithms.

13.2   Principles

13.2.1   Fuzzy c-Means

FCM was originally a fuzzy version of *k*-means, which assigns each object to one and only one cluster; that is, FCM allows one object to belong to two or more clusters. FCM originated from the fuzzy *k*-partitions and the fuzzy ISODATA algorithm (Dunn, 1973, 1974; Bezdek, 1974, 1976, 1981). FCM can also be considered as generalised squared-error partitional clustering. Formally, a fuzzy cluster model of a given dataset X into K clusters is defined to be optimal when it minimises the objective function given in Equation (13.1),

$$J(X, U) = \sum_{k=1}^{K}\sum_{n=1}^{N} u_{kn}^{m} D(x_n, c_k)^2 \tag{13.1}$$

where the entry of the partition matrix U, u_{kn}, represents the probabilistic membership degree of the nth object in the kth cluster. The centroid c_k is written as given in Equation (13.2).

$$c_k = \frac{\sum_{n=1}^{N} u_{kn}^m x_n}{\sum_{n=1}^{N} u_{kn}^m} \tag{13.2}$$

As mentioned in Chapter 11, in the k-means algorithm, the membership u_{kn} is binary, that is, $u_{kn} \in \{0,1\}$. Therefore, k-means is also called hard or crisp c-means. In fuzzy clustering, the probabilistic membership is relaxed to a real number in [0, 1], which is mathematically given by Equation (13.3),

$$u_{kn} = \frac{1}{\sum_{l=1}^{K} \left(\frac{D(x_n, c_k)^2}{D(x_n, c_l)^2} \right)^{\frac{1}{m-1}}} \tag{13.3}$$

which implies a constraint that the probabilistic memberships satisfy $\sum_{k=1}^{K} u_{kn} = 1$. Such a constraint clearly shows the relative character of the probabilistic membership degree. It depends not only on the distance of the object x_n to cluster k, but also on the distances between this data object to other clusters. The properties of the probabilistic membership degree u_{kn} follow the constraints given in Equation (13.4).

$$u_{kn} \in [0,1];$$
$$0 < \sum_{n=1}^{N} u_{kn} < N; \tag{13.4}$$
$$\sum_{k=1}^{K} u_{kn} = 1$$

It is worth noting that m, termed the fuzzifier, is any real number greater than 1.0. The fuzzifier was proposed to be 2.0 originally to achieve the desired fuzzification of the resulting probabilistic data partition (Dunn, 1973). The generalisation of m to $(1, \infty)$ was proposed by Bezdek (1976). With higher values for m the boundaries between clusters become softer; with lower values they get harder. Usually $m = 2.0$ is chosen.

The implementation of FCM is similar to the iterations of k-means, and is shown in Table 13.1. This iteration will stop when $\max_{kn} \left\{ \left| u_{kn}^{(t+1)} - u_{kn}^{(t)} \right| \right\} < \epsilon$, where ϵ is a small positive

Table 13.1   The procedure of fuzzy c-means

Step 1	Initialise $U^{(0)} = \left\{ u_{kn}^{(0)} \mid k = 1, \cdots, K; n = 1, \cdots, N \right\}$;
Step 2	At the step t, with the partition matrix $U^{(t)}$, calculate the centroids $\{c_k \mid k = 1, \cdots, K\}$ using Equation (13.2);
Step 3	Update the partition matrix $U^{(t+1)}$ using Equation (13.3);
Step 4	Repeat Steps 2 and 3 until $\max_{kn} \left\{ \left| u_{kn}^{(t+1)} - u_{kn}^{(t)} \right| \right\} < \epsilon$.

real number and t is the iteration index. There are many implementations of FCM in different platforms: in Fuzzy Logic Toolbox of MATLAB, function *fcm* applies the FCM method; in the package {e1071} in R, function ***cmeans*** also implements the FCM method.

13.2.2   Probabilistic c-Means

As discussed previously, FCM uses the probabilistic constraint that the memberships of a data object across clusters must sum to 1. This constraint is meant to avoid the trivial solution of all memberships being equal to 0. However, on the other hand, such constraint makes the memberships relative numbers. Although FCM has been shown to be advantageous to crisp clustering, its relative character of the probabilistic membership degrees is not desirable, and sometimes can be misleading. For example, it is not always the case that a fairly high value for the membership of an object in one cluster can lead to the impression that the object is typical for the cluster. Therefore, they are not suitable for applications in which the memberships are supposed to represent 'typicality', or compatibility with an elastic constraint.

Krishnapuram and Keller proposed PCM algorithms to restrain the clustering such that the membership value of an object in a cluster represents the typicality of the object in the cluster, that is, the possibility of the object belonging to the cluster, rather than a relative number (Krishnapuram and Keller, 1993, 1996). The membership degree u_{kn} of PCM is similar to that shown in Equation (13.4) of FCM, except that the constraint $\sum_{k=1}^{K} u_{kn} = 1$ is replaced by $\sum_{k=1}^{K} u_{kn} \leq 1$. After dropping this normalisation constraint, in order to avoid this trivial solution, a penalty term, which forces the membership degrees away from zero, is introduced into the objective function. Mathematically the objective function is written as in Equation (13.5) (Krishnapuram and Keller, 1996),

$$J(X, U) = \sum_{k=1}^{K} \sum_{n=1}^{N} u_{kn}^{m} D(x_n, c_k)^2 + \sum_{k=1}^{K} \eta_k \sum_{n=1}^{N} (1 - u_{kn})^m \tag{13.5}$$

where $\eta_k > 0$ and this cluster-specific constant is used to balance the contrary objectives expressed in the two terms of the objective function. The first term leads to a minimisation of the weighted distances, but it does not work when all memberships are zeros. The second term suppresses the trivial solution because it penalises the zero memberships by making $(1 - u_{kn})^m$ become one; in the meantime, it rewards high memberships (close to 1) that make $(1 - u_{kn})^m$ become approximately 0. Therefore, updating the membership degrees that are derived from Equation (13.5) by setting the derivative of $J(X, U)$ to zero leads to Equation (13.6).

$$u_{kn} = \frac{1}{1 + \left(\frac{D(x_n, c_k)^2}{\eta_k}\right)^{\frac{1}{m-1}}} \tag{13.6}$$

Obviously, the membership degree of an object in a cluster clearly depends only on the distance between the object and the cluster. The parameters, $\eta_k | k = 1, \ldots, K$, are crucial to the algorithm. Depending on the cluster's shape the parameters η_k have different geometrical interpretations.

Mathematically, they are expressed as in Equation (13.7) (Krishnapuram and Keller, 1993), where η_k can be either fixed or varied in each iteration.

$$\eta_k = \frac{\sum_{n=1}^{N} u_{kn}^m D(x_n, c_k)^2}{\sum_{n=1}^{N} u_{kn}^m} \qquad (13.7)$$

However, the PCM algorithms were reported to provide coincident clusters, very likely because no links exist between clusters (Barni, Cappellini and Mecocci, 1996). To address this issue, Krishnapuram and Keller proposed an improved PCM algorithm (Krishnapuram and Keller, 1996). We call this algorithm PCM2 to differentiate it from the original PCM using hybrid c-means. The objective function of PCM2 is given by Equation (13.8)

$$J(X, U) = \sum_{k=1}^{K} \sum_{n=1}^{N} u_{kn}^m D(x_n, c_k)^2 + \sum_{k=1}^{K} \eta_k \sum_{n=1}^{N} (u_{kn} \log u_{kn} - u_{kn}) \qquad (13.8)$$

and then the update equation for u_{kn} is written as in Equation (13.9).

$$u_{kn} = \exp\left\{ -\frac{D(x_n, c_k)^2}{\eta_k} \right\} \qquad (13.9)$$

13.2.3   Hybrid c-Means

Although the PCM algorithm has a limitation of producing coincident clusters, it has shed its light on fuzzy clustering, illuminating a fact that there are two main types of membership, that is, a relative type, termed probabilistic, and an absolute or possibilistic type, indicating the strength of the attribution to any cluster independent from the rest. To overcome the coincident-cluster problem of PCM there have been many attempts by using hybrid c-means methods, essentially combining probabilistic and possibilistic c-means algorithms.

The first attempt was done by Pal and colleagues, and the algorithm is called fuzzy-possibilistic c-means (FPCM), also known as the Pal–Pal–Bezdek algorithm (Pal, Pal and Bezdek, 1997). FPCM simultaneously produces both probabilistic and possibilistic membership (partition) matrices, for the sake of clarity, expressed by U^f and U^p, respectively. Thus, it leads to a mixed fuzzy-possibilistic c-means model, mathematically given by Equation (13.10),

$$J_{\text{FPCM}}\left(X, U^{(f)}, U^{(p)}\right) = \sum_{k=1}^{K} \sum_{n=1}^{N} \left(\left(u_{kn}^{(f)} \right)^m + \left(u_{kn}^{(p)} \right)^\gamma \right) D(x_n, c_k)^2 \qquad (13.10)$$

subject to the constrains $m > 1$, $\gamma > 1$, $0 \le u_{kn}^{(f)}, u_{kn}^{(p)} \le 1$, $\sum_{k=1}^{K} u_{kn}^{(f)} = 1$, and $\sum_{n=1}^{N} u_{kn}^{(p)} = 1$. The last constraint requires the normalisation over all objects in one cluster. To minimise the mixed-objective function, the update equations for $u_{kn}^{(f)}$ and $u_{kn}^{(p)}$ are written as shown in Equation (13.11),

$$u_{kn}^{(f)} = \left(\sum_{j=1}^{K} \left(\frac{D(x_n, c_k)}{D(x_n, c_j)} \right)^{\frac{2}{m-1}} \right)^{-1}, u_{kn}^{(p)} = \left(\sum_{j=1}^{N} \left(\frac{D(x_n, c_k)}{D(x_j, c_k)} \right)^{\frac{2}{\gamma-1}} \right)^{-1} \tag{13.11}$$

and the centroid c_k is updated as given by Equation (13.12).

$$c_k = \frac{\sum_{n=1}^{N} \left(\left(u_{kn}^{(f)} \right)^m + \left(u_{kn}^{(p)} \right)^\gamma \right) x_n}{\sum_{n=1}^{N} \left(\left(u_{kn}^{(f)} \right)^m + \left(u_{kn}^{(p)} \right)^\gamma \right)} \tag{13.12}$$

Pal and colleagues further improved their FPCM algorithm by relaxing the constraint of normalising the possibilistic memberships over all objects in one cluster (Pal *et al.*, 2005). The objective function of the improved FPCM is mathematically given by Equation (13.13),

$$J\left(X, U^{(f)}, U^{(p)}\right) = \sum_{k=1}^{K} \sum_{n=1}^{N} \left(a \left(u_{kn}^{(f)} \right)^m + b \left(u_{kn}^{(p)} \right)^\gamma \right) D(x_n, c_k)^2 + \sum_{k=1}^{K} \eta_k \sum_{n=1}^{N} \left(1 - u_{kn}^{(p)} \right)^\gamma \tag{13.13}$$

subject to the constrains $m > 1$, $\gamma > 1$, $0 \le u_{kn}^{(f)}, u_{kn}^{(p)} \le 1$, $\sum_{k=1}^{K} u_{kn}^{(f)} = 1$, $a > 0$ and $b > 0$. The constants a and b define the relative importance of fuzzy membership and typicality values in the objective function. Therefore, the update equations for $u_{kn}^{(f)}$ and $u_{kn}^{(p)}$ in the improved objective function are written as shown in Equation (13.14),

$$u_{kn}^{(f)} = \left(\sum_{j=1}^{K} \left(\frac{D(x_n, c_k)}{D(x_n, c_j)} \right)^{\frac{2}{m-1}} \right)^{-1}, u_{kn}^{(p)} = \frac{1}{1 + \left(\frac{b \cdot D(x_n, c_k)^2}{\eta_k} \right)^{\frac{1}{\gamma-1}}} \tag{13.14}$$

and the centroid c_k is updated as given by Equation (13.15).

$$c_k = \frac{\sum_{n=1}^{N} \left(a \cdot \left(u_{kn}^{(f)} \right)^m + b \cdot \left(u_{kn}^{(p)} \right)^\gamma \right) x_n}{\sum_{n=1}^{N} \left(a \cdot \left(u_{kn}^{(f)} \right)^m + b \cdot \left(u_{kn}^{(p)} \right)^\gamma \right)} \tag{13.15}$$

Zhang and Leung have also proposed two hybrid c-means algorithms using both probabilistic and possibilistic membership degrees (Zhang and Leung, 2004). The two hybrid c-means algorithms improve the objective functions Equations (13.5) and (13.8), proposed in Krishnapuram and Keller (1993) and Krishnapuram and Keller (1996), respectively. For the sake of clarity, the improved version of Equation (13.5) is called *IPCM1,* and the improved version of Equation (13.8) is called *IPCM2*. The objective function of IPCM1 is mathematically expressed as given in Equation (13.16).

$$J_{\text{IPCM1}}\left(X, U^{(f)}, U^{(p)}\right) = \sum_{k=1}^{K} \sum_{n=1}^{N} \left(u_{kn}^{(f)} \right)^m \left(u_{kn}^{(p)} \right)^\gamma D(x_n, c_k)^2 + \sum_{k=1}^{K} \eta_k \sum_{n=1}^{N} \left(u_{kn}^{(f)} \right)^m \left(1 - u_{kn}^{(p)} \right)^\gamma$$

$$\tag{13.16}$$

To minimise J_{IPCM1}, the update equations for $u_{kn}^{(f)}$ and $u_{kn}^{(p)}$ in the improved objective function are written as shown in Equation (13.17),

$$u_{kn}^{(f)} = \sum_{j=1}^{K} \left(\frac{\left(u_{kn}^{(p)}\right)^{\frac{\gamma-1}{2}} D(\boldsymbol{x}_n, \boldsymbol{c}_k)}{\left(u_{jn}^{(p)}\right)^{\frac{\gamma-1}{2}} D(\boldsymbol{x}_n, \boldsymbol{c}_j)} \right)^{-\frac{2}{m-1}} , u_{kn}^{(p)} = \frac{1}{1 + \left(\frac{D(\boldsymbol{x}_n, \boldsymbol{c}_k)^2}{\eta_k} \right)^{\frac{1}{\gamma-1}}} \tag{13.17}$$

and the centroid $\boldsymbol{c}_k$ is updated as given by Equation (13.18).

$$\boldsymbol{c}_k = \frac{\sum_{n=1}^{N} \left(u_{kn}^{(f)}\right)^m \left(u_{kn}^{(p)}\right)^{\gamma} \boldsymbol{x}_n}{\sum_{n=1}^{N} \left(u_{kn}^{(f)}\right)^m \left(u_{kn}^{(p)}\right)^{\gamma}} \tag{13.18}$$

The scale parameter η_k is generalised in order to incorporate both the possibilistic memberships and fuzzy memberships, as shown in Equation (13.19).

$$\eta_k = \frac{\sum_{n=1}^{N} \left(u_{kn}^{(f)}\right)^m \left(u_{kn}^{(p)}\right)^{\gamma} D(\boldsymbol{x}_n, \boldsymbol{c}_k)^2}{\sum_{n=1}^{N} \left(u_{kn}^{(f)}\right)^m \left(u_{kn}^{(p)}\right)^{\gamma}} \tag{13.19}$$

The objective function of IPCM2 is mathematically expressed as given in Equation (13.20)

$$\begin{aligned}
J_{\text{IPCM2}}\left(\boldsymbol{X}, \boldsymbol{U}^{(f)}, \boldsymbol{U}^{(p)}\right) &= \sum_{k=1}^{K} \sum_{n=1}^{N} \left(u_{kn}^{(f)}\right)^m \left(u_{kn}^{(p)}\right) D(\boldsymbol{x}_n, \boldsymbol{c}_k)^2 \\
&\quad + \sum_{k=1}^{K} \eta_k \sum_{n=1}^{N} \left(u_{kn}^{(f)}\right)^m \left(u_{kn}^{(p)} \log u_{kn}^{(p)} - u_{kn}^{(p)} + 1\right)
\end{aligned} \tag{13.20}$$

To minimise J_{IPCM2}, the update equations for $u_{kn}^{(f)}$ and $u_{kn}^{(p)}$ in the improved objective function are written as shown in Equation (13.21),

$$u_{kn}^{(f)} = \sum_{j=1}^{K} \left(\frac{\eta_k \left(1 - e^{-D(\boldsymbol{x}_n, \boldsymbol{c}_k)^2/\eta_k}\right)}{\eta_j \left(1 - e^{-D(\boldsymbol{x}_n, \boldsymbol{c}_j)^2/\eta_j}\right)} \right)^{-\frac{2}{m-1}} , u_{kn}^{(p)} = e^{-\frac{D(\boldsymbol{x}_n, \boldsymbol{c}_k)^2}{\eta_k}} \tag{13.21}$$

and the centroid $\boldsymbol{c}_k$ is updated as given by Equation (13.22).

$$\boldsymbol{c}_k = \frac{\sum_{n=1}^{N} \left(u_{kn}^{(f)}\right)^m \left(u_{kn}^{(p)}\right) \boldsymbol{x}_n}{\sum_{n=1}^{N} \left(u_{kn}^{(f)}\right)^m \left(u_{kn}^{(p)}\right)} \tag{13.22}$$

13.2.4   Gustafson–Kessel Algorithm

Gustafson and Kessel developed a fuzzy clustering algorithm, which replaces the Euclidean distance by the Mahalanobis distance, to consider cluster-specific geometrical shapes (Gustafson and Kessel, 1978). The Gustafson–Kessel (GK) algorithm is designed for iteratively minimising the objective function, which is expressed as given in Equation (13.23),

$$J_{GK}(X, U, A) = \sum_{k=1}^{K}\sum_{n=1}^{N} u_{kn}^{m} D(x_n, c_k, A_k)^2 \tag{13.23}$$

where the distance metric $D(x_n, c_k, A_k)$ is the Mahalanobis distance, which is written as shown in Equation (13.24),

$$D(x_n, c_k, A_k) = \sqrt{(x_n - c_k)^T A_k (x_n - c_k)} \tag{13.24}$$

where A_k is obtained by application of Equation (13.25),

$$A_k = \left(\rho_k |\Sigma_k|\right)^{1/M} \Sigma_k^{-1} \tag{13.25}$$

where $\rho_k = \det(\Sigma_k)$ is the size of the cluster and Σ_k is the fuzzy covariance matrix of the kth cluster, given by Equation (13.26).

$$\Sigma_k = \frac{\sum_{n=1}^{N} u_{kn}^{m} (x_n - c_k)(x_n - c_k)^T}{\sum_{n=1}^{N} u_{kn}^{m}} \tag{13.26}$$

The rest of the GK algorithm is the same as that for FCM. The advantage of GK over FCM is that its geometrical shape of the clusters is not restricted to a hyper-spherical shape like FCM. However, the GK algorithm exhibits higher computational demands due to the matrix inversions. Moreover, the GK algorithm may suffer from some numerical problems also because of the matrix inversions.

In order to address these issues, Babuška and colleagues improved the GK algorithm to be an adaptive clustering algorithm (Babuška, van der Veen and Kaymak, 2002). The improved GK algorithm employs the new covariance matrix update as given in Equation (13.27),

$$\Sigma_k' = (1 - \beta)\Sigma_k + \beta \det(\Sigma_0)^{1/n} I \tag{13.27}$$

where I is an identity matrix with the dimension of M, and $\beta \in [0,1]$ is a tuning parameter.

13.2.5   Gath–Geva Algorithm

The Gath–Geva (GG) algorithm is a fuzzy maximum-likelihood estimate clustering algorithm proposed by Gath and Geva (1989). The GG algorithm employs the probability as distance metric, expressed as shown in Equation (13.28),

$$D(x_n, c_k, \Sigma_k) = \frac{(2\pi)^{M/2}[\det(\Sigma_k)]^{1/2}}{\alpha_k} \exp\{(x_n - c_k)^T \Sigma_k^{-1}(x_n - c_k)/2\} \qquad (13.28)$$

where $\alpha_k = \frac{1}{N}\sum_{n=1}^{N} u_{kn}$. The update equations for u_{kn} are written as Equation (13.29).

$$u_{kn} = \frac{1/D(x_n, c_k, \Sigma_k)}{\sum_{j=1}^{K} 1/D(x_n, c_j, \Sigma_j)} \qquad (13.29)$$

The fuzzy covariance matrix Σ_k is expressed identically as in Equation (13.26).

13.2.6 Fuzzy c-Shell

Up to now, all clustering algorithms, which we discussed above, search for solid interior clusters. They are also called squared-error clustering algorithms, where the sum of weighted distances of the objects from the cluster prototypes is minimised. However, for characterisation and detection of clusters with hollow interiors, the prototype definition has to be changed (Dave, 1990, 1992). There have been a number of algorithms proposed to employ elliptical surfaces as prototypes (Dave, 1990, 1992; Dave and Bhaswan, 1992; Frigui and Krishnapuram, 1996; Klawonn, Kruse and Timm, 1997). These elliptical surfaces are called shells, and these algorithms are categorised as FCS algorithms (hard algorithms are viewed as special cases of fuzzy algorithms). Here, we only sketch the main idea of FCS rather than all the details of FCS and its variants.

First of all, the elliptical shell prototype is defined as given in Equation (13.30),

$$P(c, A, r) = \{x | (x - c)^T A(x - c) = r^2\} \qquad (13.30)$$

where r is the radius of the shell and A is an $M \times M$ positive definite matrix. The distance D_{kn} of the nth object x_n from the kth prototype $P(c_k, A_k, r_k)$ is given by Equation (13.31).

$$D_{kn}^2 = \left\{ \left[(x_n - c_k)^T A_k(x_n - c_k)\right]^{1/2} - r_k \right\}^2 \qquad (13.31)$$

Therefore, FCS is designed to minimise the objective function given by Equation (13.32).

$$J_{FCS} = \sum_{k=1}^{K}\sum_{n=1}^{N} u_{kn}^m D_{kn}^2, \text{ subject to } \sum_{k=1}^{K} u_{kn} = 1 \qquad (13.32)$$

The update equation for u_{km} is written as Equation (13.33).

$$u_{kn} = \frac{1}{\sum_{j=1}^{K} \left[\frac{D_{kn}^2}{D_{jn}^2}\right]^{\frac{1}{m-1}}} \qquad (13.33)$$

The update equation for A_k is given by Equation (13.34),

$$A_k = \left(\rho_k |\mathbf{\Sigma}_k|\right)^{1/M} \mathbf{\Sigma}_k^{-1} \tag{13.34}$$

where Equation (13.35) holds

$$\mathbf{\Sigma}_k = \sum_{n=1}^{N} u_{kn}^m \frac{D_{kn}^2}{d_{kn}^2} (\mathbf{x}_n - \mathbf{c}_k)(\mathbf{x}_n - \mathbf{c}_k)^T, d_{kn}^2 = (\mathbf{x}_n - \mathbf{c}_k)^T \mathbf{\Sigma}_k^{-1}(\mathbf{x}_n - \mathbf{c}_k) \tag{13.35}$$

13.2.7 FANNY

In fuzzy clustering, each object is assigned to various clusters and the degree of belongingness of an object to different clusters is quantified by means of membership coefficients, which range from 0 to 1, with the stipulation that the sum of their values is one. This is called a fuzzification of the cluster configuration. It has the advantage that it does not force every object into a specific cluster. It has the disadvantage that there is much more information to be interpreted. Fuzzy analysis (FANNY) aims at the minimisation of the objective function shown in Equation (13.36) (Kaufman and Rousseeuw, 1990),

$$J_{\text{FANNY}} = \sum_{k=1}^{K} \frac{\sum_{i,j=1}^{N} u_{ik}^2 u_{jk}^2 d(i,j)}{2 \sum_{j=1}^{N} u_{jk}^2} \tag{13.36}$$

where $d(i,j)$ represents the distance between data objects i and j, u_{ik} is the unknown membership of object i to the cluster k. The membership functions are subject to the following constraints:

1. $u_{ik} \geq 0$ for all $i = 1, \ldots, N$ and all $k = 1, \ldots, K$;
2. $\sum_{k=1}^{K} u_{ik} = 1$, for all $i = 1, \ldots, N$.

These constraints imply that membership cannot be negative and that each object has a certain total membership distributed over different clusters. By convention, this total membership is normalised to 1. The objective function is minimised numerically by means of an iterative algorithm, taking into account the above constraints. When each object has equal membership in all clusters, the clustering is entirely fuzzy. On the other hand, when each object has a membership of one in some cluster and zero membership in all other clusters; the clustering is entirely hard.

13.2.8 Other Fuzzy Clustering Algorithms

There are also some other fuzzy clustering algorithms in the clustering literature. For example, fuzzy Kohonen clustering networks (FKCN), also known as FSOM, was proposed by Tsao,

Table 13.2 Summary of the publicly accessible resources of fuzzy clustering algorithms

Algorithm Name	Platform	Package	Function
Fuzzy c-means	R/MATLAB	e1071	cmeans/fcm
FANNY	R	Cluster	Fanny
Fuzzy c-shell	R	e1071	cshell
Fuzzy cluster indices	R	e1071	fclustIndex

Bezdek and Pal (1994), because KCN (Kohonen clustering network) suffered from several major problems. FKCN combined the ideas of fuzzy membership values for learning rates, the parallelism of fuzzy c-means, and the structure and update rules of KCNs. Another example is that Horng *et al.* proposed a FAHC algorithm for clustering documents (Horng *et al.*, 2005). Horng *et al.* constructed fuzzy logic rules based on the document clusters and the centres of these document clusters. The constructed fuzzy logic rules were applied to modify the users' query for query expansion and to guide the information-retrieval system to retrieve documents relevant to the user's request. The proposed document-retrieval method was implemented based on FAHC.

13.3 Discussion

Fuzzy logic theory is one of most important computational theories in the twentieth century. The concept of fuzzy logic and fuzzy set has been applied to many fields, from control theory to artificial intelligence, then to bioinformatics. In this chapter, we introduced the basic principle of fuzzy logic, together with fuzzy clustering algorithms, which applied fuzzy logic to perform soft clustering. The fuzzy clustering algorithms, which we discussed, include fuzzy c-means (Dunn, 1973, 1974; Bezdek, 1974, 1976, 1981), possibilistic c-means (Krishnapuram and Keller, 1993, 1996), hybrid c-means (Krishnapuram and Keller, 1993, 1996; Pal, Pal and Bezdek, 1997; Pal *et al.*, 2005), the GK algorithm (Gustafson and Kessel, 1978; Babuška, van der Veen and Kaymak, 2002), the GG algorithm (Gath and Geva, 1989), FCS (Dave, 1990, 1992; Dave and Bhaswan, 1992; Frigui and Krishnapuram, 1996; Klawonn, Kruse and Timm, 1997), and FANNY (Kaufman and Rousseeuw, 1990). The algorithms whose implementations are publicly available are listed in Table 13.2.

There are also many applications of fuzzy clustering in the bioinformatics field, for example DNA motifs clustering (Pickert *et al.*, 1998), microarray gene expression analysis (Datta and Datta, 2003; Dembele and Kastner, 2003; Wang *et al.*, 2003; Kim, Lee and Bae, 2006; Bandyopadhyay, Mukhopadhyay and Maulik, 2007). We will discuss more details about some applications in Chapter 19.

References

Ahmed, M.N., Yamany, S.M., Mohamed, N. *et al.* (2002). A modified fuzzy c-means algorithm for bias field estimation and segmentation of MRI data. *IEEE Transactions on Medical Imaging,* **21**(3), pp. 193–199.

Babuška, R., van der Veen, P.J. and Kaymak, U. (2002). Improved Covariance Estimation for Gustafson-Kessel Clustering. Proceedings of the 2002 IEEE International Conference on Fuzzy Systems, 12–17th May 2002. FUZZ-IEEE'02, Honolulu, HI. pp. 1081–1085.

Bandyopadhyay, S., Mukhopadhyay, A. and Maulik, U. (2007). An improved algorithm for clustering gene expression data. *Bioinformatics,* **23**(21), pp. 2859–2865.

Baraldi, A. and Blonda, P. (1999a). A survey of fuzzy clustering algorithms for pattern recognition. I. *Systems, Man, and Cybernetics, Part B: IEEE Transactions on Cybernetics,* **29**(6), pp. 778–785.

Baraldi, A. and Blonda, P. (1999b). A survey of fuzzy clustering algorithms for pattern recognition. II. *Systems, Man, and Cybernetics, Part B: IEEE Transactions on Cybernetics,* **29**(6), pp. 786–801.

Barni, M., Cappellini, V. and Mecocci, A. (1996). Comments on 'A possibilistic approach to clustering'. *IEEE Transactions on Fuzzy Systems,* **4**(3), pp. 393–396.

Bezdek, J.C. (1974). Numerical taxonomy with fuzzy sets. *Journal of Mathematical Biology,* **1**(1), pp. 57–71.

Bezdek, J.C. (1976). A physical interpretation of fuzzy ISODATA. *IEEE Transactions on Systems, Man and Cybernetics,* **SMC-6**(5), pp. 387–389.

Bezdek, J.C. (1981). *Pattern Recognition with Fuzzy Objective Function Algorithms,* Kluwer Academic Publishers, Alphen aan den Rijn.

Bezdek, J.C. and Dunn, J.C. (1975). Optimal fuzzy partitions: a heuristic for estimating the parameters in a mixture of normal distributions. *IEEE Transactions on Computers,* **100**(8), pp. 835–838.

Carpenter, G.A., Grossberg, S. and Rosen, D.B. (1991). Fuzzy ART: fast stable learning and categorization of analog patterns by an adaptive resonance system. *Neural Networks,* **4**(6), pp. 759–771.

Datta, S. and Datta, S. (2003). Comparisons and validation of statistical clustering techniques for microarray gene expression data. *Bioinformatics,* **19**(4), pp. 459–466.

Dave, R.N. (1990). Fuzzy shell-clustering and applications to circle detection in digital images. *International Journal of General Systems,* **16**(4), pp. 343–355.

Dave, R.N. (1992). Generalized fuzzy c-shells clustering and detection of circular and elliptical boundaries. *Pattern Recognition,* **25**(7), pp. 713–721.

Dave, R.N. and Bhaswan, K. (1992). Adaptive fuzzy c-shells clustering and detection of ellipses. *IEEE Transactions on Neural Networks,* **3**(5), pp. 643–662.

de Oliveira, J.V. and Pedrycz, W. (eds) (2007). *Advances in Fuzzy Clustering and its Applications,* Wiley Online Library, UK.

Dembele, D. and Kastner, P. (2003). Fuzzy C-means method for clustering microarray data. *Bioinformatics,* **19**(8), pp. 973–980.

Dougherty, E.R., Barrera, J., Brun, M. *et al.* (2002). Inference from clustering with application to gene-expression microarrays. *Journal of Computational Biology,* **9**(1), pp. 105–126.

Dunn, J.C. (1973). A fuzzy relative of the ISODATA process and its use in detecting compact well-separated clusters. *Cybernetics and Systems,* **3**(3), 32–57

Dunn, J.C. (1974). Well-separated clusters and optimal fuzzy partitions. *Journal of Cybernetics,* **4**(1), pp. 95–104.

Frigui, H. and Krishnapuram, R. (1996). A comparison of fuzzy shell-clustering methods for the detection of ellipses. *IEEE Transactions on Fuzzy Systems,* **4**(2), pp. 193–199.

Gath, I. and Geva, A.B. (1989). Unsupervised optimal fuzzy clustering. *IEEE Transactions on Pattern Analysis and Machine Intelligence,* **11**(7), pp. 773–780.

Gustafson, D.E. and Kessel, W.C. (1978). Fuzzy Clustering with a Fuzzy Covariance Matrix. 1978 IEEE Conference on Decision and Control including the 17th Symposium on Adaptive Processes, 10–12th January 1979, IEEE, San Diego, CA. pp. 761–766.

Horng, Y.-J., Chen, S.-M., Chang, Y.-C. and Lee, C.-H. (2005). A new method for fuzzy information retrieval based on fuzzy hierarchical clustering and fuzzy inference techniques., *IEEE Transactions on Fuzzy Systems,* **13**(2), pp. 216–228.

Kaufman, L. and Rousseeuw, P.J. (1990). *Finding Groups in Data: an Introduction to Cluster Analysis,* John Wiley & Sons, Inc, New York.

Kim, S.Y., Lee, J.W. and Bae, J.S. (2006). Effect of data normalization on fuzzy clustering of DNA microarray data. *BMC Bioinformatics,* **7**(1), p. 134.

Klawonn, F., Kruse, R., and Timm, H. (1997). Fuzzy shell cluster analysis. *Courses and Lectures - International Centre for Mechanical Sciences,* **382,** pp. 105–120.

Krishnapuram, R. and Keller, J.M. (1993). A possibilistic approach to clustering. *IEEE Transactions on Fuzzy Systems,* **1**(2), pp. 98–110.

Krishnapuram, R. and Keller, J.M. (1996). The possibilistic c-means algorithm: insights and recommendations. *IEEE Transactions on Fuzzy Systems,* **4**(3), pp. 385–393.

Masulli, F. and Rovetta, S. (2006). Soft transition from probabilistic to possibilistic fuzzy clustering. *IEEE Transactions on Fuzzy Systems,* **14**(4), pp. 516–527.

Pal, N.R., Pal, K. and Bezdek, J.C. (1997). A Mixed C-means Clustering Model. Proceedings of the Sixth IEEE International Conference on Fuzzy Systems, 1997, Barcelona, Spain. pp. 11–21.

Pal, N.R., Pal, K., Keller, J.M. and Bezdek, J.C. (2005). A possibilistic fuzzy c-means clustering algorithm. *IEEE Transactions on Fuzzy Systems,* **13**(4), pp. 517–530.

Pickert, L., Reuter, I., Klawonn, F. and Wingender, E. (1998). Transcription regulatory region analysis using signal detection and fuzzy clustering. *Bioinformatics,* **14**(3), pp. 244–251.

Ruspini, E.H. (1969). A new approach to clustering. *Information and Control,* **15**(1), pp. 22–32.

Ruspini, E.H. (1970). Numerical methods for fuzzy clustering. *Information Sciences,* **2**(3), pp. 319–350.

Tari, L., Baral, C. and Kim, S. (2009). Fuzzy c-means clustering with prior biological knowledge. *Journal of Biomedical Informatics,* **42**(1), pp. 74–81.

Timm, H., Borgelt, C., Döring, C. and Kruse, R. (2004). An extension to possibilistic fuzzy cluster analysis. *Fuzzy Sets and Systems,* **147**(1), pp. 3–16.

Tsao, E.C.-K., Bezdek, J.C. and Pal, N.R. (1994). Fuzzy kohonen clustering networks. *Pattern Recognition,* **27** (5), pp. 757–764.

Wang, J., Bø, T.H., Jonassen, I. *et al.* (2003). Tumor classification and marker gene prediction by feature selection and fuzzy c-means clustering using microarray data. *BMC Bioinformatics,* **4**(1), p. 60.

Zadeh, L.A. (1965). Fuzzy sets. *Information and Control,* **8**(3), pp. 338–353.

Zhang, D.-Q. and Chen, S.-C. (2004). A novel kernelized fuzzy c-means algorithm with application in medical image segmentation. *Artificial Intelligence in Medicine,* **32**(1), pp. 37–50.

Zhang, J.-S. and Leung, Y.-W. (2004). Improved possibilistic c-means clustering algorithms. *IEEE Transactions on Fuzzy Systems,* **12**(2), pp. 209–217.

14

Neural Network-based Clustering

14.1 Introduction

Clustering in the neural network literature is generally based on competitive learning (CL) model, which originated in the early 1970s through contributions of Christoph von der Malsburg (Malsburg, 1973). CL, virtually, is a neural network learning process where different neurons or processing elements compete on who is allowed to learn to represent the current input. In a CL model, a stream of input patterns to a network F_1 can train the adaptive weights that multiply the signals in the pathways from F_1 to a coding level F_2. Level F_2 is designed as a competitive network capable of choosing the node (or nodes) which receive(s) the largest total input. The winning population then triggers associative pattern learning to update the adaptive weights.

Clustering algorithms implemented by distortion-based CL techniques commonly have the prototypes corresponding to the weights of neurons, for example, the centre of their receptive field in the input feature space. A common feature of the CL clustering algorithms is a competitive stage which precedes each learning step and decides to what extent a neuron may adapt its weights to a new input pattern. The goal of CL is the minimisation of the distortion or quantisation error in vector quantisation.

Kohonen made particularly strong implementation of CL in his work on learning vector quantisation (LVQ) and self-organising maps (SOM) – also known as self-organising feature maps (SOFM) (Kohonen, 1990). Intrinsically, LVQ performs supervised learning, and is not categorised as a clustering algorithm. Nevertheless, its learning properties provide an insight to describe the potential data structure using the prototype vectors in the competitive layer. It inspires many other CL clustering algorithms, such as generalised LVQ (GLVQ) by Pal, Bezdek and Tsao (1993), Neural-gas network by Martinetz, Berkovich and Schulten (1993), and so on. SOM is one of the most popular clustering algorithms, ideally suited to exploratory

Integrative Cluster Analysis in Bioinformatics, First Edition. Basel Abu-Jamous, Rui Fa and Asoke K. Nandi.
© 2015 John Wiley & Sons, Ltd. Published 2015 by John Wiley & Sons, Ltd.

data analysis, allowing one to impose partial structure on the clusters and facilitating easy visualisation and interpretation. However, SOM also suffers the problems caused by a number of user-predefined parameters, for example the size of the lattice, the number of clusters, and so on. Additionally, trained SOM may suffer from input space density misrepresentation, where areas of low pattern density may be over-represented and areas of high density under-represented (Kohonen, 2001).

Adaptive resonance theory (ART) is another learning model growing out of the CL model, which was first introduced by Grossberg (1980) and developed by Carpenter and Grossberg (1987a, b, 1988, 1990; Carpenter, Grossberg and Reynolds, 1991; Carpenter, Grossberg and Rosen 1991a,b). ART is a cognitive and neural theory of how the brain quickly learns to categorise, recognise, and predict objects and events in a changing world. One of the key computational ideas within ART is that top-down learned expectations focus attention upon bottom-up information in a way that protects previously learned memories from being washed away by new learning, and enables new learning to be automatically incorporated into the total knowledge base of the system in a globally self-consistent way.

Most prototype-based CL algorithms employ either the winner-take-all (WTA) paradigm or the winner-take-most (WTM) paradigm. The major issue with the WTA paradigm is the possible existence of dead nodes. In such cases, some prototypes can never become a winner because of inappropriate initialisation, and therefore they have no contribution to learning. WTM decreases the dependency on the initialisation of prototype locations; however, an undesirable side effect is that since all prototypes are attracted to each input pattern, some of them are detracted from their corresponding clusters. Zhang and Liu proposed a self-splitting competitive learning (SSCL) clustering based on a distinct paradigm: one-prototype-take-one-cluster (OPTOC) (Zhang and Liu, 2002).

Aforementioned prototype-based algorithms have the advantage of being able to incorporate knowledge about the global shape or size of clusters by using appropriate prototypes and distance measures in the objective function. However, these algorithms suffer from three major drawbacks, namely, the difficulty in determining the number of clusters, the sensitivity to noise and outliers, and the sensitivity to initialisation. Rhouma and Frigui developed a self-organisation of pulse-coupled oscillators algorithm, called self-organising oscillator networks (SOON) (Rhouma and Frigui, 2001), which was used to analyse microarray gene expression data later by Salem, Jack and Nandi (2008). The SOON algorithm has its root in a biological process with an interesting physical characteristic: Fireflies flash at random when considered by themselves; however, they exhibit the characteristic of firing together when in groups that are physically close to each other. The objective of this chapter is to introduce these neural network-based algorithms.

14.2 Algorithms

14.2.1 SOM

SOM belongs to one of the categories of neural network architectures, where neighbouring cells in a neural network compete in their activities by means of mutual lateral interactions, and develop adaptively into specific detectors of different signal patterns. The objective of SOM is to represent high-dimensional input patterns with prototype vectors that can be visualised in a usually two-dimensional lattice structure.

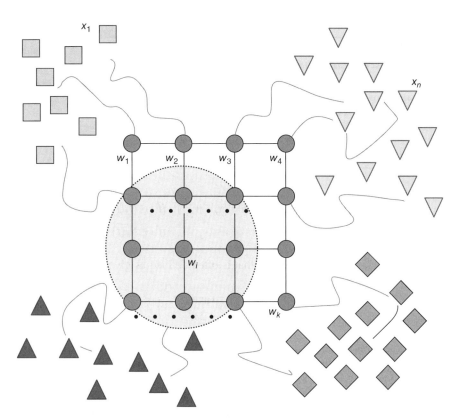

Figure 14.1 Demonstration of the strategy of SOM. Each node on the grid represents a neuron with a weight vector $w_i, i = 1, ..., K$. Only the winning neuron, for example w_i and its surrounding neurons, which are in the dotted circle, have the opportunity to update

Consider a 4×4 grid SOM depicted in Figure 14.1. Note that the arrangement of the grid is not necessary to be rectangular; it can be hexagonal, circular, and so on. The grid in Figure 14.1 has $K = 16$ neurons and each neuron has a weight vector, denoted by w_i, $i = 1, ..., K$ which is initialised randomly. Let $X = \{x_n | n = 1, ..., N\}$ be the input data. We calculate the Euclidean distances between input data X and weight vectors, and then we can obtain for each data vector x_n the best match unit (BMU) that minimises the Euclidean distance between x_n and weight vectors, expressed as shown in Equation (14.1).

$$\text{BMU} = \arg\min_j \left(x_n - w_j \right) \tag{14.1}$$

For the WTA paradigm, only BMU has the opportunity to update its weight vector. Virtually, SOM employs the WTM paradigm, in which at each learning step all neurons within a neighbourhood set around BMU can also be updated. The width or radius of this neighbourhood set is time-varying: it shrinks monotonically with time and ends with only BMU in the set. Let us define the neighbourhood set as N_s. The updating process in time t may be expressed as Equation (14.2).

Table 14.1 Summary of the clustering process of SOM by Kohonen (1990)

Step 1   Define the topology of SOM and initialise each node's weights $w_i(0)$, $i = 1, ..., K$, randomly;

Step 2   Find the best matching unit (BMU) by calculating the distance between the input vector and the weights of each node, that is $J = \arg_j \min(x - w_j)$;

Step 3   The radius of the neighbourhood around the BMU is calculated. The size of the neighbourhood decreases with each iteration;

Step 4   Each node in the BMU's neighbourhood has its weights adjusted to become more like the BMU. Nodes closest to the BMU are altered more than the nodes furthest away in the neighbourhood;

Step 5   Repeat from step 2 for enough iterations for convergence.

$$w_i(t+1) = \begin{cases} w_i(t) + h(t)[x_n - w_i(t)] & \textbf{if } i \in N_s(t) \\ w_i(t) & \textbf{if } i \notin N_s(t) \end{cases} \qquad (14.2)$$

where $h(t)$ is the neighbourhood function that is often defined as given in Equation (14.3),

$$h(t) = \alpha(t)\exp\left(-\frac{r_{\text{BMU}} - r_i^2}{2\sigma^2(t)}\right) \qquad (14.3)$$

where $\alpha(t)$ is the monotonically decreasing learning rate, r represents the position of corresponding neuron, and $\sigma(t)$ is the monotonically decreasing kernel width function. The summary of the clustering process of SOM is given in Table 14.1. In MATLAB, SOM can be performed by a combination of several functions, namely **newsom**, which initialises a SOM network, **train**, which trains the SOM network, and **sim**, which performs the clustering using the trained network. Note that the input data to the **train** function are the same as those to the **sim** function for clustering purposes. There are three packages in R providing SOM functions, namely **kohonen** package, **som** package, and **wccsom** package. In Weka, the function **SelfOrganizingMap** can be called from package **weka.clusterers**.

14.2.2   GLVQ

LVQ discovers cluster structure hidden in unlabelled data. In principle, k-means, which was introduced in Chapter 10, and LVQ are very much alike. In LVQ, the salient feature is that the LVQ network consists of two layers, namely input layer and competitive layer. Each neuron in the competitive layer has a weight vector (or prototype) attached to it. The prototypes $W = \{w_i | i = 1, ..., K\}$ are essentially a network array of cluster centres. When an input vector x_n is submitted to this network, distances are computed between each weight vector w_i and x_n. The neurons in the competitive layer compete and a winner neuron with minimum distance is found. Subsequently, the weight vector of the winner neuron is updated by using Equation (14.4).

$$w_i(t+1) = w_i(t) + \alpha(t)[x_k - w_i(t)] \qquad (14.4)$$

LVQ suffers problems which result from two causes: (1) an improper choice of initial neurons, and (2) the WTA strategy.

Pal *et al.* developed a GLVQ algorithm to circumvent these issues (Pal, Bezdek and Tsao, 1993). Let $L(X, W)$ be a loss function which measures the locally weighted mismatch of x_n with respect to the winner neuron, as given in Equation (14.5),

$$L(X, W) = \sum_{n=1}^{N} \sum_{k=1}^{K} g_{nk} \|x_n - w_k\|^2 \qquad (14.5)$$

where Equation (14.6) holds.

$$g_{nk} = \begin{cases} 1 & \text{if } k \text{ is winner} \\ \dfrac{1}{\sum_{j=1}^{K} \|x_n - w_j\|^2} & \text{otherwise} \end{cases} \qquad (14.6)$$

Therefore, for a fixed set of data points X, the problem reduces to the unconstrained optimisation problem as set out in Equation (14.7).

$$W = \arg \min_{W} L(X, W) \qquad (14.7)$$

The solution of this minimisation problem can be approximated by local gradient descent search of $L(X, W)$. Let us define $D = \sum_{j=1}^{K} \|x_n - w_j\|^2$, then the update rules are formulated as shown in Equation (14.8),

$$w_i(t+1) = w_i(t) + \alpha(t)[x_n - w_i(t)] \frac{D^2 - D + \|x_n - w_i(t)\|^2}{D^2} \qquad (14.8)$$

for the winner neuron, say i, of the data point x_n; and for other $(K-1)$ neurons Equation (14.9) holds.

$$w_j(t+1) = w_j(t) + \alpha(t)[x_n - w_j(t)] \frac{\|x_n - w_j(t)\|^2}{D^2} \qquad (14.9)$$

To avoid possible oscillations of the solution, the amount of correction should be reduced as iteration proceeds, thus $\alpha(t)$ is defined as $\alpha_0(1 - t/T)$ to satisfy the condition $t \to \infty$, $\alpha(t) \to 0$, where $\alpha_0 \in (0,1)$ and T is the maximum number of iterations.

14.2.3 Neural-gas

Neural-gas proposed by Martinetz, Berkovich and Schulten (1993) is a neural network-based clustering inspired by LVQ and SOM. The algorithm was named 'neural-gas' because it employs the dynamics of the feature vectors during the adaptation process, which distribute themselves like gas within the data space.

To avoid being confined to local minima during the adaptation procedure, like k-means, an adaptation approach called 'soft-max' is introduced into the neural-gas algorithm. Moreover, a neighbourhood ranking strategy is also employed so that not only the winning neuron but all neurons depending on their proximity to input data vector can be updated. Each time data vector x_n is presented, the neighbourhood ranking $(w_{i_0}, w_{i_1}, \ldots, w_{i_{K-1}})$ of the weight vectors is determined according to their distances from x_n, where w_{i_0} is the closest neuron to x_n, and $\|w_{i_{k-1}} - x_n\| < \|w_{i_k} - x_n\|$. Let $k_i(x_n, w)$ denote the number k associated with each vector w_i. The adaptation step for adjusting the w_i is given by Equation (14.10),

$$\Delta w_i = \epsilon \cdot h_\lambda(k_i(x_n, w)) \cdot (x_n - w_i), \quad i = 1, \dots, K \tag{14.10}$$

where the step size $\epsilon \in [0,1]$ describes the overall extent of the modification and $h_\lambda(k_i(x_n, w)) = \exp(-k_i(x_n, w)/\lambda)$.

The dynamics of the neurons obey a stochastic gradient descent on the cost function, given in Equation (14.11),

$$L(X, W) = \sum_{n=1}^{N} \sum_{i=1}^{K} \frac{1}{2C(\lambda)} p_i(x_n) h_\lambda(k_i(x_n, w)) \|w_i - x_n\|^2 \tag{14.11}$$

where $p_i(x_n)$ is the fuzzy membership of data vector x_n to cluster i, and is expressed as Equation (14.12),

$$p_i(x_n) = \frac{h_\lambda(k_i(x_n, w))}{C(\lambda)} \tag{14.12}$$

where Equation (14.13) holds.

$$C(\lambda) = \sum_{i=1}^{K} h_\lambda(k_i(x_n, w)) \tag{14.13}$$

There are two main advantages of the neural-gas algorithm: (1) it converges quickly to low distortion errors; (2) it reaches a distortion error lower than that resulting from k-means and SOM. However, the nature of the neural-gas algorithm requires knowledge of the number of neurons in advance, which limits the use of neural-gas in practical applications where the problem is to determine a suitable number of neurons a priori. Depending on the complexity of the data distribution to be modelled, very different numbers of neurons may be appropriate. Nevertheless, in light of the neural-gas algorithm, many other algorithms were proposed to circumvent this problem, for example the growing neural-gas (GNG) algorithm (Fritzke, 1995), and the growing cell structures (GCS) algorithm (Fritzke, 1994).

14.2.4   ART

ART proposes a solution of the *stability-plasticity* dilemma (Grossberg, 1976, 1980): an adequate algorithm must be capable of plasticity in order to learn about significant new events, yet it must also remain stable in response to irrelevant or often repeated events. ART can learn arbitrary input patterns in a stable, fast, and self-organising way. ART itself does not possess a neural network architecture; it is rather a learning theory in which resonance in neural circuits can trigger fast learning. Since it was invented more than two decades ago, ART has become a large family of neural network architectures with many variants, which are summarised in Table 14.2. It is not possible for us to detail all ART algorithms because of the limit of space. We focus on the basic concept of ART.

ART-1, which is the first version of ART networks for binary data, is shown in Figure 14.2. There are two major subsystems: *attentional subsystem* and *orienting subsystem*.

Table 14.2   Summary of the variants of ART

Name	Comments	References
ART-1	The simplest variety of ART networks, accepting only binary inputs	Carpenter and Grossberg (1987b)
ART-2	Extended to analog input patterns	Carpenter and Grossberg (1987a)
ART-2A	A streamlined form of ART-2 with a drastically accelerated runtime	Carpenter, Grossberg and Reynolds (1991)
ART-3	A new mechanism originating from elaborate biological processes to achieve more efficient parallel search in hierarchical structures	Carpenter and Grossberg (1990)
Fuzzy ART	Fuzzy logic was incorporated into ART's pattern recognition, thus generalising ART	Carpenter, Grossberg and Rosen (1991a)
ARTMAP	Known as Predictive ART, combines two slightly modified ART-1 or ART-2 units into a supervised learning structure	Carpenter, Grossberg and Rosen (1991b)
Fuzzy ARTMAP	ARTMAP using fuzzy ART units	Carpenter et al. (1992)
AHN	Adaptive Hamming net, a fast-learning ART 1 model without searching	Hung and Lin (1995)
Gaussian ARTMAP	A synthesis of a Gaussian classifier and ART neural network	Williamson (1996)
PART	Projective ART	Cao and Wu (2002)

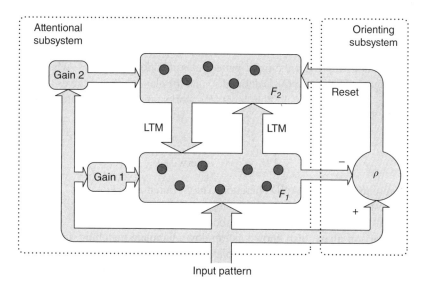

Figure 14.2   ART-1 consists of an attentional subsystem and an orienting subsystem. The attentional subsystem has two STM layers, F_1 and F_2. LTM traces between F_1 and F_2 multiply the signal in these pathways. Gain controls enable F_1 and F_2 to distinguish current stages of a running circle. The orienting subsystem generates a reset wave to F_2 when mismatches between bottom-up and top-down patterns occur at F_1

Multiple interacting memory systems are needed to monitor and adaptively react to the novelty of events. Interactions between two functionally complementary subsystems are needed to process familiar and unfamiliar events. Familiar events are processed with an attentional subsystem. This subsystem establishes precise internal representations of and responses to familiar events. It also builds up the learned top-down expectations that help to stabilise the learned bottom-up codes of familiar events. The orienting subsystem is essential for expressing whether a novel pattern is familiar and well represented by an existing recognition code or unfamiliar and in need of a new recognition code. It resets the attentional subsystem when an unfamiliar event occurs.

The attentional subsystem consists of a two-layer short-term memory (STM) structure where F_1 is called feature representation field and F_2 is called category representation field. A stream of input patterns to a network F_1 can train the adaptive weights, or long-term memory (LTM) traces that multiply the signals in the pathways from F_1 to a coding level F_2. Each neuron in F_1 is connected to all neurons in F_2 via the continuous-valued bottom-up weight matrix W^{12} and top-down weight matrix W^{21}. The prototypes of clusters are stored in layer F_2. Processing in ART-1 can be divided into four phases: recognition, comparison, search, and learn.

Recognition phase is also known as bottom-up activation. Initially, if no input pattern is applied, all recognition in F_2 is disabled and two control gains G_1 and G_2, are set to zero. This causes all F_2 elements to be zero, giving them equal chance to win the subsequent recognition competition. The gain control 2 depends only on the input pattern: $G_2 = 1$ if there is an input; $G_2 = 0$, otherwise. The gain control 1 depends on both the input pattern and the output from F_2, denoted by O_2: $G_1 = 1$, if there is an input and $O_2 = 0$; $G_1 = 0$, otherwise. Each node in F_1 whose activity is beyond the threshold ($G_1 = 1$) sends excitatory outputs to the nodes in F_2. The F_1 output pattern O_1 is multiplied by the LTM traces W^{12}. Each node in F_2 sums up all its LTM gated signals, as shown in Equation (14.14).

$$v_{2j} = \sum_i O_{1i} W_{ij}^{12} \tag{14.14}$$

These connections represent the input pattern classification categories, where each weight stores one category. The output O_{2j} is defined so that the element that receives the largest input should be the winner. As such, the layer F_2 works as a WTA strategy described by Equation (14.15),

$$O_{2j} = \begin{cases} 1 & \text{if } G_2 = 1 \cap v_{2j} = \max\{v_{2j}\} \\ 0 & \text{otherwise} \end{cases} \tag{14.15}$$

where $\cap$ represents the logical AND operation. The F_2 unit that receives the largest F_1 output is the one that best matches the input vector category, thus wins the competition. The F_2 winner node fires and simultaneously inhibits all other nodes in the layer.

In the comparison phase, also known as top-down template matching, the STM activation pattern O_2 on F_2 generates a top-down template on F_1. This pattern is multiplied by the LTM traces W^{21} and each node in F_1 obtains values shown in Equation (14.16).

$$v_{1i} = \sum_j O_{2j} W_{ij}^{21} \tag{14.16}$$

The most active recognition unit from F_2 passes a signal back to comparison layer F_1. Now G_1 is inhibited because the output from F_2 is active. If there is a good match between the top-down template and the input vector, the system becomes stable and learning then occurs; otherwise, the reset layer in the orienting subsystem then inhibits the F_2 layer. There is a threshold, called vigilance level $\rho \in (0,1]$. If the match degree is less than the vigilance level, the reset signal is then sent to F_2.

The salient difference between ART-1 and ART-2 (fuzzy ART) is that the binary AND operator in ART-1 is replaced by the fuzzy AND operator, which is defined as $A \wedge B = \min(A, B)$. The learning of the LTM traces W_j when the node j is active, is then given by Equation (14.17).

$$W_j^{new} = \beta \left(x \wedge W_j^{old} \right) + (1 - \beta) W_j^{old} \tag{14.17}$$

Many variants of ART have been developed and applied to large-scale technological and biological applications by authors around the world. Readers who are interested may be referred to a quite recent survey paper by Grossberg (2013). ART-1, ART-2, and ARTMAP are available in *R* package **RSNNS**.

14.2.5 OPTOC

OPTOC is a CL paradigm proposed by Zhang and Liu (2002). Unlike WTA and WTM paradigms, the key technique used in OPTOC is that, for each prototype, an online learning vector, asymptotic property vector (APV) is assigned to guide the learning of this prototype. With the 'help' of the APV, each prototype will locate only one natural cluster and ignore other clusters in the case that the number of prototypes is less than that of clusters. One of the well-known critical problems with CL is the difficulty in determining the number of clusters. With the OPTOC learning paradigm, SSCL algorithm starts from only a single prototype which is randomly initialised in the feature space. During the learning period, one of the prototypes (initially, the only single prototype) will be chosen to split into two prototypes based on a split validity measure. This self-splitting behaviour terminates if no more prototypes are suitable for further splitting.

Given each prototype, P_k, the key technique is that an online learning vector, APV A_k is assigned to guide the learning of this prototype. For simplicity, A_k represents the APV for prototype P_k and n_{A_k} denotes the learning counter (winning counter) of A_k. As a necessary condition of OPTOC mechanism, A_k is required to initialise at a random location, which is far from its associated prototype P_k and n_{A_k} is initially zero. Taking the input pattern x_n as a neighbour if it satisfies the condition $\langle P_k, x_n \rangle \leq \langle P_k, A_k \rangle$, where $\langle \cdot, \cdot \rangle$ is the inner product operator. To implement the OPTOC paradigm, A_k is updated online to construct a dynamic neighbourhood of P_k. The patterns 'outside' of the dynamic neighbourhood will contribute less to the learning of P_k as compared with those 'inside' patterns.

The update of A_k depends on the relative locations of the input data point x_n, prototype P_k and A_k itself, as mathematically given by Equation (14.18),

$$A_k(t+1) = A_k(t) + \frac{1}{n_{A_k}} \cdot \delta_k \cdot (x_n - A_k(t)) \cdot \Theta(P_k(t), x_n, A_k(t)) \tag{14.18}$$

where $0 < \delta_k \leq 1$, and δ_k is defined as given in Equation (14.19),

$$\delta_k = \left(\frac{\langle P_k, A_k \rangle}{\langle P_k, x_n \rangle + \langle P_k, A_k \rangle} \right)^2 \tag{14.19}$$

and $\Theta(a, b, c)$ is a general function given by Equation (14.20)

$$\Theta(a,b,c) = \begin{cases} 1 & \langle a,b \rangle \le \langle a,c \rangle \\ 0 & \text{otherwise} \end{cases} \tag{14.20}$$

For each update of A_k, its learning counter n_{A_k} is computed as shown in Equation (14.21).

$$n_{A_k} = n_{A_k} + \delta_k \cdot \Theta(P_k(t), x_n, A_k(t)) \tag{14.21}$$

Thus, it may be observed that APV A_k always attempts to move toward P_k. The update of the prototype P_k is given by Equation (14.22),

$$P_k(t+1) = P_k(t) + \alpha_k(x_n - P_k(t)) \tag{14.22}$$

where Equation (14.23) holds.

$$\alpha_k = \left(\frac{\langle P_k, A_k \rangle}{\langle P_k, x_n \rangle + \langle P_k, A_k \rangle} \right)^2 \tag{14.23}$$

In addition to the APV, there is another auxiliary vector, called distant property vector (DPV) R_k, assisting the cluster, which contains more than one prototype, to split. Let n_{R_k} denote the learning counter for R_k, which is initialised to zero. R_k will be updated to a distant location from P_k. The efficiency of splitting is improved by determining the update schedule of R_k adaptively from the analysis of the feature space. Contrary to the APV A_k, the DPV R_k always tries to move away from P_k. The update of DPV R_k is given by Equation (14.24),

$$R_k(t+1) = R_k(t) + \frac{1}{n_{R_k}} \cdot \rho_k \cdot (x_n - R_k(t)) \cdot \Theta(P_k(t), R_k(t), x_n) \tag{14.24}$$

where the learning rate of R_k, ρ_k, is given by Equation (14.25).

$$\rho_k = \left(\frac{\langle P_k, x_n \rangle}{\langle P_k, x_n \rangle + \langle P_k, R_k \rangle} \right)^2 \tag{14.25}$$

The SSCL algorithm based on the OPTOC CL paradigm is considered as a solution of two long standing critical problems in clustering, namely (1) the difficulty in determining the number of clusters, and (2) the sensitivity to prototype initialisation.

14.2.6 SOON

Rhouma and Frigui developed a self-organisation of pulse-coupled oscillators algorithm (Rhouma and Frigui, 2001), which was later named as SOON (Salem, Jack and Nandi, 2008). SOON has its root in a biological process that fireflies flash at random when considered by themselves; however, they exhibit the characteristic of firing together when in groups that are physically close to each other. Technically, SOON is an efficient synchronisation model that organises a population of integrate-and-fire oscillators into stable and structured groups. Each oscillator fires synchronically with all the others within its group, but the groups themselves fire with a constant phase difference.

Let $\mathcal{O} = \{\mathcal{O}_1, \ldots, \mathcal{O}_N\}$ be a set of N oscillators, where each oscillator $\mathcal{O}_i$ is characterised by a phase ϕ_i and a state variable s_i, given by Equation (14.26),

$$s_i = f_i(\phi_i) = \frac{1}{b}\ln\left[1 + \left(e^b - 1\right)\phi_i\right] \tag{14.26}$$

where b is a constant value used to control the curve of the oscillator. Positive values of b will make the curve concave down, while negative values of b will make the curve concave up. $\phi_i \in [0,1]$ is the phase angle of the oscillator, and determines the likelihood of the oscillator to fire, where 0 is just fired and 1 is firing. The output of the oscillator s_i is bounded in the range $[0, 1]$ for all values of $f_i(\phi_i)$ by Equation (14.27),

$$s_i^* = B\left(s_i + \epsilon_j(\phi_i)\right) \tag{14.27}$$

where $B(\cdot)$ is a limiting function as given by Equation (14.28),

$$B(x) = \begin{cases} x & \text{if } 0 \leq x \leq 1 \\ 0 & \text{if } x < 0 \\ 1 & \text{if } x > 1 \end{cases} \tag{14.28}$$

and $\epsilon_j(\phi_i)$ is the coupling strength of an oscillator j at a given phase ϕ_i, which is written as shown in Equation (14.29).

$$\epsilon_j(\phi_i) = \begin{cases} C_E\left[1 - \left(\dfrac{d_{ij}}{\delta_0}\right)^2\right], & \text{if } d_{ij} \leq \delta_0 \\ -C_I\left[\left(\dfrac{d_{ij} - \delta_0}{\delta_1 - \delta_0}\right)^2\right], & \text{if } \delta_0 < d_{ij} \leq \delta_1 \\ -C_I, & \text{otherwise} \end{cases} \tag{14.29}$$

Having decided on a limit distance δ_0, δ_1 is set to be five times δ_0. The coupling function promotes all oscillators that lie within the distance δ_0, increasing the phase value by the constant of excitation C_E, multiplied by the factor depending on the ratio of the distance between the winning oscillator and the oscillator under consideration, and δ_0. This, in turn, makes the group of oscillators more likely to synchronise with the winning oscillator in future iterations. The phase of all those with distance lying in the interval $\delta_0 < d_{ij} \leq \delta_1$ are inhibited by C_I, the coefficient of inhibition multiplied by a ratio that takes into account how close the oscillator under consideration is to the winning oscillator. All values of $d_{ij} > \delta_1$ are hard limited to $-C_I$. The value of C_E is typically relatively small, of the order of 0.1 ~ 0.2. The value of C_I is normally set to the value C_E / N (N is the number of data points), as any given data point is likely to be inhibited more often than it is likely to be excited.

Once a group of oscillators synchronise, their dynamics must remain identical in order to keep the synchronisation. Let $G = \{ \mathcal{O}_{g_1}, \ldots, \mathcal{O}_{g_n} \}$ be a set of n oscillators that have synchronised. To keep these oscillators synchronised, the coupling strength of group members should be the same [Equation (14.30)].

$$\epsilon_i(\phi_{g_1}) = \epsilon_i(\phi_{g_2}) = \cdots = \epsilon_i(\phi_{g_n}), \quad \text{or} \quad r_{ig_1} = r_{ig_2} = \cdots = r_{ig_n} = \hat{r}_{iG} \qquad (14.30)$$

There are three possible choices for $\hat{r}_{iG}$, corresponding to single linkage, complete linkage, and average linkage, respectively, and these are examined in Equation (14.31).

$$\hat{r}_{iG} = \begin{cases} \min\left(r_{ig_1}, r_{ig_2}, \cdots, r_{ig_n}\right) \\ \max\left(r_{ig_1}, r_{ig_2}, \cdots, r_{ig_n}\right) \\ \dfrac{1}{n} \sum_{j=1}^{n} r_{ig_j} \end{cases} \qquad (14.31)$$

Different choices may generate quite different results in many circumstances.

There are two SOON algorithms: SOON-1 and SOON-2 (Rhouma and Frigui, 2001; Salem, Jack and Nandi, 2008), which are summarised in Table 14.3 and Table 14.4, respectively. SOON-1 was designed for the application with only relational data available; for example, networks. SOON-2 was designed for the datasets where each data object is characterised by M numerical features. Compared with SOON-1, SOON-2 is more efficient because it uses prototypes to both represent clusters and avoid computing and storing the pairwise distances, which is a potential problem when the size of the dataset is big. SOON-2 starts with a set of K initial prototypes $\{P_k\}$, $k = 1, \ldots, K$, and each data object x_i is represented by an oscillator $\mathcal{O}_i$. The distance between oscillator i and prototype k, is denoted by $d_{ik} = d(x_i, P_k)$. Note that the distance is adjustable. If $\mathcal{O}_i$ belongs to one of the synchronised groups, then its distance will

Table 14.3 Summary of the SOON-1 algorithm

Step 1	Construct the relative distance matrix $D = \{ d_{ij} \}$; Initialise phases ϕ_i randomly;
Step 2	Identify the next oscillator to fire: $\left\{ \mathcal{O}_i : \phi_i = \max\limits_{j=1,\ldots,N} \phi_j \right\}$;
Step 3	Bring ϕ_i to threshold, and adjust other phases: $\phi_k = \phi_k + (1 - \phi_i)$ for $k = 1, \ldots, N$;
Step 4	**FOR** all oscillators $\mathcal{O}_j$, $(j \neq i)$ DO
	Compute state variable $s_k = f(\phi_k)$ using Equation (14.26);
	Compute coupling strength $\epsilon_i(\phi_k)$ using Equation (14.29);
	Adjust state variables using Equation (14.27);
	Compute new phases using $\phi_k = f^{-1}(s_k)$
	END FOR
Step 5	Identify synchronised oscillators and reset their phases;
Step 6	Adjust relations of all oscillators that have synchronised in Equation (14.31);
Step 7	Repeat Step 2 until synchronised group stabilises.

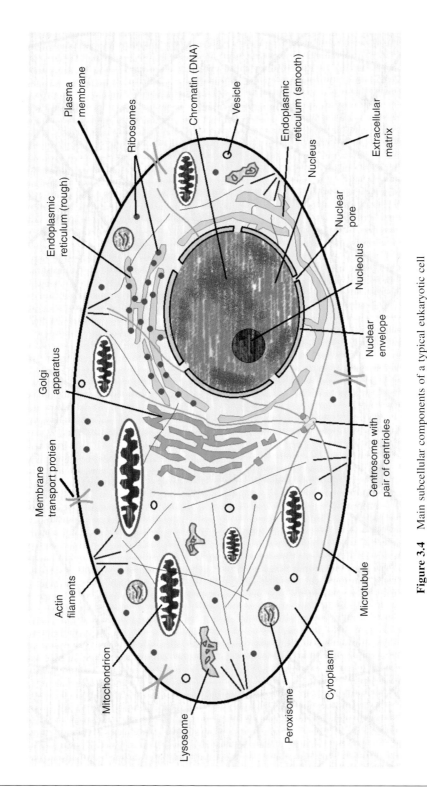

Figure 3.4 Main subcellular components of a typical eukaryotic cell

Integrative Cluster Analysis in Bioinformatics, First Edition. Basel Abu-Jamous, Rui Fa and Asoke K. Nandi.
© 2015 John Wiley & Sons, Ltd. Published 2015 by John Wiley & Sons, Ltd.

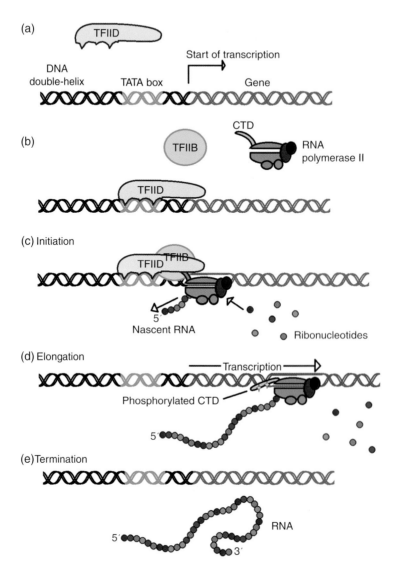

Figure 4.7 The transcription process. (a) The DNA double-helix before the initiation of transcription; the TATA box motif, the gene, and the start of transcription location are marked. (b) TFIID binds to the TATA motif. (c) Transcription initiation; TFIIB and the RNA polymerase II complex are recruited to the start of transcription site, and then transcription of a few codons takes place. (d) Transcription elongation; the CTD of the RNA polymerase II complex is phosphorylated to allow for the polymerase to slide over the gene's sequence, transcribing its codons and thus elongating the nascent RNA. (e) Transcription termination; the RNA polymerase II complex is released from the DNA and the gene is completely transcribed

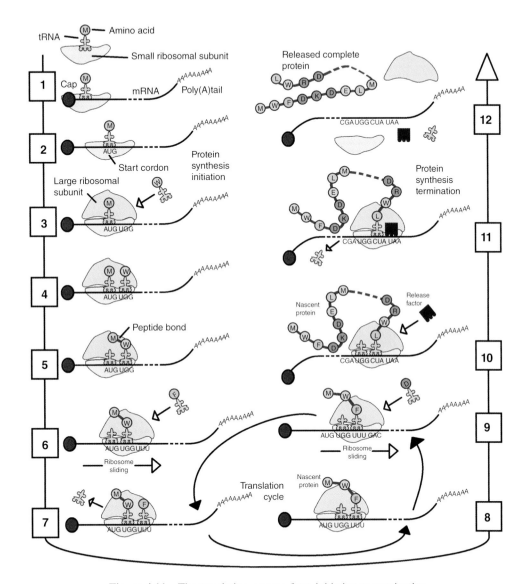

Figure 4.11 The translation process from initiation to termination

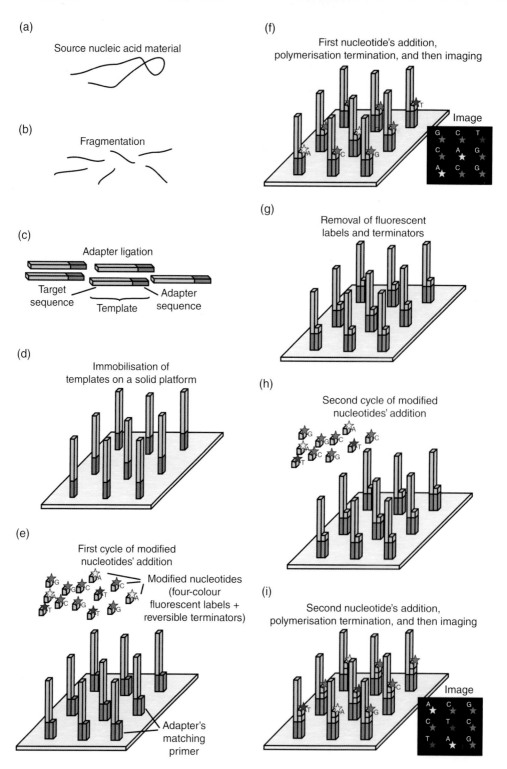

Figure 5.2 Sequencing by CRT. (a) Source nucleic acid material. (b) Fragmentation. (c) Adapter ligation. (d) Immobilisation of templates on a solid platform. (e) First cycle of modified nucleotides' addition; primers matching the adapters are bound. (f) First nucleotide's addition followed by polymerisation termination and then imaging; the image's probes' colours indicate the type of nucleotide bound to each of the templates in this cycle. (g) Removal of fluorescent labels and terminators. (h) Second cycle of modified nucleotides' addition. (i) Second nucleotide's addition followed by polymerisation termination and then imaging. The steps (h–i) are repeated iteratively until the entire template is sequenced

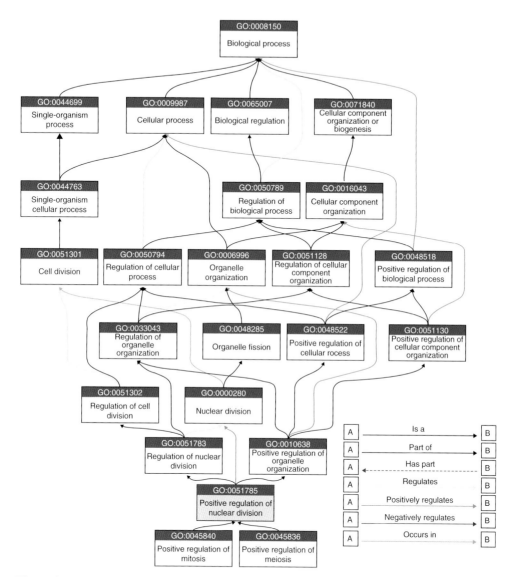

Figure 21.1 Sub-graph of GO biological process terms showing the process 'positive regulation of nuclear division' in a shaded box, its direct children, and its ancestral relations up to the root term 'biological process'. This graph can be obtained by searching the EBI QuickGO web tool (www.ebi. ac.uk/QuickGO) for the GO term ID 'GO:0051785', and then accessing the 'Ancestor Chart' tab; the direct link is (http://www.ebi.ac.uk/QuickGO/GTerm?id=GO:0051785#term=ancchart). The graph shows the names and GO term identifiers of the terms included, as well as colour-coded relations between those different biological processes

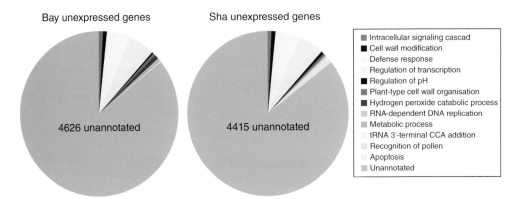

Figure 22.6 By far the majority in both accessions were unannotated genes (Richards *et al.*, 2012). Gene Ontology (GO) categories for genes that were considered 'absent' in all three replicates of at least one sample. Approximately 1/3 of the genome of both Bay-0 (8322 genes) and Sha (7948) was not detected in this study

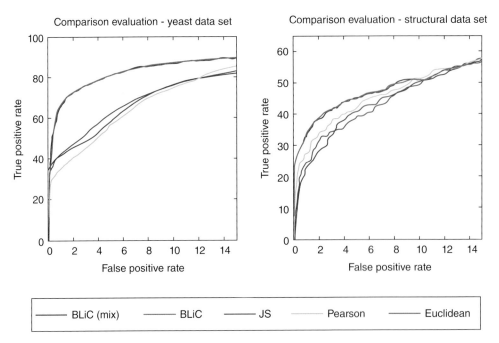

Figure 22.17 Examples of ROC from (Habib *et al.*, 2008). Left: Sensitivity and specificity of different scoring methods: Comparison of different scoring methods on the 'Yeast' dataset using a subset of motifs generated from subsets of size 35 with altered lengths (not including the full-length motifs, 685 motifs). Each similarity score was assigned an empirical statistical significance *p*-value. The ROC curve plots the true positive rate vs. the false positive rate, as computed for different *p*-value thresholds, where pairs of motifs generated from genomic binding sites that were associated with the same factor are considered true positives. The BLiC score (green, using a Dirichlet prior, or blue, using a Dirichlet-mixture prior, and blue line is overlapped with green one) outperformed all other similarity scores: Jensen–Shannon (JS) divergence (red), Euclidean distance (purple) and Pearson Correlation coefficient (cyan). Right: Sensitivity and specificity estimated by structural data: Same as the left one but using the 'Structural' dataset. Pairs of motifs from the same structural family are considered true positives

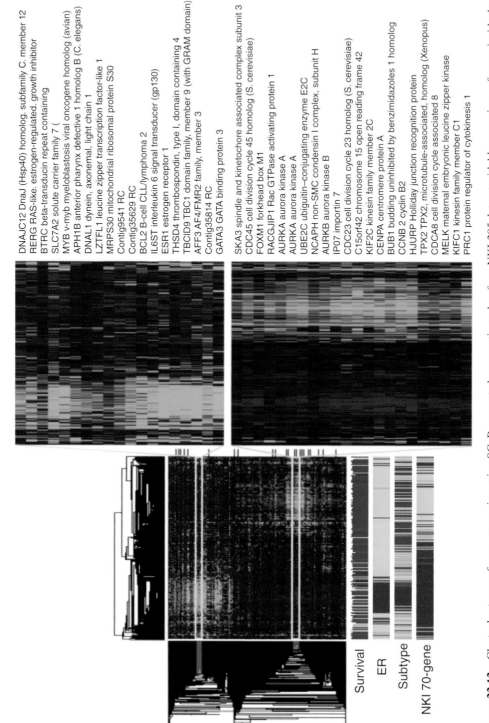

Figure 22.12 Cluster heat map of gene expression using SCoR-generated prognostic probes from NKI-295 dataset with blow-up views of genes inside the centre of poor and good prognosis gene clusters. Probes matching the NKI 70-genes are marked in black lines on the right side of the heat map (Yao et al., 2012)

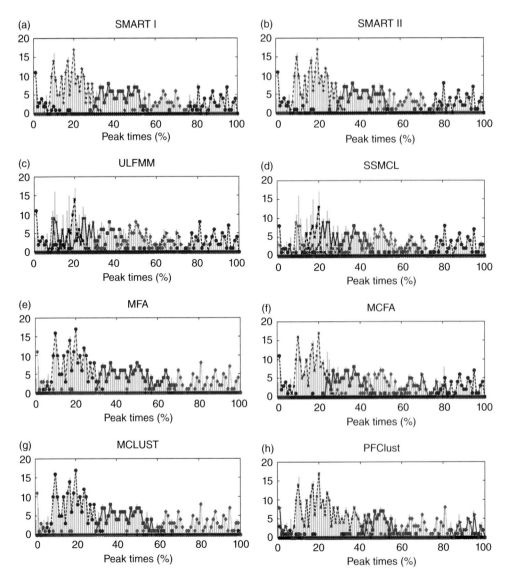

Figure 23.5 Histogram of the peak times of genes in each cluster for each algorithm in Yeast cell cycle α-38 dataset. (a) SMART-CL (SMART I), $T_{chs} = 20$ and $N_{max} = 5$, $K = 4$; (b) SMART-FMM (SMART II), $N_{max} = 5$, $K = 4$; (c) ULFMM, $k_{max} = 30$, $K = 7$; (d) SSMCL, $k_{max} = 30$, $K = 8$; (e) MFA, $q = 5$, $K = 3$; (f) MCFA, $q = 6$, $K = 5$; (g) MCLUST, $K = 3$; (h) PFClust. Sub-figures (a) and (b) show that four clusters represent reasonably good clustering since there are only few small overlap regions between clusters. Sub-figures (c) and (d) indicate that many clusters crowd and overlap in the region 5–30%, especially in Sub-figure (c), where a clustering representing peaking at 20% superimposes on another cluster, which spans over 10–30%. These overlapped clusters have to be one cluster. Sub-figures (e) and (f) show that MFA and MCFA also give reasonably good clustering results when judged by eye; however, clustering is poorer than SMART-FMM in the numerical metrics. Sub-figures (g) and (h) show the distribution of the peak times of genes based on the clustering results of MCLUST and PFClust, respectively

Table 14.4   Summary of the SOON-2 algorithm

Step 1	Select a distance measure $d(\cdot,\cdot)$; Initialise phases ϕ_i, $i = 1,\ldots,N$ randomly; Set K, and initialise the prototype P_k randomly for $k = 1,\ldots,K$;
Step 2	Identify the next oscillator to fire: $\left\{ \mathcal{O}_i : \phi_i = \max_{j=1,\ldots,N} \phi_j \right\}$;
Step 3	Identify the closest prototype to $\mathcal{O}_i$: $P^* : d(\mathcal{O}_i, P^*) = \min_{k=1,\ldots,K} d(\mathcal{O}_j, P_k)$;
Step 4	Compute $d(\mathcal{O}_j, P^*)$ for $j = 1, N$ and adjust them using Equation (14.32)
Step 5	Bring ϕ_i to threshold, and adjust other phases: $\phi_k = \phi_k + (1 - \phi_i)$ for $k = 1,\ldots,N$;
Step 6	**FOR** all oscillators $\mathcal{O}_j$, $(j \neq i)$ DO
	Compute state variable $s_k = f(\phi_k)$ using Equation (14.26);
	Compute coupling strength $\epsilon_i(\phi_k)$ using Equation (14.29);
	Adjust state variables using Equation (14.27);
	Compute new phases using $\phi_k = f^{-1}(s_k)$
	END FOR
Step 7	Identify synchronised oscillators and reset their phases;
Step 8	Update prototype $\boldsymbol{P}_k$;
Step 9	Repeat Step 2 until synchronised group stabilises.

be replaced by the average distances of all oscillators that constitute the group G_k, which is written as Equation (14.32).

$$\hat{d}_{ik} = \begin{cases} d_{ik} & \text{if } \mathcal{O}_i \notin G_k \\ \dfrac{\sum_{l \in G_k} d_{il}}{|G_k|} & \text{if } \mathcal{O}_i \in G_k \end{cases} \tag{14.32}$$

14.3   Discussion

There have been many neural network-based clustering algorithms applied in the bioinformatics field. The collections of publicly accessible resources for neural network-based clustering are shown in Table 14.5. A successful example of their application in the analysis of gene expression data was by Tamayo and colleagues (1999), where SOM was found to be significantly superior to hierarchical clustering and k-means in both robustness and accuracy. Hsu *et al.* proposed an algorithm combining unsupervised hierarchical clustering and SOM to perform class discovery and marker-gene identification in microarray gene expression data

Table 14.5   Collection of publicly accessible resources for neural network-based clustering

Algorithm name	Platform	Package	Function
SOM	*R*	kohonen/som/wccsom	som
	MATLAB		newsom/train/sim
	weka	Weka.clusterers	SelfOrganizingMap
ART	*R*	RSNNS	art1/art2/artmap

(Hsu, Tang and Halgamuge, 2003). Wang *et al.* also employed SOM as the first step of their algorithm (Wang *et al.*, 2003). The aim of applying the SOM procedure in the algorithm was to find map units that could represent the configuration of the input dataset, and at the same time to achieve a continuous mapping from the input gene space to a lattice. Chavez-Alvarez *et al.* conducted a study very recently to use SOM to cluster multiple yeast cell cycle datasets (Chavez-Alvarez, Chavoya and Mendez-Vazquez, 2014). Besides SOM, there are also many other algorithms applied in the bioinformatics field; for example, fuzzy ART was applied to analyse the time series expression data during sporulation of budding yeast (Tomida *et al.*, 2002); another variant of ART, called projective ART, was developed to select specific genes for each subtype in a cancer diagnosis marker-extraction study of soft-tissue sarcomas (Takahashi, Kobayashi and Honda, 2005; Takahashi *et al.*, 2006), and later, the analytical results were further investigated (Takahashi *et al.*, 2013). A self-splitting and merging competitive clustering algorithm based on the OPTOC paradigm was applied to identify biologically relevant groups of genes (Wu *et al.*, 2004); SOON was applied to cluster microarray data by Salem, Jack and Nandi (2008). We will discuss these applications in more detail in Chapter 19.

References

Cao, Y. and Wu, J. (2002). Projective ART for clustering data sets in high dimensional spaces. *Neural Networks,* **15**(1), pp. 105–120.

Carpenter, G.A. and Grossberg, S. (1987a). ART 2: self-organization of stable category recognition codes for analog input patterns. *Applied Optics,* **26**(23), pp. 4919–4930.

Carpenter, G.A. and Grossberg, S. (1987b). A massively parallel architecture for a self-organizing neural pattern recognition machine. *Computer Vision, Graphics, and Image processing,* **37**(1), pp. 54–115.

Carpenter, G.A. and Grossberg, S. (1988). The ART of adaptive pattern recognition by a self-organizing neural network. *Computer,* **21**(3), pp. 77–88.

Carpenter, G.A. and Grossberg, S. (1990). ART 3: Hierarchical search using chemical transmitters in self-organizing pattern recognition architectures. *Neural Networks,* **3**(2), pp. 129–152.

Carpenter, G.A., Grossberg, S. and Reynolds, J.H. (1991). ARTMAP: supervised real-time learning and classification of nonstationary data by a self-organizing neural network. *Neural Networks,* **4**(5), pp. 565–588.

Carpenter, G.A., Grossberg, S. and Rosen, D.B. (1991a). ART 2-A: an adaptive resonance algorithm for rapid category learning and recognition. *Neural Networks,* **4**(4), pp. 493–504.

Carpenter, G.A., Grossberg, S. and Rosen, D.B. (1991b). Fuzzy ART: fast stable learning and categorization of analog patterns by an adaptive resonance system. *Neural Networks,* **4**(6), pp. 759–771.

Carpenter, G.A. Grossberg, S., Markuzon, N. *et al.* (1992). Fuzzy ARTMAP: a neural network architecture for incremental supervised learning of analog multidimensional maps. *IEEE Transactions on Neural Networks,* **3**(5), pp. 698–713.

Chavez-Alvarez, R., Chavoya, A. and Mendez-Vazquez, A. (2014). Discovery of possible gene relationships through the application of self-organizing maps to DNA microarray databases. *PloS One,* **9**(4), e93233.

Fritzke, B. (1994). Growing cell structures—a self-organizing network for unsupervised and supervised learning. *Neural Networks,* **7**(9), pp. 1441–1460.

Fritzke, B. (1995). A growing neural gas network learns topologies. *Advances in Neural Information Processing Systems,* **7**, pp. 625–632.

Grossberg, S. (1976). Adaptive pattern classification and universal recoding: I. Parallel development and coding of neural feature detectors. *Biological Cybernetics,* **23**(3), pp. 121–134.

Grossberg, S. (1980). How does a brain build a cognitive code? *Psychological Review,* **87**(1), pp. 1–52.

Grossberg, S. (2013). Adaptive Resonance Theory: how a brain learns to consciously attend, learn, and recognize a changing world. *Neural Networks,* **37**, pp. 1–47.

Hsu, A.L., Tang, S.-L. and Halgamuge, S.K. (2003). An unsupervised hierarchical dynamic self-organizing approach to cancer class discovery and marker gene identification in microarray data. *Bioinformatics,* **19** (16), pp. 2131–2140.

Hung, C.-A. and Lin, S.-F. (1995). Adaptive Hamming net: a fast-learning ART 1 model without searching. *Neural Networks,* **8**(4), pp. 605–618.

Kohonen, T. (1990). The self-organizing map. *Proceedings of the IEEE,* **78**(9), pp. 1464–1480.

Kohonen, T. (2001). *Self-organizing maps*, Springer, New York.

Malsburg, C. (1973). Self-organization of orientation sensitive cells in the striate cortex. *Kybernetik,* **14**(2), pp. 85–100.

Martinetz, T.M., Berkovich, S.G. and Schulten, K.J. (1993). 'Neural-gas' network for vector quantization and its application to time-series prediction. *IEEE Transactions on Neural Networks,* **4**(4), pp. 558–569.

Pal, N.R., Bezdek, J.C. and Tsao, E.-K. (1993). Generalized clustering networks and Kohonen's self-organizing scheme., *IEEE Transactions on Neural Networks,* **4**(4), pp. 549–557.

Rhouma, M.B.H. and Frigui, H. (2001). Self-organization of pulse-coupled oscillators with application to clustering. *IEEE Transactions on Pattern Analysis and Machine Intelligence,* **23**(2), pp. 180–195.

Salem, S.A., Jack, L.B. and Nandi, A.K. (2008). Investigation of self-organizing oscillator networks for use in clustering microarray data. *IEEE Transactions on NanoBioscience,* **7**(1), pp. 65–79.

Takahashi, H., Kobayashi, T. and Honda, H. (2005). Construction of robust prognostic predictors by using projective adaptive resonance theory as a gene filtering method. *Bioinformatics,* **21**(2), pp. 179–186.

Takahashi, H., Nemoto, T., Yoshida, T. *et al.* (2006). Cancer diagnosis marker extraction for soft tissue sarcomas based on gene expression profiling data by using projective adaptive resonance theory (PART) filtering method. *BMC Bioinformatics,* **7**(1), p. 399.

Takahashi, H., Nakayama, R., Hayashi, S. *et al.* (2013). Macrophage Migration Inhibitory Factor and Stearoyl-CoA Desaturase 1: Potential Prognostic Markers for Soft Tissue Sarcomas Based on Bioinformatics Analyses. *PloS One,* **8**(10), e78250.

Tamayo, P., Slonim, D., Mesirov, J. *et al.* (1999). Interpreting patterns of gene expression with self-organizing maps: methods and application to hematopoietic differentiation. *Proceedings of the National Academy of Sciences,* **96**(6), pp. 2907–2912.

Tomida, S., Hanai, T., Honda, H. and Kobayashi, T. (2002). Analysis of expression profile using fuzzy adaptive resonance theory. *Bioinformatics,* **18**(8), pp. 1073–1083.

Wang, J., Bø, T.H., Jonassen, I. *et al.* (2003). Tumor classification and marker gene prediction by feature selection and fuzzy c-means clustering using microarray data. *BMC Bioinformatics,* **4**(1), p. 60.

Williamson, J.R. (1996). Gaussian ARTMAP: A neural network for fast incremental learning of noisy multidimensional maps. *Neural Networks,* **9**(5), pp. 881–897.

Wu, S., Liew, A.-C., Yan, H. and Yang, M. (2004). Cluster analysis of gene expression data based on self-splitting and merging competitive learning. *IEEE Transactions on Information Technology in Biomedicine,* **8**(1), pp. 5–15.

Zhang, Y.-J. and Liu, Z.-Q. (2002). Self-splitting competitive learning: a new on-line clustering paradigm. *IEEE Transactions on Neural Networks,* **13**(2), pp. 369–380.

15

Mixture Model Clustering

15.1 Introduction

In this chapter, we introduce one of most important clustering families, mixture model methods, which is also known as model-based clustering. Model-based clustering is a wide family of algorithms; the common idea behind them is to model an unknown distribution with a mixture of simpler distributions. This family of algorithms has attracted, and is still attracting, a lot of attention because of its excellent performance.

The classification of mixture model clustering can be based on the following four criteria: (i) in terms of the number of components in the mixture, they can be classified as finite mixture model (parametric) and infinite mixture model (non-parametric); (ii) in terms of the clustering kernel, there are several classes, namely, the multivariate normal models or Gaussian mixture models (GMMs), the hidden Markov mixture models, and other mixture models based on non-Gaussian distributions like the Student t-distribution; (iii) in terms of the estimation method, they can be classified to non-Bayesian methods (maximum likelihood (ML) criterion) and Bayesian methods, and within Bayesian methods, they can be further classified into Markov Chain Monte Carlo (MCMC) methods and variational approximation methods; (iv) in terms of dealing with dimensionality, there are classes of factorising algorithms, for example, mixture of factor analysers (MFA), MFA with common factor loadings, mixture of probabilistic principal component analysers, mixture of independent component analysers, and so on.

For the sake of clarity and simplicity, the hierarchy of this chapter is shown in Figure 15.1. We classify the mixture models into two large groups in terms of the number of components in the mixture, namely finite mixture models and infinite mixture models. Finite mixture models

Integrative Cluster Analysis in Bioinformatics, First Edition. Basel Abu-Jamous, Rui Fa and Asoke K. Nandi.
© 2015 John Wiley & Sons, Ltd. Published 2015 by John Wiley & Sons, Ltd.

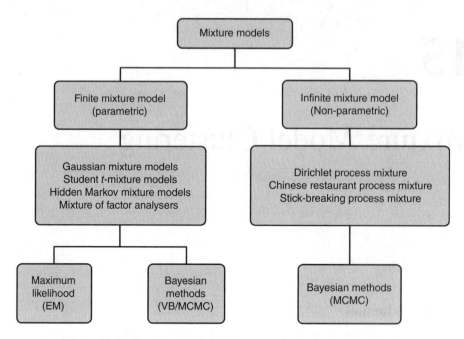

Figure 15.1   The hierarchy of the family of mixture models clustering

assume that the number of mixture components is correctly specified; while infinite mixture models do not rely on this assumption, and consider a countably infinite number of components. Since infinite mixture models may bypass the overfitting problem naturally, they have received more and more attention recently.

There are several types of finite mixture models either in terms of the clustering kernel, namely, the multivariate normal models or Gaussian mixture model, the hidden Markov model (HMM), and other non-Gaussian models like Student t-distribution, or in terms of dealing with dimensionality, MFA, MFA with common factor loadings, mixture of probabilistic principal component analysers, and mixture of multivariate t-distributions. In most of the finite mixture models, there are mainly two subgroups in terms of whether the prior is employed or not in the estimation methods. The models that do not consider the prior are called non-Bayesian models, and are based on an ML method, that is the expectation maximisation (EM) algorithm. The models that consider the prior are called Bayesian models in which either variational Bayes (VB) approximation methods or MCMC methods are employed.

The infinite mixture model is also known as Bayesian non-parametric mixture model, which has become extremely popular in recent years, with applications in diverse fields, particularly bioinformatics. Infinite mixture models can be classified into several types in terms of the different assumptions imposed on the prior of the mixing distribution. The well known Dirichlet process mixture (DPM) model, which is one of the Bayesian non-parametric mixture models, may trace back to the 1970s and was developed by Ferguson and Antoniak (Ferguson, 1973; Antoniak, 1974). In this chapter, we will only introduce DPM, Chinese restaurant process (CRP) mixture, and stick-breaking process (SBP) mixture.

15.2   Finite Mixture Models

In this section, we mainly focus on finite mixture models, which have been widely used to cluster and model the distributions of a variety of random phenomena. Comprehensive technical details have been given in many books (McLachlan and Peel, 2000; Frühwirth-Schnatter, 2006; McLachlan and Krishnan, 2007).

Suppose that we have a dataset $X = \{x_n \in \mathbb{R}^{M \times 1} | n = 1, \ldots, N\}$, where x_n is an M-dimensional vector of feature variables. Generally speaking, finite mixture model methods model the density of X as a mixture of a number G of multivariate component distributions [Equation (15.1)],

$$f(x_n | \boldsymbol{\Theta}) = \sum_{g=1}^{G} \tau_g f(x_n | \boldsymbol{\theta}_g) \tag{15.1}$$

where $\boldsymbol{\Theta} = \{\tau_g, \boldsymbol{\theta}_g | g = 1, \ldots, G\}$ denotes the parameter set, $\tau_g \in [0,1]$ is the probability of membership of the component g, and $\sum_{g=1}^{G} \tau_g = 1$. Essentially, $f(x_n | \boldsymbol{\theta}_g)$ is a generic symbol representing the density of the multivariate random variable x_n with parameters $\boldsymbol{\theta}_g$.

15.2.1   Various Mixture Models

In this section, the main task is to introduce various finite mixture models and leave estimation methods to the subsequent two subsections.

15.2.1.1   Gaussian Mixture Models

In model-based clustering, the Gaussian mixture model, which is also known as the mixture of multivariate normal distributions, is most commonly used (Banfield and Raftery, 1993; Dempster *et al.*, 1977; Yeung *et al.*, 2001; Fraley and Raftery, 2002), because of its advantage over conventional Euclidean distance-based algorithms, say *k*-means and hierarchical clustering. The covariance structure of the Gaussian mixture model potentially accounts for correlations between features within an object. Moreover, the covariance can also be flexible subject to different constraints (Yeung *et al.*, 2001). The density of a finite Gaussian mixture model is given by Equation (15.2),

$$f(x_n | \boldsymbol{\Theta}) = \sum_{g=1}^{G} \tau_g p(x_n | \boldsymbol{\mu}_g, \boldsymbol{\Sigma}_g) \tag{15.2}$$

where $p(x_n | \boldsymbol{\mu}_g, \boldsymbol{\Sigma}_g)$ is a density of a multivariate Gaussian random variable x_n with mean $\boldsymbol{\mu}_g$ and covariance matrix $\boldsymbol{\Sigma}_g$, and $\boldsymbol{\Theta} = \{(\tau_g, \boldsymbol{\mu}_g, \boldsymbol{\Sigma}_g) | g = 1, \cdots, G\}$. The density function of a multivariate Gaussian distribution can be written as shown in Equation (15.3).

Table 15.1 Ten covariance structures are characterised based on the parameterisations

ID	Model	Distribution	Volume	Shape	Orientation
EI	λI	Spherical	Equal	Equal	NA
VI	$\lambda_g I$	Spherical	Variable	Equal	NA
EEE	$\lambda D A D^T$	Ellipsoidal	Equal	Equal	Equal
EEV	$\lambda D_g A D_g^T$	Ellipsoidal	Equal	Equal	Variable
VEV	$\lambda_g D_g A D_g^T$	Ellipsoidal	Variable	Equal	Variable
VVV	$\lambda_g D_g A_g D_g^T$	Ellipsoidal	Variable	Variable	Variable
EEI	λA	Diagonal	Equal	Equal	Coordinate axes
VEI	$\lambda_g A$	Diagonal	Variable	Equal	Coordinate axes
EVI	λA_g	Diagonal	Equal	Variable	Coordinate axes
VVI	$\lambda_g A_g$	Diagonal	Variable	Variable	Coordinate axes

$$p(x|\mu,\Sigma) = \frac{1}{\sqrt{(2\pi)^M |\Sigma|}} \exp\left(-\frac{1}{2}(x-\mu)^T \Sigma^{-1}(x-\mu)\right) \qquad (15.3)$$

Considering that each covariance matrix is parameterised by eigenvalue decomposition in the form shown in Equation (15.4),

$$\Sigma_g = \lambda_g D_g A_g D_g^T \qquad (15.4)$$

where λ_g is a scalar, A_g is a diagonal matrix whose elements are proportional to the eigenvalues of Σ_g, and D_g is the orthogonal matrix of eigenvectors (Fraley and Raftery, 1999; Yeung et al., 2001). These three parameters control three characteristics of the covariance matrix, namely volume (λ_g), shape (A_g) and orientation (D_g), respectively. The parameterisation unifies many well known models, for example, the uniform spherical variance, constant variance, unconstrained variance and so on. Ten covariance structures are shown in Table 15.1. The model identifiers code geometric characteristics of the model. For example 'EEV' denotes the model in which the volumes of all components are equal (E), the shapes of all components are equal (E), and the orientation is allowed to vary (V); 'EI' denotes equal volume spherical model (both A_g and D_g are identity matrices).

15.2.1.2 Hidden Markov Mixture Models

A HMM describes an unobservable stochastic process consisting of a series of states, each of which is related to another stochastic process that emits observable symbols (Rabiner, 1989). A complete HMM can be specified by a set of the following elements:

1. A set of L unobservable states $S = \{S_1, \ldots, S_L\}$;
2. A set of D observable symbols $O = \{O_1, \ldots, O_D\}$;
3. A state transition probability distribution $A = \{a_{ij} | i, j = 1, \ldots, L\}$, where a_{ij} represents the transition probability from state S_i at time t, to state S_j at time $t + 1$ [Equation (15.5)],

$$a_{ij} = P\left(S_j^{(t+1)} | S_i^{(t)}\right), \quad \text{and} \quad \sum_{j=1}^{L} a_{ij} = 1 \tag{15.5}$$

4. A symbol emission probability distribution $B = \{b_i(d) | i = 1, \ldots, L; d = 1, \ldots, D\}$, where $b_i(d)$ represents the emission probability of the symbol x_d at time t, in the state S_i [Equation (15.6)],

$$b_i(d) = P\left(o_d^{(t)} | S_i^{(t)}\right), \quad \text{and} \quad \sum_{d=1}^{D} b_i(d) = 1 \tag{15.6}$$

5. An initial state distribution $\gamma = \{\gamma_i | i = 1, \ldots, L\}$, where $\gamma_i = P\left(S_i^{(0)}\right)$.

The complete set of parameters of an HMM is $\vartheta = \{A, B, \gamma\}$. A graph demonstration of an HMM with three states and four observable symbols is shown in Figure 15.2.

Considering the clustering framework of finite mixture models shown in Equation (15.1), we ought to cluster a set of N sequences $X = \{x_1, \ldots, x_N\}$ into G clusters, where x_n consists of M observation symbols. The mixture probability density can be written as Equation (15.7),

$$f(x_n | \Theta) = \sum_{g=1}^{G} \tau_g f(x_n | \vartheta_g) \tag{15.7}$$

where $f(x_n | \vartheta_g)$ is the density of an HMM with parameters $\vartheta_g = \{A_g, B_g, \gamma_g\}$ including state transition probabilities, observation emission probabilities and initial state probabilities.

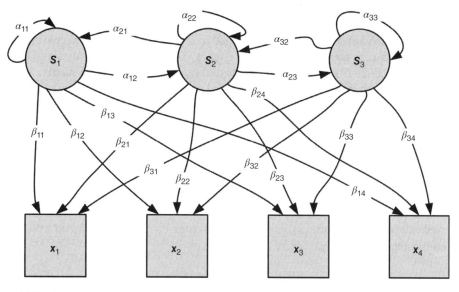

Figure 15.2   Graph demonstration of an HMM. In this demonstration, a set of states $S = \{S_1, S_2, S_3\}$ and an observation set $x = \{x_1, x_2, x_3, x_4\}$ are shown. A state transition probability distribution $A = \{\alpha_{ij}\}$ and a symbol emission probability distribution $B = \{\beta_{il}\}$ are also shown

15.2.1.3 Student *t*-Distribution Mixture Models

The *t*-distribution, also known as Student's *t*-distribution, is a family of continuous probability distributions. An *M*-dimensional random variable x follows a multivariate *t*-distribution with mean, μ, positive definite symmetric and real covariance matrix Σ, and degrees of freedom $\nu \in [0, \infty)$ when, given a weight u, the variable x has the multivariate normal distribution with mean μ and covariance matrix Σ/u as given in Equation (15.8),

$$x|\mu, \Sigma, \nu, u \sim \mathcal{N}(\mu, \Sigma/u) \tag{15.8}$$

where u follows a Gamma distribution parameterised by ν which is denoted as Ga(.,.) [Equation (15.9)].

$$u \sim \text{Ga}(\nu/2, \nu/2) \tag{15.9}$$

The density function of a multivariate *t*-distribution is given by Peel and McLachlan (2000) and McLachlan, Bean and Peel (2002) as Equation (15.10),

$$p(x|\mu, \Sigma, \nu) = \frac{\Gamma\left(\frac{\nu+M}{2}\right)|\Sigma|^{-1/2}}{(\pi\nu)^{M/2}\Gamma\left(\frac{\nu}{2}\right)\left\{1 + \frac{\delta(x,\mu;\Sigma)}{\nu}\right\}^{(\nu+M)/2}} \tag{15.10}$$

where μ is mean vector, Σ is positive definite covariance matrix, $\nu \in [0, \infty)$ is degrees of freedom, and Equation (15.11)

$$\delta(x,\mu;\Sigma) = (x-\mu)^T \Sigma^{-1} (x-\mu) \tag{15.11}$$

denotes the Mahalanobis distance between x and μ. If $\nu > 1$, μ is the mean vector of x, and if $\nu > 2$, $\nu/(\nu-1)\Sigma$ is its covariance matrix. As ν tends to infinity, the multivariate *t*-distribution becomes multivariate Gaussian distribution with mean, μ, and covariance matrix, Σ. As the multivariate *t*-distributions provide a heavy-tailed density function alternative to the normal distribution, they have a sound mathematical basis for a robust method of mixture estimation and clustering.

Therefore, the mixture of *t*-distributions has the following form [Equation (15.12)],

$$f(x_n|\Theta) = \sum_{g=1}^{G} \tau_g f\left(x_n|\mu_g, \Sigma_g, \nu_g\right) \tag{15.12}$$

where the degrees of freedom $\{\nu_g|g = 1, ..., G\}$ can either be fixed or be inferred from the data for each component thereby.

15.2.1.4 Mixture of Factor Analysers

Factor analysis (FA) is a statistical method for modelling the covariance structure of high dimensional data using a small number of latent variables. Suppose that each independent and identically distributed (i.i.d.) M dimensional data vector $\boldsymbol{x}_n$ in dataset X follows a q-factor model as shown in Equation (15.13),

$$\begin{cases} \boldsymbol{x}_n = A\boldsymbol{y}_n + \boldsymbol{\mu} + \boldsymbol{\varepsilon}_n \\ \boldsymbol{y}_n \sim \mathcal{N}(0, \boldsymbol{I}), \ \boldsymbol{\varepsilon}_n \sim \mathcal{N}(0, \boldsymbol{\Psi}) \end{cases} \tag{15.13}$$

where $\boldsymbol{\mu}$ is an M-dimensional mean vector, A is an $M \times q$ factor loading matrix, $\boldsymbol{y}_n$ is a q-dimensional latent factor vector satisfying multivariate Gaussian distribution with zero mean and covariance matrix as identity matrix, $\boldsymbol{I}$ is an identity matrix, and $\boldsymbol{\Psi} = \text{diag}\{\psi_1, \ldots, \psi_M\}$ is a positive diagonal matrix. According to this factor model, the observation vector $\boldsymbol{x}_n$ satisfies the multivariate Gaussian distribution $\mathcal{N}(\boldsymbol{\mu}, \boldsymbol{\Sigma})$, where $\boldsymbol{\Sigma} = AA^T + \boldsymbol{\Psi}$.

The MFA is a mixture of G FA models with mixture proportions $\boldsymbol{\tau} = \{\tau_g | g = 1, \ldots, G\}$. The mixture probability density can be written as Equation (15.14),

$$f(\boldsymbol{x}_n | \boldsymbol{\Theta}) = \sum_{g=1}^{G} \tau_g f(\boldsymbol{x}_n | \boldsymbol{\mu}_g, A_g, \boldsymbol{\Psi}_g) \tag{15.14}$$

where the conditions given in Equation (15.15) hold,

$$\begin{cases} \boldsymbol{x}_n | g = A_g \boldsymbol{y}_{ng} + \boldsymbol{\mu}_g + \boldsymbol{\varepsilon}_{ng} \\ \boldsymbol{y}_{ng} \sim \mathcal{N}(0, \boldsymbol{I}), \ \boldsymbol{\varepsilon}_{ng} \sim \mathcal{N}(0, \boldsymbol{\Psi}_g) \end{cases} \tag{15.15}$$

which indicates that the observation vector $\boldsymbol{x}_n$ generated from the gth FA model has parameters $(\boldsymbol{\mu}_g, A_g, \boldsymbol{\Psi}_g)$.

MFA is also known as parsimonious Gaussian mixture model (PGMM (McNicholas and Murphy, 2008)). Similarly to the GMM, in terms of covariance matrix structure, eight different PGMMs have been provided based on a full range of constraints, as shown in Table 15.2. As we mentioned the covariance structure as being given by $\boldsymbol{\Sigma}_g = A_g A_g^T + \boldsymbol{\Psi}_g$, there are constraints that can be imposed on the covariance matrix, namely loading matrix, error covariance and isotropic. The model identifier code shows the characteristics of these constraints; for example, 'CCU' means that the loading matrix is constrained to be the same for all FA models, and error covariance is also constrained; however, the isotropic characteristics are unconstrained and its

Table 15.2 Eight covariance structures are characterised based on the constraints

ID	Model	Loading matrix	Error variance	Isotropic	Covariance parameters
CCC	$(A, \psi I)$	Constrained	Constrained	Constrained	$Mq - q(q-1)/2 + 1$
CCU	$(A, \boldsymbol{\Psi})$	Constrained	Constrained	Unconstrained	$Mq - q(q-1)/2 + M$
CUC	$(A, \psi_g I)$	Constrained	Unconstrained	Constrained	$Mq - q(q-1)/2 + G$
CUU	$(A, \boldsymbol{\Psi}_g)$	Constrained	Unconstrained	Unconstrained	$Mq - q(q-1)/2 + GM$
UCC	$(A_g, \psi I)$	Unconstrained	Constrained	Constrained	$G(Mq - q(q-1)/2) + 1$
UCU	$(A_g, \boldsymbol{\Psi})$	Unconstrained	Constrained	Unconstrained	$G(Mq - q(q-1)/2) + M$
UUC	$(A_g, \psi_g I)$	Unconstrained	Unconstrained	Constrained	$G(Mq - q(q-1)/2 + 1)$
UUU	$(A_g, \boldsymbol{\Psi}_g)$	Unconstrained	Unconstrained	Unconstrained	$G(Mq - q(q-1)/2 + M)$

shape is ellipsoidal rather than spherical. Therefore, the number of parameters in the covariance matrix for this case is $(Mq - q(q-1)/2 + M)$.

15.2.1.5 Other Mixtures

There are also many other mixture models in the literature; for example, mixture of common factor analysers (Baek, McLachlan and Flack, 2010), mixture of principal component analysers (Tipping and Bishop, 1997, 1999), mixture of t-factor analysers (Andrews and McNicholas, 2011), and mixture of common t-factor analysers (Baek and McLachlan, 2011). For more finite mixture models, the interested reader may be referred to the book by McLachlan and Peel (2000).

15.2.2 Non-Bayesian Methods

15.2.2.1 Maximum Likelihood (ML)

ML has been the most commonly used approach to the fitting of mixture distributions. In this subsection, we first define ML estimation in general, and will dive deep into EM algorithms in the subsequent subsections.

With the ML approach to the estimation of the parameter set Θ in a postulated density function $f(x_n | \Theta)$, an estimate $\widehat{\Theta}$ is provided in regular solution by maximising the likelihood (or log-likelihood) function, which is mathematically given by Equation (15.16),

$$\widehat{\Theta}_{ML} = \arg\max_{\Theta} \mathcal{L}(\Theta) \text{ or } \widehat{\Theta}_{ML} = \arg\max_{\Theta} \{\log \mathcal{L}(\Theta)\} \tag{15.16}$$

where $\mathcal{L}(\Theta)$ is the likelihood function, written as shown in Equation (15.17).

$$\mathcal{L}(\Theta) = \prod_{n=1}^{N} f(x_n | \Theta) \tag{15.17}$$

Thus, $\widehat{\Theta}_{ML}$ can be obtained by solving Equation (15.18).

$$\frac{\partial \mathcal{L}(\Theta)}{\partial \Theta} = 0, \text{ or equivalently } \frac{\partial \log \mathcal{L}(\Theta)}{\partial \Theta} = 0 \tag{15.18}$$

15.2.2.2 Expectation Maximisation (EM) Algorithms

EM algorithms are iterative methods for finding ML estimates of parameters in a wide variety of situations best described as incomplete-data problems. The EM algorithm was originally proposed by Dempster *et al.* (1977). Following this pioneering paper, a large number of applications of EM algorithms have appeared in the literature. It was named expectation maximisation because each iteration of the algorithm consists of an expectation step (E-step) and a maximisation step (M-step).

We now present a simple characterisation of the EM algorithm which can usually be applied. Suppose that $\Theta^{(t)}$ denotes the current value of Θ after t cycles of the algorithm. The next cycle can be described in two steps, as follows:

- E-step: Compute the expected $\mathcal{L}(\Theta)$ with given X and $\Theta^{(t)}$ by use of Equation (15.19) where $E(\cdot)$ denotes the mathematical expectation.

$$Q\left(\Theta|\Theta^{(t)}\right) = \mathbb{E}\left(\mathcal{L}|X,\Theta^{(t)}\right) \tag{15.19}$$

- M-step: Maximise $Q\left(\Theta|\Theta^{(t)}\right)$ with respect to Θ.

The EM algorithms for different mixture models are essentially extended from the above basic idea of E and M steps.

EM Algorithm for GMM

Suppose that the true cluster membership $z_n = \{z_{1n},\ldots,z_{Gn}\} \in \{0,1\}$ for the nth data object, where G is the number of models, is known; the complete data are $y_n = \{(x_n,z_n)|n=1,\ldots,N\}$, where z_n is defined by Equation (15.20).

$$z_{gn} = \begin{cases} 1 \text{ if } x_n \text{ belongs to group g} \\ 0 \text{ otherwise} \end{cases} \tag{15.20}$$

Therefore, it is easy to derive the complete-data likelihood as shown in Equation (15.21)

$$\mathcal{L}(\Theta) = \prod_{n=1}^{N}\prod_{g=1}^{G}\left\{\tau_g f\left(x_n|\mu_g,\Sigma_g\right)\right\}^{z_{gn}} \tag{15.21}$$

or equivalently Equation (15.22).

$$\log\mathcal{L}(\Theta) = \sum_{n=1}^{N}\sum_{g=1}^{G}z_{gn}\log\left\{\tau_g f\left(x_n|\mu_g,\Sigma_g\right)\right\} \tag{15.22}$$

To proceed with the EM algorithm, first of all, a model for Σ_g has to be specified based on Table 15.1; then an initial setting of z_n must be given. A hierarchical agglomerative clustering was used to initialise the EM algorithm (Ghosh and Chinnaiyan, 2002). In all clustering problems, the missing data are z_n representing the cluster assignment. Therefore in the E-step, given $\Theta^{(t)} = \left\{(\tau_g,\mu_g,\Sigma_g)^{(t)}|g=1,\ldots,G\right\}$, we can estimate $\mathbb{E}\left(z_n|X,\Theta^{(t)}\right)$ by Equation (15.23).

$$\hat{z}_{gn}^{(t)} = \frac{\tau_g^{(t)}f\left(x_n|\mu_g^{(t)},\Sigma_g^{(t)}\right)}{\sum_{j=1}^{G}\tau_j^{(t)}f\left(x_n|\mu_j^{(t)},\Sigma_j^{(t)}\right)}, \quad g=1,\ldots,G; n=1,\ldots,N \tag{15.23}$$

The estimated membership $\hat{z}_n$ is then substituted into Equations (15.21) or (15.22) in the M-step. The likelihood or log-likelihood of the complete data can be maximised as a function of Θ. Thus, we may obtain the parameters updated by Equation (15.24).

$$n_g^{(t+1)} = \sum_{n=1}^{N} \hat{z}_{gn}^{(t)}$$

$$\tau_g^{(t+1)} = \frac{n_g^{(t+1)}}{N} \tag{15.24}$$

$$\hat{\mu}_g^{(t+1)} = \frac{\sum_{n=1}^{N} \hat{z}_{gn}^{(t)} x_n}{n_g^{(t+1)}}$$

and $\hat{\Sigma}_g^{(t+1)}$ depends on the chosen covariance model (Celeux and Govaert, 1995; Fraley and Raftery, 1998).

EM Algorithm for Mixture of HMMs

The basic procedure for clustering using a mixture of HMMs was discussed by Smyth (1997). It was suggested that the mixture of HMMs could be described as a single composite HMM with a composite transition matrix [Equation (15.25)],

$$A = \begin{bmatrix} A_1 & \cdots & 0 \\ \vdots & \ddots & \vdots \\ 0 & \cdots & A_G \end{bmatrix} \tag{15.25}$$

where A_g is the transition matrix for the gth component where $g = 1, \ldots, G$. Therefore, the clustering procedure is summarised as follows (Xu and Wunsch, 2008):

1. Model each M-symbol (it is not necessary for them to be of equal length) sequence $x_n | x_{nm} \in O$, $m = 1, \ldots, M$ and $n = 1, \ldots, N$, with an L-state HMM;
2. Calculate the log-likelihood $\log\{f(x_n | \vartheta_n)\}$ of each sequence x_n with respect to a given model ϑ_n, $n = 1, \ldots, N$;
3. Cluster the sequences into G clusters in terms of the distance measure based on the log-likelihood;
4. Model each cluster with an HMM, and initialise the overall composite HMM using the derived G HMMs;
5. Train the composite HMM with the Baum–Welch algorithm.

The EM algorithm for a single HMM, also known as the Baum–Welch algorithm, is used to find the unknown parameters of HMM (Rabiner, 1989), expressed by Equation (15.26).

$$\vartheta = \arg\max_{\vartheta} P(x | \vartheta), \quad \vartheta = \{A, B, \gamma\} \tag{15.26}$$

The Baum–Welch algorithm employs a forward-backward procedure to estimate ϑ iteratively, as shown in the following steps:

1. Initialise the state sequence $\mathcal{S} = \left\{ s^{(1)}, \ldots, s^{(M)} \mid s^{(t)} \in S \right\}$, the observation sequence $\mathcal{O} = \left\{ o^{(1)}, \ldots, o^{(M)} \mid o^{(t)} \in \mathcal{O} \right\}$ and set ϑ with random initial conditions;
2. E-step: Calculate the joint probability of the observations and state sequence for a given model ϑ [Equation (15.27)].

$$P(\mathcal{O}, \mathcal{S} \mid \vartheta) = P(\mathcal{O} \mid \mathcal{S}, \vartheta) P(\mathcal{S} \mid \vartheta)$$
$$P(\mathcal{O} \mid \vartheta) = \sum_{\mathcal{S}} P(\mathcal{O}, \mathcal{S} \mid \vartheta) P(\mathcal{S} \mid \vartheta) \tag{15.27}$$

In the forward procedure, considering $\alpha_i^{(t)} = P\left(o^{(1)}, o^{(2)}, \ldots, o^{(t)}, s^{(t)} = s_i \mid \vartheta \right)$, we can have

$$\alpha_i^{(1)} = \gamma_i b_i \left(o^{(1)} \right);$$
$$\alpha_j^{(t)} = \left[\sum_{i=1}^{L} \alpha_i^{(t-1)} \hat{a}_{ij} \right] \hat{b}_j \left(o^{(t)} \right), j = 1, \ldots, L; t = 2, \ldots, M \tag{15.28}$$

In the backward procedure, considering $\beta_i^{(t)} = P\left(o^{(t+1)}, o^{(t+2)}, \ldots, o^{(M)} \mid s^{(t)} = s_i, \vartheta \right)$, we can obtain the results set out in Equation (15.29).

$$\beta_i^{(M)} = 1;$$
$$\beta_i^{(t)} = \sum_{j=1}^{L} \hat{a}_{ij} \hat{b}_j \left(o^{(t+1)} \right) \beta_j^{(t+1)}, j = 1, \ldots, L; t = 2, \ldots, M \tag{15.29}$$

We can now calculate the temporary variables, as shown in Equation (15.30):

$$\delta_i^{(t)} = \frac{\alpha_i^{(t)} \beta_i^{(t)}}{\sum_{j=1}^{L} \alpha_j^{(t)} \beta_j^{(t)}};$$
$$\zeta_{ij}^{(t)} = \frac{\alpha_i^{(t-1)} \hat{a}_{ij} \hat{b}_j \left(o^{(t)} \right) \beta_j^{(t)}}{\sum_{k=1}^{L} \sum_{l=1}^{L} \alpha_k^{(t-1)} \hat{a}_{kl} \hat{b}_l (o^{(t)}) \beta_l^{(t)}}, i, j = 1, \ldots, L \tag{15.30}$$

3. M-step: $\widehat{\vartheta}$ is updated to maximise the log-likelihood as follows [Equation (15.31)]:

$$\widehat{\vartheta}_{\text{new}} = \arg \max_{\vartheta} P(\mathcal{O} \mid \vartheta) \tag{15.31}$$

Therefore, we can individually optimise each parameter of HMM as follows [Equation (15.32)]:

$$\widehat{\gamma}_i = \delta_i^{(t)}; \hat{a}_{ij} = \frac{\sum_{t=2}^{M} \zeta_{ij}^{(t)}}{\sum_{t=2}^{M} \delta_i^{(t)}}; \hat{b}_j = \frac{\sum_{t=1}^{M} 1_{o^{(t)} = k} \delta_i^{(t)}}{\sum_{t=1}^{M} \delta_i^{(t)}} \tag{15.32}$$

where $1_{o^{(t)} = k}$ is an indicator, which is one if $o^{(t)} = k$, otherwise, 0.

For the new model $\widehat{\boldsymbol{\vartheta}}_{\text{new}}$, $P\left(\mathcal{O}|\widehat{\boldsymbol{\vartheta}}_{\text{new}}\right) \geq P(\mathcal{O}|\boldsymbol{\vartheta})$.

EM Algorithm for Student t-Distribution Mixture Models

In the EM framework, the complete data vector is (X, Z, u), where $X = \{x_n | n = 1, \ldots, N\}$ is the set of observations, $Z = \{z_n | n = 1, \ldots, N\}$ is the component label vector, and $u = \{u_n | n = 1, \ldots, N\}$ is the weighting vector. Suppose that x_n belongs to the gth component; then Equations (15.33) and (15.34) hold.

$$x_n | u_n, z_{gn} = 1 \sim \mathcal{N}\left(\boldsymbol{\mu}_g, \boldsymbol{\Sigma}_g / u_n\right) \tag{15.33}$$

and

$$u_n | z_{gn} = 1 \sim \text{Ga}\left(\frac{\nu_g}{2}, \frac{\nu_g}{2}\right) \tag{15.34}$$

Therefore the complete data log-likelihood Equation (15.35) holds,

$$\log \mathcal{L}(\boldsymbol{\Theta}) = \sum_{n=1}^{N} \sum_{g=1}^{G} z_{gn} \log\left\{ \tau_g f\left(x_n | \boldsymbol{\mu}_g, \boldsymbol{\Sigma}_g, u_n\right) \text{Ga}\left(\frac{\nu_g}{2}, \frac{\nu_g}{2}\right) \right\}$$

$$= \log \mathcal{L}_1(\tau) + \log \mathcal{L}_2\left(\boldsymbol{\mu}_g, \boldsymbol{\Sigma}_g\right) + \log \mathcal{L}_3(\nu) \tag{15.35}$$

where Equations (15.36) hold,

$$\log \mathcal{L}_1(\tau) = \sum_{n=1}^{N} \sum_{g=1}^{G} z_{gn} \log \tau_g \log \mathcal{L}_2\left(\boldsymbol{\mu}_g, \boldsymbol{\Sigma}_g\right) = \sum_{n=1}^{N} \sum_{g=1}^{G} z_{gn}\left\{ -\frac{M}{2}\log 2\pi - \frac{1}{2}\log|\boldsymbol{\Sigma}_g| - \frac{1}{2}u_n \delta_{gn} \right\}$$

$$\log \mathcal{L}_3(\nu) = \sum_{n=1}^{N} \sum_{g=1}^{G} z_{gn}\left\{ -\log \Gamma\left(\frac{\nu_g}{2}\right) + \frac{\nu_g}{2}\log\left(\frac{\nu_g}{2}\right) + \frac{\nu_g}{2}(\log u_n - u_n) - \frac{1}{2}\log u_n \right\} \tag{15.36}$$

where $\delta_{gn} = \left(x_n - \boldsymbol{\mu}_g\right)^T \boldsymbol{\Sigma}_g^{-1}\left(x_n - \boldsymbol{\mu}_g\right)$. Therefore, the E-step calculates the expected complete data log-likelihood $Q(\boldsymbol{\Theta}) = Q_1(\tau) + Q_2\left(\boldsymbol{\mu}_g, \boldsymbol{\Sigma}_g\right) + Q_3(\nu)$, where Equation (15.37) holds.

$$Q_1(\tau) = \sum_{n=1}^{N} \sum_{g=1}^{G} \hat{z}_{gn} \log \tau_g, Q_2\left(\boldsymbol{\mu}_g, \boldsymbol{\Sigma}_g\right)$$

$$= \sum_{n=1}^{N} \sum_{g=1}^{G} \hat{z}_{gn}\left\{ -\frac{M}{2}\log 2\pi - \frac{1}{2}\log|\boldsymbol{\Sigma}_g| - \frac{1}{2}\mathbb{E}\left[u_n | x_n, \hat{z}_{gn} = 1\right]\delta_{gn} \right\} \tag{15.37}$$

To this end, we need to calculate Equations (15.38) and (15.39),

$$\hat{z}_{gn} = \frac{\tau_g f\left(x_n | \mu_g, \Sigma_g, \nu_g\right)}{\sum_{j=1}^{G} f\left(x_n | \mu_j, \Sigma_j, \nu_j\right)}, \hat{\tau}_g = \frac{1}{N}\sum_{n=1}^{N} \hat{z}_{gn} \tag{15.38}$$

$$\mathbb{E}\left[u_n | x_n, \hat{z}_{gn} = 1\right] = \frac{\nu_g + M}{\nu_g + \delta_{gn}}, \mathbb{E}\left[\log u_n | x_n, \hat{z}_{gn} = 1\right] = \psi\left(\frac{\nu_g + M}{2}\right) - \log\left\{\frac{1}{2}\left[\nu_g + \delta_{gn}\right]\right\} \tag{15.39}$$

where $\psi(x) = d\log\Gamma(x)/dx$. In the M-step, maximising the log-likelihood of the complete data yields the update equations of the respective mixture model parameters [Equation (15.40)].

$$\mu_g = \frac{\sum_{n=1}^{N} \hat{z}_{gn} u_{gn} x_n}{\sum_{n=1}^{N} \hat{z}_{gn} u_{gn}}, \hat{\Sigma}_g = \frac{\sum_{n=1}^{N} \hat{z}_{gn} u_{gn}\left(x_n - \mu_g\right)\left(x_n - \mu_g\right)^T}{\sum_{n=1}^{N} \hat{z}_{gn}} \tag{15.40}$$

The degrees of freedom for each component are computed as the solution to Equation (15.41).

$$\log\left(\frac{\nu_g}{2}\right) - \psi\left(\frac{\nu_g}{2}\right) + 1 - \log\left(\frac{\nu_g + M}{2}\right) + \frac{\sum_{n=1}^{N} \hat{z}_{gn}\left(\log u_{gn} - u_{gn}\right)}{\sum_{n=1}^{N} \hat{z}_{gn}} + \psi\left(\frac{\nu_g + M}{2}\right) = 0 \tag{15.41}$$

EM Algorithm for MFA

The EM algorithm was adopted to fit FA and MFA by Ghahramani and Hinton (1996). Let us consider the single factor analyser first of all, as shown in Equation (15.13). The random vector $\left(x_n^T, y_n^T\right)^T$ has a multivariate normal distribution with mean $(\mu^T, 0^T)^T$ and covariance matrix Equation (15.42).

$$\begin{bmatrix} AA^T + \Psi & A \\ A^T & I_q \end{bmatrix} \tag{15.42}$$

It thus follows that the conditional distribution of y_n given x_n is given by Equation (15.43),

$$y_n | x_n \sim \mathcal{N}\left(\beta^T\left(x_n - \mu\right), I_q - \beta^T A\right) \tag{15.43}$$

where $\beta = \left(AA^T + \Psi\right)^{-1} A$. The complete data log-likelihood for the single factor analyser is mathematically given by Equation (15.44).

$$\begin{aligned}
\log\mathcal{L}(X,Y) &= \sum_{n=1}^{N}\log[f(x_n | y_n)f(y_n)] \\
&= C - \frac{N}{2}\log|\Psi| - \frac{1}{2}\text{tr}\left\{\Psi^{-1}\sum_{n=1}^{N}(x_n - \mu)(x_n - \mu)^T\right\} \\
&\quad + \sum_{n=1}^{N}(x_n - \mu)^T\Psi^{-1}Ay_i - \frac{1}{2}\text{tr}\left\{A^T\Psi^{-1}A\sum_{n=1}^{N}y_n y_n^T\right\}
\end{aligned} \tag{15.44}$$

Therefore, in the E-step, we may obtain the expected complete data log-likelihood Q as shown in Equation (15.45).

$$Q(A,\Psi) = C - \frac{N}{2}\log|\Psi| - \frac{1}{2}\mathrm{tr}\left\{\Psi^{-1}\sum_{n=1}^{N}(x_n-\mu)(x_n-\mu)^T\right\} + \sum_{n=1}^{N}(x_n-\mu)^T\Psi^{-1}A\mathbb{E}(y_n|x_n)$$

$$- \frac{1}{2}\mathrm{tr}\left\{A^T\Psi^{-1}A\mathbb{E}(y_n y_n^T|x_n)\right\} \tag{15.45}$$

The expected value of y_n and $y_n y_n^T$ conditional on x_n are $\beta^T(x_n-\mu)$ and $I_q-\beta^T A+\beta^T$ $(x_n-\mu)(x_n-\mu)^T\beta$ respectively. Substituting them in Equation (15.45) yields Equation (15.46),

$$Q(A,\Psi) = C - \frac{N}{2}\log|\Psi| - \frac{N}{2}\mathrm{tr}\{\Psi^{-1}S\} + N\mathrm{tr}\{\Psi^{-1}A\beta^T S\} - \frac{N}{2}\mathrm{tr}\{A^T\Psi^{-1}A\Omega\} \tag{15.46}$$

where $\Omega = I_q - \beta^T A + \beta^T S\beta$.

In the M-step, the factor loading matrix A and the error covariance matrix Ψ can be re-estimated by solving following Equations (15.47) and (15.48).

$$\frac{\partial Q(A,\Psi)}{\partial A} = N\Psi^{-1}S^T\beta - \frac{N}{2}\left[\Psi^{-1}A\Omega^T + \Psi^{-1}A\Omega\right] = 0 \rightarrow \hat{A} = S^T\beta\Omega^{-1} \tag{15.47}$$

$$\frac{\partial Q(A,\Psi)}{\partial \Psi^{-1}} = \frac{N}{2}\Psi - \frac{N}{2}S^T + NS^T\beta A^T - \frac{N}{2}A\Omega^T A^T = 0 \rightarrow \hat{\Psi} = \mathrm{diag}\{S - A\beta^T S\} \tag{15.48}$$

The complete data log-likelihood $\log\mathcal{L}(X|\mu,A,\Psi)$ for the mixture density can be derived as shown in Equation (15.49),

$$\log\mathcal{L}(X|\mu,A,\Psi) = \log\left\{\prod_{n=1}^{N}\prod_{g=1}^{G}\left\{\tau_g f\left(x_n|\mu_g,A_g,\Psi_g\right)\right\}^{z_{gn}}\right\}$$

$$= \sum_{n=1}^{N}\sum_{g=1}^{G}z_{gn}\log\left\{\tau_g f\left(x_n|\mu_g,A_g,\Psi_g\right)\right\} \tag{15.49}$$

where $f\left(x_n|\mu_g,A_g,\Psi_g\right)$ is given by Equation (15.50).

$$f\left(x_n|\mu_g,A_g,\Psi_g\right) = \frac{1}{(2\pi)^{M/2}|\Psi_g|^{1/2}}\exp\left\{-\frac{1}{2}(x_n-\mu_g-A_g y_n)^T\Psi_g^{-1}(x_n-\mu_g-A_g y_n)\right\} \tag{15.50}$$

The alternating expectation conditional maximisation (AECM) algorithm was proposed for fitting this model (Meng and Dyk, 1997). This algorithm is an extension of the EM algorithm that uses different definitions of missing data at different stages. In the E-step, we find the expected complete-data log-likelihood is of the form shown in Equation (15.51),

$$Q(\boldsymbol{\mu}, \boldsymbol{\tau}) = \sum_{g=1}^{G}\sum_{n=1}^{N} z_{gn}\log\tau_g - \frac{NM}{2}\log 2\pi - \sum_{g=1}^{G}\frac{1}{2}\sum_{n=1}^{N} z_{gn}\log\left|\boldsymbol{A}_g\boldsymbol{A}_g^T + \boldsymbol{\Psi}_g\right|$$
$$+ \sum_{g=1}^{G}\sum_{n=1}^{N} z_{gn}\mathrm{tr}\left\{\boldsymbol{S}_g\left(\boldsymbol{A}_g\boldsymbol{A}_g^T + \boldsymbol{\Psi}_g\right)^{-1}\right\}$$

(15.51)

Where Equation (15.52) holds.

$$\hat{z}_{gn} = \frac{\hat{\tau}_g f\left(\boldsymbol{x}_n|\hat{\boldsymbol{\mu}}_g, \hat{\boldsymbol{A}}_g, \hat{\boldsymbol{\Psi}}_g\right)}{\sum_{j=1}^{G}\hat{\tau}_j f\left(\boldsymbol{x}_n|\hat{\boldsymbol{\mu}}_j, \hat{\boldsymbol{A}}_j, \hat{\boldsymbol{\Psi}}_j\right)}$$

(15.52)

and $n_g = \sum_{n=1}^{N} z_{gn}$. Subsequently, in the M-step, maximising the expected complete data log-likelihood with respect to $\boldsymbol{\mu}_g$ and τ_g yields Equations (15.53).

$$\hat{\boldsymbol{\mu}}_g = \frac{\sum_{n=1}^{N}\hat{z}_{gn}\boldsymbol{x}_n}{\sum_{n=1}^{N}\hat{z}_{gn}}$$
$$\hat{\tau}_g = \frac{n_g}{N}$$
$$\boldsymbol{S}_g = \frac{1}{n_g}\sum_{n=1}^{N}\hat{z}_{gn}\left(\boldsymbol{x}_n - \boldsymbol{\mu}_g\right)\left(\boldsymbol{x}_n - \boldsymbol{\mu}_g\right)^T$$

(15.53)

At the second stage of the M-step in the AECM algorithm, when estimating $\boldsymbol{A}_g$ and $\boldsymbol{\Psi}_g$, both the cluster labels $\boldsymbol{Z}$ and the latent factor $\boldsymbol{Y}$ are considered as the missing data. The complete data log-likelihood is mathematically written as Equation (15.54).

$$\log\mathcal{L}(\boldsymbol{X},\boldsymbol{Z},\boldsymbol{Y}) = \sum_{n=1}^{N}\sum_{g=1}^{G} z_{gn}\left[\log\tau_g + \log\{f\left(\boldsymbol{x}_n|\boldsymbol{\mu}_g, \boldsymbol{A}_g, \boldsymbol{\Psi}_g\right)\} + \log f(\boldsymbol{y}_n)\right]$$
$$= C + \sum_{g=1}^{G} n_g\log\tau_g - \sum_{g=1}^{G}\frac{n_g}{2}\log\left|\boldsymbol{\Psi}_g\right| - \sum_{g=1}^{G}\frac{n_g}{2}\mathrm{tr}\left\{\boldsymbol{\Psi}_g^{-1}\boldsymbol{S}_g\right\}$$
$$+ \sum_{g=1}^{G}\sum_{n=1}^{N} z_{gn}\left(\boldsymbol{x}_n - \boldsymbol{\mu}_g\right)^T\boldsymbol{\Psi}_g^{-1}\boldsymbol{A}_g\boldsymbol{y}_n - \sum_{g=1}^{G}\frac{1}{2}\mathrm{tr}\left\{\boldsymbol{A}_g^T\boldsymbol{\Psi}_g^{-1}\boldsymbol{A}_g\sum_{n=1}^{N} z_{gn}\boldsymbol{y}_n\boldsymbol{y}_n^T\right\}$$

(15.54)

It follows that the expected complete data log-likelihood evaluated with $\hat{\boldsymbol{\mu}}_g$ and $\hat{\tau}_g$ is given by Equation (15.55),

$$Q(\boldsymbol{A}_g, \boldsymbol{\Psi}_g) = C + \sum_{g=1}^{G} n_g\log\tau_g + \sum_{g=1}^{G}\frac{n_g}{2}\log\left|\boldsymbol{\Psi}_g^{-1}\right| - \sum_{g=1}^{G}\frac{n_g}{2}\mathrm{tr}\left\{\boldsymbol{\Psi}_g^{-1}\boldsymbol{S}_g\right\} + \sum_{g=1}^{G}\mathrm{tr}\left\{\boldsymbol{\Psi}_g^{-1}\boldsymbol{A}_g\boldsymbol{\beta}_g^T\boldsymbol{S}_g\right\}$$
$$- \sum_{g=1}^{G}\frac{1}{2}\mathrm{tr}\left\{\boldsymbol{A}_g^T\boldsymbol{\Psi}_g^{-1}\boldsymbol{A}_g\boldsymbol{\Omega}_g\right\}$$

(15.55)

where $\beta_g = \left(A_g A_g^T + \Psi_g \right)^{-1} A_g$, and $\Omega_g = I_q - \beta_g^T A_g + \beta_g^T S_g \beta_g$. When we impose the constraints, which have been shown in Table 15.2, on A_g and Ψ_g, the resulting estimates can be easily derived from the expression from $Q(A_g, \Psi_g)$ given above.

15.2.3   Bayesian Methods

Although non-Bayesian methods, basically ML-based methods, have received extensive attention for their effectiveness in estimating optimal values of parameters fitting a given dataset, there are some vital problems with the ML-based methods which limit their practical use in many applications. First, they produce models that overfit the data and subsequently have suboptimal generalisation performance. Second, they cannot be used to learn the structure of graphical models (considering a graph denoting the conditional dependence structure between random variables). Third, standard parameter-estimation methods may fail because of singularity, which, in practice, arises most often for models in which the covariance is allowed to vary between components and for models in which the number of observations is less than the number of the parameters to be estimated. Bayesian methods provide a solution to the above problems in principle. They may be regarded as estimating the uncertainty of the model as a whole and the uncertainty in the estimated parameters themselves. There are three main techniques for Bayesian model-based clustering, namely EM for maximum a posteriori (MAP), variational Bayes EM (VBEM), and MCMC.

15.2.3.1   EM for MAP

Fraley and Raftery proposed an MAP estimator to replace the ML estimator and developed an EM algorithm to search for the solution (Fraley and Raftery, 2007). In this framework, a highly dispersed proper conjugate prior on the parameters was employed to eliminate the failure caused by singularity.

The Bayesian predictive density for the data is assumed to be of the form shown in Equation (15.56),

$$\mathcal{L}\left(X | \mu_g, \Sigma_g, \tau_g \right) \mathcal{P}\left(\mu_g, \Sigma_g, \tau_g | \theta \right) \tag{15.56}$$

where $\mathcal{P}\left(\mu_g, \Sigma_g, \tau_g | \theta \right)$ is a prior distribution on the parameters μ_g, Σ_g, and τ_g, which involves other hyperparameters θ. A normal prior on the mean conditional on the covariance matrix was given by Equation (15.57)

$$\mu | \Sigma \sim \mathcal{N}(\mu_\mathcal{P}, \Sigma / \kappa_\mathcal{P})$$
$$\propto |\Sigma|^{-\frac{1}{2}} \exp\left\{ -\frac{\kappa_\mathcal{P}}{2} \mathrm{tr}\left[\Sigma^{-1} (\mu - \mu_\mathcal{P})(\mu - \mu_\mathcal{P})^T \right] \right\} \tag{15.57}$$

and an inverse Wishart prior on the covariance matrix [Equation (15.58)].

$$\Sigma \sim \text{inverseWishart}(\nu_\mathcal{P}, \Lambda_\mathcal{P}) \quad \propto |\Sigma|^{-\frac{\nu_\mathcal{P}+M+1}{2}} \exp\left\{-\frac{1}{2}\text{tr}\left[\Sigma^{-1}\Lambda_\mathcal{P}\right]\right\} \tag{15.58}$$

The hyperparameters $\mu_\mathcal{P}$, $\kappa_\mathcal{P}$ and $\nu_\mathcal{P}$ are called the mean, shrinkage and degrees of freedom, respectively, of the prior distribution. The hyperparameter $\Lambda_\mathcal{P}$, which is a matrix, is called the scale of inverse Wishart prior. This is a conjugate prior for a multivariate normal distribution because the posterior can also be expressed as the product of a normal distribution and an inverse Wishart distribution. Under this prior, the EM algorithm for the MAP estimator calculates $\hat{z}_{gn}$ in the E-step as shown in Equation (15.59)

$$\hat{z}_{gn} = \frac{\tau_g f\left(x_n|\mu_g, \Sigma_g\right) f\left(\mu_g|\Sigma_g\right) f\left(\Sigma_g\right)}{\sum_{k=1}^{G} \tau_k f\left(x_n|\mu_k, \Sigma_k\right) f\left(\mu_k|\Sigma_k\right) f\left(\Sigma_k\right)} \tag{15.59}$$

where Equation (15.60) holds.

$$\tau_g = \frac{1}{N}\sum_{n=1}^{N}\hat{z}_{gn} \tag{15.60}$$

In the M-step, the update of mean $\hat{\mu}_g$ and unconstrained ellipsoidal covariance matrix $\hat{\Sigma}_k$ are given by Equation (15.61),

$$\hat{\mu}_g = \frac{n_g\bar{x}_g + \kappa_\mathcal{P}\mu_\mathcal{P}}{\kappa_\mathcal{P} + n_g}, \hat{\Sigma}_g = \frac{\Lambda_\mathcal{P} + \frac{\kappa_\mathcal{P}n_g}{\kappa_\mathcal{P}+n_g}\left(\bar{x}_g - \mu_\mathcal{P}\right)\left(\bar{x}_g - \mu_\mathcal{P}\right)^T + W_g}{\nu_\mathcal{P} + n_gM + M + 2} \tag{15.61}$$

where $n_g = \sum_{n=1}^{N}\hat{z}_{gn}$, $\bar{x}_g = \sum_{n=1}^{N}\hat{z}_{gn}x_n/n_g$ and $W_g = \sum_{n=1}^{N}\hat{z}_{gn}\left(x_n - \mu_g\right)\left(x_n - \mu_g\right)^T$. The other constrained covariance matrix structures and their derivations, together with the selection of hyperparameters $(\mu_\mathcal{P}, \kappa_\mathcal{P}, \nu_\mathcal{P}, \Lambda_\mathcal{P})$, can be found in Fraley and Raftery (2007).

However, it is worth noting that this Bayesian regularisation method did not solve the overfitting problem, and it relied on the Bayesian information criterion (BIC), which is a model selection algorithm.

15.2.3.2 MCMC Methods

Bayesian analysis of a mixture model usually leads to intractable calculations, since the posterior distribution takes into account all the partitions of the sample. MCMC techniques as approximation methods, which rely on the missing data structure of the mixture model, evaluate the posterior distribution and Bayesian estimators by sampling (Diebolt and Robert, 1994). Richardson and Green (1997) developed reversible jump MCMC (RJMCMC) to estimate the mixture models with unknown number of components using a series of combine-split moves. In Richardson and Green (1997), only a univariate normal mixture model was considered. Stephens proposed an alternative method to RJMCMC for Bayesian analysis of mixture models with unknown number of components (Stephens, 2000), which considered both univariate and bivariate normal mixture models. Dellaportas and Papageorgiou extended RJMCMC to multivariate normal mixture models (Dellaportas and Papageorgiou, 2006).

Let us start with a univariate normal mixture as studied in Diebolt and Robert (1994). The likelihood of the mixture model for the dataset X is given by Equation (15.62).

$$\mathcal{L}(X|\Theta) = \prod_{n=1}^{N}\prod_{g=1}^{G}\left\{\tau_g f\left(x_n|\mu_g,\sigma_g\right)\right\}^{z_{gn}} \tag{15.62}$$

The conjugate prior is $\mathcal{N}\left(\xi_g, \sigma_g^2/\kappa_g\right)$ for $f\left(\mu_g|\sigma_g^2\right)$, and an inverse Gamma (IG) distribution $\text{IG}\left(\nu_g, S_g^2\right)$ for σ_g^2. Let us assume that the hyperparameters are known. The posterior distribution of (μ_g, σ_g) is given in Equation (15.63).

$$\prod_{g=1}^{G}\left\{\mathcal{N}\left(\frac{\kappa_g\xi_g+n_g\bar{x}_g}{\kappa_g+n_g},\frac{\sigma_g^2}{\kappa_g+n_g}\right)\text{IG}\left(\nu_g+n_g, S_g^2+\hat{S}_g^2+\frac{\kappa_g n_g}{\kappa_g+n_g}\left(\xi_g-\bar{x}_g\right)^2\right)\right\} \tag{15.63}$$

An approximation of the posterior distribution starts with an initial value $\Theta^{(0)}$, the algorithm runs in the following way:

1. Generating $z^{(m)} \sim f\left(z|X,\Theta^{(m)}\right)$;
2. Generating $\Theta^{(m+1)} \sim f\left(\Theta|X,z^{(m)}\right)$.

Considering the conjugate priors, the mth simulation for posterior distribution $f\left(\mu_g,\sigma_g, \tau_g|X, z^{(m)}\right)$ in the second step is

i. $\tau^{(m+1)} \sim \text{Dir}(\alpha_1+n_1,...,\alpha_G+n_G)$, where $\text{Dir}(\alpha_1, ..., \alpha_G)$ is a Dirichlet distribution;
ii. For $g=1,...,G$

Generating $\sigma_g^2 \sim \text{IG}\left(\nu_g+n_g, S_g^2+\hat{S}_g^2+\frac{\kappa_g n_g}{\kappa_g+n_g}\sum_{n=1}^{N}\hat{z}_{gn}^{(m)}\left(x_n-\xi_g\right)^2\right)$;

Generating $\mu_g \sim \mathcal{N}\left(\frac{\kappa_g\xi_g+\sum_{n=1}^{N}\hat{z}_{gn}^{(m)}x_n}{\kappa_g+n_g},\frac{\left(\sigma_g^{(m+1)}\right)^2}{\kappa_g+n_g}\right)$.

However, in the above framework, the number of components was assumed known to the algorithm, which is not realistic. Richardson and Green considered a fully Bayesian analysis of mixtures of normal distributions using the RJMCMC approach (Richardson and Green, 1997). Figure 15.3 demonstrates the complete graphical model in a directed acyclic graph, which is also known as a Bayesian network. Square shapes represent parameters or latent variables; circle shapes represent hyperparameters. Arrows indicate conditional independencies between parameters. Therefore, the joint distribution of all variables is then expressed by the factorisation given in Equation (15.64),

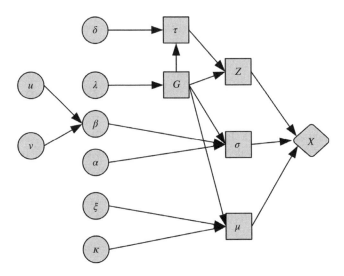

Figure 15.3 Graphical demonstration of Bayesian analysis of a normal mixture model. Squares represent parameters and latent variables; circles represent hyperparameters; diamond represents the observation

$$p(\lambda,\delta,\eta,\tau,G,Z,\Theta,X) = p(\lambda)p(\delta)p(\eta)p(G|\lambda)p(\tau|G,\delta)p(Z|G,\tau)p(\Theta|G,\eta)p(X|\Theta,Z) \quad (15.64)$$

where η is the set of hyperparameters $(\alpha, \beta, \xi, \kappa)$ and Θ denotes (μ, σ). The hyperparameter β is assume to follow a Gamma distribution with parameters u and v. In contrast with other MCMC algorithms, RJMCMC employs two extra move types: (1) splitting one mixture component into two or combining two into one, and (2) the birth or death of an empty component. Dellaportas and Papageorgiou extended RJMCMC to multivariate normal mixture models (Dellaportas and Papageorgiou, 2006).

15.2.3.3   Variational Inference

As mentioned in the last subsection, for a fully Bayesian model, in which any unknown parameters are given prior distributions and are absorbed into the set of latent variables, sometimes, especially for many models of practical interest, it is infeasible to evaluate the posterior distribution and to compute expectations with respect to this distribution. Virtually, to cope with this situation, we resort to approximation schemes. Approximation schemes primarily fall into two broad classes, according to whether they rely on stochastic or deterministic approximations: One class is stochastic techniques, such as MCMC which we discussed previously; another one is variational approximation, which is deterministic (Beal and Ghahramani, 2003; Bishop, 2006). In this subsection, we will discuss variational inference and VBEM algorithms for Bayesian analysis of mixture models.

Let us denote the set of all latent variables and parameters by Θ, the set of all observed variables by X. We can compose the log marginal probability using Equation (15.65),

$$\ln p(X) = \mathcal{L}(q) + \mathrm{KL}(q\|p) \tag{15.65}$$

where Equation (15.66) holds,

$$\mathcal{L}(q) = \int q(\boldsymbol{\Theta}) \ln\left\{\frac{p(X,\boldsymbol{\Theta})}{q(\boldsymbol{\Theta})}\right\} d\boldsymbol{\Theta}, \mathrm{KL}(q\|p) = \int q(\boldsymbol{\Theta}) \ln\left\{\frac{q(\boldsymbol{\Theta})}{p(\boldsymbol{\Theta}|X)}\right\} d\boldsymbol{\Theta} \tag{15.66}$$

where $\mathrm{KL}(q\|p)$ denotes Kullback–Leibler (KL) divergence between $q(\boldsymbol{\Theta})$ and $p(\boldsymbol{\Theta}|X)$. There-fore, maximising $\mathcal{L}(q)$ is equivalent to minimising the KL divergence between $q(\boldsymbol{\Theta})$ and $p(\boldsymbol{\Theta}|X)$. $\mathcal{L}(q)$ is also known as lower bound of log marginal probability $\ln p(X)$. Instead of cal-culating $p(\boldsymbol{\Theta}|X)$ directly, an alternative way, in which the family of distributions $q(\boldsymbol{\Theta})$ are restricted, is considered. Suppose that we partition the elements of $\boldsymbol{\Theta}$ into disjoint groups, which is denoted as $\boldsymbol{\Theta}_j, j = 1, \ldots, J$. Then the q distribution factorises as shown in Equation (15.67).

$$q(\boldsymbol{\Theta}) = \prod_{j=1}^{J} q_j(\boldsymbol{\Theta}_j) \tag{15.67}$$

Subsequently, we seek for the distributions maximising the lower bound $\mathcal{L}(q)$. To this end, we first dissect the dependence of $\mathcal{L}(q)$ on one of the factors $q_j(\boldsymbol{\Theta}_j)$ (q_j for the sake of simplicity) [Equation (15.68)].

$$\begin{aligned}
\mathcal{L}(q) &= \int \prod_j q_j \left\{ \ln p(X,\boldsymbol{\Theta}) - \sum_j \ln q_j \right\} d\boldsymbol{\Theta} \\
&= \int q_j \left\{ \int \ln p(X,\boldsymbol{\Theta}) \prod_{i \neq j} q_i d\boldsymbol{\Theta}_i \right\} d\boldsymbol{\Theta}_j - \int q_j \ln q_j d\boldsymbol{\Theta}_j + C \\
&= \int q_j \ln \widetilde{p}(X,\boldsymbol{\Theta}) d\boldsymbol{\Theta}_j - \int q_j \ln q_j d\boldsymbol{\Theta}_j + C
\end{aligned} \tag{15.68}$$

where $\ln \widetilde{p}(X,\boldsymbol{\Theta}) = \int \ln p(X,\boldsymbol{\Theta}) \prod_{i \neq j} q_i d\boldsymbol{\Theta}_i = \mathbb{E}_{i \neq j}[\ln p(X,\boldsymbol{\Theta})]$ and C is a constant. We may recog-nise that the first two items in Equation (15.68) are equal to a negative KL divergence. Maximising $\mathcal{L}(q)$ is equivalent to minimising the KL divergence between q_j and $\widetilde{p}(X,\boldsymbol{\Theta})$. Thus we obtain a general expression for the optimal solution q_j^* given by Equation (15.69).

$$q_j^* = \exp\left\{ \mathbb{E}_{i \neq j}[\ln p(X,\boldsymbol{\Theta})] \right\} \tag{15.69}$$

Here, let us consider VBEM of GMM for example. Suppose $\boldsymbol{\Theta} = \{\tau, \mu, \Sigma, z\}$, where z is the latent variables, $\tau = \{\tau_g\}, \mu = \{\mu_g\}, \Sigma = \{\Sigma_g\}, g = 1, \ldots, G$. To formulate the variational treat-ment of this model, we start with the joint distribution of all the random variables, which is given by Equation (15.70),

$$p(X,\tau,z,\mu,\Sigma) = p(X|z,\mu,\Sigma)p(z|\tau)p(\tau)p(\mu|\Sigma)p(\Sigma) \tag{15.70}$$

in which many factors are defined in Equation (15.71),

$$p(X|z,\mu,\Sigma) = \prod_{n=1}^{N}\prod_{g=1}^{G}\{\mathcal{N}(x_n|\mu_g,\Sigma_g)\}^{z_{gn}} \tag{15.71}$$

and the conditional distribution of z, given mixing coefficients τ, in the form shown in Equation (15.72).

$$p(z|\tau) = \prod_{n=1}^{N}\prod_{g=1}^{G}\tau_g^{z_{gn}} \tag{15.72}$$

Then we introduce priors over parameters τ, μ, Σ. The analysis is considerably simplified if we choose conjugate prior distribution. Therefore, we choose Dirichlet distribution over the mixing coefficients τ [Equation (15.73)],

$$p(\tau) = \mathrm{Dir}(\tau|\alpha_0) = C(\alpha_0)\prod_{g=1}^{G}\tau_g^{\alpha_0-1} \tag{15.73}$$

where $C(\alpha) = \Gamma\left(\sum_{g=1}^{G}\alpha_g\right)/\Gamma(\alpha_1)\cdots\Gamma(\alpha_g)$. We choose an independent Gaussian–Wishart prior for the mean and covariance of each Gaussian component [Equation (15.74)].

$$p(\mu,\Sigma) = p(\mu|\Sigma)p(\Sigma) = \prod_{g=1}^{G}\mathcal{N}(\mu_g|m_0,\Sigma_g/\kappa_0)\mathcal{W}\left(\Sigma_g^{-1}|W_0,v_0\right) \tag{15.74}$$

We now consider inferring the model using the VBEM algorithm. In the variational Bayes expectation (VBE) step, the update equation of the log of the optimised factor $q(z)$ is given by Equation (15.75),

$$\begin{aligned}
lnq^*(z) &= \mathbb{E}_{\mu,\Sigma,\tau}[\ln p(X,z,\mu,\Sigma,\tau)] + C \\
&= \mathbb{E}_{\tau}[\ln p(z|\tau)] + \mathbb{E}_{\mu,\Sigma}[\ln p(X|z,\mu,\Sigma)] + C \\
&= \sum_{n=1}^{N}\sum_{g=1}^{G}z_{gn}\ln\rho_{gn} + C
\end{aligned} \tag{15.75}$$

where we define $\ln\rho_{gn}$ in Equation (15.76).

$$\ln\rho_{gn} = \mathbb{E}[\ln\tau_g] - \frac{1}{2}\mathbb{E}[\ln|\Sigma_g|] - \frac{M}{2}\ln 2\pi - \frac{1}{2}\mathbb{E}_{\mu_g,\Sigma_g}\left[(x_n-\mu_g)^T\Sigma_g^{-1}(x_n-\mu_g)\right] \tag{15.76}$$

Taking the exponential of both sides of Equation (15.75), we obtain Equation (15.77).

$$q^*(z) \propto \prod_{n=1}^{N}\prod_{g=1}^{G}\rho_{gn}^{z_{gn}} \tag{15.77}$$

Requiring that this distribution be normalised, we obtain Equation (15.78),

$$q^*(z) = \prod_{n=1}^{N}\prod_{g=1}^{G} r_{gn}^{z_{gn}} \tag{15.78}$$

where $r_{gn} = \rho_{gn}\big/\sum_{j=1}^{G}\rho_{jn}$. Subsequently, we define three statistics of the observed statistics of the observed dataset, given by Equations (15.79).

$$n_g = \sum_{n=1}^{N} r_{gn}; \bar{x}_g = \frac{1}{n_g}\sum_{n=1}^{N} r_{gn}x_n; S_g = \frac{1}{n_g}\sum_{n=1}^{N} r_{gn}\left(x_n - \bar{x}_g\right)\left(x_n - \bar{x}_g\right)^T \tag{15.79}$$

In the variational Bayes maximisation (VBM) step, we consider the factorisation shown in Equation (15.80).

$$q(\tau,\mu,\Sigma) = q(\tau)\prod_{g=1}^{G} q\left(\mu_g,\Sigma_g\right) \tag{15.80}$$

Since we can have the optimised factor $q^*(\tau)$, which is given by Equation (15.81),

$$q^*(\tau) = (\alpha_0 - 1)\sum_{g=1}^{G}\ln\tau_g + \sum_{g=1}^{G}\sum_{n=1}^{N} r_{gn}\ln\tau_g + C \tag{15.81}$$

then taking the exponential of both sides, we recognise $q^*(\tau)$ as a Dirichlet distribution [Equation (15.82)].

$$q^*(\tau) = \mathrm{Dir}(\tau|\alpha), \ \alpha_g = \alpha_0 + n_k \tag{15.82}$$

Then, the optimised factor $q^*(\mu_g,\Sigma_g)$ can be obtained by using Equation (15.83),

$$q^*(\mu_g,\Sigma_g) = \mathcal{N}\left(\mu_g|m_g,\Sigma_g/\kappa_g\right)\mathcal{W}\left(\Sigma_g^{-1}|W_g,v_g\right) \tag{15.83}$$

where Equations (15.84) hold.

$$\begin{aligned} \kappa_g &= \kappa_0 + n_g \\ m_g &= \frac{1}{\kappa_g}\left(\kappa_0 m_0 + n_g\bar{x}_g\right) \\ W_g^{-1} &= W_0^{-1} + n_g S_g + \frac{\kappa_0 n_g}{\kappa_0 + n_g}\left(\bar{x}_g - m_0\right)\left(\bar{x}_g - m_0\right)^T \\ v_g &= v_0 + n_g \end{aligned} \tag{15.84}$$

Therefore, we can further calculate the expectations with respect to the variational distributions of the parameters as given in Equations (15.85).

$$\mathbb{E}_{\mu_g, \Sigma_g}\left[(x_n - \mu_g)^T \Sigma_g^{-1} (x_n - \mu_g)\right] = M\kappa_g^{-1} + v_g (x_n - m_g)^T W_g (x_n - m_g)$$

$$\ln \widetilde{\Sigma}_g^{-1} = \mathbb{E}\left[\ln |\Sigma_g^{-1}|\right] = \sum_{i=1}^{M} \psi\left(\frac{v_g + 1 - i}{2}\right) + M\ln 2 + \ln |W_g| \tag{15.85}$$

$$\ln \widetilde{\tau}_g = \mathbb{E}\left[\ln \tau_g\right] = \psi(\alpha_g) - \psi(\widehat{\alpha})$$

Finally, we can evaluate the lower bound Equation (15.66) straightforwardly as shown in Equation (15.86),

$$\mathcal{L} = \sum_n q(\tau, \mu, \Sigma, z) \ln\left\{\frac{p(X, \tau, \mu, \Sigma, z)}{q(\tau, \mu, \Sigma, z)}\right\} d\tau d\mu d\Sigma$$

$$= \mathbb{E}[\ln p(X | \mu, \Sigma, z)] + \mathbb{E}[\ln p(z | \tau)] + \mathbb{E}[\ln p(\tau)] + \mathbb{E}[\ln p(\mu, \Sigma)] \tag{15.86}$$

$$- \mathbb{E}[\ln q(z)] - \mathbb{E}[\ln q(\tau)] - \mathbb{E}[\ln q(\mu, \Sigma)]$$

where Equations (15.87)–(15.92) hold,

$$\mathbb{E}[\ln p(X | \mu, \Sigma, z)] = \frac{1}{2}\sum_{g=1}^{G} n_g \left\{\ln \widetilde{\Sigma}_g^{-1} - M\kappa_g^{-1} - v_g \text{Tr}(S_g W_g)\right\}$$

$$- \sum_{g=1}^{G} n_g v_g (x_n - m_g)^T W_g (x_n - m_g) - \frac{M}{2}\ln 2\pi \sum_{g=1}^{G} n_g \tag{15.87}$$

$$\mathbb{E}[\ln p(z | \tau)] = \sum_{n=1}^{N}\sum_{g=1}^{G} r_{gn} \ln \widetilde{\tau}_g;$$

$$\mathbb{E}[\ln p(\tau)] = \ln C(\alpha_0) + (\alpha_0 - 1)\sum_{g=1}^{G} \ln \widetilde{\tau}_g \tag{15.88}$$

$$\mathbb{E}[\ln p(\mu, \Sigma)] = \frac{1}{2}\sum_{g=1}^{G}\left\{M\ln(\kappa_0 / 2\pi) + \ln \widetilde{\Sigma}_g^{-1} - \frac{M\kappa_0}{\kappa_g} - \kappa_0 v_k (m_g - m_0)^T W_g (m_g - m_0)\right\}$$

$$+ G\ln B(W_0, v_0) + \frac{v_0 - M - 1}{2}\sum_{g=1}^{G}\ln \widetilde{\Sigma}_g^{-1} - \frac{1}{2}\sum_{g=1}^{G} v_g \text{Tr}(W_0^{-1} W_g) \tag{15.89}$$

$$\mathbb{E}[\ln q(z)] = \sum_{n=1}^{N}\sum_{g=1}^{G} r_{gn} \ln r_{gn} \tag{15.90}$$

$$\mathbb{E}[\ln q(\tau)] = \sum_{g=1}^{G} (\alpha_g - 1)\ln \widetilde{\tau}_g + \ln C(\alpha) \tag{15.91}$$

$$\mathbb{E}[\ln p(\mu, \Sigma)] = \sum_{g=1}^{G}\left\{\frac{1}{2}\ln \widetilde{\Sigma}_g^{-1} + \frac{M}{2}\ln\left(\frac{\kappa_g}{2\pi}\right) - \frac{M}{2} - H[q(\Sigma_g)]\right\} \tag{15.92}$$

where $B(\boldsymbol{W}_0, v_0)$ in Equation (15.89) is defined as given in Equation (15.93),

$$B(\boldsymbol{W}, v) = |\boldsymbol{W}|^{-v/2} \left(2^{\frac{vM}{2}} \pi^{\frac{M(M-1)}{4}} \prod_{i=1}^{M} \Gamma \left(\frac{v+1-i}{2} \right) \right)^{-1} \tag{15.93}$$

and $H[q(\boldsymbol{\Sigma}_g)]$ in Equation (15.92) is the entropy of the Wishart distribution, given by Equation (15.94).

$$H\left[q\left(\boldsymbol{\Sigma}_g\right)\right] = -\ln B\left(\boldsymbol{W}_g, v_g\right) - \frac{v_g - M - 1}{2} \mathbb{E}\left[\ln\left|\boldsymbol{\Sigma}_g^{-1}\right|\right] + \frac{vM}{2} \tag{15.94}$$

15.3 Infinite Mixture Models

Suppose we are interested in estimating a single distribution from an i.i.d sample $x_n | n = 1, ..., N$; non-parametric mixtures assume that observations arise from a function such as that given in Equation (15.95),

$$x_n \sim \int k(\cdot|\psi)\tau(d\psi) \tag{15.95}$$

where $k(\cdot|\psi)$ is a given parametric kernel indexed by ψ, and τ is a mixing distribution, which is assigned a flexible prior. In terms of the assumption imposed on τ, the Bayesian non-parametric mixture models may be DPM models (Rasmussen, 1999; Medvedovic and Sivaganesan, 2002; Medvedovic, Yeung and Bumgarner, 2004), CRP models (Qin, 2006), SBP models (Rodriguez and Dunson, 2011), and so on.

15.3.1 DPM Model

A DPM model is also known as an infinite Gaussian mixture model (Rasmussen, 1999). We have discussed the finite Gaussian mixture model in Section 15.2.1.1, and its formula is written in Equation (15.2), which implies that the mixture model contains G components and each component is a multivariate Gaussian distribution with parameters $\boldsymbol{\mu}_g$ and $\boldsymbol{\Sigma}_g$, $g = 1, ..., G$. All components are combined according to a mixing proportion $\{\tau_g | g = 1, ..., G\}$. The clusters then can be found by fitting the model as a function of the set of parameters $\boldsymbol{\Theta} = \{(\tau_g, \boldsymbol{\mu}_g, \boldsymbol{\Sigma}_g) | g = 1, \cdots, G\}$. We have also discussed the Bayesian framework in Section 15.2.3, which defines prior distributions for all parameters, also including some hyperparameters. A prior over the mixing proportion parameters τ is defined to be a symmetric Dirichlet distribution as the natural conjugate prior, with concentration parameter α/G as shown in Equation (15.96),

$$P(\tau|\alpha) = \frac{\Gamma(\alpha)}{\Gamma(\alpha/G)^G} \prod_{g=1}^{G} \tau_g^{\alpha/G - 1} \tag{15.96}$$

where α controls the distribution of the prior weight assigned to each cluster and $\Gamma(\cdot)$ is the Gamma function. The component label (also known as the indicator variable) z_n for each data point which takes on integer value $g \in \{1,\ldots,G\}$ corresponds to the hypothesis that the data point n belongs to the cluster g. The prior probability is proportional to the mixing proportion $P(z_n = g|\tau) = \tau_g$. Therefore, given the mixing proportions, the prior for the occupation numbers, N_g, which is the number of members in the cluster g, is multinomial and joint distribution of the labels becomes as shown in Equation (15.97).

$$P(z_1,\ldots,z_N|\tau_1,\ldots,\tau_G) = \prod_{g=1}^{G} \tau_g^{N_g} \tag{15.97}$$

The conditional probability of one label given the setting of all other labels after integrating over all possible settings of the mixing proportion parameters becomes as given in Equation (15.98).

$$P(z_n = g|z_{-n},\alpha) = \int P(z_n = g|z_{-n},\tau)P(\tau|z_{-n},\tau)d\tau = \frac{N_{-n,g} + \alpha/G}{N-1+\alpha} \tag{15.98}$$

A vague prior of inverse Gamma shape is put on the concentration parameter α to give Equation (15.99).

$$P(\alpha^{-1}) \sim \mathcal{G}(1,1) \rightarrow P(\alpha) \propto \alpha^{-\frac{3}{2}}\exp\left(-\frac{1}{2\alpha}\right) \tag{15.99}$$

Therefore, the likelihood for α may be derived as shown in Equation (15.100).

$$P(\alpha|G,N) \propto \frac{\alpha^{G-3/2}\exp\left(-\frac{1}{2\alpha}\right)\Gamma(\alpha)}{\Gamma(N+\alpha)} \tag{15.100}$$

The conditional posterior for α depends only on the number of observations N, and the number of components G.

The infinite Gaussian mixture models take the limit $G \rightarrow \infty$, and then the conditional distribution for labels is written as shown in Equation (15.101).

$$\begin{aligned} \text{Components where } N_{-n,g} > 0: \quad P(z_n = g|z_{-n},\alpha) &= \frac{N_{-n,g}}{N-1+\alpha} \\ \text{All other components}: \ P(z_n \neq z_{n'}, \forall n' \neq n|z_{-n},\alpha) &= \frac{\alpha}{N-1+\alpha} \end{aligned} \tag{15.101}$$

This shows that the probabilities are proportional to the occupation numbers $N_{-n,g}$. Using these conditional probabilities, one can Gibbs-sample from the indicator variables efficiently, even though the model has infinitely many Gaussian clusters. Having integrated out the mixing proportions, one can also Gibbs-sample from all of the remaining parameters of the model, that is $\{\mu, \Sigma\}$. The rest of the MCMC procedure can be found in Rasmussen (1999). Heller and Ghahramani proposed Bayesian hierarchical clustering (BHC), which can be interpreted as a fast bottom-up approximate inference method for a DPM model (Heller and Ghahramani, 2005).

15.3.2   CRP Mixture Model

The Chinese restaurant process (CRP) gets its name because it can be viewed as a sequential restaurant 'seating arrangement' described as follows. Assume a Chinese restaurant with an infinite number of circular tables, each with infinite capacity, and customers arrive sequentially at the restaurant and are randomly assigned to a table. A new arriving customer will be seated according to the current seating arrangement of all previous customers. Imagining that the circular tables represent partitions $\{C_1, C_2, \ldots\}$, customer $(r+1)$, $r \geq 1$, is seated according to the predic-tion rule applied to the partition $P_r = \{C_{1,r}, C_{2,r}, \ldots, C_{G(r),r}\}$, corresponding to the seating arrangement of the first r customers, that is if P_{r+1} is the partition representing the event that customer $(r+1)$ is seated at a previous table $C_{g,r}$, then P_{r+1} occurs with probability $P(P_{r+1} = P_r \cup \{r+1 \in C_{g,r}\}) = P(P_{r+1})/P(P_r) = l_{g,r}$; while if $P_{r+1} = P_r \cup C_{G(P_r)+1}$ is the event that customer $(r+1)$ is seated at a new table, then the event P_{r+1} occurs with probability $P(P_{r+1} = P_r \cup C_{G(P_r)+1}) = l_{0,r}$. The CRP is closely connected to the Dirichlet process (DP) and therefore useful in the application of Bayesian non-parametric methods.

Let us consider a univariate case, whose complete likelihood is given by Equation (15.17). The parameters Θ in (15.17) include μ, σ^2, τ, and z. The prior for cluster assignment is a CRP, where the probability of joining each existing table is proportional to the size of that table. Since the seating probability is the product of the prior and the likelihood, which compare the proper-ties of the new customer and those of the people at the table, the seating process follows a weighted CRP, and, essentially, the weight is the likelihood ratio of the customer joining the table versus he/she starts a new one. The whole process can be naturally fit into a Gibbs sampler framework, such that the memberships can be updated iteratively until convergence. CRP-based clustering approaches are able to modify the number of components increasingly or decreasingly to fit the data (Qin, 2006).

15.3.3   SBP Mixture Model

Suppose that we model dataset X using the following model [Equation (15.102)]:

$$x_n \sim F(\theta_{z_n}), \theta_{1,\ldots,\infty} \sim H(\cdot), \text{ and } z_{1,\ldots,N} \sim \text{categorical}(\tau) \qquad (15.102)$$

Therefore, the SBP defines the infinite model constructively by Equation (15.103),

$$P = \sum_{g=1}^{\infty} \tau_g \delta_{\theta_g} \qquad (15.103)$$

where τ_g, called the stick-breaking weight, is defined as $\tau_g = u_g \prod_{i<g}(1-u_i)$. Having $u_g \sim \text{Beta}(1-a, b+ga)$, for $0 \leq a < 1$ and $b > -a$ yields a generalised two-parameter model. When $a = 0$ and $b = \alpha$, the model results in the DPM. Rodriguez and Dunson (2011) proposed an approach, called the probit stick-breaking process (PSBP), to construct rich and flexible families of non-parametric priors that allow for simple computational algorithms. PSBP uses a stick-breaking construction similar to the one underlying the DP but replaces the characteristic beta distribution in the definition of the sticks by probit transformations of normal random variables. Therefore, the resulting construction for the weights of the process is reminiscent of the continuation ratio probit model popular in survival analysis.

15.4 Discussion

The literature of mixture model clustering methods is extensively rich and is still growing. In this chapter, we considered as many of the related methods as possible. A summary of mixture model clustering methods is shown in Table 15.3. We classified mixture model methods into two large groups, namely finite mixture models and infinite mixture models. Finite mixture

Table 15.3 Summary of mixture model-based clustering algorithms

Mixture models	Non-Bayesian	Bayesian
Gaussian mixture models	Dempster *et al.* (1977); Banfield and Raftery (1993); Fraley and Raftery (1999, 2002); Yeung *et al.* (2001); Ghosh and Chinnaiyan (2002)	Diebolt and Robert (1994); Richardson and Green (1997); Roberts *et al.* (1998); Stephens (2000); Frühwirth-Schnatter (2001); Beal and Ghahramani (2003); Teschendorff *et al.* (2005); Dellaportas and Papageorgiou (2006); Fraley and Raftery (2007); Geweke (2007)
Mixture of HMMs	Smyth (1997); Schliep, Schönhuth and Steinhoff (2003); Schliep *et al.* (2005)	
Mixture of *t*-distributions	Peel and McLachlan (2000); McLachlan, Bean, and Peel (2002); Andrews and McNicholas (2011)	Archambeau and Verleysen (2007)
Mixture of factor analysers	Ghahramani and Hinton (1996); McLachlan, Peel and Bean (2003); McNicholas and Murphy (2008); Zhao, Yu and Jiang (2008); Zhao and Yu (2008); McNicholas and Murphy (2010)	Ghahramani and Beal (1999); Utsugi and Kumagai (2001)
Mixture of common factor analysers	Baek, McLachlan and Flack (2010);	
Mixture of common *t*-factor analysers	Baek and McLachlan (2011)	
Mixture of principal component analysers	Tipping and Bishop (1997, 1999)	
Infinite mixture models		Rasmussen (1999); Medvedovic and Sivaganesan (2002); Medvedovic, Yeung and Bumgarner (2004); Qin (2006); Rodriguez and Dunson (2011)

Table 15.4 Collection of publicly accessible resources for mixture model-based clustering

Algorithm name	Platform	Package	Function
GMM-EM	*R*	mclust	mclust
GMM-VBEM	MATLAB	Chen (2012)	vbgm
MFA-EM	MATLAB	Ghahramani (2000)	mfa
MFA-VBEM	MATLAB	Ghahramani (2000)	vbmfa
BHC	*R*	BHC	bhc
Bayesian non-parametric	*R*	DPpackage	

models were further classified into Bayesian and non-Bayesian (or frequentist). Infinite mixture models are also known as Bayesian non-parametric models. Therefore, this chapter covered three very important categories of statistical model clustering, including frequentist parametric models, Bayesian parametric models, and Bayesian non-parametric models. We spent a large proportion of space to introduce finite mixture models, which does not mean that finite mixture models are more important than infinite mixture models; on the contrary infinite mixture models are receiving more and more attention. Infinite mixture models have a similar representation to finite mixture models except for the limit $G \rightarrow \infty$. Almost all Bayesian mixture models can be solved using the MCMC algorithm. Thus, we avoid the redundancy by only introducing the infinite mixture models briefly. Interested readers are strongly referred to respective references.

A collection of publicly accessible resources for mixture model-based clustering is listed in Table 15.4. The applications of mixture model-based clustering in bioinformatics will be discussed in Chapter 19.

References

Andrews, J.L. and McNicholas, P.D. (2011). Extending mixtures of multivariate t-factor analyzers. *Statistics and Computing, 21*(3), pp. 361–373.

Antoniak, C.E. (1974). Mixtures of Dirichlet processes with applications to Bayesian nonparametric problems. *Annals of Statistics, 2*(6), pp. 1152–1174.

Archambeau, C. and Verleysen, M. (2007). Robust Bayesian clustering. *Neural Networks, 20*(1), pp. 129–138.

Baek, J. and McLachlan, G.J. (2011). Mixtures of common t-factor analyzers for clustering high-dimensional microarray data. *Bioinformatics, 27*(9), pp. 1269–1276.

Baek, J., McLachlan, G.J. and Flack, L.K. (2010). Mixtures of factor analyzers with common factor loadings: applications to the clustering and visualization of high-dimensional data. *IEEE Transactions on Pattern Analysis and Machine Intelligence, 32*(7), pp. 1298–1309.

Banfield, J.D. and Raftery, A.E. (1993). Model-based Gaussian and non-Gaussian clustering. *Biometrics, 49*(3), pp. 803–821.

Beal, M.J. and Ghahramani, Z. (2003). The variational Bayesian EM algorithm for incomplete data: with application to scoring graphical model structures. In: J. Bernardo, *et al.* (eds) *Bayesian Statistics.* Oxford University Press, Oxford, pp. 453–464.

Bishop, C.M. (2006). *Pattern Recognition and Machine Learning,* Springer, New York.

Celeux, G. and Govaert, G. (1995). Gaussian parsimonious clustering models. *Pattern Recognition, 28*(5), pp. 781–793.

Chen, M. (2012) *Mathworks File Exchange Center,* http://www.mathworks.co.uk/matlabcentral/fileexchange/35362-variational-bayesian-inference-for-gaussian-mixture-model/content/vbgm.m (accessed 23 July 2014).

Dellaportas, P. and Papageorgiou, I. (2006). Multivariate mixtures of normals with unknown number of components. *Statistics and Computing,* **16**(1), pp. 57–68.

Dempster, A.P., Laird, N.M., Rubin, D.B. *et al.* (1977). Maximum likelihood from incomplete data via the EM algorithm. *Journal of the Royal Statistical Society,* **39**(1), 1–38.

Diebolt, J. and Robert, C.P. (1994). Estimation of finite mixture distributions through Bayesian sampling. *Journal of the Royal Statistical Society. Series B (Methodological),* **56**(2) pp. 363–375.

Ferguson, T.S. (1973). A Bayesian analysis of some nonparametric problems. *Annals of Statistics,* **1**(2) pp. 209–230.

Fraley, C. and Raftery, A.E. (1998). How many clusters? Which clustering method? Answers via model-based cluster analysis. *The Computer Journal,* **41**(8), pp. 578–588.

Fraley, C. and Raftery, A.E. (1999). MCLUST: software for model-based cluster analysis. *Journal of Classification,* **16**(2), pp. 297–306.

Fraley, C. and Raftery, A.E. (2002). Model-based clustering, discriminant analysis, and density estimation. *Journal of the American Statistical Association,* **97**(458), pp. 611–631.

Fraley, C. and Raftery, A.E. (2007). Bayesian regularization for normal mixture estimation and model-based clustering. *Journal of Classification,* **24**(2), pp. 155–181.

Frühwirth-Schnatter, S. (2001). Markov chain Monte Carlo estimation of classical and dynamic switching and mixture models. *Journal of the American Statistical Association,* **96**(453), pp. 194–209.

Frühwirth-Schnatter, S. (2006). *Finite Mixture and Markov Switching Models: Modeling and Applications to Random Processes,* Springer, New York.

Geweke, J. (2007). Interpretation and inference in mixture models: simple MCMC works. *Computational Statistics and Data Analysis,* **51**(7), pp. 3529–3550.

Ghahramani, Z. (2000). MATLAB codes for EM for Mixture of Factor Analyzers and Variational Bayesian Mixture of Factor Analysers, Online resources, http://mlg.eng.cam.ac.uk/zoubin/software.html (accessed 23 July 2014).

Ghahramani, Z. and Beal, M.J. (1999). *Variational Inference for Bayesian Mixtures of Factor Analysers,* NIPS, Denver, CO, pp. 449–455.

Ghahramani, Z. and Hinton, G.E. (1996). The EM Algorithm for Mixtures of Factor Analyzers, University of Toronto. Technical Report CRG-TR-96-1.

Ghosh, D. and Chinnaiyan, A.M. (2002). Mixture modelling of gene expression data from microarray experiments. *Bioinformatics,* **18**(2), pp. 275–286.

Heller, K.A. and Ghahramani, Z. (2005). Bayesian Hierarchical Clustering. Proceedings of the 22nd International Conference on Machine Learning, 7–11 August, 2005, Bonn, German. ACM, New York, NY, pp. 297–304.

McLachlan, G. and Krishnan, T. (2007). *The EM Algorithm and Extensions,* John Wiley & Sons, Inc, Hoboken.

McLachlan, G. and Peel, D. (2000). *Finite Mixture Models,* John Wiley & Sons, Inc, New York.

McLachlan, G.J., Bean, R. and Peel, D. (2002). A mixture model-based approach to the clustering of microarray expression data. *Bioinformatics,* **18**(3), pp. 413–422.

McLachlan, G.J., Peel, D. and Bean, R. (2003). Modelling high-dimensional data by mixtures of factor analyzers. *Computational Statistics and Data Analysis,* **41**(3), pp. 379–388.

McNicholas, P.D. and Murphy, T.B. (2008). Parsimonious Gaussian mixture models. *Statistics and Computing,* **18**(3), pp. 285–296.

McNicholas, P.D. and Murphy, T.B. (2010). Model-based clustering of microarray expression data via latent Gaussian mixture models. *Bioinformatics,* **26**(21), pp. 2705–2712.

Medvedovic, M. and Sivaganesan, S. (2002). Bayesian infinite mixture model based clustering of gene expression profiles. *Bioinformatics,* **18**(9), pp. 1194–1206.

Medvedovic, M., Yeung, K.Y. and Bumgarner, R.E. (2004). Bayesian mixture model based clustering of replicated microarray data. *Bioinformatics,* **20**(8), pp. 1222–1232.

Meng, X.-L. and Dyk, D.V. (1997). The EM algorithm—an old folk-song sung to a fast new tune. *Journal of the Royal Statistical Society: Series B (Statistical Methodology),* **59**(3), pp. 511–567.

Peel, D. and McLachlan, G.J. (2000). Robust mixture modelling using the t distribution. *Statistics and Computing,* **10**(4), pp. 339–348.

Qin, Z.S. (2006). Clustering microarray gene expression data using weighted Chinese restaurant process. *Bioinformatics,* **22**(16), pp. 1988–1997.

Rabiner, L. (19890. A tutorial on hidden Markov models and selected applications in speech recognition. *Proceedings of the IEEE,* **77**(2), pp. 257–286.

Rasmussen, C.E. (1999). *The Infinite Gaussian Mixture Model.* NIPS, Denver, CO, pp. 554–560.

Richardson, S. and Green, P.J. (1997). On Bayesian analysis of mixtures with an unknown number of components (with discussion). *Journal of the Royal Statistical Society: Series B (Statistical Methodology),* **59**(4), pp. 731–792.

Roberts, S.J., Husmeier, D., Rezek, I. and Penny, W. (1998). Bayesian approaches to Gaussian mixture modeling. *IEEE Transactions on Pattern Analysis and Machine Intelligence,* **20**(11), pp. 1133–1142.

Rodriguez, A. and Dunson, D.B. (2011). Nonparametric Bayesian models through probit stick-breaking processes. *Bayesian Analysis (Online),* **6**(1), 145–178.

Schliep, A., Schönhuth, A. and Steinhoff, C. (2003). Using hidden Markov models to analyze gene expression time course data. *Bioinformatics,* **19**(suppl 1), pp. i255–i263.

Schliep, A., Costa, I.G., Steinhoff, C. and Schonhuth, A. (2005). Analyzing gene expression time-courses. *IEEE/ACM Transactions on Computational Biology and Bioinformatics,* **2**(3), pp. 179–193.

Smyth, P. (1997). Clustering sequences with hidden Markov models. *Advances in Neural Information Processing Systems,* **9**, pp. 648–654.

Stephens, M. (2000). Bayesian analysis of mixture models with an unknown number of components – an alternative to reversible jump methods. *Annals of Statistics,* **28**(1) pp. 40–74.

Teschendorff, A.E., Wang, Y., Barbosa-Morais, N.L., *et al.* (2005). A variational Bayesian mixture modelling framework for cluster analysis of gene-expression data. *Bioinformatics,* **21**(13), pp. 3025–3033.

Tipping, M.E. and Bishop, C.M. (1997). Mixtures of Principal Component Analysers. Proceedings of the IEE Fifth International Conference on Artificial Neural Networks, Cambridge UK, 7–9 July, 1997. pp. 13–18.

Tipping, M.E. and Bishop, C.M. (1999). Mixtures of probabilistic principal component analyzers. *Neural Computation,* **11**(2), pp. 443–482.

Utsugi, A. and Kumagai, T. (2001). Bayesian analysis of mixtures of factor analyzers. *Neural Computation,* **13**(5), pp. 993–1002.

Xu, R. and Wunsch, D. (2008). *Clustering,* John Wiley & Sons, Inc, Hoboken.

Yeung, K.Y., Fraley, C., Murua, A. *et al.* (2001). Model-based clustering and data transformations for gene expression data. *Bioinformatics,* **17**(10), pp. 977–987.

Zhao, J.-H. and Yu, P.L.H. (2008). Fast ML estimation for the mixture of factor analyzers via an ECM algorithm. *IEEE Transactions on Neural Networks,* **19**(11), pp. 1956–1961.

Zhao, J.H., Yu, P.L.H. and Jiang, Q.B. (2008). ML estimation for factor analysis: EM or non- EM? *Statistics and Computing,* **18**(2), pp. 109–123.

16

Graph Clustering

16.1 Introduction

Graph clustering, also known as graph theoretic clustering or graph-based clustering, has attracted a great deal of research interest in the bioinformatics field because a wealth of information is available on interactions involving proteins, genes and metabolites, that is small molecules. In studies of molecular and cellular biology, the graph representation has been widely used in the analysis of protein–protein interaction networks, gene regulatory networks and metabolic networks. Figure 16.1 shows an example of metabolic network of worm *Caenorhabditis elegans* (Jeong *et al.*, 2000).

In network science, a cluster is also called a community, and graph clustering algorithms are also known as the algorithms for community detection. Traditionally, hierarchical clustering and partitional clustering were employed in the graph partitioning. However, considering that the problems of graph clustering are much different from those of other clusterings, some specific graph clustering algorithms other than traditional and general clustering algorithms are highly desired. The objective of this chapter is to overview the family of graph clustering algorithms, including their problems, principles and applications. Since the field of network science has a rich literature and is still growing rapidly, it is impossible for us to include every algorithm in the field in this single chapter. We will focus on basic concepts and methodologies, briefly review some applications in bioinformatics, and introduce some available online resources for graph clustering and network analysis.

Integrative Cluster Analysis in Bioinformatics, First Edition. Basel Abu-Jamous, Rui Fa and Asoke K. Nandi.
© 2015 John Wiley & Sons, Ltd. Published 2015 by John Wiley & Sons, Ltd.

Figure 16.1 Example of metabolic network of the worm *Caenorhabditis elegans*. This dataset was originally reported by Jeong *et al.* (2000), and was analysed by Duch and Arenas (2005)

16.2 Basic Definitions

16.2.1 Graph and Adjacency Matrix

A graph G is a pair of sets (V, E), where V is a set of vertices or nodes and E contains the pairwise edges of vertices in V. The elements in E are called edges or links, and the two vertices connected by an edge are called endpoints. In an *undirected* graph, each edge is an unordered pair $\{v, w\}$. If each edge is an ordered pair of vertices, the graph is called a *directed* graph. A graph is called a *weighted* graph if a real number (weight) is assigned to each edge; on the other hand, it is an unweighted graph if all edges are equal. If the set of vertices contains two subsets $\mathcal{A}$ and $\mathcal{B}$, and there is no interaction between $\mathcal{A}$ and $\mathcal{B}$, the graph is called a *bipartite* graph.

The number of vertices and edges of a graph are $N_V = |V|$ and $N_E = |E|$, respectively. A graph $G' = (V', E')$ is a sub-graph of $G = \{V, E\}$ if $V' \subset V$ and $E' \subset E$. A partition of the vertex set V in two subsets V' and $(V - V')$ is called a cut. The maximum size of a graph equals the total number of unordered pairs of vertices, $N_V(N_V - 1)/2$. If $N_E = N_V(N_V - 1)/2$, the graph is a *clique* (or *complete* graph). Two vertices are neighbours or adjacent if they are connected by an edge.

The whole topology of a graph of size N_V is entailed in an *adjacency matrix* A, which is an $N_V \times N_V$ matrix whose element a_{ij} equals w_{ij} representing the weight of the edge joining vertices i and j. Generally, the weight may have two entirely opposite connotations; for example, the weight may be the distance between two vertices; on the other hand, the weight may be the similarity of two vertices. Throughout this chapter, we specify that the weight that we refer to measures how similar two vertices are and how closely they are connected. The larger the weight is, more similar the two vertices are and the closer they are connected. If w_{ij} is zero, it means that vertices i and j are not connected at all. Mathematically, the adjacency matrix A is given by Equation (16.1),

$$a_{ij} = \begin{cases} w_{ij} & \text{if } i \text{ is adjacent to } j \\ 0 & \text{otherwise} \end{cases} \tag{16.1}$$

which implies that A is a symmetric matrix because the semantic interpretation of *adjacency* is a symmetric concept. Therefore, the adjacency matrix is not suitable to describe the directed graph and bipartite graph. An *incident* matrix B was introduced to describe the asymmetric graph (Newman, 2010), which is mathematically expressed as Equation (16.2).

$$b_{ij} = \begin{cases} w_{ij} & \text{if } i \text{ is incident to } j \\ 0 & \text{otherwise} \end{cases} \tag{16.2}$$

The incident matrix for the directed graph is also an $N_V \times N_V$ matrix, whose rows, however, represent the vertices at which the edges begin and whose columns represent the vertices at which the edges end. For a bipartite graph, the incident matrix is an $N_V^{(A)} \times N_V^{(B)}$ matrix, whose rows represent the vertices in subset $\mathcal{A}$ and columns represent the vertices in subset $\mathcal{B}$.

16.2.2   Measures and Metrics

The *degree* of a vertex i, denoted by d_i, is the number of its neighbours, which is written as shown in Equation (16.3).

$$d_i = \sum_{j=1}^{N_V} a_{ij} \tag{16.3}$$

For a directed graph, one distinguishes two types of degree for a vertex i: the *in-degree* d_i^I and the *out-degree* d_i^O. The in-degree is the number of edges beginning at the vertex i, and the out-degree is the number of edges ending at the vertex i. They can be written as shown in Equations (16.4).

$$d_i^I = \sum_{j=1}^{N_V} b_{ji}, \ d_i^O = \sum_{j=1}^{N_V} b_{ij} \tag{16.4}$$

A path is a sub-graph $\wp = (V(\wp), E(\wp))$, in which $V(\wp) \subset V$ and $E(\wp) \subset E$. Suppose that $V(\wp)$ contains $\{v_1, v_2, \ldots, v_l\}$, l vertices, and then $E(\wp)$ has $\{v_1 v_2, v_2 v_3, \ldots, v_{l-1} v_l\}$. The length of a path $\wp$ is the sum of the reciprocal of weight of all the edges in $E(\wp)$, which is given by Equation (16.5),

$$L(\wp) = \sum_{i,j \in V(\wp)} \frac{1}{w_{ij}^\alpha} \tag{16.5}$$

where α is a positive tuning parameter to be set according to the application and has the value 1 for general cases. Paths allow one to define the concept of connectivity and distance in graphs. A graph is connected if there is at least one path connecting any given pair of vertices in the graph. A *shortest path* $s(i,j)$, also known as a *geodesic*, between two vertices i and j of a graph, is a path of minimal length. The shortest path $s(i,j)$ can be written as given in Equation (16.6),

$$s(i,j) = \min\left\{ L\left(\wp_{ij} \right) \right\}$$ (16.6)

where $\wp_{ij}$ denote all paths connecting vertices i and j. Such a minimal length is the *distance* between two vertices. The diameter of a connected graph is the maximal distance between two vertices. A *cycle graph* or *cyclic graph* is a graph that consists of some number of vertices connected in a closed chain. A graph is called a *forest* if it has no cycle in it. A connected forest is called a *tree*. The tree concept is very important in graph theory. There is only one path from a vertex to any other vertex in a tree because if there were at least two paths between the same pair of vertices they would form a cycle, which is against the acyclic definition of a tree. Moreover, the number of edges of a tree with N_V vertices is $(N_V - 1)$. Therefore, a tree is a minimally connected, maximally acyclic graph of a given size: If any edge of a tree is removed, it would become disconnected in two parts; if a new edge is added, there would be at least one cycle. Every connected graph contains a spanning tree; that is, a tree sharing all vertices of the graph. On weighted graphs, one can define a *minimum (maximum) spanning tree*; that is, a spanning tree such that the sum of the weights on the edges is minimal (maximal).

Suppose that we are given a subgraph G' of graph G, with $|G'| = n_c$, and $|G| = N_V$ vertices. The *intra-cluster density* is defined as a ratio between the number of internal edges of G' and the number of all possible internal edges [Equation (16.7)],

$$\delta_{\text{int}}(G') = \frac{e_c}{n_c(n_c - 1)/2}$$ (16.7)

where e_c is the number of internal edges of subgraph G'. The inter-cluster density of a given clustering of a graph G in K clusters $\{C_1, \ldots, C_K\}$ is given by Equation (16.8).

$$\delta_{\text{int}}(G|C_1, \ldots, C_K) = \frac{1}{K} \sum_{k=1}^{K} \delta_{\text{int}}(C_k)$$ (16.8)

Similarly, the *inter-cluster density* can be defined as shown in Equation (16.9),

$$\delta_{\text{ext}}(G') = \frac{e_{\bar{c}}}{n_c(n - n_c)}$$ (16.9)

where $e_{\bar{c}}$ is the number of edges running from the vertices of G' to the rest of the graph $\bar{G}'$. The value of $\delta_{\text{ext}}(G')$ reflects the ratio between the number of edges running from the vertices of G' to $\bar{G}'$ and the maximum possible number of inter-cluster edges.

Centrality of a vertex reflects its importance within a network. It has been the key statistics in the analysis of networks; for instance, a high centrality measure may indicate that the given vertex either is very influential and active in the network or is located in a central position in the network and is highly connected with others. There are many classic centrality measures, including degree, closeness and betweenness (Newman, 2010). Degree centrality is the simplest indicator, which is the degree of the vertex. A generalised degree centrality for a weighted network, which combines both degree and strength, is given by Equation (16.10),

$$C_D(i) = k_i \times \left(\frac{d_i}{k_i}\right)^{\alpha} \tag{16.10}$$

where k_i is the number of edges connected with the ith vertex, d_i is the degree of the ith vertex, and α is a positive tuning parameter to be set according to the application. Both the closeness and betweenness centrality measures rely on the measure of shortest path among nodes. The closeness centrality is given by Equation (16.11).

$$C_C(i) = \left[\sum_j^{N_V} s(i,j)\right]^{-1} \tag{16.11}$$

Betweenness centrality is mathematically written as given in Equation (16.12),

$$C_B(i) = \sum_{j \neq i \neq k} \frac{g_{jk}(i)}{g_{jk}} \tag{16.12}$$

where $g_{jk}(i)$ is the number of shortest paths between the jth and kth vertices thought the ith vertices and g_{jk} is the total number of shortest paths between the jth and kth vertices. Note that $g_{jk}(i)/g_{jk}$ has to be set to zero if g_{jk} is zero.

A *degree matrix*, D, is a diagonal matrix, whose element d_{ii} is the degree of the vertex i, denoted as diag$\{d_1,\ldots,d_{N_V}\}$. A very important matrix is the *Laplacian matrix* $L = D - A$. This form of Laplacian is usually referred to as *unnormalised Laplacian*. A *normalised Laplacian* usually has two main form: $L_{\text{sym}} = D^{-1/2} L D^{-1/2}$ and $L_{\text{rw}} = D^{-1} L = I - D^{-1} A$. The matrix L_{sym} is symmetric, while L_{rw} is asymmetric. The Laplacian is one of most studied matrices in graph theoretical research. The fact that the sum of the elements of each row of the Laplacian is zero implies that it always has at least one zero eigenvalue, corresponding to the eigenvector with all ones. Interestingly, L (both unnormalised and normalised) has as many zero eigenvalues as the number of connected components in the graph. Eigenvectors of Laplacian matrices have been used in *spectral clustering*, which we will discuss in the next section. In particular, the eigenvector corresponding to the second smallest eigenvalue, called *Fiedler vector*, is commonly used in graph partitioning.

Modularity, proposed by Newman and Girvan (2004), is the most popular quality function for graph clustering or community detection. The job of community detection is to divide network vertices into groups within which the network connections are dense, but between which they are sparser. Among all community detection algorithms, modularity-optimisation methods have attracted considerable attention due to its computational efficiency and practical applicability. The logic behind the modularity methods is that dividing a network into reasonably good communities will give a high value of the benefit function Q, called modularity. Conceptually, the modularity is the value of the fraction of edges within communities minus the expected fraction of such edges. The modularity in an undirected unipartite graph is given by Equation (16.13),

$$Q = \frac{1}{2N_E} \sum_{ij} \left[a_{ij} - \frac{d_i d_j}{2N_E} \right] \delta_{ij} \qquad (16.13)$$

where δ_{ij} is Kronecker delta symbol, which is one if the vertices i and j are in the same community; otherwise it is zero. The matrix expression of modularity leads to another concept, namely modularity matrix M, defined with each element as shown in Equation (16.14).

$$M_{ij} = a_{ij} - \frac{d_i d_j}{2N_E} \qquad (16.14)$$

Therefore, the modularity Q is given by Equation (16.15),

$$Q = \frac{1}{2N_E} \text{Tr}\{U^T M U\} \qquad (16.15)$$

where U is an $N_V \times N_c$ index matrix, whose every element is either one, indicating if the vertex belongs to the community, or zero otherwise. N_c is the number of communities. The modularity for a directed unipartite graph can be mathematically written as given in Equation (16.16).

$$Q = \frac{1}{N_E} \sum_{ij} \left[a_{ij} - \frac{d_i^I d_j^O}{N_E} \right] \delta_{ij} \qquad (16.16)$$

16.2.3   Similarity Matrices

Another important point worth noting is about the difference between the similarity matrix and the adjacency matrix. Conceptually, they are different: the elements of a similarity matrix are pairwise similarity between two data objects, which is a quantitative concept; while the elements of an adjacency matrix indicate the connectivity between vertices and the strength of the connections, which is a qualitative concept. However, sometimes, the adjacency matrix can be represented by the similarity matrix when the connectivity information among data points is not available but the feature data are available, in other words, the adjacency matrix is inferred from the similarity matrix. Many clustering algorithms, for example, spectral clustering algorithm and affinity propagation (AP) algorithm, are developed based on the similarity matrix. Graph partitioning and modularity-based algorithms virtually work on the adjacency matrix. There are several popular construction techniques to transform a given set of data points $X = \{x_n | n = 1, \ldots, N\}$ with pairwise similarities s_{ij} or pairwise distances d_{ij} into a graph G. When constructing similarity graphs the goal is to model the local neighbourhood relationships between the data points.

   The ε-neighbourhood graph: all data points whose pairwise distances are smaller than ε are connected. Each data point can be viewed as a vertex. As the distances between all connected vertices are roughly of the same scale (at most, ε), weighting the edges would not incorporate more information about the data to the graph. Hence, the ε-neighbourhood graph is usually considered as an unweighted graph.

   The k-nearest neighbour graph: the goal is to connect vertex i with vertex j if vertex j is among the k-nearest neighbours of vertex i. However, this definition leads to a directed graph, as the neighbourhood relationship is not symmetric. There are two ways of making this graph undirected. The first way is to simply ignore the directions of the edges; that is, we connect vertices i and j with an undirected edge if vertex i is among the k-nearest neighbours of vertex j or if vertex j is among the k-nearest neighbours of vertex i. The resulting graph is what is usually called the k-nearest neighbour graph. The second choice is to connect vertices i and j if both i is among the k-nearest neighbours of j and vertex j is among the k-nearest neighbours of vertex i. The resulting graph is called the mutual k-nearest neighbour graph. In both cases, after connecting the appropriate vertices we weight the edges by the similarity of their endpoints.

The fully connected graph: Here we simply connect all points with positive similarity with each other, and we weight all edges by s_{ij}. As the graph should represent the local neighbourhood relationships, this construction is only useful if the similarity function itself models local neighbourhoods. An example for such a similarity function is the Gaussian similarity function $(x_i, x_j) = \exp\{-||x_i - x_j||^2/(2\sigma^2)\}$, where the parameter controls the width of the neighbourhoods. This parameter plays a similar role as the parameter ε in case of the ε-neighbourhood graph.

16.3   Graph Clustering

As we mentioned before, the objective of graph clustering is to look for the community structure, within which the interconnections are dense, but between which they are sparser. No quantitative definition of the community, however, is universally accepted, and as a matter of fact, the definition is highly dependent on the specific problem at hand.

16.3.1   Graph Cut Clustering

The number of edges across communities is called *cut size*. The goal in graph partitioning is to minimise the number of edges that cross from one sub-group of vertices to another, usually posing limits on the number of groups as well as to the relative size of the groups; in other words, to minimise the cut size, which is mathematically expressed as given in Equation (16.17),

$$\text{cut}(C_a, C_b) = \sum_{u \in C_a, v \in C_b} w(u,v) \tag{16.17}$$

where $w(u, v)$ is the weight of edge linking vertices u and v, which belong to C_a and C_b, respectively. Wu and Leahy proposed a graph clustering algorithm based on minimum cut criterion (Wu and Leahy, 1993). The Wu–Leahy algorithm partitions a graph into k sub-graphs such that the total cut size across the sub-graphs is minimised. This problem can be solved efficiently by recursively finding the minimum cuts that bisect the existing sub-graphs. However, the Wu–Leahy algorithm often results in a skewed cut; that is, a very small sub-graph is cut away. Therefore, to circumvent the problem, many constraints have been introduced; for example, the normalised cut (Ncut) (Shi and Malik, 2000). The Ncut is mathematically written as shown in Equation (16.18),

$$\text{ncut}(C_a, C_b) = \frac{\text{cut}(C_a, C_b)}{\text{conn}(C_a, C_a \cup C_b)} + \frac{\text{cut}(C_a, C_b)}{\text{conn}(C_b, C_a \cup C_b)} \tag{16.18}$$

where $\mathrm{conn}(C_a, C_a \cup C_b) = \sum_{u \in C_a, v \in C_a \cup C_b} w(u,v)$ is the total connection from nodes in C_a to all nodes in both C_a and C_b. The Ncut is used as a partition criterion. Because minimising Ncut exactly is non-deterministic polynomial-time (NP)-complete, an approximate discrete solution can be found by embedding the criterion in the real value domain.

Suppose that we partition a graph G into C_a and C_b subgraphs, where $|G| = N_V$. Let $d_i = \sum_j w(i,j)$ denote the degree of vertex i, and x be an N_V-dimensional indicator vector where $x_i = 1$ if vertex i is in A, and $x_i = -1$ otherwise. Let D be an $N_V \times N_V$ diagonal matrix with d on its diagonal, and A be an $N_V \times N_V$ adjacency matrix. The Ncut can be rewritten as in Equation (16.19),

$$
\begin{aligned}
\mathrm{ncut}(C_a, C_b) &= \frac{\sum_{x_i > 0, x_j < 0} -w_{ij} x_i x_j}{\sum_{x_i > 0} d_i} + \frac{\sum_{x_i < 0, x_j > 0} -w_{ij} x_i x_j}{\sum_{x_i < 0} d_i} \\
&= \frac{(1+x)^T (D-A)(1+x)}{k 1^T D 1} + \frac{(1-x)^T (D-A)(1-x)}{(1-k) 1^T D 1} \qquad (16.19) \\
&= \frac{[(1+x) - b(1-x)]^T (D-A)[(1+x) - b(1-x)]}{b 1^T D 1}
\end{aligned}
$$

where $k = \sum_{x_i > 0} d_i / \sum_i d_i$, and $b = k/(1-k)$. Let $y = (1+x) - b(1-x)$, thus $\mathrm{ncut}(C_a, C_b)$ becomes defined as in Equation (16.20),

$$
\mathrm{ncut}(C_a, C_b) = \frac{y^T (D-A) y}{y^T D y} \qquad (16.20)
$$

and the problem turns to the minimisation of $\mathrm{ncut}(C_a, C_b)$ as in Equation (16.21).

$$
\min_x \mathrm{ncut}(C_a, C_b) = \min_y \frac{y^T (D-A) y}{y^T D y} \qquad (16.21)
$$

Then we can minimise the Ncut by solving the generalised eigenvalue system shown in Equation (16.22).

$$
(D-A) y = \lambda D y \rightarrow D^{-\frac{1}{2}} (D-A) D^{-\frac{1}{2}} z = \lambda z, \quad \text{where } z = D^{\frac{1}{2}} y \qquad (16.22)
$$

As we introduced previously, $D^{-\frac{1}{2}} (D-A) D^{-\frac{1}{2}}$ is the normalised Laplacian matrix L_{sym}. Thus, the second smallest eigenvector of L_{sym}, which is known as the Fiedler vector, is the real-valued solution to the Ncut problem. Then the real valued solution has to be converted into a discrete form. The algorithm is summarised in Table 16.1.

16.3.2   Spectral Clustering

Spectral clustering is an algorithm which is very close to the graph cut clustering algorithm. It requires the computation of the first k eigenvectors of a Laplacian matrix (Ng, Jordan and

Table 16.1   Summary of the normalised cut graph partitioning algorithm

Step 1.   Reform the dataset into a weighted graph $G = (V, E)$, and the adjacency matrix A whose each element denotes the weight on the edge connecting two nodes set to be a measure of the similarity between the two nodes;

Step 2.   Solve $D^{-\frac{1}{2}}(D - A)D^{-\frac{1}{2}}z = \lambda z$ for eigenvectors with the smallest eigenvalues;

Step 3.   Use the eigenvector with the second smallest eigenvalue to bisect the graph;

Step 4.   Decide if the current partition should be subdivided and recursively repartition the segmented parts if necessary;

Step 5.   Recursively bisect the partitioned parts if necessary.

Table 16.2   Summary of the spectral clustering algorithm by Ng, Jordan and Weiss (2001)

Step 1.   Reform the dataset into a weighted graph $G = (V, E)$, and the adjacency matrix A whose each element denotes the weight on the edge connecting two nodes set to be a measure of the similarity between the two nodes;

Step 2.   Construct $T = D^{-\frac{1}{2}}AD^{-\frac{1}{2}}$, where D is the degree matrix;

Step 3.   Find the k eigenvector of T corresponding to k largest eigenvalues $U \in \mathfrak{R}^{N \times k}$;

Step 4.   Form the matrix Y from U by normalising each row of Y to have unit length $Y_{ij} = U_{ij} / \left(\sum_j U_{ij}^2 \right)^{1/2}$;

Step 5.   Treating each row of Y as a data point and cluster Y into k clusters via k-means or any other algorithms;

Step 6.   Finally, assign the original data point i to the cluster j if and only if the ith point in the matrix Y was assigned to the cluster j.

Weiss, 2001; Luxburg, 2007). If the graph is large, an exact computation of the eigenvectors is impossible, as it would require time of the order $O(N_V^3)$. Fortunately there are approximate techniques, like the power method or Krylov subspace techniques like the Lanczos method (Golub and Van, 2012), whose speed depends on the size of the gap between eigenvalues $|\lambda_{k+1} - \lambda_k|$, where λ_k and λ_{k+1} are the kth and the $(k+1)$th smallest eigenvalue of the matrix.

The algorithm proposed by Ng, Jordan and Weiss (2001) is summarised in Table 16.2. It is worth noting that Ng *et al.* employed the normalised adjacency matrix $D^{-1/2}AD^{-1/2}$, rather than the normalised Laplacian matrix. In principle, spectral clustering methods require a Laplacian matrix to be used in the applications (Luxburg, 2007). Furthermore, the choice of Laplacian matrix matters: if the graph vertices have the same or similar degrees, there is no substantial difference between the unnormalised and the normalised Laplacian; if there is large inhomogeneity among the vertex degrees, on the other hand, the choice of the Laplacian considerably affects the results. In general, a normalised Laplacian is more promising because the corresponding spectral clustering techniques implicitly impose a double optimisation on the set of partitions, such that the intra-cluster edge density is high and, at the same time, the inter-cluster edge density is low. On the other hand, the unnormalised Laplacian is related to the inter-cluster edge density only. Moreover, unnormalised spectral clustering does not always converge, and sometimes yields trivial partitions in which one or more clusters consist of a single vertex. Of the normalised Laplacian, L_{rw} is more reliable than L_{sym} because the eigenvectors of

L_{rw} corresponding to the lowest eigenvalues are cluster indicator vectors; that is, they have equal non-vanishing entries in correspondence to the vertices of each cluster, and zero elsewhere, if the clusters are disconnected. The eigenvectors of L_{sym}, instead, are obtained by (left-) multiplying those of L_{rw} by the matrix $D^{1/2}$. In this way, eigenvector components corresponding to vertices of the same cluster are no longer equal; in general, a complication that may induce artefacts in the spectral clustering procedure.

16.3.3 AP Clustering

The AP clustering algorithm was developed by Frey and Dueck (2007). The AP algorithm recursively transmits real-valued messages along edges of the network until a good set of exemplars and corresponding clusters emerges. The AP algorithm takes as input a real-valued similarity matrix, whose element $s(i, k)$ indicates how well the data point with index k is suited to be the exemplar for data point i, expressed as $s(i,k) = -||x_i - x_k||^2$.

One of the features of the AP algorithm is that it does not require the number of clusters. Instead, AP takes as input a real number $s(k, k)$ for each data point k so that data points with larger values of $s(k, k)$ are more likely to be chosen as exemplars. These values are referred to as 'preferences'. Initially, all data points are equally suitable as exemplars, and the preferences are set to a common value. This shared value could be the median of the input similarities.

The AP algorithm clusters data by passing messages between data points. Two kinds of messages are passed between data points. One is called 'responsibility' $r(i, k)$, sent from data point i to candidate exemplar point k, which reflects the accumulated evidence for how well suited point k is to serve as the exemplar for point i, taking into account other potential exemplars. Another is called 'availability' $a(i, k)$, sent from candidate exemplar point k to point i, which reflects the accumulated evidence for how appropriate it would be for point i to choose point k as its exemplar, taking into account the support from other points that point k should be an exemplar.

The AP algorithm starts with initial settings of two types of messages that availabilities are set to zeros, $a(i,k) = 0$, and the responsibilities are computed using the rule shown in Equation (16.23).

$$r(i,k) \leftarrow s(i,k) - \max_{k' \neq k} \{a(i,k') + s(i,k')\} \tag{16.23}$$

For $k = i$, the self-responsibility $r(k, k)$ is set to the input preference that point k be chosen as an exemplar $s(k, k')$, minus the largest of the similarities between point k and all other candidate exemplars. To gather evidence from data points as to whether each candidate exemplar would make a good exemplar, the availability is updated by use of Equation (16.24).

$$a(i,k) \leftarrow \min\left\{0, r(k,k) + \sum_{i' \neq i, k} \max\{0, r(i',k)\}\right\} \tag{16.24}$$

It implies that the availability $a(i, k)$ is set to the self-responsibility $r(k, k)$ plus the sum of the positive responsibilities candidate exemplar k receives from other points. The self-availability $a(k, k)$ is updated differently by use of Equation (16.25),

$$a(k,k) \leftarrow \sum_{i' \neq k} \max\{0, r(i',k)\} \tag{16.25}$$

which reflects accumulated evidence that point k is an exemplar, based on the positive responsibilities sent to candidate exemplar k from other points. The above update rules require only simple, local computations that are easily implemented, and messages only are exchanged between pairs of points with known similarities. At any point during the AP algorithm, availabilities and responsibilities can be combined to identify exemplars. For point i the value of k that maximises $\{a(i,k) + r(i,k)\}$ either identifies point i as an exemplar if $k = i$, or identifies the data point that is the exemplar for point i. The message-passing procedure may be terminated after a fixed number of iterations, after changes in the messages fall below a threshold, or after the local decisions stay constant for some number of iterations.

To evaluate the AP algorithm, we employ a 2-D four-clusters dataset for a demonstrating example and the clustering results are shown in Figure 16.2. We notice that although the AP algorithm does not require the number of clusters to be pre-set, its accuracy of estimating the correct number of clusters is limited. Obviously, in the case shown in this Figure, the 4-cluster dataset was over-clustered into 10 clusters. Nevertheless, the AP algorithm is still of interest to many fields of research because it reveals the affinity between data points.

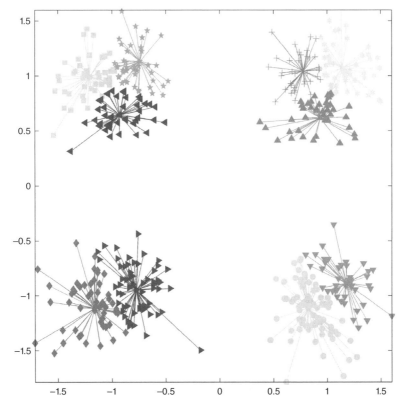

Figure 16.2 Example of 2-D four-clusters data using the AP algorithm. Obviously, the data are over-clustered by the AP algorithm into 10 clusters.

16.3.4   Modularity-based Clustering

Modularity, which was originally introduced to define a stopping criterion for the graph clustering algorithm by Newman and Girvan (2004), has rapidly become an essential element of graph clustering methods. Modularity is by far the most used and best known quality function. In this section, we shall focus on some basic modularity-based graph clustering techniques that attempt to optimise modularity directly or indirectly.

16.3.4.1   Greedy Techniques

A greed method of Newman (2004a) was the first method devised to maximise the modularity of the graph. This algorithm falls in the general category of agglomerative hierarchical clustering methods. It starts with a state in which every vertex is a community with a sole member; that is, there are N_V communities. These communities are repeatedly joined together in pairs, choosing at each step the connection that results in the greatest increase (or smallest decrease) in modularity function, which was defined in Equation (16.13). Like a hierarchical clustering algorithm, the algorithm produces a dendrogram, which shows the order of the joins. Cuts through the dendrogram at different levels give divisions of the network into larger or smaller numbers of communities. The best cut that results in the maximal value of modularity, therefore, can be found.

Let e_{ij} be one-half of the fraction of edges in the network that connect vertices in group i to those in group j. Let g_i be the fraction of all ends of edges that attached to vertices in group i, where g_i can be calculated straightforwardly by $g_i = \sum_j e_{ij}$. Thus, the modularity can be rewritten as in Equation (16.26).

$$Q = \sum_i (e_{ii} - g_i)^2 \tag{16.26}$$

Only those pairs of vertices between which there are edges are considered in the joining process. The change in Q upon joining two communities is given by Equation (16.27).

$$\Delta Q = 2(e_{ij} - g_i g_j) \tag{16.27}$$

The entire algorithm runs in the worst case in time $\mathcal{O}(N_V(N_E + N_V))$ or $\mathcal{O}(N_V^2)$ in sparse graph, much faster than a previous algorithm based on betweenness, which runs in time $\mathcal{O}(N_E N_V^2)$ or $\mathcal{O}(N_V^3)$.

16.3.4.2   Spectral Techniques

Modularity can be optimised using eigenvalues and eigenvectors of a spectral matrix, known as a modularity matrix, whose element was written as shown in Equation (16.14) (Newman, 2006). Suppose that a given vector s be the vector representing any partition of the graph in two clusters C_a and C_b: $s_i = 1$ if vertex i belongs to C_a; $s_i = -1$, if vertex j belongs to C_b. Thus Equation (16.15) can be rewritten as Equation (16.28).

$$Q = \frac{1}{4N_E} \sum_{ij} \left(a_{ij} - \frac{d_i d_j}{2N_E} \right) (s_i s_j + 1) = \frac{1}{4N_E} s^T M s \tag{16.28}$$

In the above expression the vector s can be decomposed on the basis of eigenvectors u_i of the modularity matrix $s = \sum_i \alpha_i u_i$; thus we can further rewrite modularity function Q as shown in Equation (16.29),

$$Q = \frac{1}{4N_E} \sum_{i=1}^{N_V} (u_i^T s)^2 \beta_i \qquad (16.29)$$

where β_i is the eigenvalue of M corresponding to the eigenvector u_i and the eigenvalues are in decreasing order, $\beta_1 \geq \beta_2, \ldots, \geq \beta_{N_V}$. Maximising the modularity by choosing an appropriate division of the network is equivalent to that by choosing the value of the index vector s parallel to the leading eigenvector u_1. There is a constraint on the problem imposed by the restriction of the elements of s to value ± 1, which means that we cannot choose s parallel to u_1. Nevertheless, we can choose s as close to parallel as possible: the vertices with positive components in u_1 are all in one group, the others in the other group, that is, set $s_i = +1$ if the corresponding element of u_1 is positive and $s_i = -1$ otherwise. However, it may happen that there is no positive eigenvalue of the modularity matrix. In this case, the leading eigenvector is an all-one vector, corresponding to all vertices in a single group together. This is precisely correct since in this case the algorithm tells the fact that there is no division of the network resulting in positive modularity.

The standard approach to cluster the network into more than two clusters is to repeat the above division of a large network into two small sub-networks. In doing so it is crucial to note that it is not correct, after first dividing a network in two, to simply delete the edges falling between the two parts and then apply the algorithm again to each sub-graph. This is because the degrees appearing in the definition [Equation (16.13)] of the modularity will change if edges are deleted, and any subsequent maximisation of modularity would thus maximise the wrong quantity. Instead, the correct approach is to write the additional contribution ΔQ to the modularity upon further dividing a group g of size n_g in two as given in Equation (16.30),

$$\Delta Q = \frac{1}{4N_E} s_g^T M_g s_g \qquad (16.30)$$

where $M_g(i,j) = M_{ij} - \delta_{ij} \sum_{k \in g} M_{ik}$, where $i, j \in g$, M_g is an $n_g \times n_g$ matrix, and s_g is an $n_g \times 1$ index vector. If the leading eigenvalue is zero, then the sub-graph is indivisible. This feature can be conveniently employed as a termination rule. The complete algorithm is summarised in Table 16.3.

Table 16.3 Summary of spectral modularity optimisation algorithm by Newman (2006)

Step 1. Construct the modularity matrix as shown in Equation (16.14);

Step 2. Find its leading (most positive) eigenvalue and the corresponding eigenvector u_1;

Step 3. Divide the network into two parts according to the signs of the elements of this vector;

Step 4. Repeat the process in Steps 2 and 3 for each of the parts, using the generalised modularity matrix as shown in Equation (16.30);

Step 5. If at any stage a proposed split makes a zero or negative contribution to the total modularity, leave the corresponding sub-graph undivided. When the entire network has been decomposed into indivisible sub-graphs in this way, the algorithm ends.

The community detected by Newman (2006) can be further fine-tuned by the vertex-moving method to reach the best possible modularity value. The whole procedure is repeated to subdivide the network until every remaining sub-graph is indivisible, and no further improvement in the modularity is possible.

16.3.4.3 Modifications

Since there are many types of networks, for example weighted networks, directed networks and bipartite networks, several modifications and extensions of modularity optimisation algorithms have been developed for various specific networks.

For weighted networks, a modularity optimisation algorithm can be easily extended to weighted edges (Newman, 2004b). The modularity in Equation (16.13) can be rewritten as Equation (16.31),

$$Q_w = \frac{1}{2W} \sum_{ij} \left(a_{ij} - \frac{d_i d_j}{2W} \right) \delta_{ij} \tag{16.31}$$

where W is the sum of the weights of all the edges, $a_{ij} = w_{ij}$, and $d_i = \sum_j w_{ij}$, $d_j = \sum_i w_{ij}$. In this way, one could derive the corresponding weighted modularity and use it to detect communities, with a potentially better exploitation of the structural information of the graph as compared with standard modularity.

The modularity optimisation algorithm has also been straightforwardly extended to directed networks (Leicht and Newman, 2008). Because if an edge is directed, the probability that it will be oriented in either of the two possible directions depends on the in- and out-degrees of the end vertices. Therefore, the expression of modularity for directed networks is expressed as Equation (16.16), and the modularity for weighted directed networks is expressed as Equation (16.32).

$$Q_{wd} = \frac{1}{W} \sum_{ij} \left(a_{ij} - \frac{d_i^I d_j^O}{W} \right) \delta_{ij} \tag{16.32}$$

Modularity for bipartite networks has also been developed (Barber, 2007; Guimerà, Sales-Pardo and Amaral, 2007). In Barber (2007), suppose that there are two classes of vertices with p and q vertices, respectively; Barber developed a modularity matrix for bipartite networks as $M_b = \tilde{B} - \tilde{P}$, where $\tilde{B}$ has a block off-diagonal form of the type shown in Equation (16.33),

$$\tilde{B} = \begin{bmatrix} 0_{p \times p} & B_{p \times q} \\ B_{p \times q}^T & 0_{p \times p} \end{bmatrix} \tag{16.33}$$

where $B_{p \times q}$ is the incident matrix, $0_{p \times p}$ is the all-zero matrix. Similarly, $\tilde{P}$ is also a block off-diagonal matrix as shown in Equation (16.34),

$$\tilde{P} = \begin{bmatrix} 0_{p \times p} & P_{p \times q} \\ P_{p \times q}^T & 0_{p \times p} \end{bmatrix} \tag{16.34}$$

where $\boldsymbol{P}_{p \times q}$ is the expected probability matrix in the null model. However, when the number of clusters is unknown, the performance of spectral optimisation of modularity worsens. Therefore, Barber proposed a different optimisation technique, called bipartite recursively induced modules (BRIM), based on the bipartite nature of the graph. The vertices of a bipartite network can be partitioned into two disjoint sets such that no two vertices within the same set are adjacent; that is, two sets can be assigned one of two colours, say red and blue, with no adjacent vertices bearing the same colour. The algorithm is based on the special expression of modularity for the bipartite case, for which once the partition of the red or the blue vertices is known, it is easy to get the partition of the other vertex class that yields the maximum modularity. Therefore, one starts from an arbitrary partition in c clusters of, say, the blue vertices, and recovers the partition of the red vertices, which is in turn used as input to get a better partition of the blue vertices, and so on until modularity converges. BRIM does not predict the number of clusters c of the graph, but one can obtain good estimates for it by exploring different values with a simple bisection approach. Another interesting extension of modularity for bipartite networks was introduced by Guimerà, Sales-Pardo and Amaral (2007). Let us call the two classes of vertices actors and teams, and indicate with t_i the degree of actor i and m_a the degree of team a. The null model graphs are random graphs with the same expected degrees for the vertices, as usual. The bipartite modularity for a partition $\wp$ is given by Equation (16.35),

$$Q_b(\wp) = \sum_{c=1}^{n_c} \left(\frac{\sum_{i \neq j \in c} a_{ij}}{\sum_a m_a(m_a - 1)} - \frac{\sum_{i \neq j \in c} t_{ij}}{\left(\sum_a m_a \right)^2} \right) \tag{16.35}$$

where n_c is the number of communities.

There are also many other modifications and extensions. Because of the limitations on space, we cannot detail them all here. Readers who are interested may be referred to the more comprehensive review (Fortunato, 2010).

16.3.4.4   Limitations

The first important question is about the value of the maximum modularity. The modularity maximisation algorithms are all based on a statement that the existence of modules in complex networks may result in a large enough modularity, which implies that random graphs have low modularity. However, it has been reported by Guimerà, Sales-Pardo and Nunes (2004) that a large value for the modularity maximum does not necessarily mean that a graph has community structure. Random graphs are supposed to have no community structure, while they may have partitions that themselves have large modularity values. This is due to fluctuations in the distribution of edges in the graph, which may be inhomogeneous even if the linking probability is constant, as in Erdős–Rényi graphs. Therefore, the maximum modularity of a graph reveals a significant community structure only if it is appreciably larger than the maximum modularity of random graphs of the same size and expected degree sequence. The significance of the maximum modularity for a graph can be calculated by testing it with the maximum modularity for many realisations of a null model, which is obtained from the original graph by randomly rewiring its edges. The average modularity $\bar{Q}$ and the standard deviation σ_Q of the results can be calculated. The statistical significance of the maximum modularity $Q_{\max}$ is indicated by the distance of Q_{max} from the null model average modularity $\bar{Q}$ in units of standard deviation σ_Q; that is, by the z-score, calculated using Equation (16.36).

$$z = \frac{Q_{max} - \bar{Q}}{\sigma_Q} \qquad (16.36)$$

If $z \gg 1$, Q_{max} indicates strong community structure.

Another issue, raised by Fortunato and Barthelemy (2007), is more fundamental since it affects the capability of modularity as a measure to assess the quality of partitions. In Fortunato and Barthelemy (2007), it was pointed out that modularity optimisation has a resolution limit that may prevent it from detecting clusters which are comparatively small with respect to the graph as a whole. The resolution limit comes from the very definition of modularity, in particular from its null model. The weak point of the null model is the implicit assumption that each vertex can interact with every other vertex, which implies that each part of the graph knows about everything else. This is certainly not true for large systems. A possible way to go around the resolution-limit problem could be to perform further subdivisions of the clusters obtained from modularity optimisation, in order to eliminate possible artificial mergers of communities.

16.3.5   Multilevel Graph Partitioning and Hypergraph Partitioning

Karypis and Kumar proposed a multilevel graph partition algorithm, called METIS[1] (Karypis and Kumar, 1995, 1998). The basic idea behind the multilevel graph partition algorithms implemented in METIS is that the graph is coarsened down to a small portion of vertices in the first place, then a bisection of this much smaller graph is done, and finally the partition is projected back towards the original graph by iteratively refining the partition, which is called the uncoarsening phase. During the coarsening phase, a sequence of smaller graphs $G_l = (V_l, E_l)$ is constructed from the original graph $G_0 = (V_0, E_0)$ such that $|V_l| < |V_{l-1}|$. Graph G_l is constructed from G_{l-1} by finding a maximal matching of G_l and collapsing together the vertices that are incident on each edge of the matching. When vertices $v, u \in V_{l-1}$ are collapsed to find vertex $w \in V_l$, the weight of vertex w is set equal to the sum of the weights of vertices v and u, while the edges incident on w are set equal to the union of the edges incident on v and u minus the edge (u,v). If there is an edge that is incident to both on v and u, then the weight of this edge is set equal to the sum of the weights of these edges. Therefore, the weights of both vertices and edges increase during the coarsening process. Matching not only is essential for the coarsening level but also greatly affects the quality of the bisection and the time required during the uncoarsening phase. Maximal matching can be computed in many different ways. We only briefly describe one matching scheme implemented in METIS, called heavy-edge matching (HEM). Readers who are interested in other matching schemes may be referred to Karypis and Kumar (1998). HEM computes a matching M_l, such that the weight of edges in M_l is high. The HEM is computed using a randomised algorithm. The vertices are visited in a random order. HEM matches a vertex with the unmatched vertex that is connected with the heavier edge. The second phase of METIS is to compute a minimum edge-cut bisection of the coarse graph G_l such that each part contains

---

[1] 'Metis' is the Greek word for wisdom. Metis was a titaness in Greek mythology.

roughly half the vertex weight of the original graph. A partition of G_l can be obtained using various algorithms such as spectral bisection or graph cut, which we have discussed previously. Since the size of the coarse graph G_l is small, the computation in this phase is light. In the last uncoarsening phase, the partition of the coarse graph G_l is projected back to the original graph by going through G_{l-1}, G_{l-2}, ..., G_0. After projecting a partition, a partition-refinement algorithm is used. The basic purpose of a partition-refinement algorithm is to select two subsets of vertices, one from each part such that the resulting partition has a smaller edge-cut when swapping them.

Karypis *et al.* further extended METIS to solve the hypergraph partitioning problem (Karypis *et al.*, 1999). The hypergraph partitioning problem is to partition the vertices of a hypergraph into k roughly equal parts, such that the number of hyperedges connecting vertices in different parts is minimised. The extended multilevel hypergraph partitioning algorithm is named HMETIS.

16.3.6   Markov Cluster Algorithm

The Markov clustering (MCL) algorithm was designed specifically for simple graphs and weighted graphs (van Dongen, 1998). It was extended to TRIBE-MCL, which was used to detect protein families in large databases based on domain architecture or the presence of sequence motifs (Enright, van Dongen and Ouzounis, 2002). Natural clusters in a graph are characterised by the presence of many edges between the members of that cluster, and one expects that the number of 'higher-length' paths between two arbitrary nodes in the cluster is high. Particularly, this number should be higher than the number of node pairs lying in different natural clusters. The MCL algorithm found cluster structure in graphs by a mathematical bootstrapping procedure. The process deterministically computed the probabilities of random walks through the similarity graph, and used two operators transforming one set of probabilities into another. It did so by capturing the mathematical concept of random walks on a graph. The MCL algorithm simulated random walks within a graph by alternation of two operators called expansion and inflation. Expansion coincided with taking the power of a stochastic matrix using the normal matrix product. Inflation corresponded with taking the Hadamard power of a matrix, followed by a scaling step, such that the resulting matrix is stochastic; that is, the matrix elements correspond to probability values. It did not contain high-level procedural rules for the assembling, splitting or joining of groups.

16.4   Resources

There are many platforms and tools existing to analyse and visualise the graphs and networks. First of all, we summarise the file formats used to store graphs and networks in Table 16.4. For visualisation tools, besides the software tools listed in Table 16.4, namely Pajek, Graphviz and UCINET, there are some other software tools, which are widely used for complex network analysis. We recommend **Gephi**, which is an open-source and free network analysis and visualisation software package written in Java on the NetBeans platform, initially developed by students of the University of Technology of Compiègne in France. Gephi supports most formats listed in Table 16.4 except XGMML and GXL. **Cytoscape** is another option, which is an open-source bioinformatics software platform for visualising molecular interaction networks and integrating

Table 16.4 Summary of file formats as a storage of graphs and network

Formats	Filename extension	Comments
GraphML	.graphml	An XML-based file format for graphs.
XGMML	.xml	(The eXtensible Graph Markup and Modeling Language) is an XML application based on GML which is used for graph description.
Pajek NET	.net	Pajek (Slovene word for Spider) is a program, for Windows, for analysis and visualisation of large networks.
Graphlet GML	.gml	Graphlet is GML graph data format. GML is an acronym derived from Graph Modelling Language.
Graphviz DOT	.dot	DOT is the text file format of the suite GraphViz.
CSV	.csv	A comma-separated values (CSV) (also sometimes called character-separated values) file stores tabular data in plain-text form.
UCINET DL	.dl	UCINET DL format is the most common file format used by the UCINET package. UCINET 6 for Windows is a software package for the analysis of social network data.
GXL	.gxl	GXL (Graph eXchange Language) is designed to be a standard exchange format for graphs.
Text	.txt	Delimited text table.

with gene expression profiles and other state data. Cytoscape is also written in Java and it supports GML, XGMML and text files (including CSV and delimited text table).

For analysing the numerical network data, besides Gephi, Cytoscape and their plugins, which are open source, there are some resources available online. The first is the MIT Network Analysis Toolbox for MATLAB, which is downloadable from the link http://strategic.mit.edu/downloads.php?page=matlab_networks. The second is Lev Muchnik's Complex Network Toolbox for MATLAB, which is available from http://www.levmuchnik.net/Content/Networks/ComplexNetworksPackage.html. However, MATLAB itself is a commercial software, which requires a licence. There is another commercial software, called MATHEMATICA, which also has a package to deal with complex networks.

16.5 Discussion

Graph clustering is an extremely important clustering family in the literature. As we introduced before, data objects can be organised into a matrix (either adjacency matrix or similarity matrix), which can be presented as a graph. To cluster such a graph, there are many different types of graph clustering methods, namely graph cut methods, spectral-based methods, affinity propagation methods, modularity-based methods, and hypergraph methods. In these methods, both feature-based data and relation-based data can be clustered, unlike other clustering methods which are unable to cluster relational data. Therefore, graph clustering has been widely used in network analysis, including gene co-expression networks (Xu, Olman and Xu, 2002; Wilkinson and Huberman, 2004), gene regulatory networks (Basso *et al.*, 2005), protein-interaction networks (Rives and Galitski, 2003; Spirin and Mirny, 2003; Dunn, Dudbridge and Sanderson, 2005; Altaf-Ul-Amin *et al.*, 2006; Chen and Yuan, 2006;

Table 16.5 Summary of graph clustering algorithms mentioned in this chapter

Name	Year	Reference	Software
Minimum cut criterion	1993	Wu and Leahy (1993)	
METIS	1995	Karypis and Kumar (1995, 1998)	
Normalised cut	2000	Shi and Malik (2000)	
MCL	2000	van Dongen (1998)	C (van Dongen, 2012)
Spectral clustering	2001	Ng, Jordan and Weiss (2001); Luxburg (2007)	
Greed method of Newman	2004	Newman (2004a)	MATLAB (MIT, 2011)
Spectral modularity optimisation	2006	Newman (2006)	MATLAB (MIT, 2011)
Affinity propagation	2007	Frey and Dueck (2007)	MATLAB/R (Frey, 2011)
Directed networks	2008	Leicht and Newman (2008)	MATLAB (MIT, 2011)
Overlapping community	2013	Gopalan and Blei (2013)	

Farutin *et al.*, 2006; Dittrich *et al.*, 2008), and metabolic networks (Guimerà and Nunes, 2005). In Table 16.4 we present a summary of file formats used to store graphs and network.

In Table 16.5, we summarise some graph clustering algorithms mentioned in this chapter. Some of them have publicly accessible software. We will discuss more about these applications in Chapter 19.

References

Altaf-Ul-Amin, M., Shinbo, Y., Mihara, K. *et al.* (2006). Development and implementation of an algorithm for detection of protein complexes in large interaction networks. *BMC Bioinformatics*, **7**(1), p. 207.

Barber, M. (2007). Modularity and community detection in bipartite networks. *Physical Review E*, **76**(6), p. 066102.

Basso, K., Margolin, A.A., Stolovitzky, G. *et al.* (2005). Reverse engineering of regulatory networks in human B cells. *Nature Genetics*, **37**(4), pp. 382–390.

Chen, J. and Yuan, B. (2006). Detecting functional modules in the yeast protein - protein interaction network. *Bioinformatics*, **22**(18), pp. 2283–2290.

Dittrich, M.T., Klau, G.W., Rosenwald, A. *et al.* (2008). Identifying functional modules in protein - protein interaction networks: an integrated exact approach. *Bioinformatics*, **24**(13), pp. i223–i231.

van Dongen, S. (1998). *A New Cluster Algorithm for Graphs*, CWI (Centre for Mathematics and Computer Science), Amsterdam.

van Dongen, S. (2012). *MCL - A Cluster Algorithm for .Graphs*, http://micans.org/mcl/ (accessed 24 July 2014).

Duch, J. and Arenas, A. (2005). Community detection in complex networks using extremal optimization. *Physical Review E*, **72**(2), p. 027104.

Dunn, R., Dudbridge, F. and Sanderson, C.M. (2005). The use of edge-betweenness clustering to investigate biological function in protein interaction networks. *BMC Bioinformatics*, **6**(1), p. 39.

Enright, A.J., van Dongen, S. and Ouzounis, C.A. (2002). An efficient algorithm for large-scale detection of protein families. *Nucleic Acids Research*, **30**(7), pp. 1575–1584.

Farutin, V., Robison, K., Lightcap, E. *et al.* (2006). Edge-count probabilities for the identification of local protein communities and their organization. *Proteins: Structure, Function, and Bioinformatics*, **62**(3), pp. 800–818.

Fortunato, S. (2010). Community detection in graphs. *Physics Reports*, **486**(3–5), pp. 75–174.

Fortunato, S. and Barthelemy, M. (2007). Resolution limit in community detection. *Proceedings of the National Academy of Sciences*, **104**(1), pp. 36–41.

Frey, B.J. (2011). *Frey Lab*, http://genes.toronto.edu/index.php?q=affinity%20propagation (accessed 24 July 2014).

Frey, B.J. and Dueck, D. (2007). Clustering by passing messages between data points. *Science,* **315**(5814), pp. 972–976.

Golub, G.H. and Van, C.F. (2012). *Matrix Computations*, JHU Press, Baltimore.

Gopalan, P.K. and Blei, D.M. (2013). Efficient discovery of overlapping communities in massive networks. *Proceedings of the National Academy of Sciences,* **110**(36), pp. 14534–14539.

Guimerà, R. and Nunes, L.A. (2005). Functional cartography of complex metabolic networks. *Nature,* **433**(7028), pp. 895–900.

Guimerà, R., Sales-Pardo, M. and Nunes, L.A. (2004). Modularity from fluctuations in random graphs and complex networks. *Physical Review E,* **70**(2), p. 025101.

Guimerà, R., Sales-Pardo, M. and Amaral, L. (2007). Module identification in bipartite and directed networks. *Physical Review E,* **76**(3), p. 036102.

Jeong, H., Tombor, B., Albert, R. *et al.* (2000). The large-scale organization of metabolic networks. *Nature,* **407**(6804), pp. 651–654.

Karypis, G. and Kumar, V. (1995). *Metis-Unstructured Graph Partitioning and Sparse Matrix Ordering System, Version 2.0*. Department of Computer Science, University of Minnesota, Minneapolis.

Karypis, G. and Kumar, V. (1998). A fast and high quality multilevel scheme for partitioning irregular graphs. *SIAM Journal on Scientific Computing,* **20**(1), pp. 359–392.

Karypis, G., Aggarwal, R., Kumar, V. and Shekhar, S. (1999). Multilevel hypergraph partitioning: applications in VLSI domain. *IEEE Transactions on Very Large Scale Integration (VLSI) Systems,* **7**(1), pp. 69–79.

Leicht, E. and Newman, M.E.J. (2008). Community structure in directed networks. *Physical Review Letters,* **100**, p. 118703.

Luxburg, U.V. (2007). A tutorial on spectral clustering. *Statistics and Computing,* **17**(4), pp. 395–416.

MIT (2011). *Matlab Tools for Network Analysis*, http://strategic.mit.edu/downloads.php?page=matlab_networks (accessed 24 July 2014).

Newman, M.E.J. (2004a). Fast algorithm for detecting community structure in networks. *Physical Review E,* **69**(6), p. 066133.

Newman, M.E.J. (2004b). Analysis of weighted networks. *Physical Review E,* **70**(5), p. 056131.

Newman, M.E.J. (2006). Modularity and community structure in networks. *Proceedings of the National Academy of Sciences of the United States of America,* **103**(23), pp. 8577–8582.

Newman, M.E.J. (2010). *Networks: an Introduction*, Oxford University Press, Oxford.

Newman, M.E.J. and Girvan, M. (2004). Finding and evaluating community structure in networks. *Physical Review E,* **69**(2), p. 026113.

Ng, A.Y., Jordan, M.I. and Weiss, Y. (2001). On spectral clustering: analysis and an algorithm. *Proceedings of Advances in Neural Information Processing Systems* (eds T.G. Dietterich, S. Becker and Z. Ghahramani), MIT Press, Cambridge, MA, Volume **14**, pp. 849–856.

Rives, A.W. and Galitski, T. (2003). Modular organization of cellular networks. *Proceedings of the National Academy of Sciences,* **100**(3), pp. 1128–1133.

Shi, J. and Malik, J. (2000). Normalized cuts and image segmentation. *IEEE Transactions on Pattern Analysis and Machine Intelligence,* **22**(8), pp. 888–905.

Spirin, V. and Mirny, L.A. (2003). Protein complexes and functional modules in molecular networks. *Proceedings of the National Academy of Sciences,* **100**(21), pp. 12123–12128.

Wilkinson, D.M. and Huberman, B.A. (2004). A method for finding communities of related genes. *Proceedings of the National Academy of Sciences,* **101**(suppl 1), pp. 5241–5248.

Wu, Z. and Leahy, R. (1993). An optimal graph theoretic approach to data clustering: theory and its application to image segmentation. *IEEE Transactions on Pattern Analysis and Machine Intelligence,* **15**(11), pp. 1101–1113.

Xu, Y., Olman, V. and Xu, D. (2002). Clustering gene expression data using a graph-theoretic approach: an application of minimum spanning trees. *Bioinformatics,* **18**(4), pp. 536–545.

17

Consensus Clustering

17.1 Introduction

In the previous chapters, we have introduced many distinct clustering families. Those clustering algorithms have been employed in various applications in many fields. However as is known, there is no clustering method capable of correctly finding the underlying clustering structure for all datasets. Naturally, this leads to a question: given an unknown dataset, how can we decide which clustering algorithm we should use? Moreover, some stochastic clustering algorithms for the same dataset may produce different clustering results with different initialisation parameters. The question further becomes: which clustering result should we trust? In clustering analysis, one solution to the question is the use of numerical clustering validation algorithms, which assess the quality of clustering results in terms of many criteria. We will discuss numerical clustering validation in Chapter 20. Since it is also true that no single clustering validation algorithm has been claimed to impartially evaluate the results of any clustering algorithm, the use of clustering validation is not an overwhelmingly reliable solution.

An alternative solution for the above question is to combine different clustering results, also known as ensemble clustering, consensus clustering or cluster aggregation. Ensemble clustering formalises the idea that combining different clusterings into a single representative or consensus result would emphasise the common organisation in the different clustering results. Ensemble clustering has attracted a lot of interest during the past decade. Many ensemble clustering algorithms have been proposed in the literature and have thereafter been reviewed thoroughly in several good survey papers (Ghaemi *et al.*, 2009; Li, Ogihara and Ma, 2010; Vega-Pons and Ruiz-Shulcloper, 2011). Ensemble clustering is expected to possess a set of properties; for example, it has to provide more robust, more novel and more stable clustering results than the single clustering algorithms (Ghaemi *et al.*, 2009). However, it is not necessarily true for ensemble clustering to have these properties because even the natural structure or the ground truth of the dataset

Integrative Cluster Analysis in Bioinformatics, First Edition. Basel Abu-Jamous, Rui Fa and Asoke K. Nandi.
© 2015 John Wiley & Sons, Ltd. Published 2015 by John Wiley & Sons, Ltd.

may not be the *best* result (Vega-Pons and Ruiz-Shulcloper, 2011). Only one fact can be ensured, which is that the consensus of all clustering algorithms may compensate for possible errors by individual algorithms, and the final clustering result is more *statistically* reliable than any single one.

It has been widely accepted that ensemble clustering consists of two principal steps: the first is the generation step where the creation of a set of partitions of the given dataset happens, and the second is the *consensus function,* where a new partition, which is the integration of all partitions, is computed. Consensus function is the main step in ensemble clustering. The great challenge in ensemble clustering is the definition of an appropriate consensus function (Vega-Pons and Ruiz-Shulcloper, 2011). The common taxonomy is to classify ensemble clustering algorithms in terms of consensus function into two large classes: one is called *object co-occurrence* approach, and the other is called *median partition* approach (Ghaemi *et al.*, 2009; Li, Ogihara and Ma, 2010; Vega-Pons and Ruiz-Shulcloper, 2011). Nonetheless, we employ an alternative taxonomy, which may be clearer. We classify all consensus functions into four classes in terms of what is the target for comparison in the consensus process: *partition–partition* (*P–P*) *comparison, cluster–cluster* (*C–C*) *comparison, member-in-cluster* (*MIC*) *voting,* and *member– member* (*M–M*) *co-occurrence.* In this chapter, we primarily focus on the basic concepts in each class, rather than enumerating all ensemble clustering algorithms.

17.2 Overview

The diagram of the general process of ensemble clustering is depicted in Figure 17.1. Every ensemble clustering method consists of two steps: generation and consensus. Generation is the first step in ensemble clustering methods, where, in our case, R partitions, $\{Z_r | r = 1, \ldots, R\}$, are generated. Partitions are supposed to have $k^{(r)}$ clusters, and the numbers of clusters for different partitions can be either similar or different. There is no constraint about

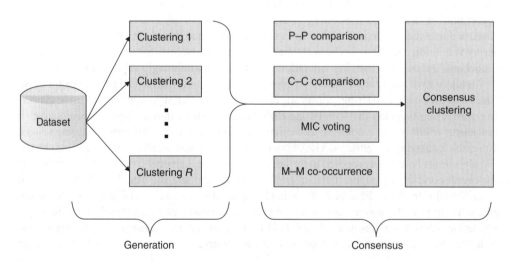

Figure 17.1 Diagram of general process of ensemble clustering. The consensus step may have four different types of consensus functions, namely P–P comparison, C–C comparison, MIC voting, M–M co-occurrence

how the partitions are obtained. They could be generated by different clustering algorithms, or the same clustering algorithm with different parameters initialisation. They even could be from different datasets for the same set of objects; for example, the same set of genes in different conditions, tissues, or organisms (Abu-Jamous *et al.*, 2013).

Topchy *et al.* reported that combining the partitions generated using weak but less expensive clustering algorithms may achieve comparable or better performance (Topchy, Jain and Punch, 2003, 2005). However, it is worth noting that Topchy *et al.* drew their conclusions based on empirical results comparison among their proposed consensus function combining weak partitions, single weak clustering algorithms and other consensus functions combining weak partitions. In other words, it is not necessarily true that combining weak partitions may provide better performance than combining better partitions. Nevertheless, the study conducted by Topchy *et al.* provided us a solution that may provide comparable or reasonable performance with low complexity.

Generally, the main challenge of ensemble clustering is the definition of consensus function. As shown in Figure 17.1, there are four classes of consensus functions in terms of what is the target for comparison in the consensus process, namely P–P comparison, C–C comparison, MIC voting, and M–M co-occurrence. This taxonomy looks different from the popular two-class taxonomy, namely object co-occurrence and median partition, in Ghaemi *et al.* (2009); Vega-Pons and Ruiz-Shulcloper (2011), although there are remarkable resemblances between them. Basically, P–P comparison is equivalent to median partition, and object co-occurrence is equivalent to the union of C–C comparison, MIC voting and M–M co-occurrence. Such four-class taxonomy may more clearly reveal the mechanism of consensus functions. In the forthcoming section, we will discuss various consensus functions.

17.3 Consensus Functions

17.3.1 P–P Comparison

P–P Comparison approaches, equivalent to median partition approaches, attempt to provide the solution of an optimisation problem, which maximises the total similarity to the given partitions. Therefore, the essential part of this class of consensus functions is the P–P similarity/ dissimilarity comparison. The optimisation problem can be written as Equation (17.1),

$$\mathbf{Z}^* = \arg\max_{\mathbf{P} \in \mathbb{P}_X} \sum_{j=1}^{R} \Gamma(\mathbf{Z}, \mathbf{Z}_j) \tag{17.1}$$

where $\Gamma(\cdot, \cdot)$ is a similarity measure between partitions. The consensus partition is defined as the partition that maximises its similarity with all partitions in the ensemble clustering. This problem implicitly appeared from as early as the late eighteenth century and the first mathematical treatment goes back to 1965 (Filkov and Skiena, 2004). The main theoretical results have been obtained for the particular case when Γ is the symmetric similarity measure. If Γ is a dissimilarity measure between partitions, the problem becomes the minimisation of the total dissimilarity with respect to the cluster ensemble.

There are a lot of similarity and dissimilarity measures between partitions that can be used in the P–P comparison. A large proportion of these measures can be found in the research of *external clustering validity index*, which we will discuss in Chapter 20; for example Rand index (RI) (Rand, 1971), adjusted Rand index (ARI) (Hubert and Arabie, 1985), normalised mutual information (NMI) (Cover and Thomas, 2006), and Jaccard index (JI) (Levandowsky and Winter, 1971). Besides the above external validity indices, the Mirkin distance is also a popular measure of partition similarity (Mirkin, 1998).

17.3.1.1 Mirkin Distance-related Methods

Mirkin distance is defined as follows: Given two partitions Z_1 and Z_2 of the same dataset X, there are four numbers defined:

- n_{00}: The number of pairs of objects that were clustered in separate clusters in Z_1 and also Z_2;
- n_{01}: The number of pairs of objects that were clustered in different clusters in Z_1, but in the same cluster in Z_2;
- n_{10}: The number of pairs of objects that were clustered in the same cluster in Z_1, but in different clusters in Z_2;
- n_{11}: The number of pairs of objects that were clustered in the same clusters in P_1 and P_2.

Mirkin distance thus is defined as $\mathcal{M}(Z_1, Z_2) = n_{01} + n_{10}$, which is the symmetric distance, representing the number of disagreements between the two partitions. Mathematically, considering two objects u and v, the measure of disagreement between clustering P_1 and P_2 on u and v is expressed as given in Equation (17.2).

$$\mathcal{M}_{uv}(Z_1, Z_2) = \begin{cases} 1 & \begin{aligned} &\text{if } (Z_1(u) = Z_1(v) \text{ and } Z_1(u) \neq Z_1(v)) \\ &\text{or } (Z_1(u) \neq Z_1(v) \text{ and } Z_1(u) = Z_1(v)) \end{aligned} \\ 0 & \qquad\qquad \text{otherwise} \end{cases} \qquad (17.2)$$

That is, Mirkin distance can be rewritten as given in Equation (17.3).

$$\mathcal{M}(Z_1, Z_2) = \sum_{u,v} \mathcal{M}_{uv}(Z_1, Z_2) \qquad (17.3)$$

The P–P comparison problem becomes as shown in Equation (17.4).

$$Z^* = \arg \min_{Z \in \mathbb{P}_X} \sum_{j=1}^{R} \mathcal{M}(Z, Z_j) \qquad (17.4)$$

With this definition of distance, the P–P comparison problem appears to be converted to a graph-based partition problem. Let us define a Mirkin distance matrix of clustering set $\{Z_1, Z_2, \ldots, Z_R\}$ as $M \in \mathbb{Z}^{N \times N}$. We also need to define the weight of edge (u, v) of any pair of clusterings as $\mathcal{M}_{uv}$ and the entry of Mirkin distance is then written as shown in Equation (17.5).

$$\mathcal{M}_{uv} = \sum_{i,j=1}^{R} \mathcal{M}_{uv}(Z_i, Z_j) \qquad (17.5)$$

Therefore, the P–P comparison problem is equivalent to *correlation clustering,* which finds a partition that minimises the score function (Gionis, Mannila and Tsaparas, 2007), as shown in Equation (17.6).

$$Z^* = \arg\min_{Z \in \mathbb{P}_X} \left(\sum_{\substack{(u,v) \\ Z(u) = Z(v)}} \mathcal{M}_{uv} + \sum_{\substack{(u,v) \\ Z(u) \neq Z(v)}} (1 - \mathcal{M}_{uv}) \right) \quad (17.6)$$

There have been many algorithms in the literature that provide solutions for this optimisation problem in a heuristic way (Filkov and Skiena, 2004; Bertolacci and Wirth, 2007; Gionis, Mannila and Tsaparas, 2007).

The easiest method is **Pick-A-Cluster,** which simply chooses a partition randomly from the given clustering set. A slightly more sophisticated method is **Best-of-k** (BOK) (Filkov and Skiena, 2004), also known as **Best-Clustering** (Bertolacci and Wirth, 2007). BOK selects the partition from the given clustering set closest to all other partitions.

Filkov and Skiena proposed three heuristic algorithms, including BOK (Filkov and Skiena, 2004). One of the other two algorithms is called simulated-annealing-one-element-move (SAOM), and the other is called best-one-element-move (BOM). SAOM follows the idea of guessing an initial partition and iteratively changing it by moving one object from one cluster to another. The initial partition can be the output of Pick-A-Cluster or BOK, or other consensus function. In SAOM heuristics, the simulated annealing meta-heuristics is applied in order to avoid the convergence to a local optimum. BOM also starts with an initial partition, which can be the output of other consensus functions. It also attempts to reach an optimal partition by moving an object from one cluster to another. The difference between SAOM and BOM is that a greedy approach is employed in BOM. Every move in BOM attempts to maximise $\Delta \mathcal{M} = \mathcal{M} - \mathcal{M}'$, where $\mathcal{M}$ is the distance before moving and $\mathcal{M}'$ denotes the distance after moving.

The Balls Algorithm was proposed by Gionis, Mannila and Tsaparas (2007) to solve the problem in a graph. Data objects are viewed as vertices in the graph. The intuition of the algorithm is to find a set of vertices that are close to each other within a cluster and far from other vertices outside the cluster. The algorithm gradually removes clusters from the graph. The difficulty lies in finding sush a set in good quality. The algorithm takes all vertices that are close to a vertex, which form a ball-shaped cluster. More formally, the procedure of the Balls algorithm is described as follows. First, the algorithm sorts the vertices in increasing order of the distance to other vertices. Then, at every step, the algorithm picks the first unclustered vertex in the ordering. A set of vertices B is formed. A parameter α is defined to be a threshold. If the average distance between the chosen vertex to the set B is smaller than or equal to α, the the vertex is considered as a member of the set; otherwise, the vertex forms a singleton cluster.

Gionis and colleagues proposed many other heuristic algorithms in Gionis, Mannila and Tsaparas (2007). One of them is ***the agglomerative algorithm,*** which is like the standard average linkage agglomerative algorithm. Another algorithm is called ***the furthest algorithm,*** which is a top-down heuristics. It starts with all objects in the same single cluster; in each step, the pair of vertices furthest apart are placed in different clusters and become the centres of the clusters. The remaining vertices are assigned to the centre that incurs the least cost. At the end of each step, the cost of the new solution is computed. If it is lower than that of the previous step, then the algorithm continues. This procedure is repeated iteratively until there is no improvement in cost reduction any more. *The local search algorithm* is an application of a local search heuristic method. It starts with an initial partition, which is iteratively updated. The cost of moving an object from one cluster to another is computed. Every move of vertex from one cluster to another yields the minimum cost. The process is iterated until there is no move that can improve the cost. In particular, the local search algorithm resembles BOM very much.

Ailon, Charikar and Newman (2008) proposed **the CC-Pivot algorithm,** which is a three-approximation algorithm for correlation clustering. The CC-Pivot algorithm attemps to solve the the problem of aggregating inconsistent information from many different sources in numerous contexts and disciplines. An instance of correlation clustering problem is represented by a graph with a vertex for each data object and an edge labelled (+) or (−) for each pair of vertices, indicating that two elements should be in the same or different clusters, respectively. The goal is to cluster the elements so as to minimise the number of (−) edges within clusters and (+) edges crossing clusters. Consensus clustering may be an application of correlation clustering, whose goal is to find a clustering that minimises the number of pairwise disagreements with the given clusterings. CC-Pivot works as follows: suppose that V contains all vertices; firstly, a random vertex i is selected as the pivot vertex and two clusters are initialised as $C = \{i\}$, $C' = \emptyset$; $\forall j \in V$, and $j \neq i$; if $(i,j) \in E^+$ then vertex j is added to C, otherwise (if $(i,j) \in E^-$), vertex j is added to C'; finally, a subgraph induced by C' is sent to CC-Pivot as input and the clustering structure can be worked out recursively.

17.3.1.2   Information Theory-based Methods

The utility function $U(C, C_r)$ evaluates the quality of a candidate median partition $C = \{C_1, ..., C_K\}$ against the given partitions $C_r = \left\{ C_1^{(r)}, ..., C_{k^{(r)}}^{(r)} \right\}$ (Topchy, Jain and Punch, 2005) [Equation (17.7)],

$$U(C, C_r) = \sum_{k=1}^{K} p(C_k) \sum_{j=1}^{k^{(r)}} p\left(C_j^{(r)} | C_k \right)^2 - \sum_{j=1}^{k^{(r)}} p\left(C_j^{(r)} \right)^2 \tag{17.7}$$

where $p(C_k) = |C_k|/N$, $p\left(C_j^{(r)} \right) = \left| C_j^{(r)} \right|/N$, and $p\left(C_j^{(r)} | C_k \right) = \left| C_j^{(r)} \cap C_k \right|/|C_k|$. The function $U(C, C_r)$ assesses the agreement between two partitions by the difference between the expected number of labels of partitions C_r that can be correctly predicted both with the knowledge of clustering C and that without the knowledge of clustering C. The overall utility of the partition C with respect to all the partitions can be measured as the sum of the pairwise agreements. Therefore, the best median partition should maximise the value of overall utility [Equation (17.8)]

$$C^* = \arg\max_C \sum_{r=1}^{R} U(C, C_r) \tag{17.8}$$

Considering the information-theoretic approach to the median partition problem, the quality of the consensus partition C^* is determined by the amount of information which it shares with the given partitions [Equation (17.9)].

$$I(C,C_r) = \sum_{k=1}^{K} \sum_{j=1}^{k^{(r)}} p\left(C_k, C_j^{(r)}\right) \log\left(\frac{p\left(C_k, C_j^{(r)}\right)}{p(C_k)p\left(C_j^{(r)}\right)}\right) \tag{17.9}$$

An optimal median partition can be found by solving the optimisation problem as shown in Equation (17.10).

$$C^* = \arg\max_C \sum_{r=1}^{R} I(P, P_r) \tag{17.10}$$

However it is not clear how to directly use these equations in a search for consensus.

17.3.2   C–C Comparison

C–C Comparison considers a cluster as a unit and compares the similarity between clusters. In other words, C–C comparison algorithms are based on clustering clusters. One typical C–C comparison algorithm was proposed by Strehl and Ghosh, called meta-clustering algorithm (MCLA) (Strehl and Ghosh, 2003). In MCLA, each cluster is represented by a hyperedge. MCLA groups and collapses related hyperedges and assigns each data object to the collapsed hyperedge in which it participates most strongly. The hyperedges that are considered related for the purposes of collapsing are determined by a graph-based clustering. Each cluster of hyperedges is referred to as a meta-cluster. The number of hyperedges is reduced from $\sum_{r=1}^{R} k^{(r)}$ to k by collapsing.

There are four steps in MCLA:

1. **Construct Meta-graph.** All the $\sum_{r=1}^{R} k^{(r)}$ indicator vectors h as vertices of another regular undirected graph, which is called a meta-graph. The edge weights are proportional to the similarity between vertices. Strehl and Ghosh employed in MCLA the binary Jaccard measure, which we will introduce in Chapter 20. Since the clusters are non-overlapping in individual partitions, there are no edges among vertices of the same partition, and thus the meta-graph is r-partite.
2. **Cluster Hyperedges.** In this step, matching labels by partitioning the meta-graph into k balanced meta-clusters are obtained. Different graph-partitioning algorithms may be employed in this step. Strehl and Ghosh used the package METIS to cluster the indicator vectors h into k meta-clusters. A meta-cluster represents a group of corresponding labels because each vertex in the meta-graph represents a distinct cluster label.
3. **Collapse Meta-clusters.** For each of the k meta-clusters, the hyperedges are collapsed into a single meta-hyperedge. Each meta-hyperedge contains an association vector, where the entry for each object describes its level of association with the corresponding meta-cluster. The level is computed by averaging all indicator vectors h of a particular meta-cluster.
4. **Compete for Objects.** Each object is assigned to its most associated meta-cluster in terms of the highest entry in the association vector.

The experiment results in Strehl and Ghosh (2003) showed that MCLA should be best suited in terms of time complexity and quality when compared with the cluster-based similarity partition algorithm (CSPA) and hypergraph-partitioning algorithm (HGPA).

17.3.3   MIC Voting

MIC Voting is a common idea of many approaches, which focuses on the relation between members and their clusters in all partitions. Via voting by data objects, the winning clusters have high probabilities to be included in the final consensus partition.

17.3.3.1   Relabelling and Voting

Combining multiple partitions has a major problem which is that those partitions generated in different situations may be totally different; that is, the clusters in these partitions are formed in totally different ways. As an example, assume that we are given two partitions, each of which contains three clusters. When combining these two partitions, we have to decide which of the clusters in the first partition matches a given cluster of the second partition. Relabelling and voting is one of the methods used to solve this problem.

Dimitriadou and colleagues proposed a relabelling and voting approach called the voting-merging (VM) algorithm (Dimitriadou, Weingessel and Hornik, 2001). Generally, VM consists of three procedures, namely the partition procedure, the voting procedure, and finally the merging procedure. These three procedures of the algorithm are applied sequentially and do not interfere with each other. The critical two procedures of VM are the voting and merging procedures. Actually, the voting procedure further consists of relabelling and voting steps. Relabelling creates a mapping between two clusters from two partitions. For example, given partitions $C^1 = \{C_1^1, \ldots, C_k^1\}$ and $C^2 = \{C_1^2, \ldots, C_k^2\}$, where C_j^i denotes the ith partition and the jth cluster. To do the relabelling, the percentage of data points assigned to C_l^1 and C_m^2 simultaneously is computed. Then the two clusters with the highest percentage of common data points are mapped together. The same action is done to the remaining clusters iteratively until all clusters in partitions are mapped (relabelled). After relabelling, the voting procedure takes place: If a data point is assigned to both clusters which have been mapped, one vote will be issued to the common cluster; if a data point is assigned to two clusters which are not mapped, only half a vote will be issued to each of the two clusters. Therefore, the resulting partition at the end of the voting procedure is a fuzzy partition of the data. Every data point may belong to more than one cluster with a certain degree of belongingness. A neighbourhood relation between two clusters can be measured in terms of how many data points in one cluster belong to the other. A merging procedure starts with many clusters and merges clusters which are closest to each other. After a single merging step, the sureness of the data points and the clusters are recalculated based on the merged clusters. The merging step repeats until some stopping criterion is met.

Bagging, also known as **Bootstrap aggregating,** is a resampling scheme, which is designed to improve the quality of clustering. Dudoit and Fridlyand proposed two resampling methods, one of which, named *BagClust1,* is to combine multiple partitions of bootstrap sample sets by voting (Dudoit and Fridlyand, 2003). Such cluster voting is used to assess the confidence of cluster assignments for individual observations. The motivation behind the bagging algorithm is to reduce variability in the partitioning results via averaging. In BagClust1, the clustering procedure is repeatedly applied to each bootstrap sample and the final partition is obtained by *plurality voting;* that is, by taking the majority cluster for each observation. For a fixed number of clusters K, a clustering procedure (any clustering algorithm) is applied to the original dataset X and cluster labels are obtained. Then B different datasets are drawn by sampling

N objects with replacement from the empirical probability distribution defined by the input dataset. The B multiple sets of objects on the resampled dataset can be considered as bootstrap replications and they are denoted by $x_n^b, 1 \leq b \leq B$. The same clustering procedure is applied to the bootstrap replicate datasets and cluster labels are obtained for all bootstrap replicate datasets. Subsequently, the cluster labels of bootstrap datasets are permuted so that there is maximum overlap with original clustering of these objects. Finally, a bagged cluster label for each object is assigned by majority voting; that is, the cluster label corresponding to x_n is the one with the maximum votes.

Ayad and Kamel (2008) proposed an idea of **cumulative voting** as a solution for the problem of aligning clustering labels generated with different number of clusters. Unlike other voting methods, a probabilistic mapping is computed in the cumulative clustering algorithm. Cumulative voting virtually is sometimes referred to as weighted voting, and maps an input k_i-partition into a probabilistic representation as a k_0-partition with cluster labels corresponding to labels of the reference clusters. A criterion was defined by Ayad and Kamel (2008) for obtaining a first summary of the ensemble as the minimum average squared distance between the mapped partitions and the optimal representation of the ensemble. Based on the maximum information content, the selection criterion is formulated for the reference clustering. Suppose that we have R clusterings with a variable number of clusters, denoted as $\mathcal{U} = \{U^1,...,U^R\}$, such that each partition U^i is represented by an $N \times k_i$ binary stochastic matrix. The optimal solution of consensus clustering (Ayad and Kamel, 2008) was found to be as shown in Equation (17.11),

$$\hat{U} = \arg\min_U \left(\frac{1}{R} \sum_{i=1}^R h\left(v_0\left(U^i\right), U\right) \right) \tag{17.11}$$

where $v_0(U^i)$ represented a mapping of partition U^i into a stochastic partition, defined with respect to a reference partition U^0 and denoted as $U^{0,i}$; that is, $v_0\left(U^i\right) : U^i \rightarrow U^{0,i}$. The mapping $v_0(U^i)$ was referred to as cumulative voting. The dissimilarity function $h(\cdot,\cdot)$ was defined as the average squared distance between the probability vectors $u_j^{0,i}$ and u_j, which is given by Equation (17.12).

$$h\left(U^{0,i}, U\right) = \frac{1}{N} \sum_{j=1}^N u_j^{0,i} - u_j^2 \tag{17.12}$$

The mapping $v_0(U^i)$ transforms the individual partitions of the ensemble into fixed stochastic partitions in the sense that for all R partitions, the cluster labels of the reference partition $U^{0,i}$ correspond to the clusters of the reference partition U^0. A categorical random variable, denoted as $C^{0,i}$, was defined over the set of reference cluster labels $\left\{c_q^0\right\}_{q=1}^{k_0}$ along the distributions given by $\hat{p}(c^{0,i}) = \hat{p}(c^0|c^i)$, where C^o and C^i denote random variables defined over $\left\{c_q^0\right\}_{q=1}^{k_0}$ and $\left\{c_l^i\right\}_{l=1}^{k_i}$, respectively. The stochastic matrix $U^{0,i}$ represents the distribution $\hat{p}(c^{0,i}|x)$. The cumulative voting method consists of two stages. First, the reference partition U^0 is selected based on the criterion given in Equation (17.13),

$$U^0 = \arg\max_{U^i} H\left(C\left(U^i\right)\right) \tag{17.13}$$

where $H(C)$ is the entropy of the partition C. Thus U^0 represents a first summary of the ensemble. Second, a formulation of a subsequent criterion is proposed for finding an optimal consensus partition. Finally, the cumulative voting map can be conducted in an unnormalised way [Equation (17.14)]

$$U^{0,i} = U^0 \left(U^{iT} U^i \right) \tag{17.14}$$

or a normalised way [Equation (17.15)],

$$U^{0,i} = U^0 \left(U^{iT} D \left(U^i \right) U^i \right) \tag{17.15}$$

where U^0 is the reference partition with k_0 clusters, U^i is the partition i with k_i clusters, and $D(U^i)$ denotes a $k_i \times k_i$ diagonal matrix whose lth diagonal element is $1/n_l^i$. The empirical distribution $\hat{p}(c,x)$, as an intermediate representation of the ensemble, can be obtained by many different means from $U^{0,i}$. The final optimal consensus clustering, in turn, can be worked out based on $\hat{p}(c,x)$.

17.3.3.2 Mixture Model-based Methods

Topchy and colleagues proposed a finite mixture model to solve the consensus problem by making use of only labels information Y delivered by the contributing clustering algorithms, without the assistance of the original patterns in X (Topchy, Jain and Punch, 2005). The main assumption is that the labels are modelled as random variables drawn from a probability distribution described as a mixture of multivariate component densities [Equation (17.16)],

$$f(y_n | \Theta) = \sum_{k=1}^{K} \alpha_k f(y_n | \theta_k) \tag{17.16}$$

where each component is parameterised by θ_k, and K components in the mixture are identified with the clusters of consensus partition P. The mixing coefficients α_k correspond to the prior probabilities of clusters. The log-likelihood function for the parameters $\Theta = \{\alpha_1, \ldots, \alpha_K, \theta_1, \ldots, \theta_K\}$ given the dataset Y is as shown in Equation (17.17).

$$\log L(\Theta | X) = \log \prod_{n=1}^{N} f(y_n | \Theta) = \sum_{n=1}^{N} \log \sum_{k=1}^{K} \alpha_k f(y_n | \theta_k) \tag{17.17}$$

The objective of consensus clustering is now formulated as a maximum-likelihood estimation problem. The best fitting mixture density for dataset Y can be obtained by maximising the log-likelihood function with respect to the unknown parameters Θ [Equation (17.18)];

$$\Theta^* = \arg\max_{\Theta} \log L(\Theta | X) \tag{17.18}$$

Thus, the original problem of clustering in the space of data X has been transformed, with the help of multiple clustering algorithms, to a space of new multivariate features Y. The ultimate goal is to make a discrete label assignment to the data X through an indirect route of density estimation of Y. The conditional probability of y_n can be represented as given in Equation (17.19).

$$f(y_n|\boldsymbol{\theta}_k) = \prod_{r=1}^{R} f^{(r)}\left(y_{nr}|\boldsymbol{\theta}_k^{(r)}\right) \tag{17.19}$$

Since the variables y_{nr} take on nominal values from a set of cluster labels in the partition P_r, it is natural to view them as the outcome of a multinomial trial [Equation (17.20)].

$$f^{(r)}\left(y_{nr}|\boldsymbol{\theta}_k^{(r)}\right) = \prod_{j=1}^{k^{(r)}} \theta_{rk}(j)^{\delta(y_{nr},j)} \tag{17.20}$$

The maximum-likelihood problem can be solved by using the expectation-maximisation (EM) algorithm. To adopt the EM algorithm, the existence of the complete data (Y,Z) is assumed. If the value of z_n is known, then the component which is used to generate the data y_n can be easily found. The E-step of the solution to the mixture model with multivariate, multinomial components is written as Equation (17.21).

$$\mathbb{E}[z_{nk}] = \frac{\alpha_k \prod_{r=1}^{R} \prod_{j}^{k^{(r)}} \theta_{rk}(j)^{\delta(y_{nr},j)}}{\sum_{g=1}^{K} \alpha_g \prod_{r=1}^{R} \prod_{j}^{k^{(r)}} \theta_{rk}(j)^{\delta(y_{nr},j)}} \tag{17.21}$$

The M-step is given in Equations (17.22) and (17.23).

$$\alpha_k = \frac{\sum_{n=1}^{N} \mathbb{E}[z_{nk}]}{\sum_{n=1}^{N} \sum_{k=1}^{K} \mathbb{E}[z_{nk}]} \tag{17.22}$$

$$\theta_{rk}(j) = \frac{\sum_{n=1}^{N} \delta(y_{nr},j)\mathbb{E}[z_{nk}]}{\sum_{n=1}^{N} \sum_{k=1}^{K} \delta(y_{nr},j)\mathbb{E}[z_{nk}]} \tag{17.23}$$

The E- and M-steps are repeated iteratively until the algorithm converges.

17.3.4   M–M Co-occurrence

The M–M co-occurrence approach converts the consensus partition problem into a co-association matrix partitioning problem. The entry of the co-association matrix is the frequency of two data objects' co-occurrence in all partitions, expressed as given in Equation (17.24),

$$A_{ij} = \frac{1}{R}\sum_{r=1}^{R} \delta\left(Z_r(x_i), Z_r(x_j)\right) \tag{17.24}$$

where $Z_r(x_i)$ represents the associated cluster label of the data object x_i in partition P_r, and $\delta(a, b)$ is 1 if $a = b$, and is 0 otherwise. There are many different algorithms to solve this problem, including tree-based methods, graph-based methods and hypergraph methods.

17.3.4.1 Tree-based Methods

Fred and Jain explored the idea of evidence accumulation for combining the results of multiple clusterings (Fred and Jain, 2005). A framework for extracting consistent clustering among given various partitions was proposed. According to the evidence-accumulation concept, each partition was viewed as an independent evidence of data organisation, and individual data partitions were combined based on a voting mechanism to generate a new $R \times R$ similarity matrix using the R partitions. In order to deal with partitions with different numbers of clusters, a measure of similarity between patterns, which benefits the combination of the clustering results, was proposed. The R partitions of N data points were mapped into a $N \times N$ co-association matrix [Equation (17.25)],

$$C(i,j) = \frac{n_{ij}}{R} \tag{17.25}$$

where n_{ij} is the number of times the pattern pair (i,j) is assigned to the same cluster among the R partitions. Different clustering algorithms could be applied to the similarity matrix. Fred and Jain explored the evidence-accumulation clustering approach with the single-link and average-link hierarchical agglomerative algorithms to extract the combined data partition. They also introduced a theoretical framework, and optimality criteria, for the analysis of clustering combination results based on the concept of mutual information, and on variance analysis using bootstrapping.

17.3.4.2 Graph-based Methods

Strehl and Ghosh proposed the use of CSPA (Strehl and Ghosh, 2003). CSPA assigns one to the entry of a similarity matrix corresponding to two objects which are in the same cluster. The co-association matrix is the sum of all similarity matrices. Mathematically, assume that the binary membership indicator matrix of the rth partition is $H^{(r)}$. The concatenated block matrix $H = \left[H^{(1)}, \ldots, H^{(R)} \right]$ defines a hypergraph with N vertices and $\sum_{r=1}^{R} k^{(r)}$ hyperedges. The co-association matrix is given by Equation (17.26).

$$A = \frac{1}{R} HH^T \tag{17.26}$$

Thus, the co-association matrix can be reclustered by any reasonable graph-partitioning algorithm. METIS, which has been introduced in Chapter 16, was used by Strehl and Ghosh (2003) because of its robust and scalable properties.

17.3.4.3   Hypergraph Methods

Strehl and Ghosh also proposed the use of HGPA (Strehl and Ghosh, 2003). As mentioned previously, the concatenated block matrix $\boldsymbol{H} = \left[\boldsymbol{H}^{(1)},...,\boldsymbol{H}^{(R)}\right]$ defines a hypergraph with N vertices and $\sum_{r=1}^{R} k^{(r)}$ hyperedges. Therefore, the consensus clustering problem is formulated as partitioning the hypergraph by cutting a minimal number of hyperedges. A hyperedge represents the relationship amongst an arbitrary number of vertices, while CSPA considers only pairwise relationships. HGPA employs the hypergraph-partitioning package, HMETIS, which has already been introduced in Chapter 16.

17.3.4.4   Resampling Method

Monti and colleagues developed a consensus clustering method in conjunction with resampling techniques (Monti *et al.*, 2003). It provided the consensus across multiple runs of a clustering algorithm and the ability to assess the stability of the discovered clusters. One of the important features of the method is that all of the information provided by the analysis of the resampled data can be graphically visualised and incorporated in the decisions about clusters' number and cluster membership.

The main motivation for the resampling-based method is the need to assess the stability of the discovered clusters; that is, the robustness of the putative clusters to sampling variability. If the data represent a sample of items drawn from distant sub-populations, and another sample was drawn from the same sub-populations, the induced cluster composition and number should not be radically different. Therefore, the more the attained clusters are robust to sampling variability, the more we are confident that these clusters represent some structure. To this end, perturbations of the original data were simulated by resampling techniques. The clustering algorithm of choice can then be applied to each of the perturbed datasets, and the consensus among multiple runs can be assessed.

Assuming a resampling scheme and a clustering algorithm had been selected, Monti and colleagues devised a method for representing and quantifying the agreement among the clustering runs over the perturbed datasets. A consensus matrix, which is an $N \times N$ matrix that stores, for each pair of items, the proportion of clustering runs in which two items are clustered together, was defined. The consensus matrix was obtained by taking the average over the connectivity matrices of every perturbed dataset. More specifically, suppose that $X^{(1)}, X^{(2)}, ..., X^{(R)}$ represents the list of R perturbed datasets from the original dataset X. Let $A^{(r)}$ denote the $N \times N$ connectivity matrix corresponding to dataset $X^{(r)}$. The entries of $A^{(r)}$ were defined as in Equation (17.27).

$$A^{(r)}(i,j) = \begin{cases} 1 & \text{if items } i \text{ and } j \text{ belong to the same cluster} \\ 0 & \text{otherwise} \end{cases} \tag{17.27}$$

Let $I^{(r)}$ be the $N \times N$ indicator matrix such that its (i,j)th entry is equal to 1 if both items i and j are present in the dataset $X^{(r)}$, and 0 otherwise. Most resampling schemes, such as bootstrapping or sub-sampling, yield datasets that do not include all items from the original dataset. The consensus matrix $\mathcal{A}$ can be defined as a properly normalised sum of the connectivity matrices of all the perturbed datasets [Equation (17.28)].

$$\mathcal{A}(i,j) = \frac{\sum_r A^{(r)}(i,j)}{\sum_r I^{(r)}(i,j)} \tag{17.28}$$

That is, the entry of $\mathcal{A}(i,j)$ records the number of times items i and j are assigned to the same cluster divided by the total number of times both items are selected. An agglomerative hierarchical tree construction algorithm is then applied to $(1-\mathcal{A})$ to yield a dendrogram of item adjacencies. Summary statistics accounting for the stability of a given cluster as well as of a cluster's members were defined based on the consensus matrix. These statistics can be used to establish a ranking of the clusters in terms of their stability, as well as to identify the more respective items within each cluster. The cluster consensus $m(k)$ was defined for each cluster $k \in K$, and item consensus $m_k(i)$ was defined for each item $e_i \in X$ and each cluster k. The cluster consensus was defined as shown in Equation (17.29),

$$m(k) = \frac{1}{N_k(N_k-1)/2} \sum_{\substack{i,j \in I_k \\ i<j}} \mathcal{A}(i,j) \tag{17.29}$$

and the item consensus was defined as in Equation (17.30),

$$m_i(k) = \frac{1}{N_k - 1\{e_i \in I_k\}} \sum_{\substack{j \in I_k \\ i \neq j}} \mathcal{A}(i,j) \tag{17.30}$$

where $1\{\textbf{condition}\}$ is the indicator function that is equal to 1 when **condition** is true, and 0 otherwise. Furthermore, an empirical cumulative distribution function (ECDF) defined over the range $[0, 1]$ is shown in Equation (17.31),

$$\text{CDF}(T) = \frac{\sum_{i<j} 1\{\mathcal{A}(i,j) \leq T\}}{N(N-1)/2} \tag{17.31}$$

where $1\{\dots\}$ denotes the indicator function. The area under the ECDF corresponding to $\mathcal{A}^K$, which denotes the consensus matrix of results with K clusters, was computed based on Equation (17.32).

$$a(K) = \sum_{i=2}^{m} [x_i - x_{i-1}] \text{CDF}(x_i) \tag{17.32}$$

Therefore a metric was defined as shown in Equation (17.33)

$$\Delta(K) = \begin{cases} a(K) & K=2 \\ \dfrac{a(K+1) - a(K)}{a(K)} & K>2 \end{cases} \tag{17.33}$$

to determine the best number of clusters.

17.4   Discussion

Consensus clustering has gained a lot of attention within the last decade, primarily because of the following reasons: first, a lot of clustering algorithms have been developed since the middle of the last century, though there is no guaranteed optimal algorithm for the clustering problem due to its unsupervised nature; secondly, even while applying the same clustering algorithm to the same dataset many times, different clustering results would be produced when different initialisation parameters are adopted, or even when the same parameters are adopted over multiple runs in the case of stochastic algorithms; and last but not the least, combining the clustering results of the same set of data points in different datasets is also an interesting, but challenging, problem; for example, the study of different clustering results collected from different gene expression datasets or experiments of the same set of genes may find the genes which are co-expressed consistently in the given datasets; that is, it may reveal that these genes are more likely to be co-regulated than the genes found co-expressed in a single experiment. This is further discussed in Chapter 24 where a recently proposed framework of consensus clustering that tackles such aspects is presented, namely the *unification of clustering results from multiple datasets using external specifications (UNCLES)* framework and method.

In this chapter, we introduced the fundamental principles of consensus clustering, which could be classified into four categories in terms of consensus functions, namely P–P comparison, C–C comparison, MIC voting, and M–M co-occurrence. P–P comparison-type consensus functions compare the similarity or dissimilarity between two partitions, and the objective function is to maximise the average similarity (or minimise the average dissimilarity) between given partitions and the optimal consensus partitions, where the distance metric can be the Mirkin distance or an information theoretical distance metric. C–C comparison-type consensus functions consider the similarity or dissimilarity between a pair of clusters from different

Table 17.1   Summary of consensus clustering algorithms mentioned in this chapter

Name	Year	Reference	Software
VM	2001	Dimitriadou, Weingessel and Hornik (2001)	
Resampling methods	2003	Monti *et al.* (2003)	
MCL (MetaClustering)	2003	Strehl and Ghosh (2003)	MATLAB
CSPA			(Strehl 2011).
HGPA			
BOK	2004	Filkov and Skiena (2004)	
SAOM			
BOM (Mirkin distance)			
Clustering ensembles weak	2005	Topchy, Jain and Punch (2005)	
Evidence accumulation	2005	Fred and Jain (2005)	
Graph Consensus Clustering (GCC)	2007	Yu, Wong and Wang (2007)	
Clustering aggregation	2007	Gionis, Mannila and Tsaparas (2007)	
Consensus clustering	2010	Brannon *et al.* (2010); Seiler *et al.* (2010b)	Python (Seiler *et al.*, 2010a)

clustering results and form a graph structure, which is called a meta-graph, whose vertices are clusters and weighted links are similarities of pairwise clusters; the meta-graph is then clustered by using graph-based clustering methods. MIC voting-type consensus functions relabel the clusters in terms of the common members over all clusters and count the votes of each data object in each cluster from all clustering results. M–M co-occurrence-type consensus functions generate an association matrix in terms of the frequency of co-occurrence of every pair of data points in all clustering results, thus the rest of the clustering problem can be solved by either tree-based algorithms like hierarchical clustering algorithms or graph- or hypergraph-based algorithms. We summarise these consensus clustering algorithms discussed in this chapter in Table 17.1. The applications of using consensus clustering in the bioinformatics field are also rich. We will discuss them in detail in Chapter 19.

References

Abu-Jamous, B., Fa, R., Roberts, D.J. and Nandi, A.K. (2013). Paradigm of tunable clustering using binarization of consensus partition matrices (Bi-CoPaM) for gene discovery. *Plos One,* **8**(2), e56432.

Ailon, N., Charikar, M. and Newman, A. (2008). Aggregating inconsistent information: ranking and clustering. *Journal of the ACM (JACM),* **55**(5), Article No. 23.

Ayad, H.G. and Kamel, M.S. (2008). Cumulative voting consensus method for partitions with variable number of clusters.. *IEEE Transactions on Pattern Analysis and Machine Intelligence,* **30**(1), pp. 160–173.

Bertolacci, M. and Wirth, A. (2007). Are Approximation Algorithms for Consensus Clustering Worthwhile? Proceedings of the Seventh SIAM International Conference on Data Mining, 26–28, April 2007, SIAM, Minneapolis, MN.

Brannon, A.R., Reddy, A., Seiler, M. *et al.* (2010). Molecular stratification of clear cell renal cell carcinoma by consensus clustering reveals distinct subtypes and survival patterns. *Genes and Cancer,* **1**(2), pp. 152–163.

Cover, T.M. and Thomas, J.A. (2006). *Elements of Information Theory,* 2nd edn, Wiley-Interscience, New York.

Dimitriadou, E., Weingessel, A. and Hornik, K. (2001). Voting-merging: an ensemble method for clustering, in *Artificial Neural Networks–ICANN 2001,* Springer, New York, pp. 217–224.

Dudoit, S. and Fridlyand, J. (2003). Bagging to improve the accuracy of a clustering procedure. *Bioinformatics,* **19**(9), pp. 1090–1099.

Filkov, V. and Skiena, S. (2004). Integrating microarray data by consensus clustering. *International Journal on Artificial Intelligence Tools,* **13**(04), pp. 863–880.

Fred, A.L. and Jain, A.K. (2005). Combining multiple clusterings using evidence accumulation. *IEEE Transactions on Pattern Analysis and Machine Intelligence,* **27**(6), pp. 835–850.

Ghaemi, R., Sulaiman, N., Ibrahim, H. and Mustapha, N. (2009). A survey: clustering ensembles techniques. *Proceedings of World Academy of Science: Engineering and Technology, Volume,* **50**.

Gionis, A., Mannila, H. and Tsaparas, P. (2007). Clustering aggregation. *ACM Transactions on Knowledge Discovery from Data (TKDD),* **1**(1), Article No. 4.

Hubert, L. and Arabie, P. (1985). Comparing partitions. *Journal of Classification,* **2**(1), pp. 193–218.

Levandowsky, M. and Winter, D. (1971). Distance between sets. *Nature,* **234**(5323), pp. 34–35.

Li, T., Ogihara, M. and Ma, S. (2010). On combining multiple clusterings: an overview and a new perspective. *Applied Intelligence,* **33**(2), pp. 207–219.

Mirkin, B. (1998). *Mathematical Classification and Clustering: From How to What and Why,* Springer, New York.

Monti, S., Tamayo, P., Mesirov, J. and Golub, T. (2003). Consensus clustering: a resampling-based method for class discovery and visualization of gene expression microarray data. *Machine Learning,* **52**(1–2), pp. 91–118.

Rand, W.M. (1971). Objective criteria for the evaluation of clustering methods. *Journal of the American Statistical Association,* **66**(336), pp. 846–850.

Seiler, M., Huang, C.C., Szalma, S. and Bhanot, G. (2010a). *Consensus Clustering Software Package*, http://code.google.com/p/consensus-cluster/ (accessed 24 July 2014).

Seiler, M., Huang, C.C., Szalma, S. and Bhanot, G. (2010b). Consensus Cluster: a software tool for unsupervised cluster discovery in numerical data. *OMICS A Journal of Integrative Biology,* **14**(1), pp. 109–113.

Strehl, A. (2011). *Cluster Analysis and Cluster Ensemble Software*, http://strehl.com/download/cluster-PackV20.zip (accessed November 2014).

Strehl, A. and Ghosh, J. (2003). Cluster ensembles - a knowledge reuse framework for combining multiple partitions. *Journal of Machine Learning Research,* **3**, pp. 583–617.

Topchy, A., Jain, A.K. and Punch, W. (2003). Combining Multiple Weak Clusterings. ICDM 2003. Third IEEE International Conference on Data Mining, 19–22 November 2003, Melbourne, FL, IEEE, Rhode Island, pp. 331–338.

Topchy, A., Jain, A.K. and Punch, W. (2005). Clustering ensembles: models of consensus and weak partitions. *IEEE Transactions on Pattern Analysis and Machine Intelligence,* **27**(12), pp. 1866–1881.

Vega-Pons, S. and Ruiz-Shulcloper, J. (2011). A survey of clustering ensemble algorithms. *International Journal of Pattern Recognition and Artificial Intelligence,* **25**(03), pp. 337–372.

Yu, Z., Wong, H.-S. and Wang, H. (2007). Graph-based consensus clustering for class discovery from gene expression data. *Bioinformatics,* **23**(21), pp. 2888–2896.

18

Biclustering

18.1 Introduction

The concept of simultaneously clustering a dataset in both dimensions can be traced back to 1972, proposed by Hartigan (1972) to analyse the voting data of the percentage Republican vote for President of the United States in the southern states over the years 1900–1968. It did not attract much attention until Cheng and Church (2000) developed a biclustering algorithm to analyse gene expression data. We have mentioned that the rationale behind clustering gene expression data is that the co-expressed genes may have a higher possibility to be co-regulated. Most clustering algorithms can do the job of grouping co-expressed genes. However, sometimes the case is that some genes are co-regulated only in some conditions rather than over the conditions of the all of the samples. General clustering algorithms hardly can do this job.

Biclustering has gained much interest since the first biclustering algorithm was proposed by Cheng and Church (CC). Now there are more than 30 different biclustering algorithms in the literature, many survey and performance-comparison papers (Madeira and Oliveira, 2004; Prelić *et al.*, 2006; Tchagang *et al.*, 2011; Eren *et al.*, 2013; Oghabian *et al.*, 2014), and many toolboxes in many different platforms available (Barkow *et al.*, 2006; Kaiser and Leisch, 2008; Eren, 2012). In this chapter, we will discuss the basic concept of biclustering and introduce the types of bicluster. We will also borrow the taxonomy introduced by Oghabian *et al.* (2014), and detail the most typical algorithms in each class, as it is impossible to detail every biclustering algorithm in a single chapter.

Integrative Cluster Analysis in Bioinformatics, First Edition. Basel Abu-Jamous, Rui Fa and Asoke K. Nandi.
© 2015 John Wiley & Sons, Ltd. Published 2015 by John Wiley & Sons, Ltd.

18.2   Overview

18.2.1   Statement of the Biclustering Problem

A biological system's interaction with its environment is complex and gene regulation is multi-factorial. Gene expression is influenced by the cell type, cell phase, external signals, and other factors. Therefore, genes might not be co-regulated across all experimental conditions observed in any comprehensive set of transcript or protein levels. In this case, the clustering task is to find a set of genes co-regulated only in a subset of the samples. This defines the problems of biclustering as follows

1. A cluster of genes should be defined with respect to only a subset of conditions; a cluster of conditions should be defined with respect to only a subset of the genes;
2. A gene/condition should be able to belong to more than one cluster or no cluster at all and be grouped using a subset of conditions/genes.

Assume that the gene expression dataset $X \in \mathbb{R}^{N \times M}$ is an N by M data matrix where N denotes the number of genes and M denotes the number of samples. Let us define the set of genes as g and the set of samples as s. Therefore, the matrix $X_{g,s} = (g, s)$ denotes full dataset X. Considering that $I \subseteq g$ and $J \subseteq s$, the matrix $X_{I,J} = (I, J)$ is a sub-matrix of X with the subset of genes I and the subset of samples J, where $N' = |I|$ and $M' = |J|$. A bicluster is a subset of rows that exhibit similar behaviour across a subset of columns, which thus can be defined as $X_{I,J} = (I, J)$.

18.2.2   Types of Biclusters

The objective of biclustering is to find biclusters. However, there are many different types of biclusters, which leads to a fact that some biclustering algorithms may only find few types of biclusters but not all of them. We can identify four major types:

1. Biclusters with constant values in a gene expression matrix describe subsets of genes with equal expression values within a subset of experimental conditions. Mathematically, considering a noise free case, they can be modelled as Matrix (18.1).

$$X_{I,J} = (I, J) = \begin{bmatrix} x & \cdots & x \\ \vdots & \ddots & \vdots \\ x & \cdots & x \end{bmatrix} \qquad \text{(Matrix 18.1)}$$

It is obvious that the variance of such a noise-free bicluster is zero;
2. Biclusters with constant values on rows or columns: Biclusters with constant values on rows indicate a subset of genes with expression levels that do not change across a subset of samples, as shown in Matrix (18.2).

$$X_{I,J} = (I, J) = \begin{bmatrix} x_1 & x_1 & \cdots & x_1 \\ x_2 & x_2 & \cdots & x_2 \\ \vdots & \vdots & \ddots & \vdots \\ x_{N'} & x_{N'} & \cdots & x_{N'} \end{bmatrix} \qquad \text{(Matrix 18.2)}$$

Biclusters with constant values on columns indicate a subset of samples with expression levels that do not change across a subset of genes, modelled as in Matrix (18.3).

$$X_{I,J} = (I, J) = \begin{bmatrix} x_1 & x_2 & \cdots & x_{M'} \\ x_1 & x_2 & \cdots & x_{M'} \\ \vdots & \vdots & \ddots & \vdots \\ x_1 & x_2 & \cdots & x_{M'} \end{bmatrix}$$ (Matrix 18.3)

This type of bicluster can follow either an additive model; that is, $x_i = \mu + \alpha_i$, or a multiplicative model, that is, $x_i = \mu \alpha_i$;

3. Biclusters with coherent values identify a subset of genes that are up-regulated and down-regulated coherently across subsets of conditions; that is, with the same magnitude and same direction across experimental conditions. Mathematically, an example of a noise-free bicluster with coherent values can be modelled as in Matrix (18.4) or Matrix (18.5).

$$X_{I,J} = (I, J) = \begin{bmatrix} \alpha_1 + \beta_1 & \alpha_2 + \beta_1 & \cdots & \alpha_{M'} + \beta_1 \\ \alpha_1 + \beta_2 & \alpha_2 + \beta_2 & \cdots & \alpha_{M'} + \beta_2 \\ \vdots & \vdots & \ddots & \vdots \\ \alpha_1 + \beta_{N'} & \alpha_2 + \beta_{N'} & \cdots & \alpha_{M'} + \beta_{N'} \end{bmatrix}$$ (Matrix 18.4)

or

$$X_{I,J} = (I, J) = \begin{bmatrix} \alpha_1\beta_1 & \alpha_2\beta_1 & \cdots & \alpha_{M'}\beta_1 \\ \alpha_1\beta_2 & \alpha_2\beta_2 & \cdots & \alpha_{M'}\beta_2 \\ \vdots & \vdots & \ddots & \vdots \\ \alpha_1\beta_{N'} & \alpha_2\beta_{N'} & \cdots & \alpha_{M'}\beta_{N'} \end{bmatrix}$$ (Matrix 18.5)

4. Biclusters with coherent evolutions: Unlike biclusters with coherent values, biclusters with coherent evolutions identify subsets of genes that are up-regulated or down-regulated coherently across subsets of conditions irrespective of their actual values; that is, in the same directions but with varying magnitude. Coherent-evolution biclusters are difficult to model using a mathematical equation. But, depending on how coherent evolution is defined, several merit functions can be defined for their statistical validation.

18.2.3 Classification of Biclustering

Biclustering has attracted a lot of interest since it was first used in the analysis of microarray gene expression data (Cheng and Church, 2000). There are tens of biclustering algorithms in the literature. To study such enriched literature, we employ the taxonomy by Oghabian *et al.* (2014), to categorise biclustering algorithms into four classes in terms of the criteria of searching biclusters:

1. **Variance-minimisation biclustering methods (VMB)**: VMB searches for biclusters in which the expression values have low variance throughout the selected genes, conditions, or the whole sub-matrix;

2. **Correlation-maximisation biclustering methods (CMB)**: CMB mines for subsets of genes and samples where the expression values of the genes correlate highly among the samples;
3. **Two-way clustering methods (TWC)**: TWC discovers the homogeneous subsets of genes and samples; that is, biclusters, by iteratively performing one-way clustering on the genes and samples;
4. **Probabilistic and generative methods (PGM)**: PGM employs probabilistic techniques to discover genes (or, respectively, samples) that are similarly expressed across a subset of samples (or, respectively, genes) in the data matrix.

In the next section, we will introduce the most typical algorithms in each class.

18.3 Biclustering Methods

18.3.1 Variance-minimisation Biclustering Methods

18.3.1.1 CC

CC proposed a deterministic greedy algorithm to seek biclusters with low variance, which is the first biclustering algorithm, applied in analysis of microarray gene expression data (Cheng and Church, 2000). A variance measure, called mean squared residue (MSR), was defined: if I and J are sets of rows and columns of a bicluster, respectively, MSR then is mathematically expressed as shown in Equation (18.1),

$$\text{MSR} = \frac{1}{|I||J|} \sum_{i \in I, j \in J} \left(x_{ij} - \bar{x}_{iJ} - \bar{x}_{Ij} + \bar{x}_{IJ} \right)^2 \tag{18.1}$$

where x_{ij} is the data element at row i and column j. $\bar{x}_{iJ}$, $\bar{x}_{Ij}$, and $\bar{x}_{IJ}$ are the mean values of the expression values in row i, column j, and the whole bicluster, respectively, for $i \in I$ and $j \in J$, which are expressed as Equations (18.2).

$$\bar{x}_{iJ} = \frac{1}{|J|} \sum_{j \in J} x_{ij}$$

$$\bar{x}_{Ij} = \frac{1}{|I|} \sum_{i \in I} x_{ij} \tag{18.2}$$

$$\bar{x}_{IJ} = \frac{1}{|I||J|} \sum_{i \in I, j \in J} x_{ij} = \frac{1}{|I|} \sum_{i \in I} \bar{x}_{iJ} = \frac{1}{|J|} \sum_{j \in J} \bar{x}_{Ij}$$

The CC algorithm starts with the whole data matrix and removes the rows and columns that have high residues gradually. Once the MSR of the bicluster reaches a given threshold, δ, the rows and columns that produce a smaller residue than the bicluster residue are added back to the bicluster. The found biclusters are masked with random values and then the process repeats until no biclusters can be found. The whole CC algorithm consists of four sub-routines to complete a piece of task.

1. **Algorithm 0** computes the score MSR for each possible row/column addition/deletion and chooses the action that decreases MSR the most. If no action will decrease MSR, or MSR $\leq \delta$, return X_{IJ}.

2. **Algorithm 1**, whose task is single-node deletion, firstly computes $\bar{x}_{iJ}$ for $\forall i \in I$, $\bar{x}_{Ij}$ for $\forall j \in J$, $\bar{x}_{IJ}$, and MSR(I, J). If MSR$(I, J) \leq \delta$, return X_{IJ}. Secondly, it finds the row $i \in I$ with the largest $d(i) = \frac{1}{|J|}\sum_{j \in J}\left(x_{ij} - \bar{x}_{iJ} - \bar{x}_{Ij} + \bar{x}_{IJ}\right)^2$, and finds the column $j \in J$ with the largest $d(i) = \frac{1}{|I|}\sum_{i \in I}\left(x_{ij} - \bar{x}_{iJ} - \bar{x}_{Ij} + \bar{x}_{IJ}\right)^2$. Then it removes the row or column which induces the largest d value.

3. **Algorithm 2**, whose task is multiple-node deletion, firstly computes $\bar{x}_{iJ}$ for $\forall i \in I$, $\bar{x}_{Ij}$ for $\forall j \in J$, $\bar{x}_{IJ}$, and MSR(I, J). If MSR$(I, J) \leq \delta$, return X_{IJ}. Secondly, it removes the rows $i \in I$ with $\frac{1}{|J|}\sum_{j \in J}\left(x_{ij} - \bar{x}_{iJ} - \bar{x}_{Ij} + \bar{x}_{IJ}\right)^2 > \alpha \text{MSR}(I, J)$, and recomputes $\bar{x}_{iJ}, \bar{x}_{Ij}, \bar{x}_{IJ}$, and MSR$(I, J)$. Then it removes the columns $j \in J$ with $\frac{1}{|I|}\sum_{i \in I}\left(x_{ij} - \bar{x}_{iJ} - \bar{x}_{Ij} + \bar{x}_{IJ}\right)^2 > \alpha \text{MSR}(I, J)$. Finally, if nothing has been removed in the iteration, it switches to Algorithm 1.

4. **Algorithm 3**, whose task is node addition, first computes $\bar{x}_{iJ}, \bar{x}_{Ij}, \bar{x}_{IJ}$, and MSR$(I, J)$, $\forall i, j$. Secondly, it adds the columns $j \notin J$ with $\frac{1}{|I|}\sum_{i \in I}\left(x_{ij} - \bar{x}_{iJ} - \bar{x}_{Ij} + \bar{x}_{IJ}\right)^2 \leq \text{MSR}(I, J)$, and recomputes $\bar{x}_{iJ}, \bar{x}_{Ij}, \bar{x}_{IJ}$, and MSR$(I, J)$. Then it adds the rows $i \notin I$ with $\frac{1}{|J|}\sum_{j \in J}\left(x_{ij} - \bar{x}_{iJ} - \bar{x}_{Ij} + \bar{x}_{IJ}\right)^2 \leq$ MSR(I, J), and for each row, $i \notin I$, adds its inverse of $\frac{1}{|J|}\sum_{j \in J}\left(-x_{ij} + \bar{x}_{iJ} - \bar{x}_{Ij} + \bar{x}_{IJ}\right)^2 \leq$ MSR(I, J). Finally, if nothing is added, return X_{IJ}.

Thus, the CC algorithm combines all the above algorithms to discover the biclusters as follows:

1. Apply *Algorithm 2* on X, δ, and α. If the size of X is small, the multiple nodes deletion can be skipped. After *Algorithm 2*, we may obtain X';
2. Apply *Algorithm 1* on X' and δ. After *Algorithm 1*, we may obtain X'';
3. Apply *Algorithm 3* on X and X'', we may obtain X''';
4. Return X''', and replace the elements in X and also in X'' with random values, repeat 1–4, until no more bicluster is produced.

18.3.1.2 Spectral Biclustering

An assumption in tumour classification is that samples drawn from a population containing several tumour types have similar expression profiles if they belong to the same type. Under this assumption, the data matrix X could be organised in a checkerboard-like structure with blocks of high-expression levels and low-expression levels. The task of spectral biclustering is to uncover the checkerboard structure through solving an Eigen problem. In doing so, spectral biclustering consists of several processing steps: (i) simultaneous normalisation of genes and conditions, (ii) post-processing the Eigenvectors to find partitions, and (iii) probabilistic interpretation.

The microarray data can be viewed as a bipartite graph, where one set of nodes in this graph represents the genes, and the other represents experimental conditions. A link connecting a gene and condition represents the level of over-expression (or under-expression) of this gene under this condition. Spectral biclustering includes the normalisation of rows and columns as an integral part of the algorithm. It attempts to simultaneously normalise both genes and conditions. This can be achieved by repeating independent scaling of rows and columns iteratively until convergence. This process, which is called *bistochatisation,* results in a rectangular matrix X' that has a doubly stochastic-like structure. All rows of X' sum to a constant and all columns sum to a different constant. X' can be written as a product $X' = D_1 X D_2$, where D_1 and D_2 are diagonal matrices. Generally, X' can be computed by repeated normalisation of rows and columns. The normalisation matrices R^{-1} and C^{-1} are computed as $R = \mathrm{diag}(X \cdot 1_N)$, where 1_N denotes the all-one vector with the length of N, and $C = \mathrm{diag}(1_M \cdot X)$. D_1 and D_2 then will represent the product of all these normalisations. Once D_1 and D_2 are found, singular value decomposition (SVD) is applied to X' without further normalisation. After SVD, a set of gene and condition eigenvectors and eigenvalues is produced. The largest eigenvalue is discarded because it is trivial in the sense that its eigenvectors make a trivial constant to the matrix, and therefore carry no partitioning information. The terminology 'largest eigenvalue' means the largest non-trivial eigenvalue. A common practice in spectral clustering is to perform a final clustering step to the data projected to a small number of eigenvectors, instead of clustering each eigenvector individually. The final clustering algorithm can be either k-means or the normalised cuts method. Furthermore, the degrees of membership of genes and conditions to the respective bicluster according to the actual values in the partitioning-sorted eigenvectors are ranked.

18.3.2   Correlation-maximisation Biclustering Methods

18.3.2.1   BiMine

BiMine is a typical CMB method proposed by Ayadi, Elloumi and Hao (2009). It relies on a new evaluation function called average Spearman's rho (ASR), which is used to guide the exploration of the search space effectively. BiMine uses a new tree structure called bicluster enumeration tree (BET) to represent the different biclusters discovered during the enumeration process. BiMine also introduces a parametric rule that allows the enumeration process to cut those tree branches that cannot lead to good biclusters. Different from MSR, which is deficient in assessing the quality of biclusters with coherent evolutions, ASR is proposed based on Spearman's rank correlation, shown in Equation (18.3),

$$\rho_{ij} = 1 - \frac{6 \sum_{k=1}^{m} \left(r_k^i(x_k^i) - r_k^j(x_k^j) \right)^2}{m(m^2 - 1)} \qquad (18.3)$$

where $r_k^i(x_k^i)$ is the rank of x_k^i, and m is the size of the data vector. Therefore, ASR can be expressed as Equation (18.4)

$$\mathrm{ASR}(I,J) = 2 \cdot \max \left\{ \frac{\sum_{i \in I} \sum_{j \geq i+1, j \in I} \rho_{ij}}{|I|(|I|-1)}, \frac{\sum_{k \in J} \sum_{l \geq k+1, l \in J} \rho_{kl}}{|J|(|J|-1)} \right\} \qquad (18.4)$$

and $\mathrm{ASR}(I,J) \in [-1,1]$. A high value (close to 1) indicates that the data are strongly correlated between two vectors. BiMine consists of three steps: pre-processing the data, conducting a BET

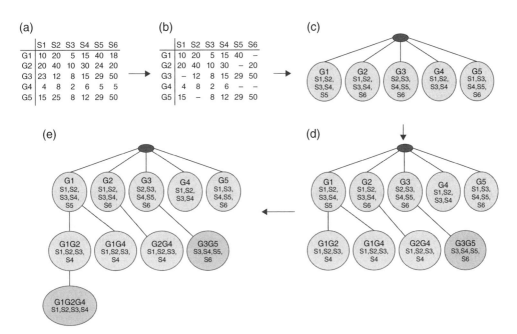

Figure 18.1 An example of BiMine. (a) The original datasets, where columns represent samples denoted by letter 'S' and rows represent genes denoted by letter 'G'. (b) The pre-processed data matrix, whose remaining elements are significant. (c) The first level of BET: the BET is initialised with a tree whose leaves are single genes with all significant samples or conditions. (d) The second level of BET. (e) The final level of BET

associated with X, and identifying the best biclusters. An example of BiMine is depicted in Figure 18.1. The pre-processing step aims to remove irrelevant expression values of the data matrix that do not contribute in obtaining apposite results. A value x_{ij} of X is considered to be insignificant if condition (18.5) holds,

$$\frac{|x_{ij}-\mathrm{avg}_i|}{\mathrm{avg}_i}\leq\delta \tag{18.5}$$

where avg_i is the average over the non-missing values in the ith row, and δ is a fixed threshold. The step of conducting a BET starts with an initial tree, whose root is an empty bicluster and leaves are all possible biclusters containing one gene with all significant conditions. Subsequently, the BET-tree creates recursively the BET and generates the set of best biclusters. In doing so, the ith child of a node is made up, on one hand, of the union of the genes of the father node and the genes of the ith uncle node, starting from the right side of the father; on the other hand, it is made up of the intersection of the conditions of the father and those of the ith uncle. If the ASR value associated with the ith child is smaller than or equal to the given threshold, then this child will be ignored and only the child with a considerably high ASR value, which has not been included in the bicluster, is kept. Finally, good biclusters can be identified from the leaves which have to be (i) without children, (ii) not included in any other bicluster, and (iii) with a good number of genes and samples.

18.3.2.2 Bimax

Bimax is a divide-and-conquer algorithm. It assumes that there are only two expression levels, on and off, of each gene. In this case, the gene expression data must first be binarised. Thresholding is used: expression values higher than the predefined threshold are set to ones; otherwise, to zeros. The threshold for binarisation is based on the mean of the data. Therefore, Bimax is expected to find only up-regulated biclusters. A bicluster (I_k, J_k) defines a sub-matrix of a matrix B whose elements are supposed to be one. Furthermore, all biclusters are found inclusion maximal; that is, they are not entirely contained in any other biclusters. Bimax employs an incremental procedure, which is based on work by Alexe *et al.* (2004). Alexe *et al.*'s method finds all inclusion maximal cliques in general graphs. Each node in the input graph is visited, and all maximal cliques are found that contain the visited node. A visit-to-a-node operation comprises an iteration through all other nodes of the graph as well, and each newly found bicluster is globally extended to its maximality. Bimax has an additional step that extends newly created biclusters to their maximality and an additional absorption check operation is made.

18.3.2.3 Robust Biclustering Algorithm

Tchagang and Tewfik proposed a robust biclustering algorithm (ROBA) (Tchagang and Tewfik, 2006). Their objective was to discover all types of biclusters in a given dataset of any type defined by the user in a timely manner. The ROBA is different from previous algorithms in several ways: first, the ROBA can be used to find the exact number of all valid perfect biclusters in each type and identify all of them in a timely manner; secondly, the ROBA uses basic linear algebra and arithmetic tools and avoids the need for heuristic cost functions of prior approaches that may miss some pertinent biclusters; thirdly, the ROBA relies on the manipulation of elementary binary matrices with entries equal to 0 and 1; lastly, the ROBA allows users to discover biclusters under any specific experimental conditions.

The ROBA consists of three stages of processing. The first stage includes a series of pre-processing procedures; namely, recovering missing values and quantising the data. In Tchagang and Tewfik (2006), Tchagang and Tewfik replaced each missing value by zero and quantised the data into L intervals $\{\alpha_1, \ldots, \alpha_L\}$. The second stage of the ROBA decomposes the gene expression matrix into the sum of the products of each of its distinct elements with a corresponding elementary matrix. Each elementary matrix consists only of binary entries. Suppose that X_Q denotes the quantised gene expression data. X_Q can be expressed as shown in Equation (18.6),

$$X_Q = \sum_{l=1}^{l=L} \alpha_l X_l \tag{18.6}$$

where X_l are binary matrices. The third stage of the ROBA is to identify all types of biclusters from the gene expression matrix. From the gene expression matrix decomposition in stage two, such matrices can be obtained by analysing each elementary matrix X_l separately to obtain sub-groups of genes that have constant expression level α_l under different conditions. Such matrices will therefore correspond to a sub-group of matrices of each elementary matrix whose elements are only the binary number 1. Based on this idea, Tchagang and Tewfik developed many approaches to discover all types of biclusters, namely biclusters with constant values, biclusters with constant values on columns or rows, biclusters with coherent values, and biclusters with coherent evolution. We will not detail all of them here and the interested readers are referred to Tchagang and Tewfik (2006).

18.3.3 Two-way Clustering Methods

18.3.3.1 Coupled Two-way Clustering

Getz *et al.* developed a coupled two-way clustering (CTWC) approach for microarray gene expression data analysis (Getz, Levine and Domany, 2000). Considering that the number of possible sub-matrices grows exponentially with the size of the original matrix, CTWC provides an efficient heuristic to generate pairs of gene and sample subsets (I_k, J_k) by an iterative process that severely restricts those possible candidates for such subsets. The iterative process is initialised with full matrix containing all genes (g^0) and all samples (s^0) to perform standard two-way clustering. Stable clusters and samples found in this step are denoted by g_i^1 and s_i^1. Then, every pair (g_i^m, s_j^n), where m, n may be either 0 or 1, defines a sub-matrix of expression data. These sub-matrices are further clustered and the resulting stable gene (or sample) clusters are denoted by g_k^2 and s_k^2. Thus new clusters (g_i^m, s_j^n), together with parent clusters, are obtained. These steps are iterated further, using pairs of all previously found clusters. Every pair is used only once. The process is terminated when no new clusters that satisfy some criteria are found. The output of CTWC provides a broad list of gene and sample clusters.

18.3.3.2 Interrelated Two-way Clustering

Interrelated Two-way Clustering (ITWC) is one of the TWC algorithms, proposed by Tang *et al.* (2001). It consists of five steps within each iteration during the clustering process:

Step 1 Clustering in the gene dimension: N genes in the dataset are clustered into K clusters $\{C_k | k = 1, \ldots, K\}$ by any clustering algorithm, such as k-means and SOM;

Step 2 Clustering in the sample dimension: Based on gene clusters generated by Step 1, each cluster is split into two sub-clusters in samples independently, represented by $\{C_k^a, C_k^b\}$;

Step 3 Clustering-results combination: the clustering results from Steps 1 to 2 are combined, as $C_i' = \left(C_k^\alpha, C_l^\beta\right)$, where α, β are either a or b, and $k, l \in [1, K]$, and $k \neq l$. There are 2^K possible combinations for C_i';

Step 4 Find *heterogeneous groups:* a heterogeneous group is defined as a pair of clusters whose samples are never grouped in any clusters.

Step 5 Sorting and reducing: the vector-cosine distance measures between members of heterogeneous groups and binary sample membership pattern are calculated. Then these distances are sorted in a descending order and only (first) one third of genes are selected to be a cluster. The remaining genes are sent to the next iteration.

The steps above are repeated until the termination conditions are satisfied. An *occupancy ratio* is defined to control the termination: if the occupancy ratio reaches a predefined threshold, the clustering procedure then is terminated.

18.3.4   Probabilistic and Generative Methods

18.3.4.1   Plaid Method

The Plaid method, whose name implies the appearance of a colour image plot of microarray gene expression data, is one of the typical algorithms in the PGM class. The biclustering problem is the same as the idea of reordering the array and producing an image with some number K of rectangular blocks on a diagonal. Every gene in gene-block k is expressed within and only within those samples in sample-block k. There are many models representing different types of biclusters, which can be written as Equations (18.7)–(18.10),

$$X_{ij} = \mu_0 + \sum_{k=1}^{K} \mu_k \rho_{ik} \kappa_{jk} \tag{18.7}$$

$$X_{ij} = \mu_0 + \sum_{k=1}^{K} (\mu_k + \alpha_{ik}) \rho_{ik} \kappa_{jk} \tag{18.8}$$

$$X_{ij} = \mu_0 + \sum_{k=1}^{K} (\mu_k + \beta_{jk}) \rho_{ik} \kappa_{jk} \tag{18.9}$$

$$X_{ij} = \mu_0 + \sum_{k=1}^{K} (\mu_k + \alpha_{ik} + \beta_{jk}) \rho_{ik} \kappa_{jk} \tag{18.10}$$

where μ_0 is a background colour, μ_k describes the colour in the block k, ρ_{ik} is 1 if gene i is in the block k (zero otherwise), and κ_{jk} is 1 if sample j is in the block k (zero otherwise). α_{ik} represents the identical response of gene i in a subset of samples and β_{jk} represents the identical response of sample j across a subset of genes. Essentially, Equation (18.6) is equivalent to biclusters with constant values; (18.7) and (18.8) are equivalent to biclusters with constant values on rows and columns respectively; (18.9) is equivalent to biclusters with coherent values with additive model. A more general model by mixing layer types can be expressed as shown in Equation (18.11),

$$X_{ij} = \sum_{k=0}^{K} \theta_{ijk} \rho_{ik} \kappa_{jk} \tag{18.11}$$

where θ_{ij0} describes the background layer. θ_{ijk} is used to represent μ_k, $\mu_k + \alpha_{ik}$, $\mu_k + \beta_{jk}$, $\mu_k + \alpha_{ik} + \beta_{jk}$, as needed. Each bicluster can be viewed as a layer, which represents the presence of a set of biological processes or conditions.

The plaid model that we are seeking is the one minimising the following cost function [Equation (18.12)].

$$\frac{1}{2}\sum_{i=1}^{N}\sum_{j=1}^{M}\left(X_{ij}-\theta_{ij0}-\sum_{k=1}^{K}\theta_{ijk}\rho_{ik}\kappa_{jk}\right)^2 \tag{18.12}$$

To simplify matters, assuming that $(K-1)$ layers are available, the layer K is to minimise the sum of squared error as given in Equation (18.13),

$$Q=\frac{1}{2}\sum_{i=1}^{N}\sum_{j=1}^{M}\left(Y_{ij}^{K-1}-\theta_{ijK}\rho_{iK}\kappa_{jK}\right)^2 \tag{18.13}$$

where Equation (18.14) shows the residual from the $(K-1)$ layer.

$$Y_{ij}^{K-1}=X_{ij}-\theta_{ij0}-\sum_{k=1}^{K-1}\theta_{ijk}\rho_{ik}\kappa_{jk} \tag{18.14}$$

Therefore, the problem is converted into a minimisation problem on a single layer and the subscript K can be dropped for simplicity [Equation (18.15)],

$$Q=\frac{1}{2}\sum_{i=1}^{N}\sum_{j=1}^{M}\left(Y_{ij}-\left(\mu+\alpha_i+\beta_j\right)\rho_i\kappa_j\right)^2 \tag{18.15}$$

subject to identifying conditions $\sum_{i=1}^{N}\rho_i^2\alpha_i=\sum_{j=1}^{M}\kappa_j^2\beta_j=0$. Straightforward Lagrange multiplier arguments show that conditions (18.16)–(18.18) hold.

$$\mu=\frac{\sum_i\sum_j\rho_i\kappa_jY_{ij}}{\left(\sum_i\rho_i^2\right)\left(\sum_j\kappa_j^2\right)} \tag{18.16}$$

$$\alpha_i=\frac{\sum_i\left(Y_{ij}-\mu\rho_i\kappa_j\right)\kappa_j}{\rho_i\sum_j\kappa_j^2} \tag{18.17}$$

$$\beta_i=\frac{\sum_i\left(Y_{ij}-\mu\rho_i\kappa_j\right)\rho_i}{\kappa_j\sum_i\rho_i^2} \tag{18.18}$$

Then, the next step is to update ρ_i and κ_j. Given values for θ_{ij} and κ_j, the values for ρ_i that minimise Q are given by Equation (18.19),

$$\rho_i=\frac{\sum_j\theta_{ij}\kappa_jY_{ij}}{\sum_j\theta_{ij}^2\kappa_j^2} \tag{18.19}$$

and similarly, given θ_{ij} and ρ_i, the minimising values for κ_j are given by Equation (18.20).

$$\kappa_j = \frac{\sum_i \theta_{ij}\rho_j Y_{ij}}{\sum_i \theta_{ij}^2 \rho_i^2} \tag{18.20}$$

The quantities ρ_i, κ_j and θ_{ij} are updated iteratively.

18.3.4.2 Bayesian Plaid

The Plaid model was further extended into a Bayesian frame to be Bayesian Plaid. All parameters in the original plaid algorithm are assumed to follow distributions: μ, α, and β follow Gaussian distribution, expressed as Equations (18.21)–(18.23),

$$\mu_0 \sim \mathcal{N}\left(0, \sigma_\mu^2 \sigma^2\right) \tag{18.21}$$

$$\alpha_{ik} \sim \mathcal{N}\left(0, \sigma_\alpha^2 \sigma^2\right) \tag{18.22}$$

$$\beta_{ik} \sim \mathcal{N}\left(0, \sigma_\beta^2 \sigma^2\right) \tag{18.23}$$

where σ_μ^2, σ_α^2, and σ_β^2 are scalar hyper-parameters specified by the user. The binary membership variables ρ and κ follow Binomial distribution as shown by Equations (18.24) and (18.25).

$$\rho_{\cdot k} \sim \text{Binomial}(N, \pi_k) \tag{18.24}$$

$$\kappa_{\cdot k} \sim \text{Binomial}(N, \lambda_k) \tag{18.25}$$

The parameters π_k and λ_k are the probability of a gene to belong to a bicluster k and the probability of a condition to belong to bicluster k, respectively. They follow a Beta distribution, which is the conjugate prior of the Binomial distribution [Equations (18.26) and (18.27)],

$$\pi_k \sim \text{Beta}\left(\sigma_\rho^{(k)}, \gamma_\rho^{(k)}\right) \tag{18.26}$$

$$\lambda_k \sim \text{Beta}\left(\sigma_\kappa^{(k)}, \gamma_\kappa^{(k)}\right) \tag{18.27}$$

where $\sigma_\rho^{(k)}, \gamma_\rho^{(k)}$, $\sigma_\kappa^{(k)}$, and $\gamma_\kappa^{(k)}$ are hyper-parameters. $P(X|\rho,\kappa)$ is the conditional probability density function of X, which is written as Equation (18.28),

$$P(X|\rho,\kappa) = \int P\left(X|\rho,\kappa,\Theta,\sigma^2\right) P(\Theta)P\left(\sigma^2\right) d\Theta d\sigma^2 \tag{18.28}$$

where Θ is the set of parameters as $\Theta = [\mu_0, \alpha_{\cdot 1}, \beta_{\cdot 1}, \ldots, \alpha_{\cdot K}, \beta_{\cdot K}]^T$. A collapsed Gibbs sampler for inferring the posterior distribution of the bicluster membership variables was derived. The interested readers are suggested to refer to Caldas and Kaski (2008) for more details.

18.3.4.3 CMonkey

CMonkey is an integrative biclustering algorithm proposed by Reiss, Baliga, and Bonneau (2006), which combines gene expression data, DNA-sequence data and associated network data to produce biclusters based on a probabilistic model. CMonkey models each bicluster via the Markov chain process, in which the bicluster is iteratively optimised, and its state is updated based upon condition probability distributions computed using the cluster's previous state. Thus, the probabilities that each gene or sample belongs to the bicluster, conditioned upon the current state of the bicluster, are defined. The components of the conditional probability are modelled independently as p-values based upon individual data likelihoods, which are then combined into a regression model to derive the full conditional probability. Three major distinct data types are used for three components: the expression component, the sequence component, and the network component.

Each bicluster begins as a 'seed' that is iteratively optimised by adding/removing genes and samples to/from the cluster. By sampling from the conditional probability distribution using a Monte Carlo procedure, it prevents premature convergence. Additional clusters are seeded and optimised until a given number ($k_{\max}$) of clusters have been generated or significant optimisation is no longer possible. Each bicluster k is defined as a sub-matrix of X, with the rows I_k and J_k, $I_k \subset I$, and $J_k \subset J$. The variance in the measured levels of condition j is $\sigma_j^2 = |I|^{-1} \sum_{i \in I} (x_{ij} - \bar{x}_j)^2$, $\bar{x}_j = \sum_{i \in I} x_{ij}/|I|$. The mean expression level of condition j over the bicluster genes I_k is $\bar{x}_{jk} = \sum_{i \in I_k} x_{ij}/|I_k|$. Then, the likelihood of an arbitrary measurement x_{ij} relative to this mean expression level is given by Equation (18.29),

$$p\left(x_{ij}\right) = \frac{1}{\sqrt{2\pi\left(\sigma_j^2 + \varepsilon^2\right)}} \exp\left[-\frac{1}{2} \cdot \frac{\left(x_{ij} - \bar{x}_{jk}\right)^2 + \varepsilon^2}{\sigma_j^2 + \varepsilon^2}\right] \tag{18.29}$$

where ε is an unknown systematic error in simple j. The likelihood for the measures of an arbitrary gene i among the conditions in bicluster k are $p(x_i) = \prod_{j \in J_k} p(x_{ij})$, and similarly the likelihood of a condition j's measurements are $p\left(x_j'\right) = \prod_{i \in I_k} p(x_{ij})$. Then the two tails of normal distribution in Equation (18.28) are integrated to derive co-expression p-values for gene i, r_{ik}, and for each condition j, r_{jk}', relative to bicluster k. With the help of the MEME algorithm, the motif p-values, s_{ik}, for each gene i relative to bicluster k can be calculated. Independently, the p-values, $q_{ik}^{\mathcal{N}}$, for each gene i and each network $\mathcal{N}$ are computed for bicluster k. Therefore, a procedure of combining all p-values into a single joint likelihood by a multiparametric logistic regression is carried out. The joint likelihood is written as shown in Equation (18.30),

$$\pi_{ik} = p(\mathcal{Y}_{ik} = 1 | X_k, S_i, \mathcal{N}) \propto \exp\left[\beta_0 + r_0 \log(r_{ik}) + s_0 \log(s_{ik}) + \sum_{\mathcal{N} \in I} q_0^{\mathcal{N}} \log\left(q_{ik}^{\mathcal{N}}\right)\right] \tag{18.30}$$

where β_0, r_0, s_0, and q_0 are four independent variables.

The CMonkey iterative procedure starts with a cluster-seeding method. There are many different seeding methods available, including (i) single-gene seeds, (ii) random or semi-correlated seeds using a pre-specified distribution of cluster sizes, and (iii) seeding on the basis of co-expressed edges in association networks. In principle, any seeding method may be used, but we have to note that the algorithm may be more or less sensitive to the starting points. Once a newly seeded bicluster is obtained, it is iteratively improved with respect to the joint likelihood derived above. At each iteration, significant motifs of sequence are detected (using MEME), and the joint membership probabilities π_{ik} for each gene or condition are computed. Moves including adding genes or samples to the bicluster and dropping genes or samples from the bicluster are performed in terms of the membership probability using a simulated annealing algorithm. They are parameterised by an annealing temperature T as shown in Equation (18.31).

$$p(\text{add}|\pi_{ik}) = e^{-\pi/T}; p(\text{drop}|\pi_{ik}) = e^{-1(1-\pi_{ik})/T} \qquad (18.31)$$

There are some additional constraints inside the CMonkey algorithm to restrict the biclusters from being either changed dramatically or dragged into local minima. We will not go further into the algorithm, and interested readers are referred to Reiss, Baliga and Bonneau (2006). One particular point with regard to CMonkey is that it is not clear as to what extent combining the sequence component and the network component may benefit condition clustering, although surely it may benefit gene clustering, which is only one aspect of biclustering.

18.4   Discussion

A comprehensive collection of biclustering algorithms is listed in Table 18.1, including their references, classification, and software packages (if available). Unlike most clustering families, biclustering is dedicated to solving the problem in gene expression data analysis. Therefore, most biclustering algorithms have been applied in the analysis of gene expression datasets. All these datasets are collected from either yeast or human cells, which represent the expression level under specific conditions. Several biclustering algorithms were used to analyse yeast cell cycle datasets (Cho et al., 1998; Spellman et al., 1998), yeast stress datasets (Gasch et al., 2000, 2001), yeast compendium (Hughes et al., 2000), and yeast galactose utilization (Ideker et al., 2001). Other algorithms were used for human breast cancer (Miller et al., 2005; Pawitan et al., 2005; Loi et al., 2007), lymphoma (Alizadeh et al., 2000), leukaemia (Golub et al., 1999), and micro-RNA (miRNA)-mRNA functional modules (Bryan et al., 2014).

There are two very recent comparative studies of biclustering algorithms for gene expression data: Eren et al. (2013) and Oghabian et al. (2014). Eren et al. compared 12 biclustering algorithms, namely CC (Cheng and Church, 2000), Plaid (Lazzeroni and Owen, 2002), OPSM (Ben-Dor et al., 2003), ISA (Bergmann, Ihmels and Barkai, 2003), Spectral (Kluger et al., 2003), xMOTIFs (Murali and Kasif, 2003), Bimax (Prelić et al., 2006), BBC (Gu and Liu, 2008), COALESCE (Huttenhower et al., 2009), CPB (Bozdağ, Parvin and Catalyurek, 2009), QUBIC (Li et al., 2009) and FABIA (Hochreiter et al., 2010). Eight real gene expression datasets, together with a synthetic dataset, were used to test these biclustering algorithms. Oghabian et al. compared 13 biclustering algorithms, namely CTWC (Getz, Levine and Domany, 2000), FLOC (Yang et al., 2005), SAMBA (Tanay, Sharan and Shamir, 2002), BiMine (Ayadi, Elloumi and Hao, 2009), R/MSBE (Liu and Wang, 2007), CC, Bimax, factor

Table 18.1 Summary of biclustering algorithms

Bicluster method	Class	Year	Availability
δ-Clustering (Hartigan, 1972)		1972	
CC (Cheng and Church, 2000)	VMB	2000	Barkow et al. (2006); Kaiser and Leisch (2008); Eren et al. (2013)
δ-jk (Califano, Stolovitzky and Tu, 2000)		2000	
CTWC (Getz, Levine and Domany, 2000)	TWC	2000	
ITWC (Tang et al., 2001)	TWC	2001	
DCC (Busygin et al., 2002; Busygin, Prokopyev and Pardalos, 2008)	TWC	2002	
SA (Ihmels et al., 2002; Ihmels, Bergmann, and Barkai, 2004)	TWC	2002	Barkow et al. (2006)
Plaid (Lazzeroni and Owen, 2002; Turner, Bailey and Krzanowski, 2005)	PGM	2002	Kaiser and Leisch (2008); Eren et al. (2013)
SAMBA (Tanay, Sharan and Shamir, 2002)	PGM	2002	
δ-P-Clustering (Wang et al., 2002)	VMB	2002	
Gibbs Clustering (Sheng, Moreau and Moor, 2003)	PGM	2003	
ISA (Bergmann, Ihmels and Barkai, 2003)	TWC	2003	Eren et al. (2013)
OP-Clustering (Liu and Wang, 2003; Liu, Wang and Yang, 2004)	CMB	2003	
OPSM (Ben-Dor et al., 2003)	CMB	2003	Barkow et al. (2006); Eren et al. (2013)
Spectral (Kluger et al., 2003)	VMB	2003	Kaiser and Leisch (2008); Eren et al. (2013)
xMOTIF (Murali and Kasif, 2003)	VMB	2003	Barkow et al. (2006); Kaiser and Leisch (2008); Eren et al. (2013)
GEMS (Wu et al., 2004; Wu and Kasif, 2005)	PGM	2004	
FLOC (Yang et al., 2005)	CMB	2005	
SA (Bryan, Cunningham and Bolshakova, 2005)	CMB	2005	
ZBDD (Yoon et al., 2005)	VMB	2005	
ROBA (Tchagang and Twefik, 2005)	CMB	2005	
Bimax (Prelić et al., 2006)	CMB	2006	Barkow et al., 2006
CMonkey (Reiss, Baliga and Bonneau, 2006)	PGM	2006	
SEBI (Divina and Aguilar-Ruiz, 2006)	VMB	2006	
MOEB (Mitra and Banka, 2006)	CMB	2006	
UBCLUST (Li et al., 2006)	CMB	2006	
R/MSBE (Liu and Wang, 2007)	VMB	2007	
Bayesian Biclustering (Gu and Liu, 2008)	PGM	2008	Eren et al., 2013
ACV (Teng and Chan, 2008)	CMB	2008	
Bayesian Plaid (Caldas and Kaski, 2008)	PGM	2008	
BiMine (Ayadi, Elloumi and Hao, 2009)	CMB	2009	
CPB (Bozdağ, Parvin and Catalyurek, 2009)	CMB	2009	Eren et al., 2013
COALESCE (Huttenhower et al., 2009)	CMB	2009	Eren et al., 2013
QUBIC (Li et al., 2009)	VMB	2009	Eren et al., 2013
FABIA and FABIAS (Hochreiter et al., 2010)	PGM	2010	Eren et al., 2013
TreeBic (Caldas and Kaski, 2010)	PGM	2010	

analysis for bicluster acquisition or factor analysis for bicluster acquisition with sparseness projection (FABIA/FABIAS), ISA, OPSM, Plaid, and QUBIC. The interested readers are referred to those studies. We will introduce some typical applications in Chapter 19.

References

Alexe, G., Alexe, S., Crama Y. *et al.* (2004). Consensus algorithms for the generation of all maximal bicliques. *Discrete Applied Mathematics*, **145**(1), pp. 11–21.

Alizadeh, A.A., Eisen, M.B., Davis, R.E. *et al.* (2000). Distinct types of diffuse large B-cell lymphoma identified by gene expression profiling. *Nature*, **403**(6769), pp. 503–511.

Ayadi, W., Elloumi, M. and Hao, J.-K. (2009). A biclustering algorithm based on a bicluster enumeration tree: application to DNA microarray data. *BioData Mining*, **2**, p. 9.

Barkow, S., Bleuler, S., Prelic, A. *et al.* (2006). BicAT: a biclustering analysis toolbox. *Bioinformatics*, **22**(10), pp. 1282–1283.

Ben-Dor, A., Chor, B., Karp, R. and Yakhini, Z. (2003). Discovering local structure in gene expression data: the order-preserving submatrix problem. *Journal of Computational Biology*, **10**(3–4), pp. 373–384.

Bergmann, S., Ihmels, J. and Barkai, N. (2003). Iterative signature algorithm for the analysis of large-scale gene expression data. *Physical Review E*, **67**(3), p. 031902.

Bozdağ, D., Parvin, J.D. and Catalyurek, U.V. (2009). A biclustering method to discover co-regulated genes using diverse gene expression datasets. In: *Bioinformatics and Computational Biology* (ed. S. Rajasekaran), Springer, New York, pp. 151–163.

Bryan, K., Cunningham, P. and Bolshakova, N. (2005). Biclustering of Expression Data using Simulated Annealing. Proceedings of the 18th IEEE Symposium on Computer-Based Medical Systems, 23–24 June 2005, Dublin, Ireland, pp. 383–388.

Bryan, K., Terrile, M., Bray, I.M. *et al.* (2014). Discovery and visualization of miRNA-mRNA functional modules within integrated data using bicluster analysis. *Nucleic Acids Research*, **42**(3), e17.

Busygin, S., Jacobsen, G., Krämer, E. and Ag, C. (2002). Double Conjugated Clustering Applied to Leukemia Microarray Data. In 2nd SIAM ICDM, Workshop on Clustering High Dimensional Data, Arlington, Virginia.

Busygin, S., Prokopyev, O. and Pardalos, P.M. (2008). Biclustering in data mining. *Computers and Operations Research*, **35**(9), pp. 2964–2987.

Caldas, J. and Kaski, S. (2008). Bayesian Biclustering with the Plaid Model. IEEE Workshop on Machine Learning for Signal Processing, MLSP 2008, Cancún, Mexico. pp. 291–296.

Caldas, J. and Kaski, S. (2010). Hierarchical Generative Biclustering for MicroRNA Expression Analysis (ed. B. Berger), Proceedings of the 14th Annual International Conference on Research in Computational and Molecular Biology, 25–28 April 2010, Lisbon, Portugal. pp. 65–79.

Califano, A., Stolovitzky, G. and Tu, Y. (2000). *Analysis of Gene Expression Microarrays for Phenotype Classification*. ISMB, Boston, MA, pp. 75–85.

Cheng, Y. and Church, G.M. (2000). *Biclustering of Expression Data*. ISMB, Boston, MA, pp. 93–103.

Cho, R.J., Campbell, M.J., Winzeler, E.A. *et al.* (1998). A genome-wide transcriptional analysis of the mitotic cell cycle. *Molecular Cell*, **2**(1), pp. 65–73.

Divina, F. and Aguilar-Ruiz, J.S. (2006). Biclustering of expression data with evolutionary computation., *IEEE Transactions on Knowledge and Data Engineering*, **18**(5), pp. 590–602.

Eren, K. (2012). *Application of Biclustering Algorithms to Biological Data*, The Ohio State University, Columbus.

Eren, K., Deveci, M., Küçüktunç, O. and Çatalyürek, Ü.V. (2013). A comparative analysis of biclustering algorithms for gene expression data. *Briefings in Bioinformatics*, **14**(3), pp. 279–292.

Gasch, A.P., Spellman, P.T., Kao, C.M. *et al.* (2000). Genomic expression programs in the response of yeast cells to environmental changes. *Molecular Biology of the Cell*, **11**(12), pp. 4241–4257.

Gasch, A.P., Huang, M., Metzner, S. *et al.* (2001). Genomic expression responses to DNA-damaging agents and the regulatory role of the yeast ATR homolog Mec1p. *Molecular Biology of the Cell*, **12**(10), pp. 2987–3003.

Getz, G., Levine, E. and Domany, E. (2000). Coupled two-way clustering analysis of gene microarray data. *Proceedings of the National Academy of Sciences*, **97**(22), pp. 12079–12084.

Golub, T.R., Slonim, D.K., Tamayo, P. *et al.* (1999). Molecular classification of cancer: class discovery and class prediction by gene expression monitoring. *Science*, **286**(5439), pp. 531–537.

Gu, J. and Liu, J.S. (2008). Bayesian biclustering of gene expression data. *BMC Genomics*, **9**(Suppl 1), S4.

Hartigan, J.A. (1972). Direct clustering of a data matrix. *Journal of the American Statistical Association*, **67**(337), pp. 123–129.

Hochreiter, S., Bodenhofer, U., Heusel, M. *et al.* (2010). FABIA: factor analysis for bicluster acquisition. *Bioinformatics*, **26**(12), pp. 1520–1527.

Hughes, T.R., Marton, M.J., Jones, A.R. *et al.* (2000). Functional discovery via a compendium of expression profiles. *Cell*, **102**(1), pp. 109–126.

Huttenhower, C., Mutungu, K.T., Indik, N. *et al.* (2009). Detailing regulatory networks through large scale data integration. *Bioinformatics*, **25**(24), pp. 3267–3274.

Ideker, T., Thorsson, V., Ranish, J.A. *et al.* (2001). Integrated genomic and proteomic analyses of a systematically perturbed metabolic network. *Science*, **292**(5518), pp. 929–934.

Ihmels, J., Friedlander, G., Bergmann, S. *et al.* (2002). Revealing modular organization in the yeast transcriptional network. *Nature Genetics,* **31**(4), pp. 370–377.

Ihmels, J., Bergmann, S. and Barkai, N. (2004). Defining transcription modules using large-scale gene expression data. *Bioinformatics*, **20**(13), pp. 1993–2003.

Kaiser, S. and Leisch, F. (2008). A toolbox for bicluster analysis in R, *Proceedings in Computational Statistics* (ed. P. Brito), Physica Verlag, Heidelberg, pp. 201–208.

Kluger, Y., Basri, R., Chang, J.T. and Gerstein, M. (2003). Spectral biclustering of microarray data: coclustering genes and conditions. *Genome Research*, **13**(4), pp. 703–716.

Lazzeroni, L. and Owen, A. (2002). Plaid models for gene expression data. *Statistica Sinica*, **12**(1), 61–86.

Li, H., Chen, X., Zhang, K. and Jiang, T. (2006). A general framework for biclustering gene expression data. *Journal of Bioinformatics and Computational Biology,* **4**(04), pp. 911–993.

Li, G., Ma, Q., Tang, H. *et al.* (2009). QUBIC: a qualitative biclustering algorithm for analyses of gene expression data. *Nucleic Acids Research*, **37**(15), e101.

Liu, J. and Wang, W. (2003). Op-cluster: Clustering by Tendency in High Dimensional Space. Third IEEE International Conference on Data Mining. ICDM 2003, Melbourne, FL. pp. 187–194.

Liu, X. and Wang, L. (2007). Computing the maximum similarity bi-clusters of gene expression data. *Bioinformatics,* **23**(1), pp. 50–56.

Liu, J., Wang, W. and Yang, J. (2004). Gene Ontology Friendly Biclustering of Expression Profiles. Proceedings of IEEE Computational Systems Bioinformatics Conference, CSB 2004, Stanford, CA. pp. 436–447.

Loi, S., Haibe-Kains, B., Desmedt, C. *et al.* (2007). Definition of clinically distinct molecular subtypes in estrogen receptor - positive breast carcinomas through genomic grade. *Journal of Clinical Oncology*, **25**(10), pp. 1239–1246.

Madeira, S.C. and Oliveira, A.L. (2004). Biclustering algorithms for biological data analysis: a survey. *IEEE/ACM Transactions on Computational Biology and Bioinformatics,* **1**(1), pp. 24–45.

Miller, L.D., Smeds, J., George, J. *et al.* (2005). An expression signature for p53 status in human breast cancer predicts mutation status, transcriptional effects, and patient survival. *Proceedings of the National Academy of Sciences of the United States of America*, **102**(38), pp. 13550–13555.

Mitra, S. and Banka, H. (2006). Multi-objective evolutionary biclustering of gene expression data. *Pattern Recognition*, **39**(12), pp. 2464–2477.

Murali, T. and Kasif, S. (2003). Extracting Conserved Gene Expression Motifs from Gene Expression Data. Pacific Symposium on Biocomputing, Lihue, HI, pp. 77–88.

Oghabian, A., Kilpinen, S., Hautaniemi, S. and Czeizler, E. (2014). Biclustering methods: biological relevance and application in gene expression analysis. *Plos One*, **9**(3), e90801.

Pawitan, Y., Bjöhle, J., Amler, L. *et al.* (2005). Gene expression profiling spares early breast cancer patients from adjuvant therapy: derived and validated in two population-based cohorts. *Breast Cancer Research*, **7**(6), R953.

Prelić, A., Bleuler, S., Zimmermann, P. *et al.* (2006). A systematic comparison and evaluation of biclustering methods for gene expression data. *Bioinformatics*, **22**(9), pp. 1122–1129.

Reiss, D.J., Baliga, N.S. and Bonneau, R. (2006). Integrated biclustering of heterogeneous genome-wide datasets for the inference of global regulatory networks. *BMC Bioinformatics*, **7**(1), p. 280.

Sheng, Q., Moreau, Y. and Moor, B.D. (2003). Biclustering microarray data by Gibbs sampling. *Bioinformatics-Oxford*, **19**(2), pp. 196–205.

Spellman, P.T., Sherlock, G., Zhang, M.Q. *et al.* (1998). Comprehensive identification of cell cycle – regulated genes of the yeast Saccharomyces cerevisiae by microarray hybridization. *Molecular Biology of the Cell*, **9**(12), pp. 3273–3297.

Tanay, A., Sharan, R. and Shamir, R. (2002). Discovering statistically significant biclusters in gene expression data. *Bioinformatics*, **18**(suppl 1), pp. S136–S144.

Tang, C., Zhang, L., Zhang, A. and Ramanathan, M. (2001). Interrelated Two-Way Clustering: an Unsupervised Approach for Gene Expression Data Analysis. Proceedings of the IEEE 2nd International Symposium on Bioinformatics and Bioengineering Conference, 2001, Bethesda, MD. pp. 41–48.

Tchagang, A.B. and Twefik, A.H. (2005). Robust Biclustering Algorithm (ROBA) for DNA Microarray Data Analysis. IEEE 13th Workshop on Statistical Signal Processing, 2005, Bordeaux, pp. 984–989.

Tchagang, A.B. and Tewfik, A.H. (2006). DNA microarray data analysis: a novel biclustering algorithm approach. *EURASIP Journal on Advances in Signal Processing*, **2006**, p. 059809.

Tchagang, A.B., Pan, Y., Famili, F. *et al.* (2011). Biclustering of DNA microarray data: theory, evaluation, and applications. *Handbook of Research on Computational and Systems Biology: Interdisciplinary Applications* (**1** Vol) (eds L.A. Liu, D. Wei Y.X. Li and H.M. Lei). Medical Information Science Reference, Hershey, PA, p. 148.

Teng, L. and Chan, L. (2008). Discovering biclusters by iteratively sorting with weighted correlation coefficient in gene expression data. *Journal of Signal Processing Systems*, **50**(3), pp. 267–280.

Turner, H., Bailey, T. and Krzanowski, W. (2005). Improved biclustering of microarray data demonstrated through systematic performance tests. *Computational Statistics and Data Analysis*, **48**(2), pp. 235–254.

Wang, H., Wang, W., Yang, J. and Yu, P.S. (2002). Clustering by Pattern Similarity in Large Data Sets. Proceedings of the 2002 ACM SIGMOD International Conference on Management of Data, Madison, WI, pp. 394–405.

Wu, C.-J. and Kasif, S. (2005). GEMS: a web server for biclustering analysis of expression data. *Nucleic Acids Research,* **33**(suppl 2), pp. W596–W599.

Wu, C.-J., Fu, Y., Murali, T. and Kasif, S. (2004). Gene expression module discovery using Gibbs sampling. *Genome Informatics*, **15**(1) pp. 239–248.

Yang, J., Wang, H., Wang, W. and Yu, P.S. (2005). An improved biclustering method for analyzing gene expression profiles. *International Journal on Artificial Intelligence Tools*, **14**(05), pp. 771–789.

Yoon, S., Nardini, C., Benini, L. and Micheli, G.D. (2005). Discovering coherent biclusters from gene expression data using zero-suppressed binary decision diagrams. *IEEE/ACM Transactions on Computational Biology and Bioinformatics (TCBB)*, **2**(4), pp. 339–354.

19

Clustering Methods Discussion

19.1   Introduction

In the previous 9 Chapters, we have introduced the technical details of clustering methods comprehensively. The principles have been described; many individual algorithms have been presented; many resources including publicly accessible open source software packages have been collected. In this chapter, we will switch to another flavour in the sense that we start to focus on the applications of clustering algorithms rather than the techniques; that is, *what* is the information that we may obtain from massive biological data using clustering algorithms rather than *how*. As an essential part of the integrative cluster analysis pipeline, to choose a proper clustering algorithm from loads of existing clustering algorithms for a given dataset is not trivial. Therefore, the objectives of this chapter are two-fold: on one hand, we may learn more by studying others' successful examples; for instance, what did other research obtain, by using which clustering algorithm, and why? On the other hand, we also like to point out that some clustering algorithms, which have not yet been utilised in the bioinformatic field, or which have been evaluated in real biological datasets but have not been applied in analytical studies of newly collected data, still have possibilities to be applied in future developments. This chapter is organised such that each clustering family takes a section.

19.2   Hierarchical Clustering

In this section, we will discuss the applications of hierarchical clustering algorithms. As introduced in Chapter 12, hierarchical clustering is one of most widely used clustering algorithms. It has been applied in yeast cell cycle gene expression data analysis (gene clustering), human breast cancer classification (sample clustering), human lymphoma classification (sample clustering), and others.

Integrative Cluster Analysis in Bioinformatics, First Edition. Basel Abu-Jamous, Rui Fa and Asoke K. Nandi.
© 2015 John Wiley & Sons, Ltd. Published 2015 by John Wiley & Sons, Ltd.

19.2.1   Yeast Cell Cycle Data

Two pioneering papers from a group at Stanford University (Eisen *et al.*, 1998; Spellman *et al.*, 1998), which opened a new field of clustering analysis in high-throughput gene expression studies, employed hierarchical clustering to study genome-wide regulation in the yeast cell cycle (Cho *et al.*, 1998). The hierarchical clustering algorithm used in these two studies was based on the average-linkage method (Sneath and Sokal, 1973). A dendrogram that assembled all genes into a single tree was computed. Spellman and colleagues analysed yeast cell cycle datasets produced using different synchronisation methods; for example, elutriation, alpha pheromone, cdc-15 (Spellman *et al.*, 1998), and cdc-28 (Cho *et al.*, 1998). Datasets elutriation, alpha pheromone, and cdc-15 are publicly available, with the GEO accession numbers GSE22, GSE23 and GSE24, respectively. An example of clustering the cdc-28 dataset using hierarchical clustering is depicted as a heatmap in Figure 19.1. In this example, the subset of 800 genes in Spellman *et al.* (1998) is further filtered and those genes with missing data are removed. The remaining dataset has 591 genes. To obtain the final clustering result, the tree can be cut based on different requirements; for example, in terms of the between-cluster distance, the tree in Figure 19.1 can be cut into three clusters, four clusters or five clusters.

19.2.2   Breast Cancer

Perou and colleagues conducted a study of molecular portraits of human breast cancer tumours (Perou *et al.*, 2000). Gene expression patterns in 65 surgical specimens of human breast tumours from 42 individuals, employing cDNA microarrays representing 8102 genes, were clustered using hierarchical clustering. Forty-two individuals were considered, including 36 infiltrating ductal carcinomas, 2 lobular carcinomas, 1 ductal carcinoma *in situ*, 1 fibroadenoma and 3 normal

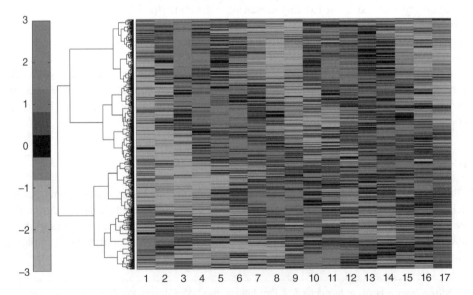

Figure 19.1   An example of hierarchal clustering for yeast cell cycle dataset cdc-28

breast samples. Twenty of the forty breast tumours were sampled twice. After an open surgical biopsy to obtain the 'before' sample, each of these patients was treated with doxorubicin for an average of 16 weeks, followed by resection of the remaining tumour. To interpret the variation in expression patterns, 17 cultured cell lines were also included. In total, 84 cDNA microarray experiments were carried out. The raw microarray dataset can be obtained from NCBI website with the series number GSE61. First of all, out of 8102 genes, 1753 genes (about 22%), whose transcripts varied in abundance by at least four-fold from their median abundance in this sample set in at least three of the samples, were chosen. The first experiment remarkably showed that the tumours have great variation in their gene expression, and that the gene expression patterns have a pervasive order reflecting relationships among genes, relationships among the tumours, and connections between specific genes and specific tumours. As the main goal of this study was to classify tumours on the basis of their gene expression patterns, the first experiment was not perfect for this purpose. In the second experiment, a smaller subset (496) of genes, called the intrinsic gene subset, was selected and these genes have significantly greater variation in expression between different tumours than between paired samples from the same tumour. The cluster analysis of the intrinsic gene subset revealed that the tissue samples were divided into two subgroups. It is also worth noting that there is extensive residual variation in expression patters within each of these two broad subgroups. Four subgroups of samples were identified to be related to different molecular features of mammary epithelial biology; that is, estrogen receptor positive(ER+)/luminal-like, basal-like, Erb-B2+ and normal breast. This study concluded that the tumours could be classified into subtypes distinguished by pervasive differences in their gene expression patterns.

Sørlie and colleagues continued the study of classifying breast carcinomas based on variations in gene expression patterns derived from cDNA microarrays, thereby correlating tumour characteristics with clinical outcome (Sørlie et al., 2001). A larger number of tumours were analysed and the clinical value of the subtypes was explored to refine the previous work (Perou et al., 2000). A total of 78 breast carcinomas including 71 ductal, 5 lobular and 2 ductal carcinomas in situ were obtained from 77 different individuals; two independent tumours from one individual were diagnosed at different times. In this study, hierarchical clustering was employed to analyse a total of 85 cDNA microarray experiments representing 78 cancerous, three fibroadenomas and four normal breast tissues. Similarly to the second experiment in Perou et al. (2000), the intrinsic gene set of 456 cDNA clones was selected to identify the intrinsic characteristics of breast tumours optimally. Cluster analysis using hierarchical clustering revealed that the tumours were separated into two main branches: the first branch contained three subgroups that all were characterised by low to absent gene expression of the ER, which has been reported previously (Perou et al., 2000); the second branch, which had been defined as a luminal/ER+ cluster, could be separated into two or probably three distinct subgroups. The group of 32 tumours, which was termed luminal subtype A, demonstrated the highest expression of the ER α gene, GATA binding protein 3, X-box binding protein 1, trefoil factor 3, hepatocyte nuclear factor 3 α, and estrogen-regulated LIV-1. The second group of tumours positive for luminal-enriched genes could be broken into 2 smaller units, a small group of 5 tumours called luminal subtype B, and a group of 10 tumours called luminal subtype C. The striking facts are: (i) luminal subtype C was distinguished from luminal subtypes A and B by high expression of a set of genes whose coordinated function was unknown, which was shared with basal-like and ERBB21 subtypes; (ii) luminal subtype A was distinguished from luminal subtypes B and C by high expression of a set of genes labelled with group F. A second hierarchical clustering analysis was conducted using the intrinsic gene set and the subset of 51 carcinomas, three benign tumours and four normal breast samples to examine the robustness of the observed clusters in the first experiment. The results presented that the same major subtypes were discovered, except that the position of the five luminal subtype B tumours changed to be grouped with the Erb-B2+ subtype, although luminal subtype B tumours do not over-express Erb-B2. All these results suggested that the groupings into the four subtypes, namely Luminal A, Luminal B + C, HER2-enriched, and Basal-like are reasonably robust with most of the tumour samples staying together in the same groups when using different sample sets for the analysis. The light shed on subtype classification of breast carcinomas by these two studies made it clear that there are four molecular 'intrinsic' subtypes of breast cancer.

Herschkowitz and colleagues conducted a study to classify a large set of mouse mammary tumour models and human breast tumours using gene expression analysis. As genomic studies evolve, further sub-classification of breast tumours into new molecular entities is expected to occur and a new breast cancer intrinsic subtype, known as Claudin-low, was identified in human tumours, again by using hierarchical clustering (Herschkowitz et al., 2007).

19.2.3   Diffuse Large B-Cell Lymphoma

Diffuse large B-cell lymphoma (DLBCL), the most common subtype of non-Hodgkin's lymphoma, was identified as being clinically heterogeneous: more than half of patients succumbed to the disease while 40% of patients responded well to the therapy and had prolonged survival, according to a report in 2000 (Alizadeh et al., 2000). It motivated Alizadeh and colleagues to investigate the unrecognised molecular heterogeneity in tumours underneath this variabilty in natural history. A systematic characterisation of gene expression in B-cell malignancies was conducted by using DNA microarrays.

Two molecularly distinct forms of DLBCL, which have gene expression patterns indicative of different stages of B-cell differentiation, were identified by Alizadeh and colleagues (2000). One type expressed genes characteristic of germinal centre B-cells ('germinal centre B-like DLBCL'); the second type expressed genes normally induced during in vitro activation of peripheral blood B-cell ('activated B-like DLBCL'). Patients with germonal centre B-like DLBCL had a significantly better overall survival than did those with activated B-like DLBCL.

The genome-wide gene expression dataset, with the GEO accession numbers GSE60, was obtained from 128 microarray analyses of 96 samples of normal and malignant lymphocyptes. Hierarchical clustering was used to group tumours and cell samples on the basis of similarity in their genes, rather than to group genes on the basis of similarity in the pattern with which their expression varied over all samples as what Eisen and colleagues did in Eisen et al. (1998). Distinct clones representing the same gene were typically clustered in adjacent rows in the dendrogram produced by hierarchical clustering, indicating that these genes have characteristic and individually distinct patterns of expression. The clusters of coordinately expressed genes were defined as good expression 'signatures'. A gene expression signature was named by either the cell type in which its component genes were expressed (T-cell signature) or the biological process in which its component genes are known to function (the proliferation signature). Alizadeh and colleagues reclustered the DLBCL cases with hierarchical clustering using only expression pattern of the genes that define the germinal centre B-cell signature. Two large branches of tumour samples were clearly shown: one branch was defined as germinal centre B-like and the other was defined as activated B-like DLBCL.

With the advent of more advanced clustering methods, in many newer studies, hierarchical clustering was not used as a solo analytical tool; on the contrary, it was employed together with other methods to produce more precise results (Shipp *et al.*, 2002; Monti *et al.*, 2005).

19.3   Fuzzy Clustering

The greatest advantage of fuzzy clustering over hard (crisp) clustering is that fuzzy clustering calculates for each data point a defined degree of belongingness to every suggested cluster. It suggests that assigning a data point to multiple clusters is possible. In this section, we will discuss some applications that employed fuzzy clustering in the bioinformatics field.

19.3.1   DNA Motifs Clustering

Functional transcription factor binding sites generally appear in clusters, which represent promoters or enhancers. In contrast, isolated sequence elements normally are not biologically significant even when they match a consensus pattern perfectly. Therefore, searching for clustered potential transcription factor binding sites may help to filter insignificant items from the output lists of sequence-scanning routines. To this end, fuzzy clustering algorithms were applied to a two-dimensional cluster analysis because the two-dimensional analysis allows one to consider the scoring of the potential sites in addition to the position along the DNA. Fuzzy approaches are best suited to deal with vague data such as scored binding sites (Pickert *et al.*, 1998). Two fuzzy clustering algorithms, FCM and the Gath–Geva algorithm, were employed in this study. FCM is very fast and it searches for spherical clusters. The Gath–Geva algorithm extends the cluster shapes to hyper-elliptic. Using fuzzy clustering, 19 binding sites were suggested, where 13 binding sites, which were all sites that have been proven experimentally, were included. Remarkably, 49 apparent false-positive matches were removed from the original output lists of sequence-scanning routines.

19.3.2   Microarray Gene Expression

Many proteins serve different roles depending on the demands of the organism, and therefore the corresponding genes are often co-expressed with different groups of genes under different situations. Therefore many genes are similarly expressed to multiple, distinct groups of genes. Because most commonly used analytical methods cannot appropriately represent these relationships, fuzzy clustering is a candidate to identify overlapping clusters of conditionally co-regulated genes. The performance of fuzzy clustering algorithms in real microarray gene expression datasets has been investigated in many different studies (Datta and Datta, 2003; Dembele and Kastner, 2003). For example, in Dembele and Kastner (2003), FCM has been investigated in many real microarray gene expression datasets, including gene expression in response to serum concentration in human fibroblasts (Iyer *et al.*, 1999), yeast cell cycle cdc28 dataset (Cho *et al.*, 1998), and human cancer cell lines (Ross *et al.*, 2000). Furthermore, Dembele and Kastner also proposed a method to compute an upper bound value for the fuzziness parameter m. They pointed out that m was commonly fixed to two, but FCM often failed in microarray datasets when $m = 2$. The upper bound value for m, which was defined as m_{ub}, was estimated. Therefore, the choice of fuzziness parameter m is determined as $m = 1 + m_0$, where $m_0 = 1$ if $m_{ub} \geq 10$; and $m_0 = m_{ub}/10$ if $m_{ub} < 10$. Dembele and Kastner proved that fuzzy clustering is a convenient method to select genes exhibiting tight association to given clusters. In the case of yeast dataset, genes which are likely to be relevant for biological significance of the cluster are indeed retained preferentially following threshold-based selection.

FCM clustering was used by Gasch and Eisen (2002) to identify overlapping clusters yeast genes based on published gene expression data following the response of yeast cells to environment changes, for example zinc starvation (Lyons *et al.*, 2000), phosphate limitation (Ogawa, DeRisi and Brown, 2000), DNA-damaging agents (Gasch *et al.*, 2001), and a variety of other stressful environment conditions (Gasch *et al.*, 2000). Yeast cells must have a precise mechanism to mediate the synthesis and function of proteins in the cell to respond to diverse and frequently changing environmental conditions. Many regulatory factors, which are essential in the overall genomic expression program, act under specific conditions, and together they govern the expression of overlapping sets of genes. The analytical study using fuzzy clustering by Gasch and Eisen (2002) suggested that the condition-specific regulation of overlapping sets of yeast genes is a prevalent theme in the regulation of yeast gene expression. A large fraction of yeast genes is expressed in patterns that are similar to different groups of genes in response to different subsets of the experiments. Moreover, a substantial number of these genes contain multiple transcription factor binding sites in their promoters, consistent with the idea that they are conditionally regulated by multiple, independent regulatory systems. Another benefit of the fuzzy clustering algorithm is that it identifies continuous memberships of genes. This allows each cluster to be expanded or collapsed to view genes of varying similarity in expression. While the genes of highest membership in a given cluster are often tightly correlated in terms of biochemical function and regulation, expanding the cluster can identify genes that are similarly expressed in only subsets of the experimental conditions.

Kim and colleagues studied the effect of data normalisation on fuzzy clustering of DNA microarray (Kim, Lee and Bae, 2006). To identify the effect of data normalisation, three normalisation methods, including two common scale and location transformations and lowess normalisation methods, were used to normalise three microarray datasets. The effects of normalisation were evaluated. The comparative analysis showed that the clustering performance depends on the normalisation method used. Particularly, the selection of the fuzzification parameter value for the FCM method was sensitive to the normalisation method used for datasets with large variations across samples. Lowess normalisation is more robust for clustering genes from general microarray data than the two common scale and location adjustment methods when samples have varying expression patterns or are noisy. In particular, the FCM method slightly outperformed the hard clustering methods when the expression patterns of genes overlapped and was advantageous in finding co-regulated genes.

Wang and colleagues developed an algorithm employing two models for tumour classification and target gene prediction. First, gene expression profiles are summarised by optimally selected self-organising map (SOM), followed by tumour sample classification by FCM clustering (Wang *et al.*, 2003). The aim of applying the SOM procedure in the algorithm was to find map units that could represent the configuration of the input dataset, and at the same time to achieve a continuous mapping from the input gene space to a lattice. The FCM was employed in order to assign fuzzy membership values that could serve as a confidence measure in tumour classification. Furthermore, the Fisher's linear discriminant is a general method in discrimination analysis, which searches for good separation between groups by finding the maximal ratio of the between-group-sum of squares to the within-group-sum of squares. The pair-wise Fisher's linear discriminant was combined into the algorithm to accomplish the cross validation of the selected features. The algorithm was tested on four published datasets, including leukaemia (Golub *et al.*, 1999), colon cancer (Alon *et al.*, 1999), brain tumours (Pomeroy *et al.*, 2002), and NCI60 cancer cell lines (Ross *et al.*, 2000).

19.4   Neural Network-based Clustering

SOM is one of the most famous and popular neural network-based clustering algorithms. It has been widely used in many different applications since it was invented by Kohonen in the 1980s (1990, 1997). Tamayo and colleagues first applied SOM in the analysis of gene expression data (Tamayo *et al.*, 1999). SOM constructs a geometry of neurons as a rectangular grid (could be another shape) and each neuron is mapped into the data space corresponding to a prototype. Neurons are updated iteratively towards their respective prototypes. Therefore, SOM imposes structure on the data, with neighbouring neurons tending to define related clusters. SOM is analogous to an entomologist's specimen drawer, with adjacent compartments holding similar insects. SOM is particularly well suited for exploratory data analysis because they expose the fundamental patterns in the data. SOM was applied to several real microarray datasets in Tamayo and colleagues's study (1999), namely yeast cell cycle cdc28 data (Cho *et al.*, 1998), macrophage differentiation in HL-60, and hematopoietic differentiation in four models including HL-60, U937, Jurkat and NB4 cells. The latter two datasets are available at http://www.broadinstitute.org/cgi-bin/cancer/datasets.cgi. Very recently, Chavez-Alvarez and colleagues employed SOM to analyse the expression level of 282 genes, which have been known to have an activity during the cell cycle, in five well-known yeast cell cycle datasets (Chavez-Alvarez, Chavoya and Mendez-Vazquez, 2014). These five datasets include cdc28 dataset (Cho *et al.*, 1998), cdc15 dataset, alpha dataset (Spellman *et al.*, 1998), alpha-30 dataset, and alpha-38 dataset (Pramila *et al.*, 2006). The study attempted to find clusters of genes with similar behaviour in the five datasets along two cell cycles.

Since SOM was found to be superior to many other clustering algorithms, it was often combined with some other clustering algorithms that are also superior to general clustering algorithms. For example, Hsu and colleagues proposed a hierarchical dynamic self-organising approach for cancel class discovery and marker gene identification (Hsu, Tang and Halgamuge, 2003). The approach integrated merits of hierarchical clustering and the robustness of SOM. It is performed in three successive phases: sample class discovery, marker gene identification, and partition refinement. The approach was evaluated with leukaemia data (Golub *et al.*, 1999) and colon cancer data (Alon *et al.*, 1999). Another example is the algorithm proposed by Wang and colleagues (2003), which has been discussed in the previous section. This algorithm combines SOM and FCM.

Another successful type of neural network based clustering algorithms, which has been widely used in the bioinformatic field, is ART and its variants. Tomida and colleagues (2002) applied fuzzy ART to analyse the time series expression data during the sporulation of budding yeast; these data were collected by Chu and colleagues (1998). In this study, 522 genes were considered in the cluster analysis. Among these 522 genes, 45 genes that have been known to have some roles in meiosis and sporulation were selected as index genes. Some of these genes were expressed at different time points, roughly labelled as 'Early', 'Middle', 'Mid-Late' and 'Late'. By using fuzzy ART, four cases of the number of clusters such as 4, 5, 6 and 7, determined by different vigilance parameter values, were examined. Finally, the clustering results were evaluated by the index genes. Projective ART is also one of the variants of ART, which was proposed by Cao and Wu (2002), and then was applied to microarray data to construct a cancer prediction model by Takahashi, Kobayashi and Honda (2005). The major difference between projective ART and original ART lies in the F_1 layer where a neuron in the F_1 layer can be active relative to some neurons in the F_2 layer, but inactive relative to other F_2 neurons for the projective ART algorithm. The aim of the projective ART algorithm is to find projected clusters, each of which consists of a subset of data points together with a subset of dimensions such that the selected data points are closely correlated in the selected dimensions. Essentially, it has the same idea as biclustering has. The projective ART algorithm was evaluated by Takahashi, Kobayashi and Honda (2005) in two well-known gene expression datasets: leukaemia data (Golub *et al.*, 1999) and lung cancer data (Bhattacharjee *et al.*, 2001). The projective ART was also applied (Takahashi *et al.*, 2006) to soft tissue sarcoma (STS) micro-array data as a gene-filtering method to select specific genes for each subtype. In this study, the results showed that the projective ART algorithm provided higher prediction accuracy than did other methods, and many selected genes are known prognostic marker genes for other tumours and could be candidate marker genes for the diagnosis of STS. The follow-up study in Takahashi *et al.* (2013) investigated the 28 selected genes from Takahashi *et al.* (2006), and found that macrophage migration inhibitory factor (MIF) and stearoyl-CoA desaturase 1 (SCD1) represent an effective diagnostic marker combination to discriminate pleomorphic malignant fibrous histiocytoma and myxofibrosarcoma.

Some other neuron network-based clustering methods have also been evaluated in real biological datasets. For example, the SSMCL algorithm based on the OPTOC paradigm (Wu *et al.*, 2004) was evaluated using yeast cell cycle cdc28 data (Cho *et al.*, 1998); in Salem, Jack and Nandi (2008), SOON was also evaluated using yeast cell cycle cdc28 data in addition to two cancer datasets, namely lymphoma data (Alizadeh *et al.*, 2000) and liver cancer data (Chen *et al.*, 2002). Although these algorithms have not been applied in the analytical studies of newly collected datasets, they can be strong candidates for being analytical tools since they have been tested in the real biological datasets.

19.5   Mixture Model-based Clustering

Mixture model clustering has been widely used in bioinformatic cluster analysis. As we have discussed in Chapter 15, mixture model-based clustering algorithms may be classified into two large categories: finite mixture models and infinite mixture models. Finite mixture models can be further classified into two subgroups in terms of statistical inference, namely maximum likelihood methods and Bayesian methods. Generally speaking, there are many more applications of finite mixture models than those of infinite mixture models, simply because infinite mixture models have been receiving more attention only very recently. After Eisen and colleagues (1998) brought the idea of clustering into high-throughput gene expression analysis, people started considering other clustering algorithms, which may do the better job. Mixture model-based clustering is one of the clustering families that were chosen to be applied to high-throughput data (Yeung *et al.*, 2001; Ghosh and Chinnaiyan, 2002; Medvedovic and Sivaganesan, 2002). We will discuss the examples of finite mixture models and infinite mixture models, respectively.

19.5.1 Examples of Finite Mixture Models

Ramoni and colleagues proposed a Bayesian method for model-based clustering of gene expression dynamics (Ramoni, Sebastiani and Kohane, 2002). Gene expression time series data are represented as a mixture of autoregressive models. The objective of the algorithm is to group the time series data into many clusters, where the time series in a particular cluster are supposed to be generated by the same stochastic process.

In the journal *Bioinformatics,* two papers about finite mixture models appeared nearly at the same time: one paper was by Yeung and colleagues (2001) and the other was by Ghosh and Chinnaiyan (2002). Both papers investigated the performance of maximisation likelihood solution of finite mixture models in gene expression data. As we have introduced in Chapter 15, the covariance of each component, say the *g*th component, can be parameterised by eigenvalue decomposition in the form shown in Equation (19.1),

$$\boldsymbol{\Sigma}_g = \lambda_g \boldsymbol{D}_g \boldsymbol{A}_g \boldsymbol{D}_g^T \tag{19.1}$$

where λ_g is a scalar, $\boldsymbol{A}_g$ is a diagonal matrix whose elements are proportional to the eigenvalues of $\boldsymbol{\Sigma}_g$, and $\boldsymbol{D}_g$ is the orthogonal matrix of eigenvectors (Fraley and Raftery, 1999). Yeung and colleagues tested the model-based algorithm in two real microarray gene expression data experiments: one was ovary data collected by Schummer and colleagues (1999) and the other was yeast cell-cycle data by Cho and colleagues (1998). Two subsets of yeast cell-cycle data, one with 384 genes and the other one with 237 genes, were investigated, respectively. Ghosh and Chinnaiyan investigated the model-based clustering algorithm in two real microarray gene expression data experiments: one is cutaneous melanoma data collected by Bittner *et al.* (2000) and the other is prostate cancer data collected by Dhanasekaran *et al.* (2001). To use the model-based clustering algorithm correctly, one has to keep two questions in mind. The first question is which covariance structure should be used (different covariance structures have been summarised in Table 14.1), since different structures may result in different numbers of parameters. For example, the algorithm with the ID 'VVV', representing elliptical distribution and variable volumes, has $(M + M(M + 1)/2)$ parameters. If the dimension M of a dataset is large, the number of parameters could be very large, which means that to estimate these parameters correctly requires a large number of observations. Furthermore, insufficient observations may result in an ill-conditioned covariance matrix, which causes numerical problems. The second question is how many clusters there are in the dataset. Since the algorithm requires the number of clusters as an input parameter, which is unknown *a priori,* the popular practice is to set a range of numbers of clusters, say $G \in [K_{min}, K_{max}]$, then to apply the algorithm to the dataset with all numbers of clusters, resulting in many different clustering results, and finally to select a clustering result among many clustering algorithms in terms of clustering validation values. Numerical clustering validation will be discussed in Chapter 20. In both studies (Yeung *et al.*, 2001; Ghosh and

Chinnaiyan, 2002), Bayesian information criterion (BIC) was employed to validate the clustering result.

McLachlan and colleagues introduced a software EMMIX-GENE for the clustering of microarray expression data, particularly of tissue samples on a large number of genes (McLachlan, Bean and Peel, 2002). EMMIX-GENE is a feasible approach which first selects a subset of the genes relevant for the clustering of the tissue samples by fitting mixtures of t distributions to rank the genes, and employs a mixture of factor analysers to reduce effectively the dimension of the feature space of genes. We have introduced both a mixture of t distributions model and a mixture factor analysers model in Chapter 15. McLachlan and colleagues investigated EMMIX-GENE with the colon tissue data by Alon and colleagues (1999) and leukaemia tissue data by Golub and colleagues (1999). Since the mixture of normal distributions requires a sufficiently large number of observations to prevent singular estimates of the component-covariance matrices, it turns out that the case of clustering tissues on the basis of gene expression levels is more difficult because it has a much larger number of points to cluster than in the case of clustering genes on the basis of their expression patterns; sometimes the number of genes is much higher than the number of tissues. Therefore, a mixture of factor analysers was applied to deal with large dimensional datasets.

McNicholas and Murphy applied a parsimonious Gaussian mixture model, which was proposed in McNicholas and Murphy (2008) and primarily based on mixture of factor analysers, to microarray gene expression data analysis (McNicholas and Murphy, 2010). As we introduced in Table 15.2 Chapter 15, there are eight covariance structures of a parsimonious Gaussian mixture model based on the constraints, similarly to Gaussian mixture models. Using the colon tissue data (Alon *et al.*, 1999) and leukaemia tissue data (Golub *et al.*, 1999), the performance of clustering results in terms of validation values were compared with many other clustering algorithms; namely, hierarchical clustering algorithms, k-means, PAM, and MCLUST. It requires the number of clusters as an input parameter, which is similar to the other model-based clustering algorithms, and additionally, it requires the specific number of factor analysers. One has to select the best results among the results with both a range of the number of clusters and a range of the number of factor analysers in terms of validation values.

Baek and McLachlan extended a mixture of common factor analysers, which was proposed by Baek, McLachlan and Flack (2010), and a mixture of multivariate t-distributions, which was developed by Andrews and McNicholas (2011), to a mixture of common t-factor analysers in gene expression data to cluster large dimensional data (Baek and McLachlan, 2011). There were two objectives of using a mixture of common t-factor analysers: the first is to reduce the number of parameters further, and the second is to be more robust against outliers. The algorithm was investigated using a breast and colon cancer dataset (Chowdary *et al.*, 2006) and a lung cancer dataset (Bhattacharjee *et al.*, 2001). Similar to the mixture model of factor analysers, this algorithm also requires both the number of clusters and the number of factor analysers.

19.5.2 Examples of Infinite Mixture Models

Heller and Ghahramani proposed Bayesian hierarchical clustering (BHC), which has several advantages over traditional distance-based agglomerative clustering algorithms (Heller and Ghahramani, 2005). First, it defines a probablistic model for the data, which can be used to compute the predictive distribution of a test point and the probability of it belonging to any of the existing clusters in the tree, while conventional hierarchical clustering algorithms do not calculate the probability for each data point. Secondly, BHC uses a model-based criterion to decide on merging clusters rather than an ad-hoc distance metric. Thirdly, Bayesian hypothesis testing is used to decide which merges are advantageous and to output the recommended depth of the tree. Lastly, the algorithm can be interpreted as a novel fast bottom-up approximate inference method for a Dirichlet process mixture model. Cooke and colleagues developed a generative BHC algorithm for microarray time series that employs Gaussian process regression to capture the structure of the data (Cooke *et al.*, 2011). Darkins and colleagues improved BHC by using a randomised algorithm (Darkins *et al.*, 2013).

Medvedovic and Sivaganesan used a clustering procedure based on the Bayesian infinite mixture model and applied it to cluster gene expression profiles (Medvedovic and Sivaganesan, 2002). Clusters of genes with similar expression patterns were identified from the posterior distribution of clusterings, which was estimated by a Gibbs sampler. The posterior distribution of clusterings was summarised by calculating posterior pairwise probabilities of co-expression. In contrast to the finite mixture models, this method does not require specifying the number of clusters, and the resulting optimal clustering is obtained by averaging over models with all possible numbers of clusters. It represents a qualitative shift in the model-based cluster analysis of expression data because it allows for incorporation of uncertainties about the true number of components. The utility of this algorithm was investigated in the analysis of the yeast cell cycle data described by Cho and colleagues (1998). Soon afterwards, Medvedovic and colleagues developed different variants of Bayesian mixture-based clustering procedures for gene expression data with experimental replicates (Medvedovic, Yeung and Bumgarner, 2004). In this approach, clusters of co-expressed genes were created from the posterior distribution of clusterings, which was estimated by a Gibbs sampler. Then infinite Bayesian mixture models with different between-replicates variance structures were defined to analyse the real-world datasets. There are some worthy points of note about using this approach: (i) if the between-replicates variability is high, the improvement in precision achieved by performing only two experimental replicates could be dramatic; (ii) precise modelling of intra-gene variability is important for accurate identification of co-expressed genes; and (iii) the overall performance of the infinite mixture model with the 'elliptical' between-replicates variance structure was better than that of any of the other tested methods; it is also capable of identifying the underlying structure of data without knowing the 'correct' number of clusters.

19.6   Graph-based Clustering

Rives and Galitski investigated the organisation of interacting proteins and protein complexes into networks of modules and a network clustering method was developed to identify modules (Rives and Galitski, 2003). The yeast protein-interaction networks were represented as graphs of vertices and edges, corresponding to proteins and interactions, respectively. The network clustering method was developed based on the following ideas: (i) the shortest path between any two vertices is likely to be the most relevant one for functional associations and information transmission; (ii) each vertex in a network has a unique profile of shortest-path distances through the network to every other vertex; and (iii) module co-members are likely to have similar shortest-path distance profiles. Each edge in the biological network was assigned a length of one. An all-pairs-shortest-path distance matrix was calculated by using standard algorithms. The all-pairs-shortest-path matrix contains the length of the shortest path between every pair of vertices in the network. Each distance in the all-pairs-shortest-path matrix was transformed into an association as $1/d^2$, where d is the shortest-path distance. The resulting associations range from 0 to 1. The association of vertices that have no connecting path was defined as 0. The network clustering algorithm simply was hierarchical agglomerative average-linkage clustering. The association matrix was clustered identically in both dimensions. This method was validated by using functionally enriched high-throughput datasets. The network of yeast filamentation proteins was analysed by the network clustering algorithm and the complexity of the network was reduced to a small number of connected units of structure and function. This simplified representation facilitates the exploration of biological system properties in terms of interactions.

Wilkinson and Huberman developed a text-mining technique for the biological literature that produced detailed results while extracting simple data from each article abstract and title (Wilkinson and Huberman, 2004). The method created a network of gene symbol co-occurrences from Medline articles and partitions this network into communities. Gene symbol mentions were first extracted from almost 12.5 million Medline article titles and abstracts. Then sets of genes found to be statistically correlated to a set of user-selected keywords were selected. Networks were subsequently created from these sets of genes. In the network, each node represents a gene, and an edge connects two genes if they co-occur in at least one article. The degree distribution of the networks follows a power law. The network partitioning was based on the process of Girvan and Newman (2002), which was shown to give very good results for variety. The genes within a community were weighted, indicating how strongly they belonged to the community. The communities produced in the case of colon cancer gave one insight into the function of the component genes.

The identification of functional modules from genome-wide information, such as transcription profiles or protein interactions, is an important goal of functional genomics. Pereira-Leal and colleagues successfully isolated 1046 functional modules from the known protein-interaction network of budding yeast involving 8046 individual pair-wise interactions by using an entirely automated and unsupervised graph-clustering algorithm (Pereira-Leal, Enright and Ouzounis, 2004). This system biology approach was able to detect many well-known protein complexes or biological processes, without reference to any additional information. Since the objective was to isolate functionally coordinated interactions, the algorithm took a graph, consisting of edges connecting nodes, and produces its associated line graph, in which edges represent nodes and nodes represent edges. The line graph generated has a number of advantages for graph clustering: (i) it does not sacrifice information content because the original bidirectional network can be recovered, (ii) it takes into account the higher-order local neighbourhood of interactions, and (iii) hence, it is more highly structured than the original graph. The protein-interaction network is derived from the yeast subset of the Database of Interacting Proteins (DIP), which is appropriate because it contains curated interactions from both small- and large-scale experimental studies. A weighted network of proteins connected by interactions was formed based on the data, where the weights qualitatively reflected the confidence that was attributed to each interaction based on the amount of experimental evidence supporting it. An algorithm, called TribeMCL, which has been introduced in Chapter 16, was used for graph clustering by graph flow simulation to cluster the interaction network and recover clusters of associated interactions. Clusters were validated by assessing the consistency of protein classifications within an individual cluster. Many detected clusters represented previously characterised functional modules. One example was a high-scoring cluster containing the mitochondrial F1-F0 ATP synthase complex. In this cluster, all known protein interactions related to the formation of this complex were recovered.

19.7 Consensus Clustering

Cluster analysis of microarray datasets may suffer from lack of inter-method consistency in assigning related gene expression profiles to clusters. Obtaining a consensus set of clusters from a number of clustering methods may improve confidence in gene expression analysis (Abu-Jamous *et al.*, 2013). Monti and colleagues developed a methodology of class discovery and clustering validation tailed to the task of analysing gene expression data (Monti *et al.*, 2003). Monti's resampling algorithm has been discussed in Chapter 17. In their study, the resampling-based consensus clustering was investigated in six real gene expression datasets, namely leukaemia dataset (Golub *et al.*, 1999), Novartis multi-tissue (Su *et al.*, 2002), St. Jude leukaemia (Yeoh *et al.*, 2002), lung cancer (Bhattacharjee *et al.*, 2001), central nervous system tumours (Pomeroy *et al.*, 2002), and normal tissue (Ramaswamy *et al.*, 2001). In general, when applied to gene expression data, consensus clustering with hierarchical clustering outperformed consensus clustering with SOM.

Swift and colleagues developed a consensus clustering algorithm, which provided the advantage of improving confidence (Swift *et al.*, 2004). The weighted-kappa metric, which was originally proposed by Cohen (1968), was used as a direct measure of similarity of partitions. A consensus strategy was applied both to produce robust and consensus clustering of gene expression data and to assign statistical significance to these clusters from known gene functions. The method is different from the resampling method (Monti *et al.*, 2003), which we discussed before, in that different clustering algorithms are used rather than perturbing the gene expression data for a single algorithm. Using consensus clustering with probabilistic measures of cluster membership derived from external validation with gene function annotations, specific transcriptionally co-regulated genes from microarray data of distinct B-cell lymphoma types (Jenner *et al.*, 2003) were identified accurately and rapidly.

Brannon and colleagues analysed gene expression microarray data using software that implements iterative unsupervised consensus clustering algorithms to identify the optimal molecular subclasses, without clinical or other classified information (Brannon *et al.*, 2010). Clear cell renal cell carcinoma (ccRCC) is the predominant RCC subtype, but even within this classification the natural history is heterogeneous and difficult to predict. ConsensusCluster was proposed by Seiler and colleagues (2010) for the analysis of high-dimensional single-nucleotide polymorphism (SNP) and gene expression microarray data. The software implemented the consensus clustering algorithm and principal component analysis (PCA) to stratify the data into a given number of robust clusters. The robustness is achieved by combining clustering results from data and sample resampling as well as by averaging over various algorithms and parameter settings to achieve accurate, stable clustering results. Several different clustering algorithms have been implemented, including *k*-means, partitioning around medoids (PAM), SOM, and hierarchical clustering methods. After clustering the data, ConsensusCluster generates a consensus matrix heat map to give a useful visual representation of cluster membership, and automatically generates a log of selected features that distinguish each pair of clusters. Such consensus clustering analysis identified two distinct subtypes of ccRCC, designated clear cell types A and B. In each subtype, logical analysis of data defined a small, highly predictive gene set that could then be used to classify additional tumours individually. The subclasses were corroborated in a validation dataset of 177 tumours and analysed for clinical outcome. Based on individual tumour assignment, tumours designated type A had markedly improved disease-specific survival compared with type B. Using patterns of gene expression based on a defined gene set, ccRCC was classified into two robust subclasses based on inherent molecular features that ultimately corresponded to marked differences in clinical outcome. This classification schema thus provided a molecular stratification applicable to individual tumours that may have implications to influence treatment decisions, define biological mechanisms involved in ccRCC tumour progression, and direct future drug discovery.

19.8   Biclustering

Although the concept of biclustering may trace back to the 1970s, it did not attract much attention until it was successfully applied in bioinformatics, particularly to cluster the gene expression datasets in both dimensions; that is, genes and samples (Cheng and Church, 2000). Thereafter, there has been a rich literature of applications for data analysis using biclustering, and most algorithms, which were listed in Table 18.1, have been applied to the applications in bioinformatics. We will only choose a few examples to discuss. Interested readers are suggested to refer to the references in Chapter 18.

Tchagang and colleagues employed a biclustering algorithm for group biomarkers identification using microarray gene expression data of ovarian cancer (Tchagang et al., 2008). Ovarian cancer is a deadly disease because it is not usually diagnosed until it has reached an advanced stage. The objective of this study was to identify group biomarkers that can be used for the diagnosis of early-stage and/or recurrent ovarian cancer. Group biomarkers were extracted from the gene expression data of over 400 normal, cancerous and diseased tissues by using robust biclustering, which was proposed Tchagang and Tewfik (2006). The gene expression matrix was organised as three sub-matrices: matrix A is a 12651×62 matrix that represents the gene expression of the 62 normal ovary tissue samples and 12651 known genes using the Affymetric GeneChip HG_U95A; matrix $B = [B_1, B_2, B_3]$ is a 12651×45 matrix that represents the gene expression of the 45 ovarian cancer tissues; and matrix C is a 12651×319 matrix that represents the gene expression of the 319 non-ovarian tissues. Biomarkers specific for ovarian cancer should be highly expressed in ovarian cancer samples and low or absent in other samples, including normal ovaries and non-ovarian tissues. For each sub-matrix, the robust biclustering algorithm was used to identify biclusters with constant values; that is, a subset of genes whose expression level stays constant across a subset of conditions or tissue samples. The receiver operating characteristics (ROC) curves were also employed to identify single or group biomarkers. In this study, both healthy ovarian and non-ovarian tissues were considered in the definition of specificity. Using these methods, the study identified many significant patterns that encode for secreted proteins, membrane proteins, and extracellular matrix proteins, which clearly discriminate between the gene expression data of ovarian cancer, normal ovary and non-ovarian tissues.

Huttenhower and colleagues proposed the combinational algorithm for expression and sequence-based cluster extraction (COALESCE) system for regulatory module prediction

(Huttenhower *et al.*, 2009). The assumptions, which are valid in unicellular systems, imply that regulation often occurs based on well defined transcriptional factor binding sites and discrete activation or regression of transcription. However, the difficulty for complex organisms is that these assumptions no long hold and predicting regulatory modules from expression data becomes an increasingly challenging problem. The objective of this study was to discover expression biclusters and putative regulatory motifs in metazoan genomes and very large microarray conditions. By using Bayesian data integration, diverse supporting data types such as evolutionary conservation or nucleosome placement can be included as a potential method to enhance the prediction ability. The basic COALESCE algorithm consumes gene expression and DNA sequence data as input to produce putative co-regulated modules as output. Each resulting module consists of a set of co-regulated genes, one or more expression conditions under which they are co-expressed, and zero or more motifs predicted to drive the co-regulation. The algorithm finds modules in a serial manner by seeding each new module with a set of co-expressed genes and iteratively refining the module to convergence. COALESCE was evaluated using both real datasets and synthetic datasets and the results showed its effectiveness in predicting regulatory modules.

Bryan and colleagues were the first who applied the technique of biclustering to model functional modules within an integrated microRNA (miRNA)-mRNA association matrix (Bryan *et al.*, 2014). MiRNAs are small non-coding RNA molecules that regulate gene expression at a post-transcriptional level through the binding of their seed region to complementary sequences in the 3' UTR of a target mRNA followed by subsequent degradation and/or translational inhibition of the mRNA transcript. Given that miRNA-directed binding of RNA-induced silencing complex may result in the degradation of the target mRNA, it is possible to determine miRNA targets by examining significant inverse correlations between miRNA and mRNA expression data. The first step of the overall analysis pipeline in the study (Bryan *et al.*, 2014) was to integrate miRNA and mRNA expression datasets and compile a two-dimensional matrix for all possible combinations of miRNA and mRNA expression. In the ideal case, the matrix contains values −1, representing direct inversely correlated interactions, +1, representing indirect positively correlated interactions, and 0, representing no correlations. A biclustering algorithm based on the simulated annealing search method was used to locate highly correlated biclusters with an miRNA-mRNA association matrix. This method was applied to investigate the interplay of miRNAs and mRNAs in integrated datasets derived from neuroblastoma and human immune cells. Results provided evidence of an extensive modular miRNA functional network and enabled characterisation of miRNA function and dysregulation in disease.

19.9   Summary

To sum up, we studied eight clustering families in Part Four and many algorithms in these families have been applied in the field of bioinformatics. In this chapter, which is the last chapter in Part Four, we discussed many examples in bioinformatics using different clustering families; for example, to discover new diseases by sample-based clustering, to discover new annotations for genes or biomarkers by gene clustering, to discover modules in a biological network by graph-based clustering, and to discover the regulatory properties by DNA sequence clustering. These clustering algorithms, as one kind of primary exploratory tool, play a significantly important role in biological data analysis. Nonetheless, it is worth noting that clustering is only one of the processing stages in the overall pipeline of data analysis. In Part Five, we will introduce other important stages in the pipeline, namely validation and visualisation.

References

Abu-Jamous, B., Fa, R., Roberts, D.J. and Nandi, A.K. (2013). Paradigm of tunable clustering using binarization of consensus partition matrices (Bi-CoPaM) for gene discovery. *Plos One*, **8**(2), e56432.

Alizadeh, A.A., Eisen, M.B., Davis, R.E. *et al.* (2000). Distinct types of diffuse large B-cell lymphoma identified by gene expression profiling. *Nature*, **403**(6769), pp. 503–511.

Alon, U., Barkai, N., Notterman, D.A. *et al.* (1999). Broad patterns of gene expression revealed by clustering analysis of tumor and normal colon tissues probed by oligonucleotide arrays. *Proceedings of the National Academy of Sciences*, **96**(12), pp. 6745–6750.

Andrews, J.L. and McNicholas, P.D. (2011). Extending mixtures of multivariate t-factor analyzers. *Statistics and Computing*, **21**(3), pp. 361–373.

Baek, J. and McLachlan, G.J. (2011). Mixtures of common t-factor analyzers for clustering high-dimensional microarray data. *Bioinformatics*, **27**(9), pp. 1269–1276.

Baek, J., McLachlan, G.J. and Flack, L.K. (2010). Mixtures of factor analyzers with common factor loadings: applications to the clustering and visualization of high-dimensional data. *IEEE Transactions on Pattern Analysis and Machine Intelligence*, **32**(7), pp. 1298–1309.

Bhattacharjee, A., Richards, W.G., Staunton, J. *et al.* (2001). Classification of human lung carcinomas by mRNA expression profiling reveals distinct adenocarcinoma subclasses. *Proceedings of the National Academy of Sciences*, **98**(24), pp. 13790–13795.

Bittner, M., Meltzer, P., Chen, Y. *et al.* (2000). Molecular classification of cutaneous malignant melanoma by gene expression profiling. *Nature*, **406**(6795), pp. 536–540.

Brannon, A.R., Reddy, A., Seiler, M. *et al.* (2010). Molecular stratification of clear cell renal cell carcinoma by consensus clustering reveals distinct subtypes and survival patterns. *Genes and Cancer*, **1**(2), pp. 152–163.

Bryan, K., Terrile, M., Bray, I.M. *et al.* (2014). Discovery and visualization of miRNA-mRNA functional modules within integrated data using bicluster analysis. *Nucleic Acids Research*, **42**(3), e17.

Cao, Y. and Wu, J. (2002). Projective ART for clustering data sets in high dimensional spaces. *Neural Networks*, **15**(1), pp. 105–120.

Chavez-Alvarez, R., Chavoya, A. and Mendez-Vazquez, A. (2014). Discovery of possible gene relationships through the application of self-organizing maps to DNA microarray databases. *PloS One*, **9**(4), e93233.

Chen, X., Cheung, S.T., So, S. *et al.* (2002). Gene expression patterns in human liver cancers. *Molecular Biology of the Cell*, **13**(6), pp. 1929–1939.

Cheng, Y. and Church, G.M. (2000). *Biclustering of Expression Data*. ISMB, Boston, MA, pp. 93–103.

Cho, R.J., Campbell, M.J., Winzeler, E.A. *et al.* (1998). A genome-wide transcriptional analysis of the mitotic cell cycle. *Molecular Cell*, **2**(1), pp. 65–73.

Chowdary, D., Lathrop, J., Skelton, J. *et al.* (2006). Prognostic gene expression signatures can be measured in tissues collected in RNAlater preservative. *Journal of Molecular Diagnostics*, **8**(1), pp. 31–39.

Chu, S., DeRisi, J., Eisen, M. *et al.* (1998). The transcriptional program of sporulation in budding yeast. *Science*, **282**(5389), pp. 699–705.

Cohen, J. (1968). Weighted kappa: nominal scale agreement provision for scaled disagreement or partial credit. *Psychological Bulletin*, **70**(4), p. 213–220.

Cooke, E.J., Savage, R.S., Kirk, P.D. *et al.* (2011). Bayesian hierarchical clustering for microarray time series data with replicates and outlier measurements. *BMC Bioinformatics*, **12**(1), p. 399.

Darkins, R., Cooke, E.J., Ghahramani, Z. *et al.* (2013). Accelerating Bayesian hierarchical clustering of time series data with a randomised algorithm. *PloS One*, **8**(4), e59795.

Datta, S. and Datta, S. (2003). Comparisons and validation of statistical clustering techniques for microarray gene expression data. *Bioinformatics*, **19**(4), pp. 459–466.

Dembele, D. and Kastner, P. (2003). Fuzzy C-means method for clustering microarray data. *Bioinformatics*, **19**(8), pp. 973–980.

Dhanasekaran, S.M., Barrette, T.R., Ghosh, D. *et al.* (2001). Delineation of prognostic biomarkers in prostate cancer. *Nature,* **412**(6849), pp. 822–826.

Eisen, M.B., Spellman, P.T., Brown, P.O. and Botstein, D. (1998). Cluster analysis and display of genome-wide expression patterns. *Proceedings of the National Academy of Sciences,* **95**(25), pp. 14863–14868.

Fraley, C. and Raftery, A.E. (1999). MCLUST: software for model-based cluster analysis. *Journal of Classification,* **16**(2), pp. 297–306.

Gasch, A.P. and Eisen, M.B. (2002). Exploring the conditional coregulation of yeast gene expression through fuzzy k-means clustering. *Genome Biology,* **3**(11), research0059.1–0059.22.

Gasch, A.P., Spellman, P.T., Kao, C.M. *et al.* (2000). Genomic expression programs in the response of yeast cells to environmental changes. *Molecular Biology of the Cell,* **11**(12), pp. 4241–4257.

Gasch, A.P., Huang, M., Metzner, S. *et al.* (2001). Genomic expression responses to DNA-damaging agents and the regulatory role of the yeast ATR homolog Mec1p. *Molecular Biology of the Cell,* **12**(10), pp. 2987–3003.

Ghosh, D. and Chinnaiyan, A.M. (2002). Mixture modelling of gene expression data from microarray experiments. *Bioinformatics,* **18**(2), pp. 275–286.

Girvan, M. and Newman, J.M.E. (2002). Community structure in social and biological networks. *Proceedings of the National Academy of Sciences,* **99**(12), pp. 7821–7826.

Golub, T.R., Slonim, D.K., Tamayo, P. *et al.* (1999). Molecular classification of cancer: class discovery and class prediction by gene expression monitoring. *Science,* **286**(5439), pp. 531–537.

Heller, K.A. and Ghahramani, Z. (2005). Bayesian Hierarchical Clustering. Proceedings of the 22nd International Conference on Machine Learning, 7–11 August 2005, Bonn, ACM, New York, pp. 297–304.

Herschkowitz, J.I., Simin, K., Weigman, V.J. *et al.* (2007). Identification of conserved gene expression features between murine mammary carcinoma models and human breast tumors. *Genome Biology,* **8**(5), R76.

Hsu, A.L., Tang, S.-L. and Halgamuge, S.K. (2003). An unsupervised hierarchical dynamic self-organizing approach to cancer class discovery and marker gene identification in microarray data. *Bioinformatics,* **19**(16), pp. 2131–2140.

Huttenhower, C., Mutungu, K.T., Indik, N. *et al.* (2009). Detailing regulatory networks through large scale data integration. *Bioinformatics,* **25**(24), pp. 3267–3274.

Iyer, V.R., Eisen, M.B., Ross, D.T. *et al.* (1999). The transcriptional program in the response of human fibroblasts to serum. *Science,* **283**(5398), pp. 83–87.

Jenner, R.G., Maillard, K., Cattini, N. *et al.* (2003). Kaposi's sarcoma-associated herpesvirus-infected primary effusion lymphoma has a plasma cell gene expression profile. *Proceedings of the National Academy of Sciences,* **100**(18), pp. 10399–10404.

Kim, S.Y., Lee, J.W. and Bae, J.S. (2006). Effect of data normalization on fuzzy clustering of DNA microarray data. *BMC Bioinformatics,* **7**, p. 134.

Kohonen, T. (1990). The self-organizing map. *Proceedings of the IEEE,* **78**(9), pp. 1464–1480.

Kohonen, T. (1997). *Self-Organizing Maps,* Springer-Verlag, New York.

Lyons, T.J., Gasch, A.P., Gaither, L.A. *et al.* (2000). Genome-wide characterization of the Zap1p zinc-responsive regulon in yeast. *Proceedings of the National Academy of Sciences,* **97**(14), pp. 7957–7962.

McLachlan, G.J., Bean, R. and Peel, D. (2002). A mixture model-based approach to the clustering of microarray expression data. *Bioinformatics,* **18**(3), pp. 413–422.

McNicholas, P.D. and Murphy, T.B. (2008). Parsimonious Gaussian mixture models. *Statistics and Computing,* **18**(3), pp. 285–296.

McNicholas, P.D. and Murphy, T.B. (2010). Model-based clustering of microarray expression data via latent Gaussian mixture models. *Bioinformatics,* **26**(21), pp. 2705–2712.

Medvedovic, M. and Sivaganesan, S. (2002). Bayesian infinite mixture model based clustering of gene expression profiles. *Bioinformatics,* **18**(9), pp. 1194–1206.

Medvedovic, M., Yeung, K.Y. and Bumgarner, R.E. (2004). Bayesian mixture model based clustering of replicated microarray data. *Bioinformatics,* **20**(8), pp. 1222–1232.

Monti, S., Tamayo, P., Mesirov, J. and Golub, T. (2003). Consensus clustering: a resampling-based method for class discovery and visualization of gene expression microarray data. *Machine Learning,* **52**(1–2), pp. 91–118.

Monti, S., Savage, K.J., Kutok, J.L. *et al.* (2005). Molecular profiling of diffuse large B-cell lymphoma identifies robust subtypes including one characterized by host inflammatory response. *Blood,* **105**(5), pp. 1851–1861.

Ogawa, N., DeRisi, J. and Brown, P.O. (2000). New components of a system for phosphate accumulation and polyphosphate metabolism in Saccharomyces cerevisiae revealed by genomic expression analysis. *Molecular Biology of the Cell*, **11**(12), pp. 4309–4321.

Pereira-Leal, J.B., Enright, A.J. and Ouzounis, C.A. (2004). Detection of functional modules from protein interaction networks. *PROTEINS: Structure, Function, and Bioinformatics*, **54**(1), pp. 49–57.

Perou, C.M., Sørlie, T., Eisen, M.B. *et al.* (2000). Molecular portraits of human breast tumours. *Nature*, **406**(6797), pp. 747–752.

Pickert, L., Reuter, I., Klawonn, F. and Wingender, E. (1998). Transcription regulatory region analysis using signal detection and fuzzy clustering. *Bioinformatics*, **14**(3), pp. 244–251.

Pomeroy, S.L., Tamayo, P., Gaasenbeek, M. *et al.* (2002). Prediction of central nervous system embryonal tumour outcome based on gene expression. *Nature*, **415**(6870), pp. 436–442.

Pramila, T., Wu, W., Miles, S. *et al.* (2006). The Forkhead transcription factor Hcm1 regulates chromosome segregation genes and fills the S-phase gap in the transcriptional circuitry of the cell cycle. *Genes and Development*, **20**(16), pp. 2266–2278.

Ramaswamy, S., Tamayo, P., Rifkin, R. *et al.* (2001). Multiclass cancer diagnosis using tumor gene expression signatures. *Proceedings of the National Academy of Sciences*, **98**(26), pp. 15149–15154.

Ramoni, M.F., Sebastiani, P. and Kohane, I.S. (2002). Cluster analysis of gene expression dynamics. *Proceedings of the National Academy of Sciences*, **99**(14), pp. 9121–9126.

Rives, A.W. and Galitski, T. (2003). Modular organization of cellular networks. *Proceedings of the National Academy of Sciences*, **100**(3), pp. 1128–1133.

Ross, D.T., Scherf, U., Eisen, M.B. *et al.* (2000). Systematic variation in gene expression patterns in human cancer cell lines. *Nature Genetics*, **24**(3), pp. 227–235.

Salem, S.A., Jack, L.B. and Nandi, A.K. (2008). Investigation of self-organizing oscillator networks for use in clustering microarray data. *IEEE Transactions on Nanobioscience*, **7**, pp. 65–79.

Schummer, M., Ng, W.V., Bumgarner, R.E. *et al.* (1999). Comparative hybridization of an array of 21 500 ovarian cDNAs for the discovery of genes overexpressed in ovarian carcinomas. *Gene*, **238**(2), pp. 375–385.

Seiler, M., Huang, C.C., Szalma, S. and Bhanot, G. (2010). ConsensusCluster: a software tool for unsupervised cluster discovery in numerical data. *OMICS: A Journal of Integrative Biology*, **14**(1), pp. 109–113.

Shipp, M.A., Ross, K.N., Tamayo, P. *et al.* (2002). Diffuse large B-cell lymphoma outcome prediction by gene-expression profiling and supervised machine learning. *Nature Medicine*, **8**(1), pp. 68–74.

Sneath, P.H. and Sokal, R.R. (1973). *Numerical Taxonomy. The Principles and Practice of Numerical Classification*, W.H.Freeman & Co Ltd, New York.

Sørlie, T., Perou, C.M., Tibshirani, R. *et al.* (2001). Gene expression patterns of breast carcinomas distinguish tumor subclasses with clinical implications. *Proceedings of the National Academy of Sciences*, **98**(19), pp. 10869–10874.

Spellman, P.T., Sherlock, G., Zhang, M.Q. *et al.* (1998). Comprehensive identification of cell cycle-regulated genes of the yeast Saccharomyces cerevisiae by microarray hybridization. *Molecular Biology of the Cell*, **9**(12), pp. 3273–3297.

Su, A.I., Cooke, M.P., Ching, K.A. *et al.* (2002). Large-scale analysis of the human and mouse transcriptomes. *Proceedings of the National Academy of Sciences*, **99**(7), pp. 4465–4470.

Swift, S., Tucker, A., Vinciotti, V. *et al.* (2004). Consensus clustering and functional interpretation of gene-expression data. *Genome Biology*, **5**(11), R94.

Takahashi, H., Kobayashi, T. and Honda, H. (2005). Construction of robust prognostic predictors by using projective adaptive resonance theory as a gene filtering method. *Bioinformatics*, **21**(2), pp. 179–186.

Takahashi, H., Nemoto, T., Yoshida, T. *et al.* (2006). Cancer diagnosis marker extraction for soft tissue sarcomas based on gene expression profiling data by using projective adaptive resonance theory (PART) filtering method. *BMC Bioinformatics*, **7**, p. 399.

Takahashi, H., Nakayama, R., Hayashi, S. *et al.* (2013). Macrophage migration inhibitory factor and stearoyl-CoA desaturase 1: potential prognostic markers for soft tissue sarcomas based on bioinformatics analyses. *PloS One*, **8**(10), e78250.

Tamayo, P., Slonim, D., Mesirov, J. *et al.* (1999). Interpreting patterns of gene expression with self-organizing maps: methods and application to hematopoietic differentiation. *Proceedings of the National Academy of Sciences*, **96**(6), pp. 2907–2912.

Tchagang, A.B. and Tewfik, A.H. (2006). DNA microarray data analysis: a novel biclustering algorithm approach. *EURASIP Journal on Advances in Signal Processing*, **2006**, 59809, pp. 1–12.

Tchagang, A.B., Tewfik, A.H., DeRycke, M.S. *et al.* (2008). Early detection of ovarian cancer using group bio-markers. *Molecular Cancer Therapeutics*, **7**(1), pp. 27–37.

Tomida, S., Hanai, T., Honda, H. and Kobayashi, T. (2002). Analysis of expression profile using fuzzy adaptive resonance theory. *Bioinformatics*, **18**(8), pp. 1073–1083.

Wang, J., Bø, T.H., Jonassen, I. *et al.* (2003). Tumor classification and marker gene prediction by feature selection and fuzzy c-means clustering using microarray data. *BMC Bioinformatics*, **4**, p. 60.

Wilkinson, D.M. and Huberman, B.A. (2004). A method for finding communities of related genes. *Proceedings of the National Academy of Sciences*, **101**(suppl 1), pp. 5241–5248.

Wu, S., Liew, A.-C., Yan, H. and Yang, M. (2004). Cluster analysis of gene expression data based on self-splitting and merging competitive learning. *IEEE Transactions on Information Technology in Biomedicine*, **8**(1), pp. 5–15.

Yeoh, E.-J., Ross, M.E., Shurtleff, S.A. *et al.* (2002). Classification, subtype discovery, and prediction of outcome in pediatric acute lymphoblastic leukemia by gene expression profiling. *Cancer Cell*, **1**(2), pp. 133–143.

Yeung, K.Y., Fraley, C., Murua, A. *et al.* (2001). Model-based clustering and data transformations for gene expression data. *Bioinformatics*, **17**(10), pp. 977–987.

Part Five

Validation and Visualisation

20

Numerical Validation

20.1 Introduction

Part Four systematically introduced and discussed many widely used clustering methods in bioinformatics. Now let us look at another important aspect of clustering analysis, namely validation. This chapter will describe numerical validation techniques and the next chapter will introduce biological validation techniques.

As we have mentioned in Part Four, clustering is also known as unsupervised learning. Therefore, there is no existing guideline to guarantee an optimal clustering and it is still an open question how to tell that a clustering algorithm or a clustering result is better. Thus, the task of assessing the results of clustering algorithms can be as important as the clustering algorithms themselves. The procedure for evaluating clustering algorithms and their results is known as cluster validation (CV) (Halkidi, Batistakis and Vazirgiannis, 2002a, b). There have been a lot of CV algorithms in the literature since the 1960s (Milligan and Cooper, 1985). Most CV algorithms can be classified into three classes (Halkidi, Batistakis and Vazirgiannis, 2002a, b), namely *external criteria, internal criteria* and *relative criteria.*

External criteria imply that the results of a clustering algorithm are evaluated based on a pre-specified structure, which is imposed on a dataset to reflect the clustering structure of the dataset. In other words, external criteria rely on the *a priori* knowledge or *ground truth* about the dataset. We will introduce four indices in this class, namely Rand index (RI) and its derivative adjusted Rand index (ARI), Jaccard index (JI) and normalised mutual information (NMI).

Internal criteria evaluate the clustering algorithms in terms of the inner structures of the datasets themselves, rather than *a priori* knowledge. The common strategy of this class of algorithms is re-sampling. This class of algorithms may have good estimates of the number of clusters in a dataset and also have good indication of the effectiveness of clustering algorithms. We will introduce figure of merit (FOM) and CLEST as criteria of this class.

Integrative Cluster Analysis in Bioinformatics, First Edition. Basel Abu-Jamous, Rui Fa and Asoke K. Nandi.
© 2015 John Wiley & Sons, Ltd. Published 2015 by John Wiley & Sons, Ltd.

Relative criteria evaluate the clustering partitions by the relative relationship between compactness and separation. Different from both external criterion and internal criterion, there are two ingredients for relative criterion, one of which is that clustering partitions, rather than clustering algorithms, are evaluated in terms of the index values, and the other is that the index values are relative values of compactness versus separation. Relative criteria can be further classified into many subclasses including model-based indices, fuzzy validity indices and crisp validity indices. We introduce minimum message length (MML), minimum description length (MDL), Bayesian information criterion (BIC) and Akaike's information criterion (AIC) in model-based indices. In fuzzy validity indices, we will introduce partition coefficient (PC), partition entropy (PE) (PC and PE are two indices that involve only the membership values), Fukuyama–Sugeno (FS) index and Xie–Beni (XB) index. Crisp validity indices include Calinski–Harabasz (CH) index, Dunn's index (DI), Davies–Bouldin (DB) index (DB is the counterpart of XB in crisp clustering), I index (II), silhouette, object-based validation (OBV-LDA) (where LDA stands for linear discriminant analysis), the geometrical index (GI) and the validity index (VI).

Suppose that N gene expression data objects are formalised as numerical vectors $x_n = \{x_{nm} | m = 1, \ldots, M; n = 1, \ldots, N\}$, where M is the number of features (here, 'features' represent different experimental conditions or different time points or different samples from different tissues/organisms) and x_{nm} is the value of the mth feature for the nth object. We define $Z = \{z_n | n = 1, \ldots, N\}$ as a partition and $z_n \in [1, \ldots, K]$, where K is the number of clusters. Partitions can also be represented as a partition matrix $U^{N \times K} = \{u_{k,n} | k = 1, \ldots, K; n = 1, \ldots, N\}$. In the crisp partitions, the nth gene belongs to the kth cluster if the binary entry $u_{k,n}$ of the partition matrix is one, otherwise the gene does not; while in fuzzy partitions, the entry $u_{k,n} \in [0,1]$ represents the degree to which the nth gene belongs to the kth cluster. Additionally, partitions can be represented as a cluster set $C = \{C_1, \ldots, C_K\}$, where C_k denotes the kth cluster and $k = 1, \ldots, K$. The number of members in the kth cluster is n_k.

20.2   External Criteria

20.2.1   Rand Index

RI was proposed by Rand (1971) to measure the similarity between two partitions. Given the ground truth as the reference partition, we may calculate the index value from which we are able to find how far the target partition is away from the ground truth. Suppose that we have the reference partition $Z^r = \{z_i^r | i = 1, \ldots, N\}$ and $z_i^r \in [1, \ldots, K^r]$, where K^r is the number of clusters of the reference partition.

Before calculating RI, we have to define a few more variables as follows:

1. **TP** denotes true positives, the number of pairs of objects that are in the same cluster in both Z^r and Z;
2. **TN** denotes true negatives, the number of pairs of objects that are in different clusters in both Z^r and Z;
3. **FN** denotes false negatives, the number of pairs of objects that are in the same cluster in Z^r but in different clusters in Z;
4. **FP** denotes false positives, the number of pairs of objects that are in different clusters in Z^r but in the same cluster in Z.

Thus, the RI is given by Equation (20.1).

$$RI = \frac{TP+TN}{TP+TN+FN+FP} = \frac{TP+TN}{\binom{N}{2}} \tag{20.1}$$

Apparently, $(TP + TN)$ is considered as the degree of agreement and $(FN + FP)$ is considered as the degree of disagreement. The RI has a value between 0 and 1, with 0 indicating that the two partitions do not agree on any pair of objects, and 1 indicating that the partitions are exactly the same.

20.2.2 Adjusted Rand Index

The ARI was proposed by Hubert and Arabie (1985) assuming that the generalised hypergeometric distribution is considered as the model of randomness.

Suppose that the reference cluster set is $C^r = \left\{ C_k^r | k = 1, ..., K^r \right\}$; we define a contingency table (Table 20.1)

Table 20.1 Contingency table

$C \backslash C^r$	C_1^r	C_2^r	$\cdots$	$C_{K^r}^r$	Sums
C_1	n_{11}	n_{12}	$\cdots$	n_{1K^r}	a_1
C_2	n_{21}	n_{22}	$\cdots$	n_{2K^r}	a_2
$\vdots$	$\vdots$	$\vdots$	$\ddots$	$\vdots$	$\vdots$
C_K	n_{K1}	n_{K2}	$\cdots$	n_{KK^r}	a_K
Sums	b_1	b_2	$\cdots$	b_{K^r}	N

where $n_{ij} = |C_i \cap C_j^r|$ and $|S|$ denotes the cardinality, which is the number of members of the set S. The ARI thus is given by Equation (20.2).

$$ARI = \frac{Index - Expected\ Index}{MaxIndex - Expected\ Index}$$

$$= \frac{\sum_{ij} \binom{n_{ij}}{2} - \left[\sum_i \binom{a_i}{2} \sum_j \binom{b_j}{2} \right] / \binom{N}{2}}{\frac{1}{2} \left[\sum_i \binom{a_i}{2} + \sum_j \binom{b_j}{2} \right] - \left[\sum_i \binom{a_i}{2} \sum_j \binom{b_j}{2} \right] / \binom{N}{2}} \tag{20.2}$$

Different from the RI, the ARI can yield negative values if the index is less than the expected index.

20.2.3   Jaccard Index

The JI, also known as Jaccard coefficient, measures the similarity between two partitions. The JI is defined as the cardinality of the intersection of two partitions divided by the cardinality of their union, which is written as Equation (20.3).

$$JI = \frac{|C^r \cap C|}{|C^r \cup C|} \tag{20.3}$$

Alternatively, one can calculate the Jaccard distance (JD), which measures the dissimilarity between two partitions, written as Equation (20.4),

$$JD = \frac{|C^r \cup C| - |C^r \cap C|}{|C^r \cup C|} \tag{20.4}$$

and the JI is equal to (1 − JD). Using MATLAB, we can obtain the JD easily by the function *pdist(…, 'jaccard')* and then JI = 1 − JD.

20.2.4   Normalised Mutual Information

Mutual information is a symmetric measure to quantify the statistical information shared between two distributions. It provides a sound indication of the shared information between a pair of partitions. Let $I(C^r, C)$ denote the mutual information between reference partition C^r and target partition C. Since $I(C^r, C)$ has no upper bound, a normalised version of $I(C^r, C)$ is desired. Let us write $I(C^r, C)$ as in Equation (20.5),

$$I(C^r, C) = H(C^r) + H(C) - H(C^r, C) \tag{20.5}$$

where $H(\cdot)$ denotes the marginal entropy and $H(\cdot, \cdot)$ is the joint entropy. Thus, the NMI is given by Equation (20.6).

$$NMI = \frac{I(C^r, C)}{\sqrt{H(C^r)H(C)}} \tag{20.6}$$

The NMI ranges from 0 to 1.

20.3   Internal Criteria

20.3.1   Adjusted Figure of Merit

The adjusted Figure of Merit (adjusted FOM) was developed based on the FOM (Yeung, Haynor and Ruzzo, 2001). It is worth noting that adjusted FOM is used to evaluate the clustering algorithms rather than the partitions. The basic idea of adjusted FOM is similar to leave-one-out cross-validation in machine learning and a small adjusted FOM value indicates a strong prediction power of a clustering algorithm. Suppose that we have a clustering algorithm to evaluate; we apply the clustering algorithm to all but one feature in the dataset and use the left-out feature to assess the predictive power of the clustering algorithm.

Suppose that the clustering algorithm is applied to the dataset with $1, \ldots, (m-1)$, $(m+1), \ldots, M$ features and leaves the mth feature out. We group the dataset into k clusters $C = \{C_1, \ldots, C_k\}$. Let $E(n, m)$ be the expression of the mth feature of the nth gene in the original dataset. We denote $\mu_{c_i}(m)$ as the average expression of the mth feature in the ith cluster. We may obtain the FOM for all genes at the mth feature with k clusters $F(m, k)$ as shown in Equation (20.7).

$$F(m,k) = \sqrt{\frac{1}{N} \sum_{i=1}^{k} \sum_{n \in C_i} \left[E(n,m) - \mu_{c_i}(m) \right]^2} \tag{20.7}$$

Thus, we obtain the FOM as an estimate of the total predictive power of the clustering algorithm over all features with k clusters in the datasets, written as given in Equation (20.8).

$$FOM(k) = \sum_{m=1}^{M} F(m,k) \tag{20.8}$$

The FOM has a critical problem, however, which is that increasing the number of clusters tends to decrease the FOM. To deal with this problem, the adjusted FOM (AFOM) was defined to correct the FOM, written as shown in Equation (20.9).

$$AFOM(k) = \frac{FOM(k)}{\sqrt{(N-k)/N}} \tag{20.9}$$

There are two strengths of the adjusted FOM: one is that it is able to evaluate the clustering algorithm in a given dataset in the fashion of cross-validation; another one is that it is accurate in estimating the number of clusters of the dataset. The adjusted FOM also has two drawbacks: one is that it is relatively complex due to its re-sampling nature; another is that since the curve of the adjusted FOM against the number of clusters has a 'knee' shape, determining the best number of clusters is somewhat subjective.

20.3.2 CLEST

CLEST was proposed by Dudoit and Fridlyand to estimate the number of clusters for microarray datasets (Dudoit and Fridlyand, 2002). It randomly and iteratively partitions the original dataset in a learning set $\mathcal{L}$ and a training set $\mathcal{T}$. The learning set $\mathcal{L}$ is partitioned by the clustering algorithm into $C(\mathcal{L})$, and the partition $C(\mathcal{L})$, in turn, is used to build a classifier and derive 'gold standard' partition of the training set $C_{\mathcal{L}}(\mathcal{T})$. That is, the classifier is assumed to be a reliable model for the data. It is then used to assess the quality of the partition of the training set $C(\mathcal{T})$ obtained by a given clustering algorithm.

Suppose that the maximum number of clusters in the given dataset is K_{max}. For each number of k, $2 \leq k \leq K_{max}$, the CLEST is performed in the following four steps:
Repeat the following B times:

1. Randomly split the given dataset into two non-overlap subsets: a learning set $\mathcal{L}^b$ and a training set $\mathcal{T}^b$.
   a. Apply a clustering algorithm to the learning set $\mathcal{L}^b$ to obtain a partition $C(\mathcal{L}^b)$.
   b. Build a classifier using the learning set $\mathcal{L}^b$ and its partition $C(\mathcal{L}^b)$.
   c. Apply the resulting classifier to the training set $\mathcal{T}^b$ to obtain a partition $C_{\mathcal{L}^b}(\mathcal{T}^b)$.
   d. Apply the clustering algorithm to the training set $\mathcal{T}^b$ to obtain a partition $C(\mathcal{T}^b)$.
   e. Take $C_{\mathcal{L}^b}(\mathcal{T}^b)$ as a reference and compute the similarity $s_{k,b}$ between $C_{\mathcal{L}^b}(\mathcal{T}^b)$ and $C(\mathcal{T}^b)$ using an external index.
2. Let $t_k = \text{median}(s_{k,1}, \ldots, s_{k,B})$ denote the observed similarity statistic for the k-partition clustering of the dataset.
3. Generate B_0 datasets under a suitable null hypothesis. For each reference dataset, repeat steps 1 and 2 above, to obtain B_0 similarity statistics $\{t_k^b | b = 1, \ldots, B_0\}$.
4. Let t_k^0 denote the average of these B_0 similarity statistics, as $1/B_0 \sum_{b=1}^{B_0} t_k^b$. Let p_k denote the proportion of $\{t_k^b | b = 1, \ldots, B_0\}$ that are as large as the observed similarity statistic t_k. Let $d_k = t_k - t_k^0$ denote the difference between the observed similarity statistic and its estimated expected value under the null hypothesis of $K = 1$.

CLEST has a very severe limitation on large datasets due to its high computational demand.

20.4   Relative Criteria

20.4.1   Minimum Description Length

The problem of model selection is one of most important problems in inductive and statistical inference. The MDL method, which was proposed by Rissanen (1978, 1986), has been widely employed as one of most popular generic solutions to the model-selection problem (Hansen and Yu, 2001; Grünwald, Myung and Pitt, 2005; Grünwald, 2007). The principle of parsimony, or so called Occam's razor, is the soul of model selection. The principle of MDL is based on the insight that finding as much regularity of data as we can enables us to compress data as much as possible. More specifically, MDL aims to choose the model that gives the shortest description of the data.

MDL is designed to correspond to probability models or distributions and to emphasise the description length interpretation of these distributions. The crucial aspect of the MDL method is found in the code or description length interpretation of probability distributions. The *description length* is defined as $L_Q = -\log Q$, where Q is a distribution. From a coding perspective, MDL coincides with the maximum likelihood (ML) principle that maximises $p(X|\theta)$ (MDL minimises $-\log p(X|\theta)$), when both transmitter and receiver know which parametric family generated the data X. In modelling applications, however, we have to transmit the parameters

θ and the description length of θ has to be added into the cost, which is written as given in Equation (20.10),

$$L = -\log p(X|\theta) + L(\theta) \tag{20.10}$$

where $L(\theta)$ is the code length needed to describe the parameters θ. The lower bound of the redundancy of the parameters corresponds to the price that one must pay for not knowing which member of the model class generated the data X. It was demonstrated that for a regular parametric family of dimension k, its redundancy amounts to at least $N_p/2 \log N$ extra bits, where N_p is the number of parameters. Any code length that achieves this lower bound qualifies as a valid description length of the model class given a data string X, and the associated model-selection criteria have good theoretical properties.

20.4.2   Minimum Message Length

The MML criterion is one of the minimum encoding length criteria (Wallace and Boulton, 1968; Wallace and Dowe, 1999), like MDL, and is used as the clustering-selection algorithm. The rationale behind minimum encoding length criteria is that if one can build a short code for any given data, it implies that the code is a good model for fitting data. The shortest code length for set $\{X\}$ is $-\log p(X|\theta)$. If $p(X|\theta)$ is fully known to both the transmitter and receiver, they can both build the same code and communication can proceed. However, if θ is *a priori* unknown, the transmitter has to start by estimating and transmitting θ. This leads to a two-part message, whose total length is given by Equation (20.11).

$$\text{Length}(X,\theta) = \text{Length}(X|\theta) + \text{Length}(\theta) \tag{20.11}$$

All minimum encoding length criteria state that the parameter estimate is the one minimising Length(X, θ). The criterion was derived to the following form [Equation (20.12)],

$$\text{Length}(X,\theta) = \frac{N_p}{2}\sum_{k=1}^{K}\log \alpha_k + \frac{N_p+1}{2}K\log N - \log p(X|\theta) + C \tag{20.12}$$

where N_p is the number of parameters which is required in each component, $\{\alpha_k | 1 \le k \le K\}$ is the mixing probability of the kth component with the constraint $\sum_{k=1}^{K}\alpha_k = 1$, and $C = (N_p + 1)K(1 - \log 12)/2$ is a constant. Note that the components with zero-probability in α_k have been eliminated and k is the number of non-zero-probability components.

20.4.3   Bayesian Information Criterion

The BIC, also known as the Schwarz criterion, was developed by Schwarz (1978) as a criterion of model selection among a finite set of models. Assuming that data distribution is in exponential family, the BIC added an asymptotic penalty of guessing the wrong model, which equivalently penalises the model with more parameters and discourages overfitting, into the log-likelihood function, written as Equation (20.13).

$$\text{BIC} \approx \log p(X|\boldsymbol{\theta}) - \frac{N_p}{2} \log N \tag{20.13}$$

Note that, coincidently, the BIC is equivalent to the MDL criterion.

20.4.4   Akaike's Information Criterion

The AIC proposed by Akaike (1974) is a measure of the quality of a statistical model with a given dataset. The AIC is defined as given in Equation (20.14),

$$\text{AIC} = -2 \log p(X|\boldsymbol{\theta}) + 2N_p \tag{20.14}$$

and the best estimate is given by the model and the ML estimates of its parameters that give the minimum AIC value.

20.4.5   Partition Coefficient

PC, which is proposed by Bezdek (1974), is one of the validation criteria for evaluating fuzzy partitions. It is defined as shown in Equation (20.15).

$$\text{PC} = \frac{1}{N} \sum_{k=1}^{K} \sum_{n=1}^{N} u_{k,n}^2 \tag{20.15}$$

20.4.6   Partition Entropy

PE is another validation criterion for evaluating fuzzy partitions proposed by Bezdek (1973), which is defined as shown in Equation (20.16).

$$\text{PE} = -\frac{1}{N} \sum_{k=1}^{K} \sum_{n=1}^{N} [u_{k,n} \log u_{k,n}] \tag{20.16}$$

20.4.7   Fukuyama–Sugeno Index

The FS index was proposed by Fukuyama and Sugeno (1989). FS exploits the concepts of compactness and separation for fuzzy partitions, which is written as Equation (20.17),

$$\text{FS}(K) = V_C(K) - V_s(K) \tag{20.17}$$

where $V_C(K)$ represents the compactness measure when the dataset is grouped into K clusters, which is given by $\sum_{n=1}^{N} \sum_{k=1}^{K} u_{k,n}^m \|x_n - c_k\|^2$, where $c_k = \sum_{n=1}^{N} u_{k,n}^m x_n / \sum_{n=1}^{N} u_{k,n}^m$ is the centroid of the kth cluster. $V_s(k)$ is the degree of separation between clusters, which is written as $\sum_{n=1}^{N}\sum_{k=1}^{K}u_{k,n}^{m}\|\mathbf{c}_k-\bar{\mathbf{c}}\|^2$, where $\bar{\mathbf{c}}=\sum_{k=1}^{K}\mathbf{c}_k/K$. In general, an optimal K^* is found by solving $\min\limits_{K\in[2,N-1]}FS(K)$ to produce the best clustering performance for the dataset X. Note that m is the fuzzifier.

20.4.8 Xie–Beni Index

The XB index proposed by Xie and Beni (1991) is also a fuzzy validation criterion which exploits the compactness and the separation. There are two differences between XB and the aforementioned FS index: first, XB considers the ratio between the compactness and the separation; second, the definition of the separation of XB is different to that of FS. The XB for a given dataset X with a partition with K clusters is mathematically written as Equation (20.18).

$$XB(K)=\frac{V_C(K)/N}{V_s(K)}=\frac{\sum_{n=1}^{N}\sum_{k=1}^{K}u_{k,n}^{m}\|\mathbf{x}_n-\mathbf{c}_k\|^2}{N\times\min\limits_{i,j}\|\mathbf{c}_i-\mathbf{c}_j\|} \tag{20.18}$$

Similar to FS, an optimal K^* is found by solving $\min\limits_{K\in[2,N-1]}XB(K)$ to produce the best clustering performance for the dataset X.

20.4.9 Calinski–Harabasz Index

The CH index was proposed by Calinski and Harabasz (1974) for crisp partitions. In Milligan and Cooper (1985), a comprehensive performance comparison of 30 indices was carried out. Among those indices, CH performed the best. Mathematically, CH is given by Equation (20.19),

$$CH(K)=\frac{B(K)/(K-1)}{W(K)/(n-K)} \tag{20.19}$$

where $B(K)$ and $W(K)$ are the between- and within-cluster sums of squares, with K clusters, which are given by $B(K)=\sum_{k=1}^{K}n_k\|\mathbf{c}_k-\mathbf{c}\|^2$ and $W(K)=\sum_{k=1}^{K}\sum_{n\in C_k}\|\mathbf{x}_n-\mathbf{c}_k\|^2$, respectively, where n_k is the number of members in the kth cluster.

20.4.10 Dunn's Index

DI is a metric proposed by Dunn (1973). It is mathematically written as shown in Equation (20.20),

$$DI(K)=\min\limits_{1\leq i\leq K}\left\{\min\limits_{1\leq j\leq K}\left\{\frac{\delta(C_i,C_j)}{\max\limits_{1\leq k\leq K}\{\Delta(C_k)\}}\right\}\right\} \tag{20.20}$$

where $\delta(C_i, C_j)$ is the minimum between-cluster dissimilarity between cluster i and cluster j, given by Equation (20.21),

$$\delta\left(C_i, C_j\right) = \min_{x_i \in C_i, x_j \in C_j} D\left(x_i, x_j\right) \tag{20.21}$$

where $D(x_i, x_j)$ is the dissimilarity between x_i and x_j. $\Delta(C_k)$ is the largest within-cluster separation of cluster k, given as Equation (20.22).

$$\Delta(C_k) = \max_{x_i, x_j \in C_k} D\left(x_i, x_j\right) \tag{20.22}$$

Similar to other indices in relative criteria, DI is a ratio of between-cluster dissimilarity and within-cluster dissimilarity and a larger DI value implies a better clustering partition.

20.4.11   Davies–Bouldin Index

DB is one of the indices in relative criteria, proposed by Davies and Bouldin (1979). It is a ratio of within-cluster dissimilarity and between-cluster dissimilarity. DB can be mathematically written as shown in Equation (20.23),

$$DB(K) = \frac{1}{K} \sum_{k=1}^{K} \left\{ \max_{1 \leq j \leq K} \frac{S_j + S_k}{M_{kj}} \right\} \tag{20.23}$$

where S_k is the average dissimilarity between objects and the cluster centroid within the kth cluster, given by $S_k = 1/N_k \sum_{i \in c_k} D(x_i, c_k)$, and M_{ij} is a between-cluster dissimilarity measure between cluster i and cluster j, given by $M_{ij} = D(c_i - c_j)$. The optimal K^* is the one that minimises $DB(K)$ where K is in the range of $(1, N]$.

20.4.12   I Index

II was proposed by Maulik and Bandyopadhyay (2002). II is written as shown in Equation (20.24),

$$II(K) = \left(\frac{1}{K} \times \frac{E_1}{E_K} \times D_K\right)^p \tag{20.24}$$

where $E_1 = \sum_j \left\|x_j - \bar{c}\right\|_2$ and $E_K = \sum_{k=1}^{K} \sum_{j \in c_k} \left\|x_j - c_k\right\|_2$, $D_K = \max_{i,j}^{K} \left\|c_i - c_j\right\|_2$ and the power p is a constant, which normally is set to be two. The optimal K^* is the one that maximises $II(K)$.

20.4.13   Silhouette

Silhouette statistic was proposed by Rousseeuw (1987) for evaluating clusters and determining the optimal number. For the ith object, let $a(i)$ be the average dissimilarity to other objects within its cluster, and $b(i)$ is the average dissimilarity to objects in the nearest cluster besides the so called 'the nearest cluster', defined by the cluster minimising the average dissimilarity. Thus the silhouette statistic for the ith object in the given partition with K clusters is defined by Equation (20.25).

$$\text{Sil}(i, K) = \left\{ \frac{b(i) - a(i)}{\max\{a(i), b(i)\}} \right\}_K \tag{20.25}$$

An object is well clustered if its silhouette statistic is large. If the task is to evaluate the whole partition, we can simply average the silhouette statistics over the whole dataset as $\text{Sil}(K) = 1/N \sum_{i=1}^{N} \text{Sil}(i, K)$. In MATLAB statistics toolbox, the *silhouette* subroutine is implemented for silhouette statistic.

20.4.14 Object-based Validation

It is worth noting that silhouette index is one of very few object based validation indices, which assign a validity value to each individual object. This capability distinguishes object based validation indices from other indices. However, silhouette index is only suitable for crisp partitions. Here, we introduce an object based validation based on linear discriminant analysis (OBV-LDA) by Fa, Abu-Jamous and Nandi (2013), which can be used to validate both crisp and fuzzy partitions.

Since clustering an individual object into a cluster is only challenged by its closest neighbouring cluster, the CV problem can be simplified to a two-class classification problem. For each object in dataset X, its validity value is given by Equation (20.26),

$$v(n, K) = \log \left\{ \frac{D_M(x_n, \mu'')}{D_M(x_n, \mu')} \right\} \tag{20.26}$$

where μ'' is the centroid of the closest neighbouring cluster of the nth object x_n, μ' is the centroid of the cluster in which x_n locates, and $D_M(a, b)$ denotes the Mahalanobis distance between a and b. Thus, we can obtain Equations (20.27) and (20.28),

$$D_M(x_n, \mu'') = \sqrt{(x_n - \mu'')^T \Sigma^{-1} (x_n - \mu'')} \tag{20.27}$$

$$D_M(x_n, \mu') = \sqrt{(x_n - \mu')^T \Sigma^{-1} (x_n - \mu')} \tag{20.28}$$

where Σ is the combined covariance matrix of the nth object-locating cluster and its closest neighbouring cluster, which is written mathematically as shown in Equation (20.29),

$$\Sigma = \frac{(L'' - 1)\Sigma' + (L' - 1)\Sigma''}{L' + L'' - 2} \tag{20.29}$$

where Σ' is the covariance matrix for the local cluster, Σ'' is the covariance matrix for the closest neighbouring cluster, and L' and L'' are the cardinality of the local cluster and the neighbouring cluster, respectively. The covariance matrix for the kth cluster is obtained from Equation (20.30).

$$\Sigma_k = \frac{1}{L_k - 1} \sum_{n=1}^{N} u_{n,k}^m (x_n - \mu_k)(x_n - \mu_k)^T \tag{20.30}$$

The quality of the whole partition can be assessed by averaging the OBV-LDA as shown in Equation (20.31),

$$V(K) = \frac{1}{N} \sum_{n=1}^{N} v(n, K) \tag{20.31}$$

or the quality of one cluster, say the kth cluster, can be also assessed by means of Equation (20.32).

$$V(k) = \frac{1}{L_k} \sum_{x_n \in C_k} v(n, K) \tag{20.32}$$

20.4.15 Geometrical Index

GI was proposed in by Lam and Yan (2007), and is expressed as shown in Equation (20.33),

$$GI(K) = \max_{1 \le k \le K} \left\{ \frac{\left(2\sum_{m=1}^{M} \sqrt{\lambda_{mk}}\right)^2}{\min_{1 \le j \le K} \|c_i - c_j\|_2} \right\} \tag{20.33}$$

where λ are the eigenvalues of the covariance matrix of the kth cluster. Note that the estimated number of clusters with the GI value closest to zero is the best one.

20.4.16 Validity Index

VI, which was proposed by Salem, Jack and Nandi (2008), is mathematically written as shown in Equation (20.34),

$$VI(K) = \frac{\sum_{i=1}^{K} Ie_i}{\sum_{i=1}^{K} Ia_i} \tag{20.34}$$

where $Ie_i = \min_{j=1, j \ne i} Ie_{ij}$, where Ie_{ij} is the largest dissimilarity of the minimum spanning tree (MST) between the ith cluster and jth cluster, representing the between-cluster separation; Ia_i is the largest dissimilarity of the MST between two data objects inside the ith cluster, representing the within-cluster scatter.

20.4.17   Generalised Parametric Validity

The Generalised Parametric Validity (GPV) index was developed by Fa and Nandi (2011). By introducing two tunable parameters α and β, the GPV index possesses better noise-resistance ability compared with other indices in relative criteria. The rationale behind the GPV index is that, in some noisy scenario, the boundaries among clusters are blurring, and, moreover, the distances between boundary areas and the centre areas are more dominant in determining the quality of a clustering result than are those only within centre areas or only within boundary areas. Thus α and β are employed to control the proportion of objects in centre areas and boundary areas, respectively.

A 2-D plane demonstration of the principle of GPV is depicted in Figure 20.1. The objects marked by 'x' and 'o' belong to two different clusters. Three spaces, namely inner space, intra outer space, and inter outer space, are defined. As shown in Figure 20.1, considering the cluster with 'x' symbol, the darkest grey area is the inner space and the set of objects in this area is denoted by A; the third darkest grey area is the intra outer space and the set of objects in this area is denoted by B; the second lightest grey area is inter outer space and the set of objects in this area is denoted by C. Correspondingly, the numbers of objects in these three spaces are defined as N_k^i, N_a^o and N_e^o, which can be expressed as Equations (20.35),

$$N_k^i = \lceil \alpha N_k \rceil, \ N_a^o = \lceil \beta N_k \rceil, \ N_e^o = \lceil \beta(N - N_k) \rceil \tag{20.35}$$

where N_k is the number of the objects in the kth cluster and $\lceil \cdot \rceil$ is the ceiling operator.

The GPV index, then, can be mathematically written as in Equation (20.36),

$$\text{GPV}(K, \alpha, \beta) = \sum_{k=1}^{K} \sum_{a=1}^{N_k^i} \left(\frac{De_k^a}{Da_k^a} \right) \tag{20.36}$$

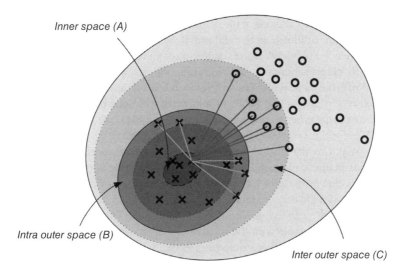

Figure 20.1   Illustration of the GPV index. Symbols 'x' and 'o' represent two clusters. Considering 'x' cluster, the darkest grey area is the inner space, labelled by 'A'; the third darkest gray is the intra outer space, labelled by 'B'; and the second lightest gray area is inter outer space, labelled by 'C'

where Da_k^a denotes a normalised intra-cluster dissimilarity, which is given by Equation (20.37),

$$Da_k^a = \frac{\sum_{b=1}^{N_a^o} D\left(a_k^a, b_k^{a,b}\right)}{N_a^o} \tag{20.37}$$

and De_k^a denotes a normalised inter-cluster dissimilarity for the ath inner end in the kth cluster, which is given by Equation (20.38),

$$De_k^a = \frac{\sum_{c=1}^{N_a^o} D\left(a_k^a, c_k^{a,c}\right)}{N_a^o} \tag{20.38}$$

where $\{a_k^a | a = 1, \ldots, N_k^i\}$ denote the inner space of the kth cluster A_k; for each object in the inner space a_k^a, we need to form subsets $B_k^a = \left\{b_k^{a,b} | b = 1, \ldots, N_a^o\right\}$ and $C_k^a = \left\{c_k^{a,c} | c = 1, \ldots, N_e^o\right\}$ for the object in the intra outer space and the inter outer space, respectively.

20.5 Discussion and Summary

In this chapter, we introduced more than 20 numerical CV algorithms and mentioned that these CV algorithms are classified into three categories. They have their capabilities and advantages to validate clustering; on the other hand, they have limitations and may fail in some circumstances.

Here we summarise these CV algorithms and their pros and cons in Table 20.2. The capabilities are listed in a form of questions about what clustering-validation jobs they can do and what requirements they need to complete the job; namely, if the ground truth is needed while validating, if the clustering algorithms can be validated, if the partitions can be validated, if the fuzzy partitions can be validated, if the crisp partitions can be validated, if the number of clusters K can be estimated, if the object validity can be provided, if the single-cluster validity can be provided; if it has noise-resistance ability, if it is probabilistic model-based criterion, and what the level of its computational complexity is. We fill the Table with 'Yes' or 'No' to indicate if the algorithm has that capability or not. There are also '?' marks in the Table, meaning that it is not a definite 'Yes' or 'No'.

Basically, the external criteria are undisputable validations: if we have ground truth at hand, the external CV algorithms can easily provide how similar the target partition is to the ground truth. However, requiring the ground truth is, on the other hand, a critical limitation, since the ground truth is not available most of the time in real-life clustering applications. Another aspect to be mentioned is that because of the nature of these partition-oriented CV algorithms, it is impossible to judge if a clustering algorithm is good or poor by the validation result using a single partition when the clustering algorithm is stochastic. This is the reason why we put '?' in Table 20.2. It is the same with the relative CV algorithms.

Internal criteria represent another group of effective CV algorithms. They are excellent in estimating the number of clusters, K, and in judging the effectiveness of the clustering algorithms. However, because they commonly employ the re-sampling strategy, they have two limitations: first, they are unable to validate the partitions; second, they cost a large amount of computational power.

Table 20.2 Summarisation of the capabilities and limitations of all CV introduced in this chapter. The capabilities that we refer to are what validation work they can do and what requirements they need; namely, (G) if the **ground truth** is needed, (A) if the clustering **algorithms** can be validated, (P) if the **partitions** can be validated, (F) if the **fuzzy** partitions can be validated, (C) if the **crisp** partitions can be validated, (K) if the number of clusters **K** can be estimated, (O) if the **object** validity can be provided, (SC) if the **single-cluster** validity can be provided, (NR) if it has **noise-resistance** ability, (MB) if it is probabilistic **model-based** criterion, and what the level of its computational complexity is. In this table, 'Yes' means that the validation algorithm has the capability and 'No' means it has not. '?' Means that it is not a definite 'Yes' or 'No'

CV algorithms	Criterion	G	A	P	F	C	K	O	SC	NR	MB	Complexity
RI	External	Yes	?	Yes	No	Yes		No	No	No	No	Low
ARI	External	Yes	?	Yes	No	Yes		No	No	No	No	Low
JI	External	Yes	?	Yes	No	Yes		No	No	No	No	Low
NMI	External	Yes	?	Yes	No	Yes		No	No	No	No	Low
AFOM	Internal	No	Yes	No	No	Yes	Yes	No	No	Yes	No	High
CLEST	Internal	No	Yes	No	No	Yes	Yes	No	No	Yes	No	High
MDL	Relative	No	?	Yes	Yes	Yes	Yes	No	No	No	Yes	Moderate
MML	Relative	No	?	Yes	Yes	Yes	Yes	No	No	No	Yes	Moderate
BIC	Relative	No	?	Yes	Yes	Yes	Yes	No	No	No	Yes	Moderate
AIC	Relative	No	?	Yes	Yes	Yes	Yes	No	No	No	Yes	Moderate
PC	Relative	No	?	Yes	Yes	No	Yes	No	No	No	No	Low
PE	Relative	No	?	Yes	Yes	No	Yes	No	No	No	No	Low
FS	Relative	No	?	Yes	Yes	Yes	Yes	No	No	No	No	Moderate
XB	Relative	No	?	Yes	Yes	Yes	Yes	No	No	No	No	Moderate
CH	Relative	No	?	Yes	No	Yes	Yes	No	No	No	No	Moderate
DI	Relative	No	?	Yes	No	Yes	Yes	No	No	No	No	Moderate
DB	Relative	No	?	Yes	No	Yes	Yes	No	No	No	No	Moderate
II	Relative	No	?	Yes	No	Yes	Yes	No	No	No	No	Moderate
Silhouette	Relative	No	?	Yes	No	Yes	Yes	Yes	Yes	No	No	Moderate
OBV-LDA	Relative	No	?	Yes	Yes	Yes	Yes	Yes	Yes	No	No	Moderate
GI	Relative	No	?	Yes	No	Yes	Yes	No	No	No	No	Moderate
VI	Relative	No	?	Yes	No	Yes	Yes	No	No	No	No	Moderate
GPV	Relative	No	?	Yes	No	Yes	Yes	No	No	Yes	No	Moderate

Relative criteria cover a large group of CV algorithms. Some of them are information the-oretic and model-based, some of them are designed for fuzzy partitions and some of them are for crisp partitions. Their common advantages are that they are simple to calculate and they are suitable for validating the partitions. However, sometimes some of them are not reliable because of their simplicity. Among these relative criteria, silhouette index is the most widely used VI for analysing microarray gene expression data (Horimoto and Toh, 2001; Hanisch *et al.*, 2002; Huang and Pan, 2006; Bandyopadhyay, Mukhopadhyay and Maulik, 2007). It is worth noting that the recently proposed GPV is claimed to possess the noise-resistant ability and to be superior to all other indices.

References

Akaike, H. (1974). A new look at the statistical model identification., *IEEE Transactions on Automatic Control*, **19**(6), pp. 716–723.

Bandyopadhyay, S., Mukhopadhyay, A. and Maulik, U. (2007). An improved algorithm for clustering gene expression data. *Bioinformatics*, **23**(21), pp. 2859–2865.

Bezdek, J. (1973). Cluster validity with fuzzy sets. *Journal of Cybernetics*, **3**(3), pp. 58–73.

Bezdek, J. (1974). Numerical taxonomy with fuzzy sets. *Journal of Mathematical Biology*, **71**, pp. 57–71.

Calinski, T. and Harabasz, J. (1974). A dendrite method for cluster analysis. *Communications in Statistics – Theory and Methods*, **3**(1), pp. 1–27.

Davies, D. and Bouldin, D. (1979). A cluster separation measure. *IEEE Transactions on Pattern Analysis and Machine Intelligence*, **1**(2), pp. 224–227.

Dudoit, S. and Fridlyand, J. (2002). A prediction-based resampling method for estimating the number of clusters in a dataset. *Genome Biology*, **3**(7), p. research0036.

Dunn, J.C. (1973). A fuzzy relative of the ISODATA process and its use in detecting compact well-separated clusters. *Journal of Cybernetics*, **3**(3), pp. 32–57.

Fa, R. and Nandi, A.K. (2011). Parametric Validity Index of Clustering for Microarray Gene Expression Data. IEEE International Workshop on Machine Learning for Signal Processing (MLSP), 18–21 September 2011, Beijing, IEEE, Piscataway, NJ, pp. 1–6.

Fa, R., Abu-Jamous, B. and Nandi, A. (2013). Object Based Validation Algorithm and Its Application to Consensus Clustering, in Proceedings of the 21st European Signal Processing (EUSIPCO) Conference, 9–13 September 2013, Marrakech, Morocco.

Fukuyama, Y. and Sugeno, M. (1989). A New Method of Choosing the Number of Clusters for the Fuzzy C-means Method, in Proceedings of the 5th Fuzzy Systems Symposium, 2–3 June 1989, Kobe, Japan.

Grünwald, P.D. (2007). *The Minimum Description Length Principle*, MIT Press, Cambridge, MA.

Grünwald, P.D., Myung, I.J. and Pitt, M.A. (2005). *Advances in Minimum Description Length: Theory and Applications*, MIT Press, Cambridge, MA.

Halkidi, M., Batistakis, Y. and Vazirgiannis, M. (2002a). Cluster validity methods: part I. *ACM Sigmod Record*, **31**(2), pp. 40–45.

Halkidi, M., Batistakis, Y. and Vazirgiannis, M. (2002b). Clustering validity checking methods: part II. *ACM Sigmod Record*, **31**(3), pp. 19–27.

Hanisch, D., Zien, A., Zimmer, R. and Lengauer, T. (2002). Co-clustering of biological networks and gene expression data. *Bioinformatics*, **18**(suppl 1), pp. S145–S154.

Hansen, M.H. and Yu, B. (2001). Model selection and the principle of minimum description length. *Journal of the American Statistical Association*, **96**(454), pp. 746–774.

Horimoto, K. and Toh, H. (2001). Statistical estimation of cluster boundaries in gene expression profile data. *Bioinformatics*, **17**(12), pp. 1143–1151.

Huang, D. and Pan, W. (2006). Incorporating biological knowledge into distance-based clustering analysis of microarray gene expression data. *Bioinformatics*, **22**(10), pp. 1259–1268.

Hubert, L. and Arabie, P. (1985). Comparing partitions. *Journal of Classification*, **2**(1), pp. 193–218.

Lam, B.S.Y. and Yan, H. (2007). Assessment of microarray data clustering results based on a new geometrical index for cluster validity. *Soft Computing*, **11**(4), pp. 341–348.

Maulik, U. and Bandyopadhyay, S. (2002). Performance evaluation of some clustering algorithms and validity indices. *IEEE Transactions on Pattern Analysis and Machine Intelligence*, **24**(12), pp. 1650–1654.

Milligan, G. and Cooper, M. (1985). An examination of procedures for determining the number of clusters in a data set. *Psychometrika*, **50**(2), pp. 159–179.

Rand, W.M. (1971). Objective criteria for the evaluation of clustering methods. *Journal of the American Statistical Association*, **66**(336), pp. 846–850.

Rissanen, J. (1978). Modeling by shortest data description. *Automatica*, **14**(5), pp. 465–471.

Rissanen, J. (1986). Stochastic complexity and modeling. *Annals of Statistics*, **14**(3) pp. 1080–1100.

Rousseeuw, P.J. (1987). Silhouettes: a graphical aid to the interpretation and validation of cluster analysis. *Journal of Computational and Applied Mathematics*, **20**, pp. 53–65.

Salem, S.A., Jack, L.B. and Nandi, A.K. (2008). Investigation of self-organizing oscillator networks for use in clustering microarray data. *IEEE Transactions on Nanobioscience*, **7**, pp. 65–79.

Schwarz, G. (1978). Estimating the dimension of a model. *Annals of Statistics*, **6**(2), pp. 461–464.

Wallace, C. and Boulton, D. (1968). An information measure for classification. *The Computer Journal*, **11**(2), 185–194.

Wallace, C. and Dowe, D. (1999). Minimum message length and Kolmogorov complexity. *The Computer Journal*, **42**(4), 270–283.

Xie, X. and Beni, G. (1991). A validity measure for fuzzy clustering. *IEEE Transactions on Pattern Analysis and Machine Intelligence*, **13**(8), pp. 841–847.

Yeung, K.Y., Haynor, D.R. and Ruzzo, W.L. (2001). Validating clustering for gene expression data. *Bioinformatics*, **17**(4), pp. 309–318.

21

Biological Validation

21.1 Introduction

The main objective of bioinformatic studies is to enhance our understanding of biology and medicine. Having said that, the results from such studies need to be valid from a biological point of view and not merely from a numerical and computational point of view. Therefore, there is a need for techniques that both embody biologically relevant knowledge and results, and guide downstream biological investigations. In the case of cluster analysis, the main computational output is a group of clusters of genes (or proteins, or metabolites, etc.), possibly with some associated meta-data such as clusters' numerical quality measures. Biological validation in this case can be performed by analysing the contents of the clusters in light of the already established biological knowledge available in the literature. Although many parts of the results are expected to be as yet undiscovered, statistical tests over the already discovered parts can provide statistical measures quantifying how strongly the cluster conforms to such established knowledge.

In this chapter, we present some of the commonly used techniques for clusters' biological validation and bioinformatic post-clustering analysis. This mainly includes gene ontology (GO) analysis, upstream sequence analysis and gene-network analysis.

21.2 GO Analysis

The Gene Ontology Consortium (GOC) runs the GO project, which assigns genes' products with terms from a controlled vocabulary of the processes in which they participate, their molecular functions and the cellular components in which they localise (www.geneontology.org)

Integrative Cluster Analysis in Bioinformatics, First Edition. Basel Abu-Jamous, Rui Fa and Asoke K. Nandi.
© 2015 John Wiley & Sons, Ltd. Published 2015 by John Wiley & Sons, Ltd.

(The Gene Ontology Consortium, 2000, 2013). Assignments are based on the literature and are updated as new studies are conducted.

The terms are structured in a hierarchical manner in which there are three root terms, namely 'biological process', 'molecular function', and 'cellular component', with the GO identifiers GO:0008150, GO:0003674 and GO:0005575, respectively. Each term has a unique identifier with the prefix 'GO:' followed by seven numerical digits.

QuickGO is a useful tool to navigate through and visualise GO terms and their relations in a graphical manner. This tool, which is available at the website www.ebi.ac.uk/QuickGO, is run by the European Bioinformatics Institutes (EBI). By searching for the biological process term 'positive regulation of nuclear division', accessing its page, and then accessing the tab 'Ancestor Chart', a graph similar to the one in Figure 21.1 is obtained. In this figure, the main relation between terms which defines the levels of hierarchy is the 'is a' relation. The relation 'A is a B', where 'A' and 'B' are terms, indicates that 'A' is a subtype of 'B'. For example, and as can be seen in Figure 21.1, the term 'cellular process' is a subtype of the more general term 'biological process', and the term 'nuclear division' is a subtype of the more general term 'organelle fission'. Indeed, by transitivity, a child term is a subtype of its ancestor terms, as well as the ancestors of its ancestors. For example, the term 'nuclear division' is a subtype of the terms 'cellular process' and 'biological process' by transitivity. A single term may have more than one direct ancestor, and an ancestor may have more than one child.

Another type of relations is the 'part of' relation, which indicates that a term is always part of another term. For example, as in Figure 21.1, 'nuclear division' is always part of 'cell division' in eukaryotes; that is, as part of performing cell division, the nucleus needs to be divided. Regulation, whether it was positive or negative, represents another type of relation between terms. This relation appears when the process represented by one term regulates, that is, controls the activity of, the process represented by its target terms. For example, and as can be intuitively inferred from its name, the process 'regulation of nuclear division' regulates the process 'nuclear division'. Some of such regulation relations are labelled as either positive or negative, while some others are not limited to one of those two options.

Similar to Figures 21.1, Figure 21.2 shows a sub-graph considering the direct children and all ancestors of the molecular function term 'mRNA binding', and Figure 21.3 shows a sub-graph considering the cellular component term 'mitochondrion' and its ancestors.

In contrast to the other relations, the relation 'has part' is directed from the parent to the child and not vice versa. If 'B' has 'A' as part of it, the term 'B' will always involve 'A', while 'A' does not always occur as part of 'B'; that is, 'A' may occur independently of 'B'. For example, the molecular function 'nucleic acid binding transcription factor activity' in Figure 21.2 has the molecular function 'nucleic acid binding' as part of it. This means that for the 'nucleic acid binding transcription factor activity' function to be performed, 'nucleic acid binding' must be performed as part of it, while 'nucleic acid binding' might be performed in other contexts unrelated to transcription factor activity, such as during DNA repair or replication. The same logic applies for the component 'mitochondrion-associated adherens complex' in Figure 21.3, which is a complex always containing a 'mitochondrion' as part of it. Note the difference between this irreversible type of relation and the aforementioned relation 'part of'.

One more type of relation is the 'occurs in' relation (not shown in the provided figures). If a process term 'A' has a relation of 'occurs in' with the component term 'B', the component 'B' represents the cellular component in which the process 'A' occurs. For example, the

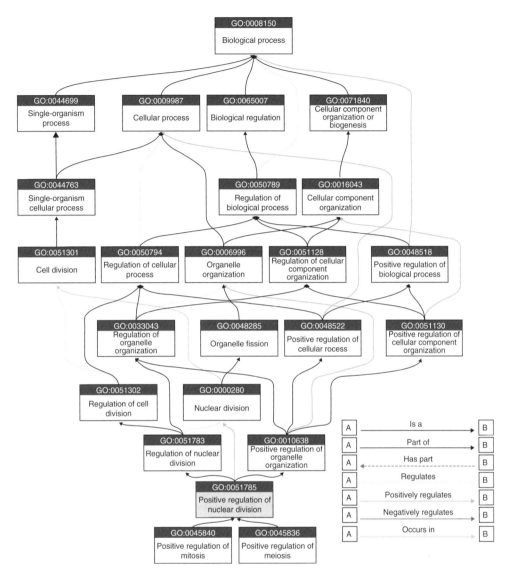

Figure 21.1 Sub-graph of GO biological process terms showing the process 'positive regulation of nuclear division' in a shaded box, its direct children, and its ancestral relations up to the root term 'biological process'. This graph can be obtained by searching the EBI QuickGO web tool (www.ebi.ac. uk/QuickGO) for the GO term ID 'GO:0051785', and then accessing the 'Ancestor Chart' tab; the direct link is (http://www.ebi.ac.uk/QuickGO/GTerm?id=GO:0051785#term=ancchart). The graph shows the names and GO term identifiers of the terms included, as well as colour-coded relations between those different biological processes (*See insert for color representation of the figure*)

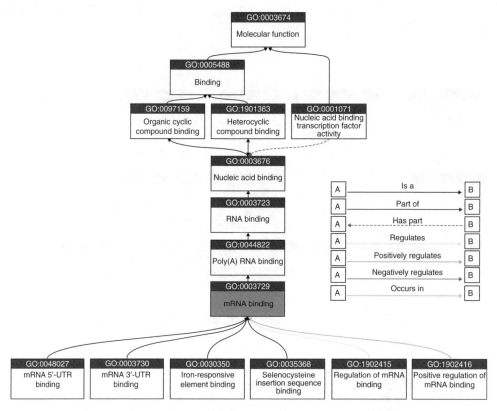

Figure 21.2   Sub-graph of GO molecular function terms showing the function 'mRNA binding' in a shaded box, its direct children, and its ancestral relations up to the root term 'molecular function'. This graph can be obtained in a similar way to the graph in Figure 21.1 while considering the GO term ID 'GO:0003729'

process 'GO:0032544 plastid translation' has a relation of 'occurs in' with the component 'GO:0009536 plastid' because plastid translation actually occurs in plastids.

21.2.1   GO Term Enrichment

GO term-enrichment analysis is the analysis of a subset of genes or genes' products (proteins) in order to identify the GO terms which are significantly represented in this subset. Many tools are available to perform GO term-enrichment analysis where some of them are specific to some species, such as the Stanford University Saccharomyces Genome Database (SGD) term-finder tool for Saccharomyces yeast at http://www.yeastgenome.org/cgi-bin/GO/goTermFinder.pl, while others are more general like the Princeton University term-finder tool at http://go.princeton.edu/cgi-bin/GOTermFinder. Most of the available tools provide similar functionalities and are used in similar ways. Therefore, the example provided below will be sufficient to introduce this type of analysis in general.

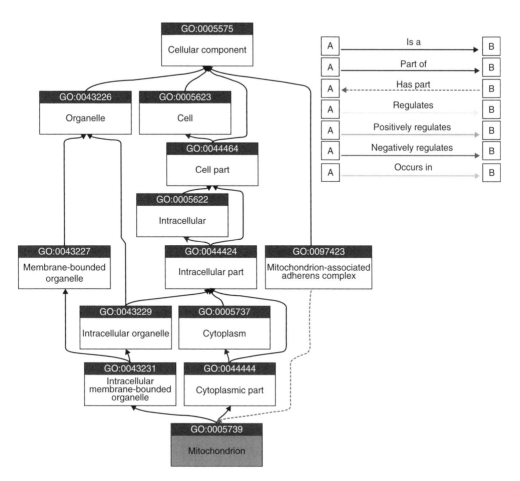

Figure 21.3   Sub-graph of GO cellular component terms showing the component 'mitochondrion' in a shaded box and its ancestral relations up to the root term 'cellular component'. This graph can be obtained in a similar way to the graph in Figure 21.1 while considering the GO term ID 'GO:0005739'. Because this term has a large number of children terms, they are not shown in this sub-graph

Clusters generated by clustering analysis are well suited subjects for GO term-enrichment analysis. To demonstrate this, we have analysed the 172 budding yeast genes contained in the cluster 'C1' generated by the Bi-CoPaM method (Abu-Jamous *et al.*, 2013a) with difference threshold binarisation (DTB) at a tightness value of $\delta = 0.9$ from the Supplementary Table S1 of the recent study of Abu-Jamous *et al.* (2013b). This cluster of genes was submitted to the Princeton University GO term-finder tool while leaving all optional variables to their default values. The tool requires specifying the type of GO terms to be investigated; that is, biological processes, molecular functions or cellular components. In this instance, we investigate biological process terms. The top part of the resulting Table is shown in Figure 21.4.

Result table

Terms from the process ontology of gene_association.sgd with p-value <= 0.01

Gene Ontology term	Cluster frequency	Genome frequency	Corrected p-value	FDR	False positives	Genes annotated to the term
DNA metabolic process	58 of 172 genes, 33.7%	553 of 7166 genes, 7.7%	1.14e-20	0.00%	0.00	YJL115W, YDL164C, YKL067W, YJR006W, YPR135W, YNL312W, YOL090W, YLL002W, YNL273W, YDR545W, YML060W, YDL003W, YNL339C, YMR048W, YBR073W, YGR296W, YFL008W, YLR383W, YLL022C, YNL082W, YJL074C, YPR018W, YMR078C, YER190W, YLR032W, YDR097C, YLR154C, YML061C, YDR279W, YNL072W, YBR275C, YBL035C, YJR043C, YNL102W, YMR075W, YDL101C, YER070W, YAR007C, YGL163C, YOR033C, YGL207W, YBR088C, YDL102W, YKL045W, YPR175W, YIL066C, YLR466W, YKL113C, YER095W, YNL262W, YLR103C, YLR466W, YKL113C, YCL061C, YHR154W, YMR076C, YPL283C, YPL153C, YNL309W
DNA-dependent DNA replication	27 of 172 genes, 15.7%	119 of 7166 genes, 1.7%	9.70e-17	0.00%	0.00	YDL164C, YDR279W, YNL072W, YBL035C, YJR006W, YJR043C, YPR135W, YNL102W, YNL312W, YMR075W, YDL101C, YAR007C, YNL273W, YGL207W, YBR088C, YDL003W, YDL102W, YKL045W, YMR048W, YPR175W, YKL113C, YMR078C, YLR103C, YNL262W, YCL061C, YPL153C, YLR154C
Cellular response to DNA damage stimulus	40 of 172 genes, 23.3%	330 of 7166 genes, 4.6%	2.24e-15	0.00%	0.00	YDL164C, YKL067W, YJR006W, YPR135W, YNL312W, YOL090W, YLL002W, YNL273W, YML060W, YDL003W, YMR048W, YBR073W, YFL008W, YLR383W, YNL082W, YMR078C, YLR032W, YDR097C, YML061C, YNL102W, YJR043C, YDR501W, YDL101C, YAR007C, YGL163C, YOR033C, YLR183C, YGL207W, YBR088C, YDL102W, YKL045W, YPR175W, YKL113C, YER095W, YNL262W, YLR103C, YCL061C, YMR076C, YHR154W, YPL153C
DNA repair	37 of 172 genes, 21.5%	280 of 7166 genes, 3.9%	2.75e-15	0.00%	0.00	YDL164C, YKL067W, YJR006W, YPR135W, YNL312W, YOL090W, YLL002W, YNL273W, YML060W, YDL003W,

Figure 21.4 Sample table of GO biological process term enrichment results. The results are for a cluster of 172 budding yeast genes representing the cluster C1 from the study (Abu-Jamous et al., 2013b) generated by the Bi-CoPaM method with the DTB binarisation technique at the tuning parameter δ = 0.9. The genes can be obtained from Supplementary Table S1 of the aforementioned study (Abu-Jamous et al., 2013b)

It can be seen in Figure 21.4 that the enriched biological process terms are ordered from the most enriched to the least enriched based on their p-values, where lower p-values indicate higher enrichment. All of the terms with p-values lower than a given threshold, whose default is 0.01, are listed in the table while the rest are not. The first column of the table shows the term. The second and the third columns show the cluster frequency and the background (genome) frequency of the term, respectively. For example, the top term, 'DNA metabolic process', is associated with 58 genes out of the 172 genes included in this cluster (33.7%), and is associated with 553 genes out of the 7166 genes included in the entire budding yeast genome (7.7%). The p-value in this case is based on the hypergeometric distribution. For example, the corrected p-value of the 'DNA metabolic process' term, that is 1.14×10^{-20}, indicates that this is the probability of having 58 or more genes annotated to this particular term in a cluster of 172 genes randomly selected from a set of 7166 genes which includes 553 genes annotated to this term. The p-value correction is a statistical adjustment for the raw hypergeometric p-value to compensate for the multiple-hypothesis testing problem. This aspect is discussed more thoroughly in Chapter 9. The remaining columns respectively show the false-discovery rate (FDR), false positives and the names of the genes in this cluster annotated to this term.

Default GO term-enrichment tools consider the entire genome (complete list of genes) in the corresponding species when performing statistical analysis. Though, if the initial list of genes considered in the study is much different, it is more accurate to provide the tool with this list as the background list instead. For example, the aforementioned cluster of 172 genes was obtained by performing clustering analysis over a selected subset of about 500 genes. Those approximately 500 genes represent the background set of this study, and would be provided to the tool for more accurate analysis. Sample results of GO term-enrichment analysis while specifying this background are shown in Figure 21.5.

By identifying the most enriched processes and/or functions and/or components of the genes (or proteins) included in the cluster, the biological context of that cluster would be elucidated. Staying with the same cluster of 172 yeast genes, it can be seen from its most enriched processes that it is enriched with processes that manipulate the DNA for its replication. Replicating the DNA is an essential step during the cell cycle and occurs at the end of the G1 stage and the beginning of the S stage of the cycle. The average temporal expression profile of this cluster is shown in Figure 21.6. This dataset includes the expression of yeast genes over 120 min covering two complete cell cycles. The number of samples is 25, and they are uniformly sampled with 5-min spacing. Cell cycle-regulated genes are expected to show cyclic profiles over the two cell cycles, and this is indeed the case of this cluster. By scrutinising the figure in light of the dataset information provided by the dataset's owners, Pramila and colleagues (2006), it can be seen that the peak expression of this cluster occurs at the G1/S stage transmission, which conforms well to the GO term-enrichment results.

It is also useful to identify the genes that are included in the cluster but have not been annotated to any GO term yet; that is, genes with unknown biological processes. Those genes are candidate subjects to downstream biological functional studies as they may be participating in the same processes in which many of their peer genes in the cluster participate. However, mere appearance in such clusters does not constitute sufficient evidence for those unknown genes to be annotated to those terms; it rather focuses and guides following functional studies.

Result table

Terms from the process ontology of gene_association.sgd with p-value <= 0.01

Gene Ontology term	Cluster frequency	Genome frequency	Corrected p-value	FDR	False positives	Genes annotated to the term
DNA metabolic process	58 of 172 genes, 33.7%	101 of 498 genes, 20.3%	6.76e-05	0.00%	0.00	YJL115W, YDL164C, YKL067W, YJR006W, YPR135W, YNL312W, YOL090W, YLL002W, YNL273W, YDR545W, YML060W, YDL003W, YNL339C, YMR048W, YBR073W, YGR296W, YFL008W, YLR383W, YLL022C, YNL082W, YJL074C, YPR018W, YMR078C, YER190W, YLR032W, YDR097C, YLR154C, YML061C, YDR279W, YNL072W, YBR275C, YBL035C, YJR043C, YNL102W, YMR075W, YDL101C, YER070W, YAR007C, YGL163C, YOR033C, YGL207W, YBR088C, YDL102W, YKL045W, YPR175W, YIL066C, YLR467W, YKL113C, YER095W, YNL262W, YLR103C, YLR466W, YCL061C, YHR154W, YMR076C, YPL283C, YPL153C, YNL309W
Mitotic sister chromatid cohesion	14 of 172 genes, 8.1%	14 of 498 genes, 2.8%	0.00015	0.00%	0.00	YER016W, YIL026C, YFL008W, YJL074C, YPR135W, YMR078C, YNL262W, YNL273W, YJL019W, YCL061C, YBR088C, YDL003W, YMR076C, YMR048W
DNA repair	37 of 172 genes, 21.5%	56 of 498 genes, 11.2%	0.00020	0.00%	0.00	YDL164C, YKL067W, YJR006W, YPR135W, YNL312W, YOL090W, YLL002W, YNL273W, YML060W, YDL003W, YMR048W, YBR073W, YFL008W, YLR383W, YNL082W, YMR078C, YLR032W, YDR097C, YML061C, YNL102W, YJR043C, YAR007C, YGL163C, YOR033C, YGL207W, YBR088C, YDL102W, YKL045W, YPR175W, YKL113C, YER095W, YNL262W, YLR103C, YCL061C, YMR076C, YHR154W, YPL153C
Cellular response to DNA damage stimulus	40 of 172 genes, 23.3%	64 of 498 genes, 12.9%	0.00059	0.00%	0.00	YDL164C, YKL067W, YJR006W, YPR135W, YNL312W, YOL090W, YLL002W, YNL273W, YML060W, YDL003W, YMR048W, YBR073W, YFL008W, YLR383W, YNL082W

Figure 21.5 Sample table of GO biological process term enrichment results with specified background list. The results are for the same cluster of genes considered in Figure 21.4 but while specifying the background list of genes to be the approximately 500 genes originally considered by the study (Abu-Jamous *et al.*, 2013b) as the starting point of analysis

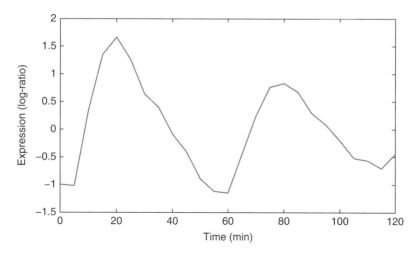

Figure 21.6 Average gene expression of the 172 genes considered in Figures 21.4 and 21.5. The horizontal axis represents time in minutes and the vertical axis represents log-ratio expression values. There are 25 samples in this signal with 5-min spacing. The data, as provided by Pramila and colleagues (Pramila *et al.*, 2006), covers two budding yeast cell cycles over the 120 min of time, and cell cycle-regulated genes are expected to show periodic signals over the two cycles; the shown average signal is clearly periodic over the two cycles

21.3 Upstream Sequence Analysis

Levels of gene expression are usually controlled by different signals including the products of other genes. A transcription factor, as discussed in Chapter 4, is a gene's product which regulates, by activation or repression, the expression of other genes, or even the producing gene itself. This regulation occurs by direct binding of the transcription factor to a specific DNA sequence in the proximity of its target gene(s). It has been found that different transcription factors have different *binding sites*, which are specific short DNA sequences that usually appear in the upstream portion of the target gene, and are specifically bound by transcription factors for regulation. Some transcription factors regulate large numbers of genes (e.g. tens or hundreds), and many transcription factors cooperate in regulating other genes.

When many genes are regulated by the same transcription-regulatory machinery (one or more cooperating transcription factors), they are expected to be *co-expressed* in many cases; that is, to show similar expression profiles to each other. This resonates well with what clustering analysis mines for, as clustering aims at identifying those subsets of genes that are co-expressed. Therefore, searching the upstream sequences of the genes included in a cluster in order to identify the DNA short sequences that significantly appear in such upstream sequences can help in identifying potential transcription factors that regulate that cluster. Hence the relevance of upstream sequence analysis to the validation and downstream analysis of clustering results.

Many tools facilitate upstream sequence analysis, such as the MEME suite, hosted at meme. nbcr.net (Bailey and Elkan, 1994; Bailey *et al.*, 2006), and the PAINT toolset, hosted at http:// www.jefferson.edu/university/jmc/departments/pathology/daniel-baugh-institute/paint.html (Vadigepalli *et al.*, 2003). The MEME suite (MEME stands for Multiple EM (Estimation Maximisation) for Motif Elicitation) provides many motif-based sequence-analysis tools.

(a)   MBF (MBP1/SWI6) transcription factor complex binding site

(b)   Most enriched motif in the cluster of 172 genes under investigation

Figure 21.7   Aligned logos for (a) the binding site of the transcription factor complex MBF (MBP1/SWI6) and (b) the most enriched motif in the upstream sequences of the 172 genes in the cluster under investigation

The main tool in this suite, also named as MEME, mines a group of provided sequences for short DNA sequences (motifs) that significantly appear in them (Bailey and Elkan, 1994; Bailey *et al.*, 2006). Another tool in this suite, the TOMTOM tool, compares a given motif with databases of motifs known in the literature, and identifies the known motifs that highly match the query motif (Tanaka *et al.*, 2011). It is facilitated within the suite to pass the discovered motifs by the MEME tool directly as inputs to the TOMTOM tool. This pipeline of two-step analysis can be semantically summarised as the discovery of the known motifs that significantly appear in the upstream sequences of the given cluster of genes. Since known motifs are usually defined as the binding sites of known transcription factors, such analysis results in identifying a list of potential regulators of the given cluster of genes.

To demonstrate the application of this type of analysis, we have submitted the upstream 300 DNA bases of the genes in a cluster of 172 genes to the MEME tool. This cluster, which is the same cluster of genes analysed in the previous section, is the cluster 'C1' generated by Abu-Jamous and colleagues using the Bi-CoPaM method with the DTB technique's parameter $\delta = 0.9$ (Abu-Jamous *et al.*, 2013b). The contents of the cluster are available in Supplementary Table S1 of that study (Abu-Jamous *et al.*, 2013b).

The most enriched sequence motif in the upstream sequences of those 172 genes is shown in Figure 21.7b. As explained in Chapter 3, DNA sequences are formed of four different nucleobases referred to by the four letters A, T, C and G. Because the motif is extracted from the upstream sequences in a statistical rather than all-inclusive way, the motif does not show the same exact sequence in all instances; it rather possesses slight variations. The relative height of any base letter at some given position in the logo reflects its frequency in the aligned motifs. The vertical axis in the logo represents the information content of that position in the motif measured by bits and calculated by the formula of Shannon's entropy. This information content measures how surprising the observation is when compared with the blind uniform selection of the four different bases. There is zero information content when the position is equally occupied by each of the four bases at different instances of aligned sequences, while the highest information content (2.0 bits) occurs when a single type of base consistently appears in all of the aligned sequences.

When the discovered motif in Figure 21.7b was submitted to the TOMTOM tool to search for known motifs that most significantly match it, the binding site of the MBF transcription factor complex was identified (Figure 21.7a). It can be seen in Figure 21.7 that there is a very strong correlation between the discovered and the previously known motifs, with a p-value of 3.86×10^{-6}, as calculated by the TOMTOM tool. Therefore, by considering that the discovered motif *is* the binding site of the MBF complex rather than merely considering it as a motif that highly matches it, the transitive inferred statement becomes that the binding site of the MBF complex highly appears in the upstream sequences of this cluster of genes.

The MBF complex constitutes two main proteins, MBP1 and SWI6. It is well known to be a key transcription factor for the G1/S stage transition during the budding yeast cell cycle (Bähler, 2005). As discussed and demonstrated in the previous section, this cluster of 172 genes is highly enriched with genes that participate in processes that are required at the G1/S transition (e.g. DNA metabolism), and its expression profile shows peak values at the G1/S transition (Figure 21.6).

Taken together, the results of upstream sequence analysis biologically validate the cluster that was generated computationally by the Bi-CoPaM method. Moreover, this biological validation gives a deeper insight regarding the biological relevance of the cluster and establishes a link that facilitates more intelligible discussions between computational and biological members of a bioinformatic collaboration.

21.4   Gene-network Analysis

Many types of relations have been defined between genes as well as gene products. For instance, and linking with the previous section, one gene's product may be a positive or negative regulator of another gene. Other types of relations include physical interactions between genes' products, genetic interactions, shared protein domains, co-localisation, co-expression and others. Indeed, some of those relations are directed, such as regulatory relations in which one gene's product regulates another gene, while some other relations are undirected, such as physical interactions in which the two products interact with each other.

Networks of genes are formed by considering genes (or their products) as nodes and the relations between them as edges. The University of Toronto-developed tool, GeneMANIA (Gene Multiple Association Network Integration Algorithm), which is hosted at www.genemania.org, is an interactive tool that builds various types of gene networks for a given subset of genes (Warde-Farley *et al.*, 2010; Zuberi *et al.*, 2013). GeneMANIA stores a massive number of gene–gene relations fetched from various high-throughput studies in the literature. Moreover, this tool can identify extra genes which have high connectivity with the query genes.

To demonstrate the use of this tool, we submitted the same cluster of 172 genes which has been considered in the two previous sections to GeneMANIA. Indeed, from the tool's list of species, we selected the *S. cerevisiae* (baker's yeast) because those 172 genes are baker's yeast genes. By keeping advanced options to their default, the network in Figure 21.8 was obtained. It is clear that there is a very large number of associations between the genes within this cluster. The grey nodes are those genes which have been predicted by the tool as related to the cluster because of their high connectivity with the cluster's genes. Although it is hard to comprehend

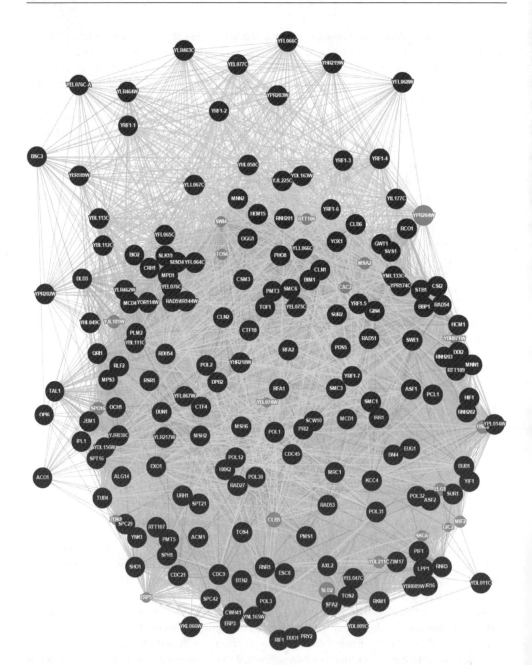

Figure 21.8   Network of multiple types of associations between the 172 yeast genes in the cluster under investigation. The black nodes represent query genes, that is the 172 genes, while the grey nodes represent predicted genes that have high connectivity with the query genes. Different types of genetic relations are shown as different edge colours

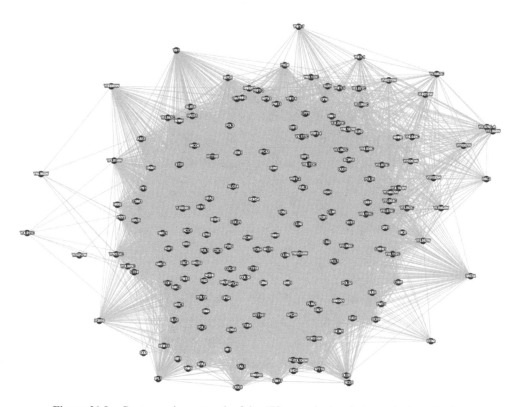

Figure 21.9   Co-expression network of the 172 genes in the cluster under investigation

all of the links in this figure, zooming and filtering options are provided by the tool for a more focused view. Additionally, the network's contents can be downloaded as a vector image as well as a tab-delimited text file, which provides the researcher with a digitised version of the visual network, allowing for more customised scrutiny and analysis.

We will examine two types of gene–gene relations captured by GeneMANIA more closely to demonstrate some types of downstream analysis. We modified the advanced options of the tool to show only 'co-expression' interactions between genes, and to show only the query genes without any extra predicted genes. The resulting network is shown in Figure 21.9. By downloading the list of interactions in a tab-delimited text format and counting the number of edges (interactions), we found that there are 8135 co-expression associations between those 172 genes. This represents 55% of the number of edges that would appear in a fully connected network of 172 genes, which is 14 706.

This percentage cannot be confirmed to be a high, moderate or low value unless it is compared with a reference. To do so, we have generated 22 different clusters, each of which includes 172 randomly selected genes, and which were subjected to the same analysis by the tool. The number of co-expression interactions in those 22 randomly generated clusters had the mean of 5538 interactions with a standard deviation of 273. By assuming a normal distribution, the probability of observing 8135 interactions or more (p-value) where the mean

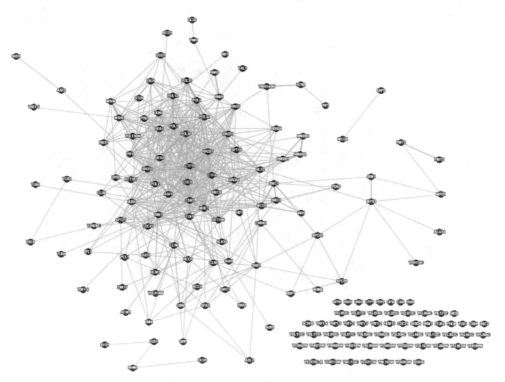

Figure 21.10 Physical interaction network of the products of the 172 genes in the cluster under investigation

and the standard deviation are as given is 9.3×10^{-22}. Thus, this subset of genes is indeed highly enriched with co-expressed genes.

It may be argued that, because the number of edges in the network represents the number of successful random binary events of the type 'does the edge exist?', it should be modelled by using a binomial distribution instead of a normal distribution. The fact that those events are highly dependent renders choosing the binomial distribution invalid. If some edges between a small subset of genes exist, the probability that some of the remaining untested edges also exist is higher than average. However, with 22 samples, normal distribution can be practically applied.

Figure 21.10 shows another network between the 172 genes in the same cluster under consideration. This network shows links between genes whose products (proteins) have been shown to interact physically. There are 518 interactions (edges) in this network, again, out of 14 706 possible edges in a fully connected network. By following the same aforementioned approach in comparing this network with a number of networks for randomly generated clusters of the same size, it can be shown that this network is indeed significantly highly connected. The average number of physical interaction edges in the tested 22 random networks is 122 with a standard deviation of 53. Therefore, the normal distribution-based p-value is 9.3×10^{-15}, which is very small, evidencing that the given cluster of genes is indeed far from being generated by mere chance when genes' products' interactions are concerned.

21.5 Discussion and Summary

It is indeed important to validate clustering results from a numerical point of view, as discussed in Chapter 20. However, when clustering is applied in bioinformatics, it is also important to evaluate those results' biological validity and relevance. Three different biological validation approaches are presented in this chapter, namely, GO term-enrichment analysis, upstream sequence analysis and gene-network analysis.

GO terms are a controlled vocabulary of terms referring to biological processes (e.g. cellular division), molecular functions (e.g. DNA binding), or cellular components (e.g. nucleus). Those terms are structured in a hierarchical manner with different types of relations between them. For instance, a term may be a subtype of another term, part of it, a regulator of it, or has it as one of its parts. Genes and their products are associated with those terms that different studies in the literature have demonstrated sufficient evidence for their association with. Such associations are being actively updated as new studies emerge. GO term-enrichment analysis of a given cluster of genes aims at identifying the terms with which a significant number of query genes are associated.

Upstream DNA sequences of genes are targets of transcription factor proteins which bind therein to regulate, that is control, the amount of expression of such target genes. Upstream sequence analysis identifies the motifs (short sequences) of DNA which repetitively appear in the upstream sequences of the genes in a considered cluster. By matching the discovered motifs with the sequences to which known transcription factors bind, candidate transcription factors would be identified as potential regulators of the cluster under investigation.

Gene networks are networks in which genes, or their products (proteins), are modelled as nodes, while known interactions between them are modelled as edges. Different types of interactions between genes, or genes' products (proteins) have been defined in the literature such as genetic interactions, co-expression, physical interactions between products, regulatory interactions, co-localisation and others. Visualising the interactions already discovered by the research community between the genes included in a cluster can be useful in evaluating the established connectivity between such a subset of genes while considering different types of interactions.

All of the aforementioned approaches can be used to validate the inclusion of the genes under consideration together in a cluster when they show significantly consistent relevance to the same GO term or transcription factor's binding site, or when they show high connectivity in a gene network. However, the usefulness of those approaches does not stop at that level. Such identified GO terms, transcription factors and networks put the cluster in its biological context such as the biological processes and pathways in which the cluster seems to participate. Furthermore, none of the databases of those approaches is absolutely complete; that is, not all of GO terms that should be associated to each single gene in each single species have been actually discovered, not all of the binding sites or the target genes of transcription factors have been identified, and not all of the existing interactions between genes or their products have been established. Consequently, identifying a GO term which is significantly associated with many genes in a cluster that has some genes with unknown associations stimulates hypothesising that the latter genes may be also associated with that same GO term. The same applies for those transcription factors that have not been identified previously as regulators of some of the genes in the cluster, or those missing links in a highly connected network of genes. Indeed, biological validation bridges the gap between numerical clustering results and the biological research community by providing them with biologically relevant hypotheses that represent starting points for their biological functional analysis. Although proving or disproving those hypotheses is the task of biological functional analysis, drawing such focused hypotheses out of the huge amounts of existing data is for sure the task of bioinformatics.

References

Abu-Jamous, B., Fa, R., Roberts, D.J. and Nandi, A.K. (2013a). Paradigm of tunable clustering using binarization of consensus partition matrices (Bi-CoPaM) for gene discovery. *Plos One*, **8**(2), e56432.

Abu-Jamous, B., Fa, R., Roberts, D.J. and Nandi, A.K. (2013b). Yeast gene CMR1/YDL156W is consistently co-expressed with genes participating in DNA-metabolic processes in a variety of stringent clustering experiments. *Journal of the Royal Society Interface*, **10**(81), 20120990.

Bähler, J. (2005). Cell-cycle control of gene expression in budding and fission yeast. *Annual Review of Genetics*, **39**, pp. 69–94.

Bailey, T.L. and Elkan, C. (1994). Fitting a Mixture Model by Expectation Maximization to Discover Motifs in Biopolymers. Proceedings of the Second International Conference on Intelligent Systems for Molecular Biology, AAAI Press, Menlo Park, CA, pp. 28–36.

Bailey, T.L., Williams, N., Misleh, C. and Li, W.W. (2006). MEME: discovering and analyzing DNA and protein sequence motifs. *Nucleic Acids Research*, **34**(Web Server), pp. W369–W373.

Pramila, T., Wu, W., Miles, S. *et al.* (2006). The Forkhead transcription factor Hcm1 regulates chromosome segregation genes and fills the S-phase gap in the transcriptional circuitry of the cell cycle. *Genes and Development*, **20**, pp. 2266–2278.

Tanaka, E., Bailey, T., Grant, C.E. *et al.* (2011). Improved similarity scores for comparing motifs. *Bioinformatics*, **27**(12), pp. 1603–1609.

The Gene Ontology Consortium (2000). Gene Ontology: tool for the unification of biology. *Nature Genetics*, **25**(1), pp. 25–29.

The Gene Ontology Consortium (2013). Gene Ontology annotations and resources. *Nucleic Acids Research*, **41**(Database), pp. D530–D535.

Vadigepalli, R., Chakravarthula, P., Zak, D.E. *et al.* (2003). PAINT: a promoter analysis and interaction network generation tool for gene regulatory network identification. *OMICS*, **7**, pp. 235–252.

Warde-Farley, D., Donaldson, S.L., Comes, O. *et al.* (2010). The GeneMANIA prediction server: biological network integration for gene prioritization and predicting gene function. *Nucleic Acids Research*, **38**(Web Server), pp. W214–W220.

Zuberi, K., Franz, M., Rodriguez, H. *et al.* (2013). GeneMANIA prediction server 2013 update. *Nucleic Acids Research*, **41**(Web Server), pp. W115–W122.

22

Visualisations and Presentations

22.1 Introduction

In the previous chapters, we discussed the whole cluster analysis pipeline in bioinformatics, including data acquisition, pre-processing, normalisation, feature selection, differential expression analysis, clustering, numerical validation and biological validation. As soon as we gather the results discovered in the cluster analysis, and validate them numerically and biologically, the ultimate goal is to publish the discovery and to see its impact on medicine development and clinical therapies. Therefore, clear and informative visualisations and presentations play an important role in helping people to understand the study.

The most widely used visualisation is a simple line plot, where the values presented by the y-axis are outputs of a function with a variable presented by the x-axis. The output values are connected by a line, which can be either solid or dashed. The line plot may give us a direct impression on what the function looks like, and what the 'trend' looks like - up, or down or periodic. Apart from the line plot, there are many other methods of visualisation and presentation. In this chapter, we discuss 15 methods of visualisation excluding the line plot, and we also present examples of these methods with real applications of cluster analysis in bioinformatics.

22.2 Methods and Examples

22.2.1 Profile Patterns

Plotting the profile patterns directly is one of the simplest methods to visualise the clustering results. Particularly, this method is appropriate for presenting the time series expression data (can be either gene or protein). This method, actually, is a display of the collection of line

Integrative Cluster Analysis in Bioinformatics, First Edition. Basel Abu-Jamous, Rui Fa and Asoke K. Nandi.
© 2015 John Wiley & Sons, Ltd. Published 2015 by John Wiley & Sons, Ltd.

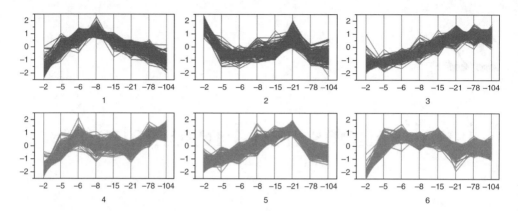

Figure 22.1 k-Means cluster analysis of differentially expressed genes (Kwekel *et al.*, 2013). These data illustrate the various biological patterns of kidney expression which exist during pre-pubertal, early adult and aged rat life stages. Differentially expressed genes were clustered into 28 k-means clusters as this is the lowest number of clusters which allows a minimum correlation coefficient of R = 0.6 between any one expression profile and its other cluster members. Only 6 out of 28 clusters are presented here. The x- and y-axes represent age in weeks and relative fold-change, respectively

plots of data points in the same cluster. The quality of clustering can be roughly judged by observing the shapes of the patterns and the tightness of the bundles. This method, however, is applicable to qualitative analysis only as it does not provide precisely quantitative measurements.

Kwekel and colleagues conducted a study on gene expression in the kidney at various life cycle stages in male F344 rats, as age is a predisposing condition for susceptibility to chronic kidney disease and progression, and age-related differences in kidney biology are a key concern in understanding drug safety and disease (Kwekel *et al.*, 2013). In this study, Agilent whole-genome rat microarrays were used to query global renal expression profiles of untreated male F344 rats at 2, 5, 6, 8, 15, 21, 78 and 104 weeks of age. An ANOVA (p <0.01), coupled with a fold-change >1.5 in relative mRNA expression, was used to identify 7683 out of a total of 41 897 array probes consisting of 3724 unique Entrez Gene IDs; that is, in this case, 3724 differentially expressed genes. A cluster analysis using k-means was performed on the 7683 differentially expressed array probes. The number of initial clusters chosen was 28 as this was the lowest number of clusters to allow a minimal correlation coefficient of 0.6 for any gene expression profile in its respective cluster. Individual gene expression patterns are allowed to cluster with genes having the same expression pattern, providing information on the various patterns that exist in the kidney and how many genes exhibit those patterns of expression during the life cycle. We arbitrarily choose six example clusters out of 28 clusters to demonstrate the expression patterns in Figure 22.1. By mere eye observation, we may find that the patterns of the six clusters are very different. Genes, which had high expression in the early period (first 2 weeks), appeared only in cluster 2 whereas genes in the other five clusters showed low expression at first 2 weeks. Furthermore, these five clusters with early low expression had also very different patterns with each other: cluster 1 had a peak at week 8; cluster 5 had a peak at week 21; cluster 4 had double peaks; cluster 3 gradually increased during the time of the study; cluster 6 sharply increased after week 2 and reached a peak at week 5.

22.2.2   Bar Chart

A bar chart or bar graph is a chart with rectangular bars with lengths proportional to the values that they represent. It uses either horizontal or vertical bars to show comparisons among categories. A vertical bar chart is sometimes called a column bar chart. One axis of the chart shows the specific categories being compared, and the other axis represents a discrete value. Some bar charts present bars clustered in groups of more than one (grouped bar charts), and others show the bars divided into subparts to show cumulative effect (stacked bar charts). Some bar charts can also be drawn in a 3D fashion, not only for fancy appearance, but also for showing the comparison of the combination of two characteristics.

Sîrbu and colleagues conducted a study to investigate the state of the art for the three technologies, namely RNA-seq, single-channel (SC) and dual-channel (DC) microarrays, with a view of assessing overlapping features, data compatibility and integration potential, in the context of gene expression time series datasets for the *Drosophila melanogaster* embryo development (Sîrbu *et al.*, 2012). Differential expression analysis was performed using *R* software; that is, the LIMMA package, for the two microarray datasets, and the DESeq package for the sequencing dataset. As one of the results presented in their study, a bar chart as depicted in Figure 22.2 was employed to show the proportion of those genes identified from each dataset, together with the different differentially expressed sets that apply at different levels. Figure 22.2 shows an example of grouped bar charts.

To understand how complex and highly connected metabolic networks are organised, Matthäus and colleagues measured the biosynthetic potential of a particular compound by determining all metabolites that can be produced from it and called this set the scope of the compound (Matthäus, Salazar and Ebenhöh, 2008). A hierarchical clustering method was applied to identify groups of compounds with similar chemical structures and appear in the same metabolic pathway. In order to measure the quality of the clustering to assure that the elements with

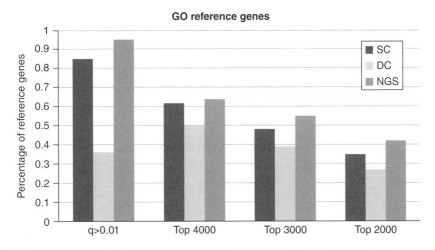

Figure 22.2   Percentage of reference genes represented in the differentially expressed sets obtained from the three datasets (Sîrbu *et al.*, 2012). The NGS dataset identifies the largest number of reference genes, and the DC dataset the lowest

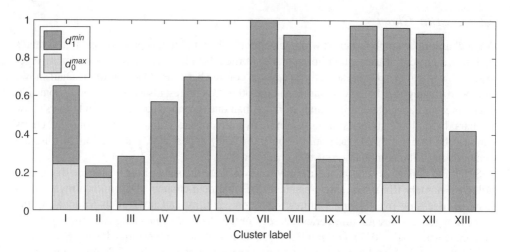

Figure 22.3 Cluster radius and cluster separation (Matthäus, Salazar and Ebenhöh, 2008). Maximum distance between compounds of a cluster to their corresponding consensus scope d_0^{max} (cluster radius), and the minimum distance of the compounds to the second nearest consensus scope d_1^{min} (cluster separation). The assignment to a cluster is good if $d_0^{max} \ll d_1^{min}$

the same cluster are similar and the clusters are well distinguishable, two metrics, namely cluster radius and cluster separation, were calculated for every cluster. Cluster radius, d_0^{max} is defined as the maximum of the distances between all data points contained in the cluster to the centroid. Cluster separation, d_1^{min}, is defined as the minimum of the distances between all data points in the cluster to the centroid of its nearest neighbouring cluster. The stacked bar chart as depicted in Figure 22.3 was used to show the cluster radius and cluster separation for all clusters. Thus, the difference between the cluster radius and cluster separation can be easily spotted and a judgement as to which cluster is better can be made simply by eye observation.

Chen and colleagues attempted to develop an effective analysis method to estimate expression-pattern similarities between different tumour tissues and their corresponding normal tissues (Chen *et al.*, 2013). A cluster analysis was conducted to characterise the gene expression pattern in gene expression level and variation. The study found that cancer-associated housekeeping genes are expressed at higher and more variable levels in cancerous conditions than in normal conditions. As shown in Figure 22.4, 3D bar charts were used to indicate the numbers of housekeeping genes in both normal and cancer tissues, and furthermore, the adjustment of the numbers of housekeeping genes when the condition changed from normal to cancerous.

22.2.3   Error Bar

Error bars commonly appear in presentations in all kinds of publications. They are able to show confidence intervals, standard errors, standard deviations or other related quantities. Error bars are valuable for understanding results in a research article and deciding whether the authors' conclusions are justified by the data. Since different types of error bars give quite different information, Cumming and colleagues suggested several simple rules to assist with effective use and

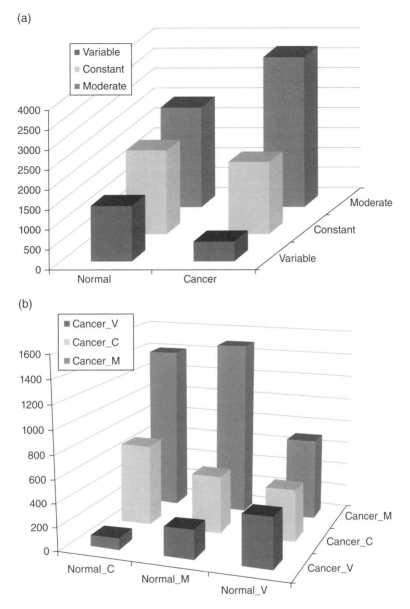

Figure 22.4 (a) Shared HK gene expression variation distribution in normal and cancer conditions. (b) Gene expression variation adjustment in shared HK genes between normal and cancer conditions (Sîrbu *et al.*, 2012). There are three gene expression variation statuses: Constant, abbreviated as suffix 'C' in (b); Moderate variable, abbreviated as Moderate in (a) and suffix 'M' in (b); and Variable, abbreviated as suffix 'V' in (b)

interpretation of error bars (Cumming, Fidler and Vaux, 2007). Although their suggestions were supposed to be given to experimental biologists, some of them are useful specifically for bioinformaticians to visualise the results of cluster analysis. For example, when showing error bars, always describe in the figure legends what they are; the value of the sample size, or

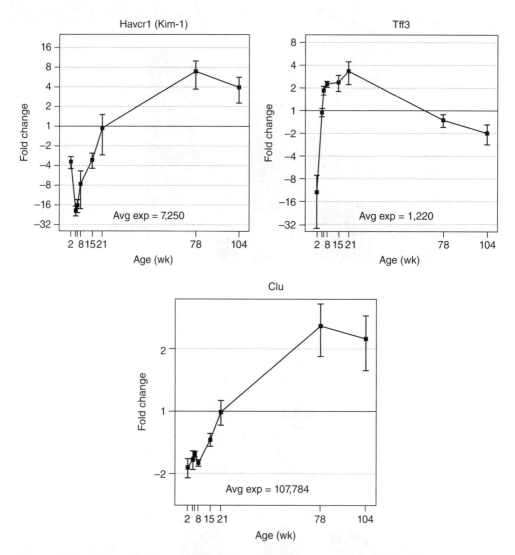

Figure 22.5 Differential life cycle expression of genes encoding qualified kidney biomarkers (Matthäus, Salazar and Ebenhöh, 2008). Genes encoding qualified renal biomarkers kidney injury molecule 1 (Kim-1), clusterin (Clu) and trefoil factor 3 (Tff3) show life cycle expression. Their microarray data and qRTPCR relative fold-changes are plotted, respectively, using error bars, which represent standard error of mean

the number of independently performed experiments, must be stated in the figure legend; error bars and statistics should only be shown for independently repeated experiments, and never for replicates.

As shown in Figure 22.5, error bars were used by Kwekel and colleagues (2013) to show the expression profiles of three qualified kidney biomarkers, namely Kim-1, Tff3 and Clu, over the life cycle of the rat.

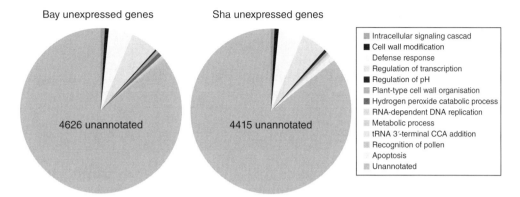

Figure 22.6   By far the majority in both accessions were unannotated genes (Richards *et al.*, 2012). Gene Ontology (GO) categories for genes that were considered 'absent' in all three replicates of at least one sample. Approximately 1/3 of the genome of both Bay-0 (8322 genes) and Sha (7948) was not detected in this study (*See insert for color representation of the figure*)

22.2.4   Pie Chart

A pie chart is a circular chart divided into sectors, illustrating numerical proportion. In a pie chart, the arc length of each sector (and consequently its central angle and area), is proportional to the quantity it represents. While it is named for its resemblance to a pie which has been sliced, there are variations on the way it can be presented. Pie charts are very widely used in all types of research articles. However, they have been criticised, and many experts recommend avoiding them, pointing out that research has shown it is difficult to compare different sections of a given pie chart, or to compare data across different pie charts. Pie charts can be replaced in most cases by other plots such as the bar chart.

Richards and colleagues assessed genome-wide gene expression patterns in two natural accessions of the model plant *Arabidopsis thaliana* (Thale Cress), namely Bay-0 and Sha, and examined the nature of transcriptional variation throughout its life cycle and gene expression correlations with natural environmental fluctuations (Richards *et al.*, 2012). In their study, a pie chart, as depicted in Figure 22.6, was used to show the unexpressed genes in both Bay-0 and Sha accessions. It was clear that the majority in both accessions were unannotated genes.

22.2.5   Box Plot

A box plot is a convenient way of graphically depicting groups of numerical data through their quartiles. Box plots may also have lines extending vertically from the boxes (whiskers) indicating variability outside the upper and lower quartiles, hence the terms box-and-whisker plot and box-and-whisker diagram. Outliers may be plotted as individual points. Box plots display differences between populations without making any assumptions of the underlying statistical distribution: they are non-parametric. The spacing between the different parts of the box helps to indicate the degree of dispersion (spread) and skewness in the data, and identify outliers. As an example depicted in Figure 22.7, the box plot was used to show that normal-unique housekeeping genes and cancer-associated housekeeping genes prefer to express more variable

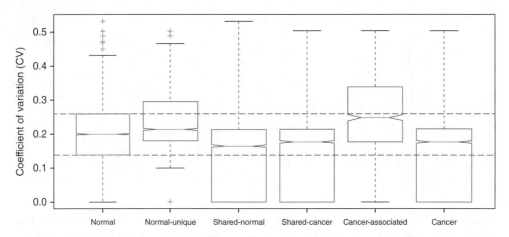

Figure 22.7   Six types of housekeeping genes Coefficient of Variation (CV) values distributions (Chen *et al.*, 2013). The up and down bars signify Q1 (one quarter) and Q3 (three quarters) of CV values, marked as constant and variable expression threshold values. 'Normal' is CV values distribution for normal housekeeping genes; 'Normal-unique' is CV values distribution calculated from specific housekeeping genes in normal condition; 'Shared-normal' is CV values distribution in nine cancer cell lines calculated from overlapped housekeeping genes in normal and cancer conditions; 'Shared-cancer' is CV values distribution in 12 normal tissues calculated from overlapped housekeeping genes in normal and cancer conditions; 'Cancer-associated' is CV values distribution calculated from specific housekeeping genes in cancer condition; 'Cancer' is CV values distribution for cancer housekeeping genes

expressed genes in 3 expression variation statuses, namely constant, moderate, and variable, compared AA with normal (the standard control) and cancer housekeeping genes, because they undertake the basic function of reacting to physiological conditions (Chen *et al.*, 2013).

22.2.6   *Histogram*

A histogram is a representation of tabulated frequencies, shown as adjacent rectangles, erected over discrete intervals (bins), with an area proportional to the frequency of the observations in the interval. The height of a rectangle is also equal to the frequency density of the interval; that is, the frequency divided by the width of the interval. The total area of the histogram is equal to the total number of data observations. A histogram may also be normalised displaying relative frequencies. It then shows the proportion of cases that fall into each of several categories, with the total area equalling to one. The categories are usually specified as consecutive, non-overlapping intervals of a variable. The categories (intervals) must be adjacent, and often are chosen to be of the same size. The rectangles of a histogram are drawn so that they touch each other to indicate that the original variable is continuous.

The first example, as depicted in Figure 22.8, is the pair of histograms that Stegmaier and colleagues used to show the distribution of ED.sqr, which is a measurement of distance involving Euclidean distance and information coverage, for inter-class, intra-class and intra-family alignments (Stegmaier *et al.*, 2013). The objective of their study was to analyse the similarities between DNA sequence motifs in a comprehensive set of vertebrate transcription factor classes.

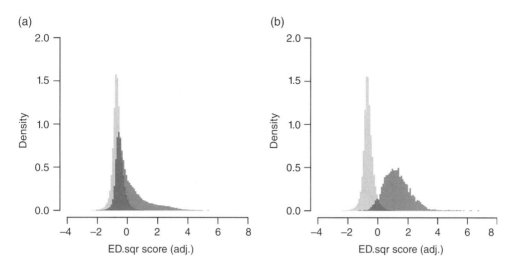

Figure 22.8 ED.sqr scores for inter-class, intra-class and intra-family alignments (Stegmaier *et al.*, 2013). (a) Histograms of adjusted ED.sqr scores for inter-class (light) and intra-class alignments (dark). (b) Histograms of adjusted ED.sqr scores for inter-class (light) and intra-family alignments (dark)

We can find that distributions of intra- and inter-class scores strongly overlapped and the intra-family distribution exhibited a smaller overlap with and a different mode than the inter-class distribution.

Caretta-Cartozo and colleagues introduced a network in which the weight of each link is a function of the phase difference between the expression peaks of two genes (Caretta-Cartozo *et al.*, 2007). The analysis of the stability of clustering through the computation of an entropy parameter reveals a structure made of four clusters, which are separated by bottleneck structures that appear to correspond to cell-cycle checkpoints. Another example of a histogram, as depicted in Figure 22.9, is the pair that Caretta-Cartozo and colleagues showed, illustrating the phase distributions and phase-difference distributions for periodic genes from the three independent studies on *Schizosaccharomyces pombe* and from the study on *Saccharomyces cerevisiae* (Figure 22.9).

22.2.7   Scatter Plot

A scatter plot is a type of mathematical diagram using Cartesian coordinates to display values for two variables for a set of data. The data are displayed as a collection of points, each having the value of one variable determining the position on the horizontal axis and the value of the other variable determining the position on the vertical axis. The scatter plot can suggest various kinds of correlations between variables with a certain confidence interval. For example, weight and height, weight would be on the x-axis and height would be on the y-axis. Correlations may be positive (rising), negative (falling) or null (uncorrelated). If the pattern of dots slopes from lower left to upper right, it suggests a positive correlation between the variables being studied. If the pattern of dots slopes from upper left to lower right, it suggests a negative correlation. One of the most powerful aspects of a scatter plot, however, is its ability to show nonlinear relationships between variables. Furthermore, if the data are represented by a mixture model of simple relationships, these relationships will be visually evident as superimposed patterns.

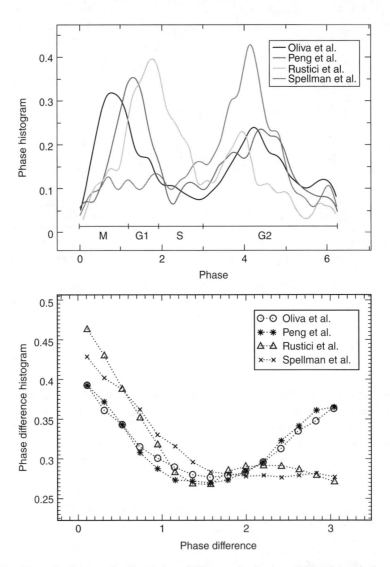

Figure 22.9 Phase distributions (Left) and phase difference distributions (Right) for periodic genes from the three independent studies on *S. pombe* and from the study on *S. cerevisiae* (Caretta-Cartozo *et al.*, 2007) (*See insert for color representation of the figure*)

A typical example of a scatter plot (Figure 22.10) comes from Chen and colleagues, who presented the scatter MDAD plot of cancer-associated housekeeping genes, where MD stands for M distance and AD stands for A distance (Chen *et al.*, 2013).

22.2.8   Venn Diagram

A Venn diagram, named after John Venn around 1880, also known as a set diagram, is a diagram that shows all possible logical relations between a finite collection of sets. A Venn diagram is constructed with a collection of simple closed and overlapping curves drawn in a

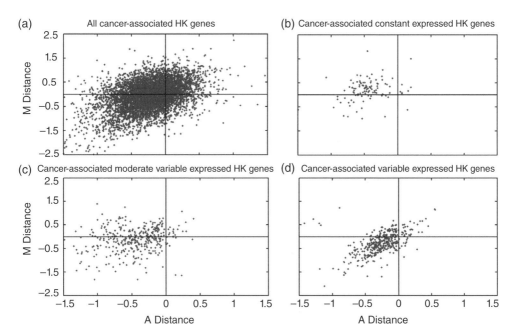

Figure 22.10　MDAD plots of cancer-associated housekeeping genes (Chen *et al.*, 2013). MD < 0 means the gene expression span width in cancer condition is larger than that in normal condition, and AD < 0 means the gene expression relative average level in cancer condition is higher than that in normal condition. According to cancer-associated housekeeping gene expression variation statuses, cancer-associated housekeeping genes are divided into three subtypes, namely constant, moderate variable and variable expressed cancer-associated housekeeping genes. Paired Wilcoxon signed-rank test is used here to measure gene expression regulation and gene expression variation status regulation in cancer. (a) All cancer-associated housekeeping genes. (b) Cancer-associated constant expressed housekeeping genes. (c) Cancer-associated moderate variable expressed housekeeping genes. (d) Cancer-associated variable expressed housekeeping genes

plane. The interior of the circle symbolically represents the elements of the set, while the exterior represents elements that are not members of the set. For instance, in a two-set Venn diagram, one circle may represent the group of all wooden objects, while another circle may represent the set of all tables. The overlapping area or intersection would then represent the set of all wooden tables. Shapes other than circles can be employed as shown below by Venn's own higher set diagrams. Venn diagrams typically represent two or three sets, but there are forms that allow for higher numbers.

Sîrbu and colleagues employed the Venn diagram, as depicted in Figure 22.11a, to show the number of differentially expressed genes in each dataset, and in overlapping areas (Sîrbu *et al.*, 2012). The Next/Second Generation Sequencing (NGS) dataset identified the largest number of genes, in agreement with previous study findings, followed by SC and DC. Datasets SC and NGS show greatest similarity for the differentially expressed sets obtained, with a large number of (mostly common) differentially expressed genes involved.

Yao and colleagues introduced an algorithm named Survival analysis using Cox proportional hazard regression and Random resampling (SCoR) to apply random resampling and clustering

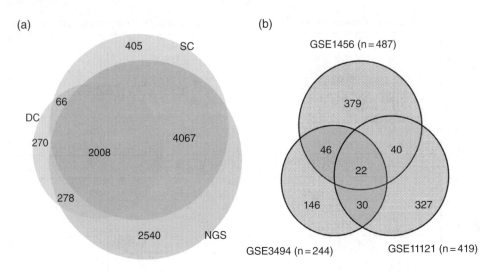

Figure 22.11   Venn diagram examples in two different studies (Sîrbu *et al.*, 2012; Yao *et al.*, 2012). (a) Differentially expressed genes. The NGS and SC datasets display the largest commonality, while the DC and SC display the smallest (Sîrbu *et al.*, 2012). (b) Overlapping genes found by SCoR in three breast cancer datasets (Yao *et al.*, 2012)

methods in identifying gene features correlated with time-to-event data (Yao *et al.*, 2012). As shown in Figure 22.11b, overlapping genes found by SCoR in three breast cancer datasets were presented in the Venn diagram neatly.

22.2.9   Tree View

Tree view, also known as a dendrogram, is a tree diagram frequently used to illustrate the arrangement of the clusters produced by hierarchical clustering. Dendrograms are often used in computational biology to illustrate the clustering of genes or samples. The bottom level of nodes represent data (individual observations), and the connected nodes represent the clusters to which the data belong, with the lines representing the distance (dissimilarity). The distance between merged clusters monotonically increases with the level of the merger: the height of each node in the plot is proportional to the value of the intergroup dissimilarity between its two daughters (the top nodes representing individual observations are all plotted at zero height). To visualise the clustering hierarchical clustering results of gene expression data, tree views are often accompanied by a heat map (discussed in the following subsection).

    An example shown in Figure 22.12 is the cluster dendrogram and heat map of gene expression using SCoR-generated prognostic probes from NKI-295 dataset in Yao and colleagues' study (Yao *et al.*, 2012). It is worth noting that in this example both dendrogram and heat map were used to illustrate the clustering results. On the left-hand side of Figure 22.12, the tree structures are shown for both rows (genes) and columns (samples or patients). In Stegmaier and colleagues' study of DNA sequence motif similarity, as depicted in Figure 22.13, the dendrogram showed the clustering result for the set of 71 non-zinc finger Jaspar motifs (Stegmaier *et al.*, 2013).

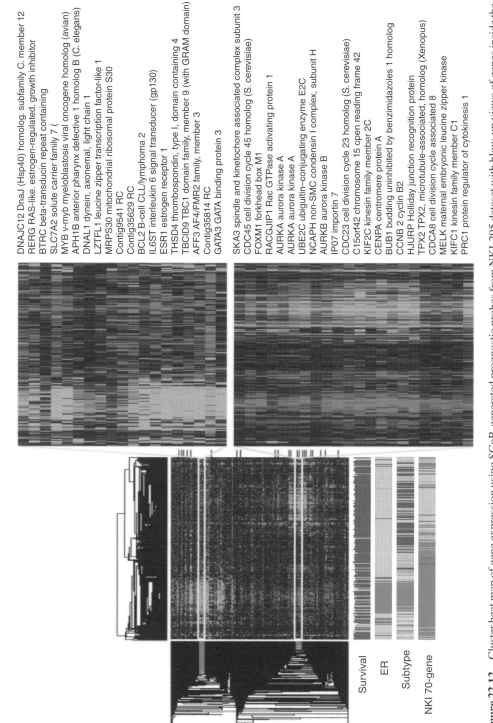

Figure 22.12 Cluster heat map of gene expression using SCoR-generated prognostic probes from NKI-295 dataset with blow-up views of genes inside the centre of poor and good prognosis gene clusters. Probes matching the NKI 70-genes are marked in black lines on the right side of the heat map (Yao *et al.*, 2012) (*See insert for color representation of the figure*)

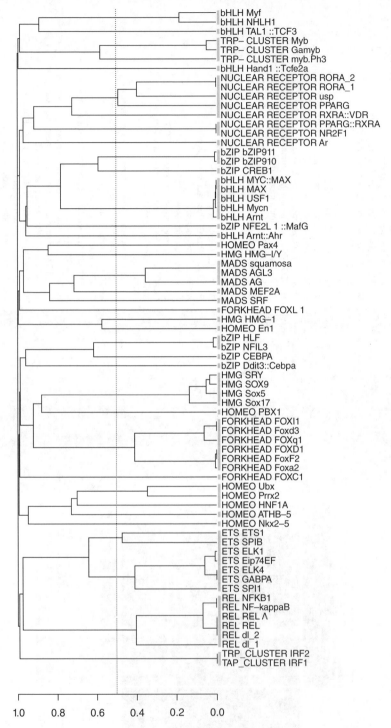

Figure 22.13 Clustering of 71 non-zinc finger motifs from Jaspar (Stegmaier *et al.*, 2013). Gray boxes between dendrogram and matrix names indicate motif clusters. The dotted line points out the 50% motif family threshold. Some clusters were merged below that threshold, because Familial Binding Profiles (FBPs) formed in the course of the clustering process provided for a better presentation of the motif family than the basic motifs

22.2.10   Heat Map

A heat map is a graphical representation of data where the individual values contained in a matrix are represented as colours. Heat maps originated in 2D displays of the values in a data matrix. Larger values were represented by small dark gray or black squares (pixels) and smaller values by lighter squares. There are many different colour schemes that can be used to illustrate the heat map, with perceptual advantages and disadvantages for each. Rainbow colour maps are often used, as humans can perceive more shades of colour than they can of gray, and this would purportedly increase the amount of detail perceivable in the image. Fractal maps and tree maps both often use a similar system of colour-coding to represent the values taken by a variable in a hierarchy. Figure 22.12 can be a very good example for the joint use of both heat map and tree view. A heat map can also be used unaccompanied. As depicted in Figure 22.14, a heat map was employed to show differential expression during flowering (Richards *et al.*, 2012). The heat

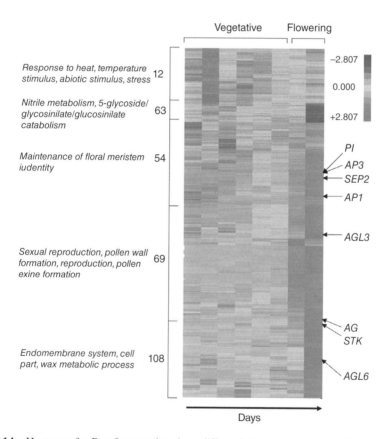

Figure 22.14   Heat map for Bay-0 genes that show differential expression during flowering (Richards *et al.*, 2012). The numbers listed to the left indicate the cluster number identified in the silhouette analyses. The four most significant (i.e., lowest *p* values) GO term categories enriched for each cluster are shown on the left. The vegetative and flowering samples are indicated at the top, and the rows corresponding to various flower-development genes are shown. Several developmental genes associated with bolting and flower development are highlighted. Scale: from brightest blue equals most down-regulated to brightest red equals most up-regulated

map presents groups of unique gene expression patterns (defined by cluster analysis) for which more than 50% of the variance was explained by flowering status in both Bay-0 and Sha.

22.2.11   Network Graph

A network graph is a representation of a set of objects where some pairs of objects are connected by links. The interconnected objects are represented by mathematical abstractions called vertices, and the links that connect some pairs of vertices are called edges. Typically, a network graph is depicted in diagrammatic form as a set of dots for the vertices, joined by lines or curves for the edges. Graphs are one of the subjects studied in discrete mathematics, but here we only consider them as a visualisation method to present networks. We have discussed many visualisation tools for graphs in Chapter 16.

Freeman and colleagues presented an approach for visualisation and analysis of transcriptional networks generated from microarray data (Freeman *et al.*, 2007). These networks consisted of nodes representing transcripts connected by virtue of their expression-profile similarity across multiple conditions. The resulting graphs, as depicted in Figure 22.15a and b, are weighted undirected graphs consisting of probes connected by their co-expression values, when the Pearson threshold is 0.95 and 0.9, respectively. Visual representation of such data is especially desirable, as it is far easier for people to infer relationships and identify structural features from a visual standpoint. Figure 22.15a hides the nodes so as to show the structure of the networks, while Figure 22.15b shows coloured nodes according to their membership of Morkov cluster (MCL) clusters.

(a)   (b)

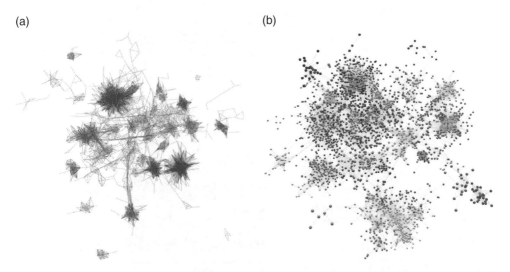

Figure 22.15   Untiled (Organic Layout) of GNF1M Network Graphs at Different Pearson Correlation (Freeman *et al.*, 2007) Thresholds Graphs show the mouse tissue transcription network graphs when the Pearson threshold is set at (a) 0.95 (2860 nodes, 201 724 edges); nodes have been hidden so as to show the structure of the networks, and (b) 0.90 (5410 nodes, 447 467 edges), nodes are shown and coloured according to their membership of MCL clusters (inflation value 1.5) (*See insert for color representation of the figure*)

22.2.12 Low-dimension Display

Low-dimension display is not an actual method of visualisation, but an idea of visualising high-dimensional data in a diagram. There are two most popular ways to do that: the first is principal component analysis (PCA) and the second is multidimensional scaling (MDS). We have discussed PCA in Chapter 8. The difference between PCA and MDS is that PCA uses two or three predominant principal components to represent all features, while MDS displays the information contained in a distance matrix and it does not matter how many features there are in the data.

Kwekel and colleagues employed PCA in their study on renal gene expression (Kwekel *et al.*, 2013). The 7683 differentially expressed probes that met the filtering criteria were used for principal component analysis. The top three principal components account for 41.3%, 23.7%, and 7.2% of the total variability in the dataset, respectively, and are plotted in three dimensions to visualise the contribution of individual animals to the global expression profiles of the differentially expressed genes. The 3D visualisation is shown in Figure 22.16. The 2-week-old group shows the greatest distance from the other ages, followed by more proximal clustering of the 5-, 6-, 8-, 15- and 21-week-old groups, with 78- and 104-week-old groups occupying a third sub-cluster. These three broad age-divisions seem to define major stages in kidney gene expression during the life cycle.

22.2.13 Receiver Operating Characteristics Curves

A receiver operating characteristic (ROC) curve has been widely used to illustrate the performance of a binary classifier system while its discrimination threshold is varying. The graphical plot presents the fraction of true positives out of the total actual positives versus the fraction of false positives out of the total actual negatives at various threshold settings. True positive rate is also known as sensitivity or recall in machine learning. The false positive rate is also known as the fall-out and can be calculated as one minus the more well-known specificity. The ROC curve is then the sensitivity as a function of fall-out. Generally, the ROC curve plots the cumulative distribution function of the detection probability in the y-axis versus the cumulative distribution function of the false alarm probability in x-axis if both of the probability distributions for detection and false alarm are known. ROC analysis enables us to select possibly optimal models and to discard suboptimal ones independently from the cost context or the class distribution. Furthermore, ROC analysis directly and naturally leads to benefit versus cost analysis of diagnostic decision making.

Habib and colleagues presented a method for comparing and merging motifs, based on Bayesian probabilistic principles (Habib *et al.*, 2008). This method, which is called Bayesian likelihood 2-component (BLiC), took into account both the similarity in positional nucleotide distributions of the two motifs and their dissimilarity to the background distribution. In this study, the sensitivity and specificity of motif similarity-scoring methods were evaluated by comparing all possible pairs of motifs from the datasets, and testing whether pairs that have high similarity indeed were generated from the same source. In the 'Yeast' dataset a pair is considered as true if the two motifs were generated from binding locations of the same transcription factor, and in the 'Structural' dataset a pair is considered as true if the motifs are

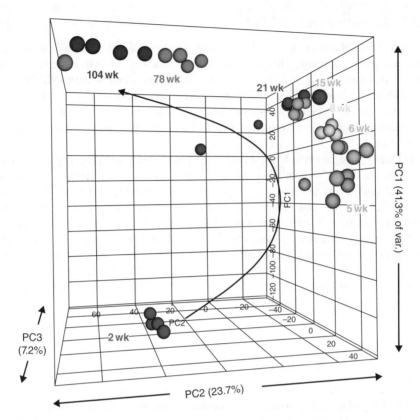

Figure 22.16 Three-dimensional principal component analysis of differentially expressed genes (Kwekel *et al.*, 2013). The 7683 differentially expressed features (ANOVA, *p* <0.01; relative fold-change >1.5) were used to assess the global view of each animal's contribution to the life cycle expression profile. Each sphere represents the composite expression profile of one animal according to the top three principal components plotted in three-dimensional space (ArrayTrack). Spheres are coloured by similar age group (N = 4 or 5) and generally cluster together according to respective age in an age-sequential pattern. Two-week animals show the most distinction from other groups, followed by the 78- and 104-week animals' separation from the remaining 5-, 6-, 8-, 15- and 21-week groups. Together, these data illustrate the relatively high reproducibility between biological replicates in a discrete and continuous linear pattern from young to old animals. It also suggests at least three general stages in kidney life cycle gene expression (*See insert for color representation of the figure*)

of factors from the same structural class. As shown in Figure 22.17, examples of ROC curves indicate that the BLiC score outperformed all other scores throughout the range of possible sensitivity/specificity tradeoffs on both datasets.

22.2.14 Kaplan–Meier Plot

A Kaplan–Meier plot is a plot of the Kaplan–Meier estimate of the survival function, which is a series of horizontal steps of declining magnitude which, when a large enough sample is taken,

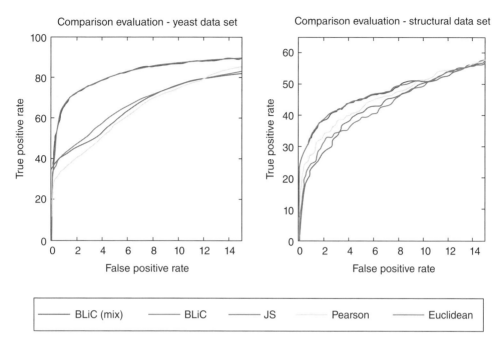

Figure 22.17 Examples of ROC from (Habib *et al.*, 2008). Left: Sensitivity and specificity of different scoring methods: Comparison of different scoring methods on the 'Yeast' dataset using a subset of motifs generated from subsets of size 35 with altered lengths (not including the full-length motifs, 685 motifs). Each similarity score was assigned an empirical statistical significance *p*-value. The ROC curve plots the true positive rate vs. the false positive rate, as computed for different *p*-value thresholds, where pairs of motifs generated from genomic binding sites that were associated with the same factor are considered true positives. The BLiC score (green, using a Dirichlet prior, or blue, using a Dirichlet-mixture prior, and blue line is overlapped with green one) outperformed all other similarity scores: Jensen–Shannon (JS) divergence (red), Euclidean distance (purple) and Pearson Correlation coefficient (cyan). Right: Sensitivity and specificity estimated by structural data: Same as the left one but using the 'Structural' dataset. Pairs of motifs from the same structural family are considered true positives (*See insert for color representation of the figure*)

approaches the true survival function for that population. The Kaplan–Meier estimator, also known as the product limit estimator, is an estimator for estimating the survival function from lifetime data. In medical research, it is often used to measure the fraction of patients living for a certain amount of time after treatment. The estimator is named after Edward L. Kaplan and Paul Meier. The value of the survival function between successive distinct sampled observations is assumed to be constant. An important advantage of the Kaplan–Meier curve is that the method can take into account some types of censored data, particularly right-censoring, which occurs if a patient withdraws from a study; that is, is lost from the sample before the final outcome is observed. On the plot, small vertical tick-marks indicate losses, where a patient's survival time has been right-censored. When no truncation or censoring occurs, the Kaplan–Meier curve is the complement of the empirical distribution function. In medical statistics, a typical application might involve grouping patients into categories; for instance, those with Gene A profile and those with Gene B profile. In the graph, patients with Gene B die much more quickly than

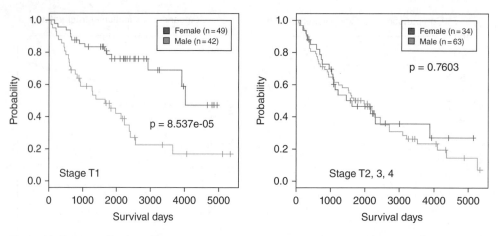

Figure 22.18 Identification of sex as a prognostic factor in one lung adenocarcinoma dataset (Yao *et al.*, 2012). Kaplan–Meier plot of patient survivals stratified by patient gender in all patients, (left) in stage T1 patients, and (right) stage T2, T3, T4 patients

those with Gene A. After 2 years, about 80% of the Gene A patients survive, but less than half of patients with Gene B.

As shown in Figure 22.18, one may easily read that the effect of patient gender on survival could only be established in stage T1 tumours (size < 3 cm), and there is no statistically significant difference in survival between male and female patients when tumour sizes exceeded 3 cm in stages T2, T3, T4 tumours even within the DCC2008-MI dataset (Yao *et al.*, 2012).

22.2.15 Block Diagram

A block diagram is usually used to describe a system in which the principal parts or functions are represented by blocks connected by lines that show the relationships among the blocks. It is typically used for a higher-level, less detailed description aimed more at understanding the overall concepts and less at understanding the details of implementation. There is an increasing use of engineering principles in biology (as well as biological principles in biology); therefore, the block diagram technique harnessed by control engineering has been used in so-called systems biology.

As shown in Figure 22.19, the hierarchical ordering of the consensus compounds is displayed in a block diagram in Matthäus, Salazar and Ebenhöh study (2008). The boxes contain a cluster representative (a compound with a scope identical to the consensus scope), the cluster label and the consensus scope size, as well as the chemical elements present in most metabolites of the corresponding cluster. In the block diagram, clusters with a large biosynthetic potential are positioned above clusters with a lower biosynthetic potential. A line between two clusters is drawn if the consensus scope of the cluster positioned below is a subset of the consensus scope of the cluster positioned above.

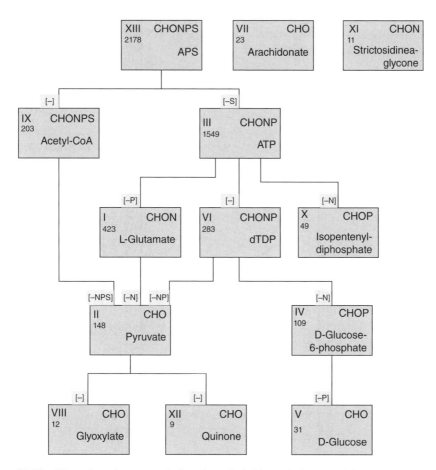

Figure 22.19 Hierarchy of compounds based on their biosynthetic potentials. Each box denotes a distinct consensus scope. On the top-left corner of each box, the cluster label and consensus scope size are shown. On the top-right corner, the chemical elements present in most cluster metabolites are given. Further, a representative metabolite of the cluster, whose scope is identical to the cluster's consensus scope, is given. Two consensus scopes are connected by an edge if the lower one is completely contained in the upper one. If the inclusion can be explained by differences in the chemical elements within the cluster members, the missing elements have been noted at the corresponding edge

22.3   Summary

We introduced 15 methods of visualisation excluding the simple line plot. The examples of these descriptive presentations can be easily found from any study in the bioinformatics field. The rule of thumb of choosing the methods of visualisation is to present the results (ideas, discoveries or experimental observations) in the clearest and simplest way. It is also worth noting that to analyse the same problem from different perspectives, we need to employ many different approaches to present the results in a single study. As we may notice in Table 22.1, all applications we

Table 22.1 Summary of the visualisation methods that have been used in different studies

Studies	LP	PP	BC	EB	PC	BP	HG	SP	VD	TV	HM	NG	LDD	ROC	KM	BD
Caretta-Cartozo et al. (2007)	✓						✓	✓	✓	✓	✓	✓				
Chen et al. (2013)	✓		✓	✓		✓	✓	✓	✓	✓	✓					
Freeman et al. (2007)	✓			✓			✓					✓				
Habib et al. (2008)			✓						✓		✓			✓		✓
Kwekel et al. (2013)		✓		✓									✓			
Matthäus, Salazar and Ebenhöh (2008)			✓		✓				✓							✓
Richards et al. (2012)	✓		✓			✓	✓	✓	✓		✓					
Sîrbu et al. (2012)	✓		✓			✓	✓	✓		✓		✓				
Stegmaier et al. (2013)	✓		✓				✓	✓	✓	✓	✓				✓	
Yao et al. (2012)	✓		✓													✓

BC, bar chart; BD, block diagram; BP, box plot; EB, error bar; HG, histogram; HM, heat map; KM, Kaplan–Meier; LP, line plot; LDD, low-dimension display; NG, network graph; PC, pie chart; PP, profile pattern; SP, scatter plot; TV, tree view; VD, Venn diagram.

mentioned in this chapter used more than one method to present their studies. Each study, at least, used three approaches, and some studies even used eight different methods to visualise their results. All in all, the ultimate goal of our research is to convince people from both inside and outside the field about our research products, hoping that our results may have a strong impact, specifically, on biological, medical and clinical research. Therefore, choosing appropriate approaches to visualise and present the research outcomes is an important part in the art of research communications.

References

Caretta-Cartozo, C., Los, P.D., Piazza, F. and Liò, P. (2007). Bottleneck genes and community structure in the cell cycle network of S. pombe. *Plos Computational Biology*, **3**(6), e103.

Chen, M., Xiao, J., Zhang, Z. *et al.* (2013). Identification of human HK genes and gene expression regulation study in cancer from transcriptomics data analysis. *Plos One*, **8**(1), e54082.

Cumming, G., Fidler, F. and Vaux, D.L. (2007). Error bars in experimental biology. *Journal of Cell Biology*, **177**(1), pp. 7–11.

Freeman, T.C., Goldovsky, L., Brosch, M. *et al.* (2007). Construction, visualisation, and clustering of transcription networks from microarray expression data. *Plos Computational Biology*, **3**(10), e206.

Habib, N., Kaplan, T., Margalit, H. and Friedman, N. (2008). A novel Bayesian DNA motif comparison method for clustering and retrieval. *Plos Computational Biology*, **4**(2), e1000010.

Kwekel, J.C., Desai, V.G., Moland, C.L. *et al.* (2013). Life cycle analysis of kidney gene expression in male F344 rats. *Plos One*, **8**(10), e75305.

Matthäus, F., Salazar, C. and Ebenhöh, O. (2008). Biosynthetic potentials of metabolites and their hierarchical organization. *Plos Computational Biology*, **4**(4), e1000049.

Richards, C.L., Rosas, U., Banta, J. *et al.* (2012). Genome-wide patterns of Arabidopsis gene expression in nature. *Plos Genetics*, **8**(4), e1002662.

Sîrbu, A., Kerr, G., Crane, M. and Ruskin, H.J. (2012). RNA-Seq vs dual-and single-channel microarray data: sensitivity analysis for differential expression and clustering. *Plos One*, **7**(12), e50986.

Stegmaier, P., Kel, A., Wingender, E. and Borlak, J. (2013). A discriminative approach for unsupervised clustering of DNA sequence motifs. *Plos Computational Biology*, **9**(3), e1002958.

Yao, J., Zhao, Q., Yuan, Y. *et al.* (2012). Identification of common prognostic gene expression signatures with biological meanings from microarray gene expression datasets. *Plos One*, **7**(9), e45894.

Part Six

New Clustering Frameworks Designed for Bioinformatics

23

Splitting-Merging Awareness Tactics (SMART)

23.1 Introduction

Clustering is one of the most difficult and challenging problems in the realm of machine learning due to the lack of universal and rigorous mathematical definition. There have been many families of clustering algorithms used in biological data analysis, including partitional clustering, hierarchical clustering, model-based clustering, fuzzy clustering and so on, as we have detailed in Part Three of this book. Results of most of the successful clustering algorithms strongly depend on the determined number of clusters, for example k-means, model-based clustering and hierarchical clustering (when the clustering memberships need to be determined). However, in many cases, *a priori* knowledge of the actual number of clusters is not available. Thus, the number of clusters has to be estimated beforehand. The problem of determining the best number of clusters needs to be addressed in another branch of research in clustering analysis, known as clustering validation, which we have also discussed in Chapter 19.

Once an appropriate clustering-validity index is selected, the general practice for determining the best number of clusters has few steps: a set of clustering results are first obtained by a clustering algorithm with a fixed number of clusters within a predetermined range $[K_{min}, K_{max}]$; then, these clustering results are evaluated by the chosen validity index; finally, depending on the chosen validity index, a maximum or minimum index value indicates the best number of clusters (in some cases if the index value has an increasing or decreasing trend against the number of clusters, the significant knee point indicates the best number of clusters). However, this solution requires an extensive search for the number of clusters and is tedious work for a large number of clusters.

Moreover, the initialisation of clustering is also a major issue. For some algorithms with deterministic initialisation, for example hierarchical clustering and k-means clustering with Kaufman approach initialisation (KA), the optimal solution is not always guaranteed. For some

Integrative Cluster Analysis in Bioinformatics, First Edition. Basel Abu-Jamous, Rui Fa and Asoke K. Nandi.
© 2015 John Wiley & Sons, Ltd. Published 2015 by John Wiley & Sons, Ltd.

algorithms sensitive to initialisation, such as k-means with random initialisation, expectation-maximisation (EM) and self-organisation map (SOM), they may get stuck at a local minimum. Addressing this problem requires running the algorithm repeatedly with the same dataset using several different initialisations. This makes such clustering algorithms more computationally unfavourable. Thus, better options would be integrative frameworks or strategies which provide an automatic and consistent clustering, so users do not have to worry about setting those data-specific parameters.

Therefore, automated clustering without employing any *a priori* knowledge of the number of clusters is always desired. To implement automated clustering, one has to integrate both clustering algorithm and clustering validation into a framework so that the product of the clustering fits the data well without *a priori* knowledge. There have been many algorithms designed for clustering without specifying the number of clusters; for example, self-splitting competitive learning (SSCL) (Zhang and Liu, 2002), self-splitting-merging competitive learning (SSMCL) (Wu *et al.*, 2004), unsupervised learning of finite mixture models (ULFMM) (Figueiredo and Jain, 2002), variational Bayesian Gaussian model (VBGM) (Teschendorff *et al.*, 2005), parameter-free clustering (PFClust) (Mavridis, Nath and Mitchell, 2013), density-based spatial clustering of applications with noise (DBSCAN) (Ester *et al.*, 1996), and so on. However, some of them still need the upper bound of the number of clusters, for instance SSMCL, ULFMM and VBGM; some of them require data-dependent parameters, for instance SSCL and DBSCAN; and some of them have poor clustering performance under noisy conditions, for instance SSCL and PFClust.

In this chapter, a recent clustering framework, called splitting-merging awareness tactics (SMART) (Fa and Nandi, 2012, 2013, 2014; Fa *et al.*, 2013; Fa, Roberts, and Nandi, 2014), is introduced. SMART integrates many crucial clustering techniques such as cluster-splitting methods, cluster-similarity measurement and clustering selection, within a framework to mimic human perception doing the sorting and grouping. The framework starts with one or two clusters and accomplishes many clustering tasks to split and merge clusters. While splitting, a merging process is also taking place to merge the clusters which meet the merging criterion. In this process, SMART has the ability to split and merge clusters automatically in iterations. Once the stop criterion is met, the splitting process terminates and then a clustering-selection method is employed to choose the best clustering from several generated ones. Moreover, the SMART framework is not restricted to a specific clustering technique.

The rest of this chapter is organised in the following sequence. The next section introduces related previous work, including SSCL, SSMCL, ULFMM, VBGM, PFClust and DBSCAN. Section 23.3 describes the philosophy of the SMART framework. Subsequently, the clustering techniques employed in the SMART framework are detailed in Section 23.4. An enhanced SMART (E-SMART) structure is introduced in Section 23.5. Finally, we present some examples and draw conclusions.

23.2   Related Work

23.2.1   SSCL

SSCL by Zhang and Liu (2002) is one of the neural network-based clustering algorithm employing a competitive learning (CL) paradigm called one-prototype-take-one-cluster (OPTOC), which was introduced in Chapter 14. SSCL employs many auxiliary vectors, including asymptotic property vector (APV), centre property vector (CPV) and distant property vector (DPV) in the learning of each prototype. After the first split happens, the prototypes start to compete for updating when a pattern is presented. Each time a prototype splits into two prototypes, one stays at the same location as the mother prototype, the other is spawned at a distant location, which is indicated by the DPV of its mother prototype. Then all the prototypes available will reset their winning counters and start to compete according to the nearest-neighbour condition. It has no dependency on the initial locations of prototypes. Moreover, it may find the right number of natural clusters via the adaptive splitting processes. However, there are two vital issues to prevent its practical uses: (1) the prototypes are easily trapped into a global centroid, especially the first few ones, and (2) the parameters for stopping both OPTOC learning and splitting are crucial to the algorithm but they are difficult to estimate reliably (Wu *et al.*, 2004).

23.2.2 SSMCL

Wu *et al.* proposed an SSMCL algorithm for clustering microarray gene expression data (Wu *et al.*, 2004), which also employed the OPTOC paradigm. Different from SSCL, SSMCL employs the OPTOC paradigm to over-cluster the dataset to a large number of partitions, say k_{max}, then it merges partitions to fewer clusters, which were closer to the natural clusters. This strategy is called splitting-then-merging (STM). The merging criterion is that when two clusters are close to each other to the extent that their joint probability density function from a unimodal structure, then it would be reasonable to merge these two clusters into one. However, there are also two critical problems of SSMCL: (1) it requires the upper bound of the number of clusters, which sometimes could be unreasonably large, say the number of data objects; (2) the merging criterion is coarse, and sometimes may fail, especially in high-dimensional datasets.

23.2.3 ULFMM

ULFMM, proposed by Figueiredo and Jain (Figueiredo and Jain, 2002), belongs to mixture model-based clustering. But, strictly speaking, it is an automatic clustering framework rather than a pure clustering algorithm. Similar to SSMCL, ULFMM employs the STM strategy, which splits the dataset into k_{max} clusters and then merges those clusters close enough to each other until only one cluster remains. The component-wise expectation maximisation (CEM), which is a variant of EM, is used as the learning method. The CEM will be detailed in Section 23.4.2. The CEM algorithm is essential in this framework in the sense that all components are updated successively and compete with each other. Strong components survive while weak components die. During the merging procedure, many clustering results are generated. A model-based clustering validation, minimum message length (MML), which was introduced in Chapter 20, is integrated to select the best clustering as the output in terms of the MML value.

23.2.4 VBGM

Variational Bayesian Gaussian mixture (VBGM) has been discussed in Chapter 15. Teschendorff *et al.* applied VBGM in the analysis of gene expression data (Teschendorff *et al.*, 2005).

The interested readers may be referred to Chapter 15, or references (Beal and Ghahramani, 2003; Bishop, 2006) for the technical details. As we mentioned before, Bayesian methods provide a solution to the over-fitting problem in principle. They may be regarded as estimating the uncertainty of the model as a whole and the uncertainty in the estimated parameters themselves. VB is one of the methods which make the evaluation of posterior distribution tractable. VBGM employs the Variational Bayesian Expectation Maximisation (VBEM) algorithm, which also requires the upper bound of the number of clusters. Remarkably, VBGM does not use any clustering validation, while it calculates the marginal probability for candidate-clustering results and selects the one with the largest value.

23.2.5   PFClust

Recently, a parameter-free clustering (PFClust) algorithm was proposed by Mavridis, Nath and Mitchell (2013). PFClust is able to cluster data and identify a suitable number of clusters automatically without requiring any parameters to be specified by the user. The algorithm partitions a dataset into a number of clusters that share some common attributions, such as variance of intra-cluster similarity. The clustering algorithm consists of two parts: the first part is the randomisation, and the second part incorporates both the threshold selection and the actual clustering. In the randomisation step, 20 thresholds are estimated by a randomisation process. These thresholds are used to cluster the data and only the best threshold is selected in the second step. The whole randomisation and threshold-selection procedure is carried out multiple times. The threshold values are estimated by multiple random clusterings of the data. A random number of clusters are chosen and data points are randomly assigned in one of clusters. Therefore, the intra-cluster mean similarities from every individual cluster across a number of randomisations are collated into a single distribution. From this distribution, 20 threshold values are retrieved from 95 to 100% significant levels. Then for each threshold value, the dataset is clustered with a similarity-based clustering. All elements are separated and no cluster is defined. The two most similar elements are placed together to form the first cluster. This procedure goes on, like an agglomerative algorithm, until the average similarity of the cluster exceeds the threshold. This process is continued iteratively until all data points have been clustered, or until remaining elements cannot form a cluster that has an expectation value of intra-cluster similarity greater than the threshold. The Silhouette width (which was introduced in Chapter 20) is the main factor used in deciding which threshold produces the best clustering.

23.2.6   DBSCAN

DBSCAN is a density-based clustering algorithm, proposed by Ester et al. (1996). It attempts to find a number of clusters starting from the estimated density distribution of corresponding data points. In principle, a data point x_i is directly density-reachable from another data point x_j if their distance is not farther away than a given distance ϵ. Data point x_i is considered density-reachable from x_j if there is a sequence $x_1, \ldots, x_n$ of data points with $x_1 = x_i$, and $x_n = x_j$, where each data point in this sequence is directly density-reachable from its neighbour points. DBSCAN is based on the concept of density-reachability. DBSCAN requires two parameters, ϵ and the minimum number of points required to form a dense region. It starts with an arbitrary starting point and the ϵ-neighbourhood of this point is retrieved. If it contains sufficiently many points, a cluster starts; otherwise, the point is labelled as noise. Once a cluster is considered as dense enough, all points that are found within the ϵ-neighbourhood are added if they are also dense. This process continues until the whole dense cluster is completely found. Then it starts with a new unvisited point for the next cluster, leading to the discovery of the whole clustering structure. DBSCAN is one of the most common clustering algorithms; however, it still requires the data-dependent parameter ϵ to control the size of the clusters.

23.3   SMART Framework

First of all, we must emphasise that SMART is a framework rather than a simple clustering algorithm, within which a number of clustering techniques are organically integrated. Thus, conceptually, SMART does not fall into any categories or families, which have been introduced in Part Three of this book. In this section, we focus on an overview of the whole framework, and describe implementation solutions and specific clustering techniques in the following sections.

As depicted in Figure 23.1, the whole clustering procedure is divided into four tasks. SMART starts with one or two clusters and the cluster needs to be initialised, which is Task 1. Subsequently, the data go through a splitting-while-merging (SWM) process, where splitting and merging are automatically conducted in iterations. In the splitting step of each iteration, which

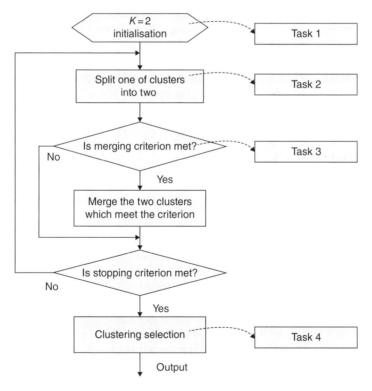

Figure 23.1   Flow chart of the SMART framework

is labelled Task 2, SMART splits one of the clusters into two. After a splitting step, the new clustering is censored by a merging criterion, which is associated with Task 3. If the condition for merging is satisfied, then one merges the two clusters, otherwise the merging step is skipped. SMART then goes through a termination-check, where a stopping criterion is applied. If the condition for termination is not satisfied, SMART goes to the next iteration and continues to split; otherwise, SMART finishes the splitting-merging process. The last step is the clustering selection (Task 4).

Note that these tasks in the SMART flow can be completed using many clustering techniques in the literature; for example, Task 1 can be done by any initialisation technique, either deterministic or random; Tasks 2 and 3 may be achieved by any splitting algorithm and merging criterion, respectively, or they may be combined into one algorithm; and Task 4 can be accomplished by any of either model order-selection algorithms or validity indices. Different techniques will make the implementation slightly different but the flow does not change. Moreover, different clustering algorithms bring different features into the framework and so SMART can be customised for different applications. In the following section, we will describe many SMART algorithms using different splitting and merging algorithms; that is, OPTOC CL, finite mixture model (FMM), and mixture of factor analysors (MFA), which are called SMART-CL (or SMART I), SMART-FMM (or SMART II) and SMART-MFA, respectively, and they have similar configurations. In particular, both use MML as their clustering-selection algorithm and use the same termination criterion in the SWM process, namely the maximum number of merges, N_{max}. The logic behind the termination criterion is that, normally, merging will not start until clustering is somehow saturated, that is, optimal clustering is nearly reached. Once N_{max} is reached, the splitting and merging will terminate automatically. All existing self-splitting-merging algorithms employ the STM strategy with different clustering paradigms; instead our SMART algorithms employ the SWM strategy. For the purposes of direct comparison with the existing STM algorithms, we propose two specific SMART algorithms. Nevertheless, it should be noted that, within the proposed SMART framework, many other algorithms can be derived for different clustering paradigms.

23.4 Implementations

23.4.1 SMART-CL

SMART with CL (Fa and Nandi, 2012) uses the CL method; more specifically, the OPTOC paradigm, as splitting algorithm. It is also called SMART-I because it is the first implementation of the SMART framework (Fa, Roberts and Nandi, 2014). In SMART-CL, OPTOC CL is employed to deal with Task 2. Given each prototype P_k, the key technique is that an online learning vector, APV A_k, is assigned to guide the learning of this prototype. For simplicity, A_k represents the APV for prototype P_k, and n_{A_k} denotes the learning counter (winning counter) of A_k. As a necessary condition of the OPTOC mechanism, A_k is required to initialise at a random location, which is far from its associated prototype P_k, and n_{A_k} is initially zero. Taking the input pattern x_i as a neighbour if it satisfies the condition $\langle P_k, x_i \rangle \leq \langle P_k, A_k \rangle$, where $\langle \cdot, \cdot \rangle$ is the inner product operator. To implement the OPTOC paradigm, A_k is updated online to construct a dynamic neighbourhood of P_k. The patterns 'outside' the dynamic neighbourhood will contribute less to the learning of P_k as compared with the 'inside' patterns.

In addition to the APV, there is another auxiliary vector, called DPV R_k, assisting the cluster, which contains more than one prototype, to split. Let n_{A_k} denote the learning counter for R_k, which is initialised to zero. Vector R_k will be updated to a distant location from P_k. The efficiency of splitting is improved by determining the update schedule of R_k adaptively from the analysis of the feature space. Contrary to the APV A_k, the DPV R_k always tries to move away from P_k.

The remarkable differences of SMART-CL from SSCL and SSMCL lie in two strategies. The first strategy is the SWM where the merging process is carried out while the dataset is split into many pieces by the OPTOC method. Every pair of clusters is measured by calculating the similarity between them by **Cohesion,** which was proposed by Lin and Chen (2005). The cohesion metric is defined as given in Equation (23.1),

$$\text{chs}(C_k, C_l) = \frac{\sum_{x \in C_k, C_l} \text{joint}(x, C_k, C_l)}{|C_k| + |C_l|} \tag{23.1}$$

where $|C_k|$ is the size of the cluster C_k. The term $\text{joint}(x, C_k, C_l)$ defines the similarity of the two clusters referring to the existence of an object x, which is defined as shown in Equation (23.2),

$$\text{joint}(x, C_k, C_l) = \min(f_k(x), f_l(x)) \tag{23.2}$$

where $f_k(x)$ and $f_l(x)$ are the probability density functions of the distributions in clusters C_k and C_l.

The second strategy is that SMART-CL employs MML as the clustering-selection criterion. MML is one of the minimum encoding-length criteria, which we have discussed in Chapter 20. Therefore, these techniques are integrated into the SMART framework to complete the whole clustering task. The pseudo-code of SMART-CL is presented in Table 23.1.

Table 23.1   The pseudo-code for SMART-CL

Task 1: Initialising SMART with $K = 1$
Randomly select P_1 and find the farthest object as A_1 and initialise $R_1 = P_1$;
terminate = 0;
WHILE !terminate
Task 2: Use the OPTOC paradigm for the learning of prototype, and the splitting of the cluster with largest variance;
IF the prototype P_k does not converge
Go back to Task 2;
ENDIF
Task 3: Calculate pairwise cohesions for all converged prototypes;
IF The maximum of cohesions is a few times larger than the median of cohesions
Merge the pair of cluster with the maximum cohesion;
Go back to Task 3 to continue merging;
ENDIF
The stage for recoding candidate clustering.
IF The number of merges is greater than or equal to N_m
terminate = 1;
ENDIF
ENDWHILE
Task 4: Calculate the length for every converged clustering; output the clustering with the minimum length.

Normally, Task 1 in SMART can be done by any initialisation algorithm, either random or deterministic, like the KA algorithm (Kaufman and Rousseeuw, 1990). In SMART-CL implementation, a simple random initialisation is used. The first prototype P_1 is randomly selected, the APV A_1 is the farthest object away from P_1, and the DPV R_1 is initialised as P_1. Thereafter, the SWM process starts. Learning with the OPTOC paradigm drags the prototype to its neighbour, which is 'inside' the range of APV, and also drags the APV towards the prototype. Task 2 will not finish until every prototype converges. Since OPTOC is an online learning algorithm, systematic errors may be introduced by the order in which data are fed into the algorithm. Thus, every time OPTOC starts, the order of input data is randomised.

Once the prototypes converge, Task 3 commences. The pairwise cohesions are calculated to measure the distance between the prototype clusters. A criterion is set to guide the merging process, stating that if the maximum of the cohesions is T_{chs}-times more than the majority of the cohesions, it reveals that two prototypes with this maximal cohesion are close enough to merge. The merging process continues until no further merge occurs. A merging counter records the number of merges. After the merging process finishes, the clustering is recorded as the candidate to output. If the merging counter exceeds the maximum number of merges N_{max}, the SWM process is terminated automatically; otherwise, it goes to Task 2 and continues splitting. Once the SWM process finishes, all the candidates are fed into the MML algorithm, which is associated with Task 4. The final clustering result is the one which minimises the MML value.

Note that there are two parameters T_{chs} and N_{max} that have to be set in SMART, but they are neutral; that is, T_{chs} is a relative number rather than an absolute one, and is a data-independent value; the reason for setting N_{max} is that normally merging occurs frequently after the natural clustering has been reached. The reasonable value for T_{chs} is 20 and that for N_{max} is 5.

23.4.2   SMART-FMM

The key technique in SMART-FMM (Fa and Nandi, 2013; Fa, Roberts and Nandi, 2014) is a modified component-wise expectation maximisation of mixtures (CEM^2) (Celeux *et al.*, 2001; Figueiredo and Jain, 2002). Since the finite mixture model (FMM) and the EM algorithm are very well known topics, we will not address their details here and readers may refer to Chapter 15. The conventional EM algorithm for mixture model has many drawbacks; for example, it is sensitive to initialisation and it is a locally greedy method that may be trapped into local minima. Therefore, the CEM^2 was proposed by Celeux *et al.* (2001) and modified by Figueiredo and Jain (2002). The greatest advantage of modified CEM^2 is that the weaker component may naturally be excluded in the iterative process, which gives the stronger ones a better chance of survival. From the merging point of view, it is a merging process combined with learning.

Unlike the conventional EM algorithm, CEM^2 updates the model parameters $\{\theta_k | k = 1, ..., K\}$ and the probabilities of components $\{\alpha_k | k = 1, ..., K\}$ sequentially, rather than simultaneously. In CEM^2, the estimation is also a two-step process, but in each iteration only one component has the opportunity to update its parameters. For the j-component, it alternates the steps:

- **CEM^2 E-step:** Compute the conditional expectation $\Gamma = \{\gamma_{k,i} | k = 1, ..., K; i = 1, ..., N\}$ of the missing labels Z for $i = 1, ..., N$ and $k = 1, ..., K$, such that Equation (23.3) holds,

$$\gamma_{k,i} \equiv \mathbb{E}\left[\hat{z}_{k,i}|X,\hat{\boldsymbol{\theta}}\right] = \frac{\hat{\alpha}_k p\left(\boldsymbol{x}_i|\hat{\theta}_k\right)}{\sum_{l=1}^{K}\hat{\alpha}_l p\left(\boldsymbol{x}_i|\hat{\theta}_l\right)} \tag{23.3}$$

- **CEM² M-step:** Set

$$\hat{\alpha}_j^* = \frac{\sum_{i=1}^{N}\gamma_{j,i}}{\sum_{l=1}^{K}\sum_{i=1}^{N}\gamma_{l,i}} \tag{23.4}$$

$$\hat{\boldsymbol{\theta}}_j^* = \arg\max_{\hat{\theta}_j}\left\{\log p(X|\boldsymbol{\theta})\right\} \tag{23.5}$$

For $l \neq j$, $\hat{\alpha}_l^* = \hat{\alpha}_l$ and $\hat{\boldsymbol{\theta}}_l^* = \hat{\boldsymbol{\theta}}_l$.

In Figueiredo and Jain (2002), the adoption of a Dirichlet-type prior for α_k results a new M-step [Equation (23.6)].

$$\hat{\alpha}_k^* = \frac{\max\left\{0, \sum_{i=1}^{N}\gamma_{j,i} - \frac{N_p}{2}\right\}}{\sum_{l=1}^{K}\left\{0, \sum_{i=1}^{N}\gamma_{l,i} - \frac{N_p}{2}\right\}}, \quad k = 1,\ldots,K \tag{23.6}$$

The corresponding components $\hat{\boldsymbol{\theta}}_k^*$ with $\hat{\alpha}_k^* = 0$ are eliminated and become irrelevant. This component annihilation can be also explained in an estimation theoretic point of view as that the estimates are not accurate unless enough samples are involved. Those estimates without enough samples are dismissed and in turn others have more chances to survive. Modified CEM² can fulfil learning and merging, which are associated with Tasks 2 (only learning part) and 3, respectively, in SMART-FMM.

Compared with SMART-CL, SMART-FMM is easier to implement since modified CEM² can do both learning and merging. In addition to the learning and merging techniques, there are two configurations different from SMART-CL. The first is that in SMART-FMM, we initially start with $K = 2$ because $K = 1$ does not need learning, but $K = 1$ is still included in the candidate list for selection in the output. The second is that the splitting process cannot be done by modified CEM² and has to be specified. Once all components converge and all zero-probability components are discounted, a new component will be injected into the framework. This new component is initialised deterministically by using the farthest object away from the closest component among all the components as the mean and averaged covariance matrix of all components' covariance matrices, as given by Equations (23.7) and (23.8),

$$\boldsymbol{\mu}_{K+1} = \arg\max_{x \in X}\left\{\min_{1 \leq k \leq K} D(\boldsymbol{x},\boldsymbol{\mu}_k)\right\} \tag{23.7}$$

$$\boldsymbol{\Psi}_{K+1} = \frac{1}{K}\sum_{k=1}^{K}\{\boldsymbol{\Psi}_k\} \tag{23.8}$$

where $D(\cdot,\cdot)$ is a distance metric, and then the clustering splits $K = (K + 1)$. The pseudo-code for SMART-FMM is given in Table 23.2. The stage for recoding the candidate clustering is after

Table 23.2   The pseudo-code for SMART-FMM

Task 1: Initialising SMART with $K = 2$

Randomly initialise $\widehat{\boldsymbol{\theta}}_k$ and $\widehat{\alpha}_k$ for $k = 1, 2$;

terminate = 0;

WHILE !terminate

Tasks 2 and 3: Use modified CEM^2 for the learning and merging based on Equations (23.3) and (23.6).

IF the prototype $\widehat{\boldsymbol{\theta}}_k$ does not converge

Go back to Tasks 2 and 3;

ENDIF

The stage for recoding candidate clustering.

Splitting: Calculate the parameters for new components (23.7) and (23.8);

IF The number of merges is greater than or equal to $N_{\max}$

terminate = 1;

ENDIF

ENDWHILE

Task 4: Calculate the length for every converged clustering; output the clustering with the minimum length.

---

all the current components have converged and all merges finish and before the splitting for a new component starts.

23.4.3   SMART-MFA

To meet the demand of clustering high-dimensional data efficiently, a component-wise expectation conditional maximisation (CW-ECM) algorithm was proposed and integrated within the SWM framework, for the mixture of factor analysers (MFA) model (Fa and Nandi, 2014). The MFA model has been discussed in Chapter 15. The SMART-MFA algorithm has two advantages over conditional MFA algorithms: it has ability to converge to the actual or close-to-actual number of clusters by an SWM strategy, and it avoids the local minima effectively and efficiently. Furthermore, the splitting strategy in the original SMART framework was improved and it may save more computational effort.

To estimate the latent label indicators $\boldsymbol{Z}$ conditional on the parameters $\boldsymbol{\Theta} = \{\boldsymbol{\Theta}_k | k = 1, \ldots, K\} = \{\boldsymbol{\mu}_k |_{k=1}^{K}, \boldsymbol{B}_k |_{k=1}^{K}, \boldsymbol{\Psi}_k |_{k=1}^{K}, \pi_k |_{k=1}^{K}\}$, the CW-ECM updates each component in parallel with a two-stage process; that is, E and CM steps. Suppose that in t-th iteration, for the kth component, it has parameters $\tilde{\Theta}^{(t,k)} = \left\{ \tilde{\Theta}_1^{(t,k)}, \ldots, \tilde{\Theta}_{k-1}^{(t,k)}, \tilde{\Theta}_k^{(t-1,k)}, \ldots, \tilde{\Theta}_K^{(t-1,k)} \right\}$ and alternates the steps as follows:

- **CW-ECM E-step:** Compute Equation (23.9) for $n = 1, \ldots, N$

$$
\gamma_{k,n} \equiv \mathbb{E}\left[\hat{z}_{k,n} | \boldsymbol{X}, \tilde{\Theta}^{(t,k)}\right] = \frac{f\left(\boldsymbol{x}_n | \tilde{\Theta}_k^{(t,k)}\right)}{\sum_{l=1}^{G} f\left(\boldsymbol{x}_i | \tilde{\Theta}_l^{(t,k)}\right)} \tag{23.9}
$$

where Equation (23.10) holds.

$$\tilde{\Theta}_l^{(t,k)} = \begin{cases} \Theta_l^{(t,k)} & \text{if } l < k \\ \Theta_l^{(t-1,k)} & \text{if } l \geq k \end{cases} \tag{23.10}$$

Thus, we may obtain Equation (23.11).

$$f\left(x_n | \tilde{\Theta}_l^{(t,k)}\right) = \begin{cases} \hat{\pi}_l^t p\left(x_n | \mu_l^{(t,k)}, B_l^{(t,k)}, \Psi_l^{(t,k)}\right) & \text{if } l < k \\ \hat{\pi}_l^{t-1} p\left(x_n | \mu_l^{(t-1,k)}, B_l^{(t-1,k)}, \Psi_l^{(t-1,k)}\right) & \text{if } l \geq k \end{cases} \tag{23.11}$$

- **CW-ECM M-step:** Set

Equation (23.12) applies.

$$\Theta_k^{(t,k)} = \arg \max_{\Theta_k^{(t-1,k)}} \left\{ \log p\left(X | \tilde{\Theta}^{(t,k)}\right) \right\} \tag{23.12}$$

More precisely, Equation (23.12) can be broken down into many individual updates [Equations (23.13)–(23.15)] as follows:

$$\hat{\pi}_k^{(t,k)} = \frac{\sum_{n=1}^N \gamma_{k,n}}{\sum_{l=1}^K \sum_{n=1}^N \gamma_{l,n}} \tag{23.13}$$

$$\mu_k^{(t,k)} = \frac{\sum_{n=1}^N \gamma_{k,n} x_n}{\sum_{n=1}^N \gamma_{k,n}} \tag{23.14}$$

$$B_k^{(t,k)} = S_k^{(t,k)} \beta_k^{(t,k)} \left(\beta_k^{(t,k)T} S_k^{(t,k)} \beta_k^{(t,k)} + \omega_k^{(t,k)}\right)^{-1} \tag{23.15}$$

where Equations (23.16)–(23.18) hold.

$$S_k^{(t,k)} = \frac{\sum_{n=1}^N \gamma_{k,n} \left(x_n - \mu_k^{(t,k)}\right)\left(x_n - \mu_k^{(t,k)}\right)^T}{\sum_{n=1}^N \gamma_{k,n}} \tag{23.16}$$

$$\beta_k^{(t,k)} = \left(B_k^{(t-1,k)} B_k^{(t-1,k)T} + \Psi_k^{(t-1,k)}\right) B_k^{(t-1,k)} \tag{23.17}$$

$$\omega_k^{(t,k)} = I_q - \beta_k^{(t,k)T} B_k^{(t-1,k)} \tag{23.18}$$

Then the updated estimate $\Psi_k^{(t,k)}$ is given by Equation (23.19).

$$\Psi_k^{(t,k)} = \text{diag}\left\{ S_k^{(t,k)} - S_k^{(t,k)} \beta_k^{(t,k)} B_k^{(t-1,k)T} \right\} \tag{23.19}$$

Note that in all EM algorithms for the MFA models, selecting the number of loading factors q is arbitrary. In our algorithm, we search a range of q in $[1, q_{max}]$.

Table 23.3   The pseudo-code for SMART-MFA

Task 1: Initialising SMART with $K = 2$
Randomly initialise $\boldsymbol{\mu}_k$, $\boldsymbol{B}_k$, $\boldsymbol{\Psi}_k$ and π_k for $k = 1, 2$;
terminate = 0;
WHILE !terminate
Learning and Merging
WHILE !converge
FOR $k = 1 : K$
Use CW-ECM for the learning and merging based on Equations (23.9) and (23.12)
IF $\hat{\pi}_k \rightarrow 0$
Get rid of k-th cluster; $K = K - 1$;
ENDIF
ENDFOR
ENDWHILE (!converge)
Splitting:
Employ the splitting strategy, $K = K + 1$;
IF The number of merges is greater than or equal to $N_{\max}$
terminate = 1;
ENDIF
ENDWHILE (!terminate)
Selecting:
Select the best clustering based on MML criterion.

The splitting strategy in the original SMART framework, which calculates all pairwise distances in the dataset and performs a KA style (Kaufman and Rousseeuw, 1990) to search for the next candidate to split, was improved (Fa and Nandi, 2014). Such calculation is huge when the size of the dataset is merely moderate, say around a thousand data objects. It is worth noticing that the likelihood of the given object allocated to the g-th cluster, which is $f(\boldsymbol{x}_n | \Theta_k)$, has been calculated during the EM algorithm, and it can be used to judge how likely the given object should be allocated in the cluster. If an object has a very small likelihood in every cluster, there may be two possibilities: one is that the object is an outlier, and another is that the object belongs to a cluster which has not been discovered. Even for an outlier, it may be an outlier near the existing clusters or near the cluster not yet having been discovered. In the CW-type EM (or ECM) algorithm, an actual cluster may survive from iteration to iteration even where the initial point is relatively far from its centre; on the other hand, the cluster may vanish if it is not an actual cluster.

Thus, we design our splitting strategy as outlined in the following few steps:

1. To create a pool to record the splitting candidates which have been selected;
2. To find a data object which has a minimum value of $\sum_{k=1}^{K} f(\boldsymbol{x}_n | \Theta_k)$ among all objects not in the pool;
3. To assign the data object as the $\boldsymbol{\mu}_{K+1}$ of the new $(K + 1)$th cluster, and generate $\boldsymbol{B}_{K+1}$, $\boldsymbol{\Psi}_{K+1} \pi_{K+1}$ randomly, record the data object in the pool, and $K = K + 1$.

To summarise, the pseudo-code of the proposed algorithm SMART-MFA is shown in Table 23.3.

23.5   Enhanced SMART

Successive processing is a very popular strategy in the signal processing and communication engineering fields, where the best example is successive interference cancellation (Fa and de Lamare, 2011; Li, de Lamare and Fa, 2011; Fa and Zhang, 2013). However, successive processing is seldom considered in clustering because of two major issues: (1) most successive processing algorithms know how many sources there are in the received signal before the processing stage, but most clustering algorithms do not know the correct number of clusters; and (2) the order of successive processing influences the performance of the outcomes greatly, so the better practice is to subtract the highest signal-to-noise ratio (SNR) from the original signal at each subtraction stage, but most clustering algorithms in the literature do not have the mechanism to order the clusters according to their quality. However, it has been found that the SMART framework has the ability to overcome these two problems, which motivates employing successive processing to enhance the performance of SMART further.

An enhanced SMART(E-SMART) framework using successive processing was proposed by Fa *et al.* (2013). Instead of selecting the best clustering from the results by using a clustering-selection criterion, we introduce a successive processing strategy into the framework to subtract clusters one by one in iterations. In doing so, the silhouette index (Rousseeuw, 1987) is employed to evaluate the intermediate clusters generated by the SWM process and order the clusters according to index values from high to low. Then we subtract the best cluster from the original data and iterate the remaining data back to the SWM process to start a new iteration. The process repeats successively and the clustering terminates automatically once no splitting happened in the SWM process. Consequently, all clusters can be obtained by iterations.

The flow chart of E-SMART is depicted in Figure 23.2. Let us first focus on the original SMART framework inside the large dotted block, which is labelled SMART. The whole clustering procedure is divided into four tasks. SMART starts with two clusters ($K = 2$), and the cluster needs to be initialised, which is Task 1. Subsequently, the data go through an SWM process, where splitting and merging are automatically conducted in iterations. In the splitting step of each iteration, which is labelled Task 2, SMART splits one of the clusters into two. After a splitting step, the new clustering is censored by a merging criterion, which is associated with Task 3. If the condition for merging is satisfied, then merge the two clusters, otherwise skip the merging step. Then SMART goes through a termination-check, where a stopping criterion is applied. If the condition for termination is not satisfied, SMART goes to the next iteration and continues to split; otherwise, SMART finishes the splitting-merging process. Particularly, if the number of mergings exceeds N_{max}, which denotes the maximum number of merging, the SWM process terminates and the intermediate clustering result is produced. Up to this point, these functioning blocks, which are coloured in grey, are kept in the E-SMART framework. The last step of the original SMART framework, which is the clustering selection according to Task 4 in the small dotted boxes, is removed.

In the new framework, a mechanism, which selects the best cluster from the intermediate clustering result generated by the SWM process and subtracts it from the remaining dataset X', is added. In each subtraction stage, the silhouette index is employed to evaluate the quality of each cluster and order the clusters based on their index values. Let us suppose that $X_k = \{x_{k,i} | 1 \le i \le N_k\}$ is the best cluster out of the remaining dataset X'. Note that in the first iteration, $X = X'$, and in the later iterations, $X' = X' - X_k$. The remaining dataset X' is sent back to the SWM process to start a new iteration. The clustering and subtracting continues until no splitting occurs in the remaining dataset. The E-SMART framework was implemented with FMM in Fa *et al.* (2013).

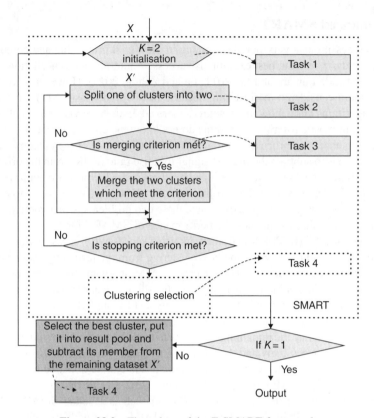

Figure 23.2   Flow chart of the E-SMART framework

23.6   Examples

First of all, a benchmark test dataset, which is a bivariate mixture model, is used to demonstrate how the SMART algorithms work. Then the SMART algorithms are investigated in some real gene expression datasets and compared with other state-of-the-art clustering algorithms.

The demonstration dataset is quadrature phase-shift keying (QPSK) data with SNR equal to 15 dB. This dataset can be viewed as a four-component Gaussian mixture. This example may clearly demonstrate how SMART-CL and SMART-FMM work, as shown in Figures 23.3 and 23.4, respectively. In both Figures 23.3 and 23.4, subfigures from (1) to (8) illustrate the proposed SWM process in the SMART framework, and subfigure (9) shows the final clustering result. The results show that the first merge of SMART I is after $K = 5$ shown in Figure 23.3-(5) and the first merge of SMART II is after $K = 5$ shown in Figure 23.4-(5). Subsequently, the merge counter measures the times of merges until the SWM process terminates.

Two real microarray gene expression datasets are studied using SMART. The performance comparisons are carried out between the SMART algorithms and other state-of-the art clustering algorithms, namely SSMCL, ULFMM, VBGM, DBSCAN, PFClust, MFA, mixture of common factor analysers (MCFA) and model-based clustering (MCLUST). The first real dataset is a subset of a leukaemia dataset (Golub *et al.*, 1999), which consists of 38 bone marrow samples obtained from acute leukaemia patients at time of diagnosis. There are

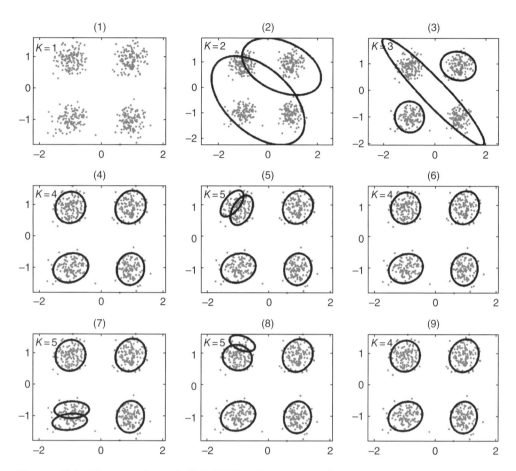

Figure 23.3 Demonstration of SMART-CL clustering in QPSK dataset. Sub-figures (1)–(8) demonstrate the procedure of SMART-CL (SWM process). It starts with $K = 1$ (sub-figure (1)), splits into K 2, $K = 3$, $K = 4$ and $K = 5$ shown in sub-figures (2)–(5) respectively, and then merges some clusters while splitting as shown in sub-figures (6)–(8). Sub-figure (9) is the final clustering result. Parameter settings: $T_{chs} = 20$ and $N_{max} = 5$

999 genes in the dataset (Monti *et al.*, 2003). The biological truth is that the samples include three groups: 11 acute myeloid leukaemia (AML) samples, 8 T-lineage acute lymphoblastic leukaemia (ALL) samples and 19 B-lineage ALL samples (Golub *et al.*, 1999; Monti *et al.*, 2003). The clustering experiments in this dataset were repeated 1000 times in order to test the ability of detecting the number of clusters for each method. A correct selection rate (CSR) of the number of clusters was defined as the ratio between the number of experiments where the true number of clusters is correctly selected, and the total number of experiments. Several clustering-validation algorithms were employed to validate the clustering results by every examined clustering algorithm.

The performance comparison is shown in Table 23.4. SSMCL and VBGM totally fail in this experiment, where SSMCL always converges to one cluster and VBGM always terminates at $k_{max} = 30$. SMART-CL has significantly better performance than ULFMM and has nearly

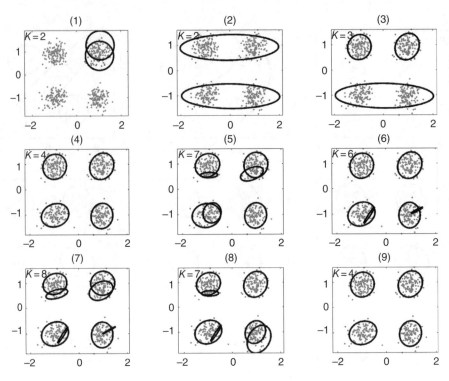

Figure 23.4 Demonstration of SMART-FMM clustering in QPSK dataset. Sub-figures (1)–(8) demonstrate the procedure of SMART-FMM (SWM process). It starts with $K = 2$ (sub-figure (1)), splits into $K = 2$, $K = 3$, $K = 4$ and $K = 7$ shown in sub-figures (2)–(5) respectively, and then merges some clusters while splitting as shown in sub-figures (6)–(8). Sub-figure (9) is the final clustering result. Parameter setting: $N_{max} = 5$

Table 23.4 Performance comparison of many metrics, including CSR, $\hat{K}$, MML, CH, SI for all algorithms in leukaemia dataset

Algorithm	$\hat{K}$ (q)	CSR (%)	MML	CH	SI
MFA	3(7)	—	4.23E + 04	6.42	0.35
MCFA	3(4)	—	4.22E + 04	6.48	0.35
SSMCL	1 ± 0	0.00	—	—	—
ULFMM	3.23 ± 0.54	69.40	3.91E ± 42.07E2	5.96 ± 0.89	0.32 ± 0.06
VBGM	30 ± 0	0.00	4.02E ± 42.27E3	0.78 ± 0.02	0.048 ± 0.013
DBSCAN	1 ± 0	0.00	—	—	—
MCLUST	2 ± 0	0.00	4.27E ± 40.0	6.73 ± 0.0	0.36 ± 0.0
PFClust	4 ± 0	0.00	4.31E ± 42.91	3.73 ± 4.3E-3	0.21 ± 2.51E-4
SMART-CL	**2.99 ± 0.13**	**99.00**	**3.89E4 ± 1.62E2**	**6.49 ± 0.3**	**0.36 ± 0.02**
SMART-FMM	**3 ± 0**	**100**	**2.9E4 ± 8.37E-3**	**6.75 ± 5.64E-5**	**0.36 ± 5.81E-8**

30% greater CSR and better performance in other metrics. In terms of mean and standard deviation of $\hat{K}$, SMART-CL has a mean closer to the true value and significantly smaller standard deviation than does ULFMM. Both MFA and MCFA have their lowest MML values with three clusters, but compared with two SMART algorithms, they show poorer performance in all metrics. SMART-FMM has the superior performance and always provides 100% CSR and best performance in all other metrics. Particularly, SMART-FMM also has very small variations in these metrics; that is, it provides consistent results even though it is randomly initialised. In this experiment, DBSCAN, MCLUST and PFClust perform poorly and do not provide the correct estimates of the true number of clusters. Furthermore, their other validation metrics are worse than the SMART-FMM algorithm.

Another real dataset is the yeast cell cycle α-38 dataset provided by Pramila *et al.* (2006). It consists of 500 genes with highest periodicity scores and each gene has 25 time samples. Additionally, their peaking times as percentages of the cell cycle have also been provided by Pramila *et al.* (2006). It is widely accepted that there are four phases in the cell cycle; namely, G1, S, G2 and M phases (Cho *et al.*, 1998; Spellman *et al.*, 1998). But there is no explicit knowledge about how many clusters should be in this dataset, so we cannot calculate CSR in this case. We obtain four clusters by using both SMART-CL and SMART-FMM, seven clusters by using ULFMM, eight clusters using SSMCL, three clusters using MFA with five factors, and five clusters using MCFA with six factors, as shown in Table 23.5. SMART-FMM has the superior performance as in other experiments. We note that VBGM fails again in this experiment as it requires a dimension reduction of the data before clustering. We do not perform a reduction in data dimensions to obtain a fair comparison.

To discern the effectiveness of the clusterings, we plot the histogram of the peak times of genes in each cluster for each algorithm, as depicted in Figure 23.5, where the grey bar plot is the histogram of the 500 genes in the dataset. Figure 23.5(a) and (b) show that four clusters represent reasonably good clustering since there are only few small overlap regions between clusters. Figure 23.5(c) and (d) indicate that many clusters crowd and overlap in the region of 5–30%, especially in Figure 23.5(c), and a clustering representing peaking at 20% superimposes on another cluster, which spans over 10–30%. These overlapped clusters have to be one cluster. Figure 23.5(e) and (f) show that MFA and MCFA also give reasonably good clustering

Table 23.5 Performance comparison of many metrics, including $\hat{K}$, MML, CH, SI, but excluding CSR, for all algorithms in yeast cell cycle α-38 dataset

Algorithm	$\hat{K}$ (q)	MML	CH	SI
MFA	3(5)	1.36E + 04	6.68	0.37
MCFA	5(6)	1.30E + 04	6.49	0.37
SSMCL	8	2.11E + 04	3.82	0.14
ULFMM	7	1.23E + 04	6.03	0.38
VBGM	20	3.97E + 04	1.98	0.17
DBSCAN	1	—	—	—
MCLUST	3	1.394	6.46	0.38
PFClust	6	1.24E + 04	3.94	0.32
SMART-CL	4	**1.26E + 04**	**6.27**	**0.37**
SMART-FMM	4	**1.16E + 04**	**6.86**	**0.39**

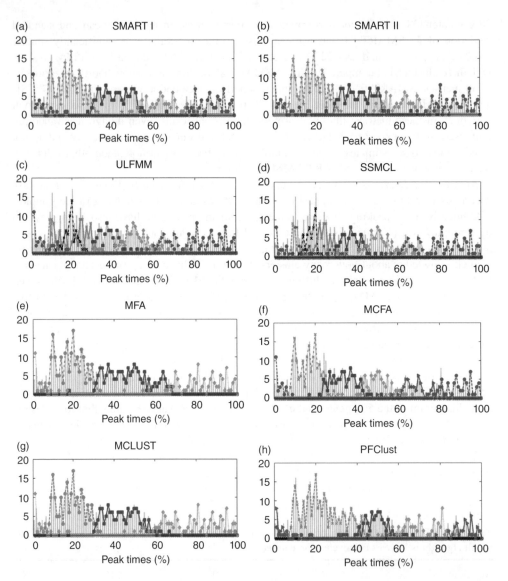

Figure 23.5 Histogram of the peak times of genes in each cluster for each algorithm in Yeast cell cycle α-38 dataset. (a) SMART-CL (SMART I), $T_{chs} = 20$ and $N_{max} = 5$, $K = 4$; (b) SMART-FMM (SMART II), $N_{max} = 5$, $K = 4$; (c) ULFMM, $k_{max} = 30$, $K = 7$; (d) SSMCL, $k_{max} = 30$, $K = 8$; (e) MFA, $q = 5$, $K = 3$; (f) MCFA, $q = 6$, $K = 5$; (g) MCLUST, $K = 3$; (h) PFClust. Sub-figures (a) and (b) show that four clusters represent reasonably good clustering since there are only few small overlap regions between clusters. Sub-figures (c) and (d) indicate that many clusters crowd and overlap in the region 5–30%, especially in Sub-figure (c), where a clustering representing peaking at 20% superimposes on another cluster, which spans over 10–30%. These overlapped clusters have to be one cluster. Sub-figures (e) and (f) show that MFA and MCFA also give reasonably good clustering results when judged by eye; however, clustering is poorer than SMART-FMM in the numerical metrics. Sub-figures (g) and (h) show the distribution of the peak times of genes based on the clustering results of MCLUST and PFClust, respectively (*See insert for color representation of the figure*)

results judged by eye, however being poorer than SMART-FMM in the numerical metrics. Figure 23.5(g) and (h) show the distribution of peak times of genes based on the clustering results of MCLUST and PFClust, respectively. MCLUST has a very similar performance to MFA. The partition provided by PFClust has a cluster (labelled by brown circles) overlapping with other clusters. The numerical metrics consistently indicate that PFClust performs poorly in the R2 dataset. Since DBSCAN and VBGM do not provide a reasonable result, we do not depict these in Figure 23.5. The results reveal that the SMART algorithms, especially SMART-FMM, provide a better representation than do other algorithms.

23.7 Discussion

A recently proposed SWM clustering framework, named SMART, has been described in this chapter. The framework employs an SWM process and intrinsically integrates many clustering techniques. SMART has the ability to split and merge the clusters automatically during the process. Once the stop criterion is met, the SWM process terminates and the optimal clustering result is selected as the final outcome by applying the selection criterion. Three main properties of the proposed SMART framework are summarised as: (1) needing no parameters dependent on the respective dataset or *a priori* knowledge about the datasets, (2) extendible to many different applications, (3) and offering superior performance compared with counterpart algorithms when tested on some benchmark and real biological datasets.

References

Beal, M.J. and Ghahramani, Z. (2003). *The Variational BayesianEM Algorithm for Incomplete Data: with Application to Scoring Graphical Model Structures*, Oxford University Press, Oxford.

Bishop, C.M. (2006). *Pattern Recognition and Machine Learning*, Springer, New York.

Celeux, G., Chrétien, S., Forbes, F. and Mkhadri, A. (2001). A component-wise EM algorithm for mixtures. *Journal of Computational and Graphical Statistics*, **10**(4), pp. 697–712.

Cho, R.J., Campbell, M.J., Winzeler, E.A. *et al.* (1998). A genome-wide transcriptional analysis of the mitotic cell cycle. *Molecular Cell*, **2**(1), pp. 65–73.

Ester, M., Kriegel, H.-P., Sander, J. and Xu, X. (1996). *A Density-Based Algorithm for Discovering Clusters in Large Spatial Databases with Noise*. KDD, Portland, OR, pp. 226–231.

Fa, R. and Lamare, R.D. (2011). Multi-branch successive interference cancellation for MIMO spatial multiplexing systems: design, analysis and adaptive implementation. *IET Communications*, **5**(4), pp. 484–494.

Fa, R. and Nandi, A. (2012). SMART: Novel Self Splitting-Merging Clustering Algorithm.), 2012 Proceedings of the 20th European Signal Processing Conference (EUSIPCO), 27–31 August 2012. Bucharest, IEEE, Piscataway, NJ, pp. 2198–2202.

Fa, R. and Nandi, A. (2013). An Enhanced Splitting-While-Merging Algorithm with Finite Mixture Models. IEEE International Conference on Acoustics, Speech and Signal Processing (ICASSP), 26–31 May 2013, Vancouver, IEEE, Piscataway, NJ, pp. 3332–3336.

Fa, R. and Nandi, A. (2014). Splitting-While-Merging Framework for Clustering High-Dimension Data With Component-Wise Expectation Conditional Maximisation, in *IEEE International on Acoustics, Speech and Signal Processing (ICASSP), 4–9 May 2013, Florence*, IEEE, Piscataway, NJ.

Fa, R. and Zhang, L. (2013). Generalised grouped minimum mean-squared error-based multi-stage interference cancellation scheme for orthogonal frequency division multiple access uplink systems with carrier frequency offsets. *IET Communications*, **7**(7), pp. 685–695.

Fa, R., Abu-Jamous, B., Roberts, D. and Nandi, A. (2013). Enhanced SMART Framework for Gene Clustering using Successive Processing. IEEE International Workshop on Machine Learning for Signal Processing (MLSP), 22–25 September 2013, *Southampton*, IEEE, Piscataway, NJ, pp. 1–6.

Fa, R., Roberts, D.J. and Nandi, A.K. (2014). SMART: unique splitting-while-merging framework for gene clustering. *Plos One*, **9**(4), e94141.

Figueiredo, M.A. and Jain, A.K. (2002). Unsupervised learning of finite mixture models. *IEEE Transactions on Pattern Analysis and Machine Intelligence*, **24**(3), pp. 381–396.

Golub, T.R., Slonim, D.K., Tamayo, P. *et al.* (1999). Molecular classification of cancer: class discovery and class prediction by gene expression monitoring. *Science*, **286**(5439), pp. 531–537.

Kaufman, L. and Rousseeuw, P.J. (1990). *Finding Groups in Data: an Introduction to Cluster Analysis*, John Wiley & Sons, Inc, New York.

Li, P., de Lamare, R.C. and Fa, R. (2011). Multiple feedback successive interference cancellation detection for multiuser MIMO systems. *IEEE Transactions on Wireless Communications*, **10**(8), pp. 2434–2439.

Lin, C.-R. and Chen, M.-S. (2005). Combining partitional and hierarchical algorithms for robust and efficient data clustering with cohesion self-merging. *IEEE Transactions on Knowledge and Data Engineering*, **17**(2), pp. 145–159.

Mavridis, L., Nath, N. and Mitchell, J.B. (2013). PFClust: a novel parameter free clustering algorithm. *BMC Bioinformatics*, **14**(1), p. 213.

Monti, S., Tamayo, P., Mesirov, J. and Golub, T. (2003). Consensus clustering: a resampling-based method for class discovery and visualization of gene expression microarray data. *Machine Learning*, **52**(1–2), pp. 91–118.

Pramila, T., Wu, W., Miles, S. *et al.* (2006). The Forkhead transcription factor Hcm1 regulates chromosome segregation genes and fills the S-phase gap in the transcriptional circuitry of the cell cycle. *Genes and Development*, **20**(16), pp. 2266–2278.

Rousseeuw, P.J. (1987). Silhouettes: a graphical aid to the interpretation and validation of cluster analysis. *Journal of Computational and Applied Mathematics*, **20**, pp. 53–65.

Spellman, P.T., Sherlock, G., Zhang, M.Q. *et al.* (1998). Comprehensive identification of cell cycle – regulated genes of the yeast Saccharomyces cerevisiae by microarray hybridization. *Molecular Biology of the Cell*, **9**(12), pp. 3273–3297.

Teschendorff, A.E., Wang, Y., Barbosa-Morais, N.L. *et al.* (2005). A variational Bayesian mixture modelling framework for cluster analysis of gene-expression data. *Bioinformatics*, **21**(13), pp. 3025–3033.

Wu, S., Liew, A.-C., Yan, H. and Yang, M. (2004). Cluster analysis of gene expression data based on self-splitting and merging competitive learning. *IEEE Transactions on Information Technology in Biomedicine*, **8**(1), pp. 5–15.

Zhang, Y.-J. and Liu, Z.-Q. (2002). Self-splitting competitive learning: a new on-line clustering paradigm. *IEEE Transactions on Neural Networks*, **13**(2), pp. 369–380.

24

Tightness-tunable Clustering (UNCLES)

24.1 Introduction

While considering a specific biological problem, most of the genes in the considered species' genome are expected to be irrelevant. Therefore, clustering the entire genome in a way that assigns each single gene to one of the output clusters does not conform to that biological fact. One commonly adopted approach to overcome this issue is to start by identifying a filtered subset of genes which are expected to be of interest through gene selection (Chapter 8) or differentially expressed genes identification (Chapter 9), and then to apply clustering to this filtered subset. Another approach, which has been proposed recently through the design of a sophisticated ensemble-clustering framework, has the crucial feature of producing clusters with *tunable tightness*. The core method in this framework is the *unification of clustering results from multiple datasets using external specifications (UNCLES)*, which is followed by a clustering-validation technique based on the *M–N scatter plots*. In this chapter we present this framework while discussing the various biological and computational problems that it tackles.

24.2 Bi-CoPaM Method

The binarisation of consensus partition matrices (Bi-CoPaM) method has been proposed recently to tackle various biology-specific clustering issues (Abu-Jamous *et al.*, 2013c). Given a set of L datasets $X_1, \ldots, X_L$ which measure the genetic expression of the same set of genes, and a set of C clustering methods $M_1, \ldots, M_C$, the Bi-CoPaM method is applied through four main steps described in the following four sub-sections.

Integrative Cluster Analysis in Bioinformatics, First Edition. Basel Abu-Jamous, Rui Fa and Asoke K. Nandi.
© 2015 John Wiley & Sons, Ltd. Published 2015 by John Wiley & Sons, Ltd.

24.2.1 Partition Generation

$R = L \times C$ partitions are generated by applying each of the C clustering methods to each of the L datasets while considering a constant number of clusters (K). The resulting R partitions are represented by the partition matrices U^1 to U^R, where each of which has K rows representing the clusters and N columns representing the genes (data points). The element $u^r_{k,i}$ represents the membership of the i-th gene (data point) in the k-th cluster according to the r-th partition. Three constraints [shown in Equations (24.1)–(24.3)] control these partition matrices:

$$u^r_{k,i} \in [0, 1], \forall r, \forall k, \forall i \tag{24.1}$$

$$\sum_{k=1}^{K} u^r_{k,i} = 1, \forall r, \forall i \tag{24.2}$$

$$0 < \sum_{i=1}^{N} u^r_{k,i} < N, \forall r, \forall k \tag{24.3}$$

The first constraint states that the membership of any given gene in any given cluster based on any given partition must be between zero and unity inclusively. If it was zero, the gene does not belong to that cluster at all; if it was equal to unity, the gene completely belongs to that cluster; while if it was between those values, the gene partially belongs to that cluster. In the special case of binary partitions, this membership value can have the value of only either zero or unity. The second constraint states that the total of the membership values of a specific gene in all of the clusters in a given partition must be unity. The third constraint states that each cluster must include some genes but not all of them.

24.2.2 Relabelling

Because clustering is unsupervised, there are no labels which map the clusters in one partition to their corresponding clusters in another partition. Therefore, the relabelling step is needed for this mapping. Given two partitions, the relabelling process can be modelled as a labelling correspondence problem, which is an NP-complete problem (Ayad and Kamel, 2010; Vega-Pons and Ruiz-Shulcloper, 2011), rendering the brute-force solution infeasible. Two techniques were considered for relabelling in Bi-CoPaM applications, namely the min-max (Abu-Jamous et al., 2013c; Abu-Jamous et al., 2013d) and the min-min approaches (Abu-Jamous et al., 2013a; Abu-Jamous et al., 2013b).

Let the function Relabel(U, U_{ref}) represent the solution to the problem of relabelling the clusters (rows) in the partition matrix U to be matched with the clusters (rows) in the reference partition matrix U^{ref}, and let $\hat{U}$ denote the resulting partition after relabelling. The first step is to produce a cluster-pairwise matrix with rows representing the clusters of U and columns representing the clusters of U^{ref}. Each element of this matrix represents the distance (dissimilarity) between the two corresponding clusters' membership vectors. Hamming distance can be used for binary clusters, while Manhattan or Euclidean distances can be generally used for fuzzy and binary clusters. An example of two pairwise similarity matrices is shown in Figure 24.1.

(b)

	1	2	3	4	5	6	7	8	9	10
1	5	7	9	9	10	7	8	6	11	10
2	0	13	8	6	8	12	7	9	8	5
3	9	1	6	8	9	6	11	7	17	8
4	4	5	8	13	16	7	9	9	8	9
5	7	9	15	6	9	9	6	12	18	7
6	4	5	10	4	11	15	8	8	9	7
7	10	11	9	6	9	18	16	1	12	6
8	6	5	8	9	15	9	8	7	9	12
9	8	6	13	2	11	7	7	6	10	8
10	12	5	7	7	9	9	10	9	7	9
Min	0	1	6	2	8	6	6	1	7	5

(a)

	1	2	3	4
1	2	7	6	7
2	5	0	3	4
3	9	5	7	1
4	6	1	8	6
Min	2	0	3	1

Figure 24.1 Sample cluster pairwise similarity matrix. (a) min-max approach. (b) min-min approach. The minimum of each column is shown in the last row and the (a) maximum or the (b) minimum of these minima is shaded

In the min-max approach, the minimum of each column in the matrix is found, and then the maximum of these minima is identified (Figure 24.1a). The pair of clusters which produced this maximum of minima are mapped to each other, the corresponding row and column are deleted (shaded in Figure 24.1a), and the process is repeated iteratively over the remaining matrix until all clusters are mapped. The min-min approach follows a similar approach but while considering the minimum of the minima rather than the maximum of them (Figure 24.1b).

The min-max approach aims at optimising the entirety of the relabelled clusters by avoiding convergence to the local minima that the min-min approach would converge to. This is because it gives priority to the clusters which do not have very strong similarity to any of the clusters in the other partition to be mapped to their relatively most similar cluster first. An example of this is the cluster which corresponds to the third column in the matrix in Figure 24.1a. In this case, if the excellent zero-distanced clusters, which are represented by the second row and the second column, were mapped to each other first, the cluster represented by the third column will be left with only relatively very dissimilar clusters to be paired with. The min-max approach tackles this problem by aiming at optimising the total of the dissimilarity values between the mapped pairs.

On the other hand, the min-min approach gives higher priority to the best matched pairs first, which is expected to produce some excellent pairs and many poor ones. This is useful when there are few clusters of interest within the results, and mapping them with their best peers should be preferably assured even if this affects the quality of the rest of the clusters. An example is shown in Figure 24.1b where there are four very good pairs of clusters out of ten. The min-min approach will result in producing those pairs of mapped clusters, namely the ordered pairs (2, 1), (3, 2), (7, 8) and (9, 4), while the six remaining pairs will be poor. In contrast, if the min-max approach were applied to this latter matrix, the first pair to be matched will be the ordered pair (2, 5). Note that the column (5) represents a cluster that is very dissimilar from all of the clusters represented by the rows. In this case, column (5) will be consuming the cluster represented by the second row, and preventing it from being mapped to its real best-match; that is, the first column. If the application requires obtaining few high-quality clusters, the min-min approach is surely the choice.

24.2.3   Fuzzy Consensus Partition Matrix Generation

Once the clusters in the partitions are relabelled, they are averaged in an element-by-element manner to produce a single fuzzy consensus partition matrix (CoPaM). This process is done by considering the first partition matrix U^1 as the first reference matrix U^{ref} and relabelling the second partition U^2 based on it. The two partitions are then combined by element-by-element averaging to produce the first intermediate fuzzy CoPaM $U^{int(1)}$. This intermediate matrix is then considered as the reference partition to relabel the following partition matrix, which is combined with it again to produce a new intermediate fuzzy CoPaM. After having relabelled and combined all of the R partition matrices, the last intermediate fuzzy CoPaM $U^{int(R)}$ is considered as the final fuzzy CoPaM U^*. This iterative process can be summarised by the following steps:

$$U^{int(1)} = U^1$$

For $r = 2$ to R

$$\hat{U}^r = \text{Relabel}\left(U^r, U^{int(r-1)}\right)$$

$$U^{int(r)} = \frac{1}{r}\sum_{r'=1}^{r}\hat{U}^{r'} = \frac{1}{r}\hat{U}^r + \frac{r-1}{r}U^{int(r-1)}$$

$$U^* = U^{int(R)}$$

24.2.4   Binarisation

The final stage of the Bi-CoPaM method is the binarisation of the fuzzy CoPaM by using one of six binarisation techniques. The result of binarisation is a binary CoPaM in which each gene's membership in each of the clusters is binary; that is, either zero (does not belong) or one (belongs). However, this method's binarisation relaxes the constraints in Equations (24.2) and (24.3), which control genes and clusters, respectively. At the level of genes, the gene can be exclusively assigned to a single cluster, which meets the constraint in Equation (24.2), but it can also be assigned to multiple clusters simultaneously or not to be assigned to any of the clusters at all. At the level of clusters, they can be complementary, as conventional clustering would produce, but they can also be wide and overlapping or tight and focused. Some clusters might be tightened to the extreme level at which they become totally empty, which does not meet the constraint in Equation (24.3).

Six binarisation techniques were proposed in Abu-Jamous *et al.* (2013c), and they belong to two main tracks as shown in Figure 24.2. The first track includes the techniques *top binarisation (TB)*, *maximum value binarisation (MVB)* and *difference threshold binarisation (DTB)*, and will be referred to hereinafter as the *TB-MVB-DTB track*. The second track includes the techniques *union binarisation (UB)*, *value threshold binarisation (VTB)* and *intersection binarisation (IB)*, and will be referred to hereinafter as the *UB-VTB-IB track*.

We start by describing the MVB technique; it assigns the gene exclusively to the cluster in which it has its largest membership value, and therefore it generates complementary clusters.

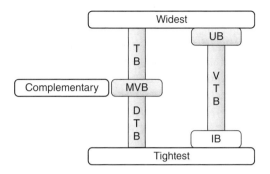

Figure 24.2   Binarisation tracks

Let the resulting binary CoPaM be B^* with K rows and N columns, where $b^*_{k,i} \in \{0, 1\}$ is an element in this matrix representing the binary membership of the ith gene in the kth cluster. Similarly, the corresponding fuzzy CoPaM is U^* with the elements $u^*_{k,i} \in [0, 1]$. Given that, the MVB technique can be expressed as given in Equation (24.4):

$$b^*_{k,i} = \begin{cases} 1, u^*_{k,i} = \max\limits_{1 \le j \le K} u^*_{j,i} \\ 0, \text{otherwise} \end{cases} \tag{24.4}$$

The TB technique moves from the MVB technique towards the wider-clusters side of the TB-MVB-DTB track. This is because it assigns the given gene to multiple clusters simultaneously if its membership values in them are not further than the value of the tuning parameter δ below its maximum membership value. The TB technique is expressed as shown in Equation (24.5):

$$b^*_{k,i} = \begin{cases} 1, u^*_{k,i} \ge \max\limits_{1 \le j \le K} u^*_{j,i} - \delta \\ 0, \text{otherwise} \end{cases} \tag{24.5}$$

On the other hand, and in a symmetric manner, the DTB technique moves from the MVB technique towards the tighter-clusters side of the track. This is by assigning a gene to the cluster in which it has its maximum membership value only if this value is far from the closest competitive cluster at least by the value of the tuning parameter δ; it is not assigned to any of the clusters otherwise. The DTB technique is expressed as shown in Equation (24.6):

$$b^*_{k,i} = \begin{cases} 1, u^*_{k,i} \ge \max\limits_{\substack{1 \le j \le K, \\ j \ne k}} u^*_{j,i} + \delta \\ 0, \text{otherwise} \end{cases} \tag{24.6}$$

When δ is equal to zero in TB or DTB, these techniques become identical to the MVB technique. When δ increases, TB or DTB start widening or tightening the clusters, respectively. The maximum value of δ is unity. When this value is reached, the TB technique reaches the extreme case of wide clusters in which each one of the clusters includes all of the genes. Also, at the

δ value of unity, the DTB produces the tightest clusters in which a gene is assigned to a cluster only if its fuzzy membership value is equal to unity in that cluster and is equal to zero in all of the other clusters; that is, if all of the R individual partitions have consensually assigned that gene to that cluster. Although the DTB would generate many empty clusters at $\delta = 1.0$, this result would not be trivial if some of the clusters still preserved some genes up to this tightest level, as opposed to the TB technique's results at such a δ value.

As for the second track, the UB technique assigns each gene to all of the clusters in which it has non-zero fuzzy membership values; that is, to all of the clusters in which at least one of the R individual partitions has assigned it. This generates wide and overlapping clusters. The UB technique is expressed as given in Equation (24.7):

$$b^*_{k,i} = \begin{cases} 1, & u^*_{k,i} > 0 \\ 0, & \text{otherwise} \end{cases} \tag{24.7}$$

On the other hand, the IB technique assigns a gene to a cluster only if all of the R individual partitions have consensually assigned that gene to it; that is, if its fuzzy membership value in it is unity while being zero elsewhere. This technique generates the tightest and most focused clusters, and is equivalent to the tightest clusters generated by the TB-MVB-DTB track, namely by the DTB technique at $\delta = 1.0$. The IB technique is expressed as given by Equation (24.8):

$$b^*_{k,i} = \begin{cases} 1, & u^*_{k,i} = 1.0 \\ 0, & \text{otherwise} \end{cases} \tag{24.8}$$

The VTB technique assigns a gene to a cluster if its membership in it is larger than or equal to the value of the tuning parameter α. When α is equal to zero, the VTB assigns each gene to all of the clusters, which is a trivial and useless result. At $\alpha = \varepsilon$, where ε is an arbitrarily small real positive number, the VTB technique becomes identical to the UB technique, and at $\alpha = 1.0$, it becomes identical to the IB technique. As the value of α increases from ε to unity, the clusters are tightened. The VTB technique is expressed as shown in Equation (24.9):

$$b^*_{k,i} = \begin{cases} 1, & u^*_{k,i} \geq \alpha \\ 0, & \text{otherwise} \end{cases} \tag{24.9}$$

24.3   UNCLES Method - Other Types of External Specifications

The general aim of the Bi-CoPaM method is to identify the subsets of genes that are consistently co-expressed in all of the given datasets. The different tunable binarisation techniques control how consistent this co-expression needs to be. Given the clustering results for the same set of genes from multiple datasets, other types of external specifications can be imposed while inferring the final consensus result. An important type of such external specifications is to combine the partitions to identify the subsets of genes that are consistently co-expressed in a subset of datasets while being poorly co-expressed in another subset of datasets. The method of the *unification of the clustering results from multiple datasets using external specifications,* abbreviated as *UNCLES,* generalises the Bi-CoPaM method by allowing various types of external specifications to be used in the process of combining partitions.

Two types of external specifications are described here:

Type A: the partitions are combined to identify the subsets of genes that are consistently co-expressed in all of the given datasets. This is equivalent to the Bi-CoPaM method.

Type B: the partitions are combined to identify the subsets of genes that are consistently co-expressed in a subset of datasets (S^+) while being poorly consistently co-expressed in another subset of datasets (S^-).

To apply UNCLES type B, the type A (Bi-CoPaM) is applied to each of the two subsets of datasets S^+ and S^- separately while considering the DTB binarisation technique with the δ values of δ^+ and δ^- respectively. Then, the genes that are included in the S^+ results and not included in the S^- results are included in the final result. The given result is referred to as the result of applying the UNCLES type B method with the tuning parameter pair (δ^+, δ^-). The parameter δ^+ controls how well co-expressed the genes should be in the S^+ datasets to be included in the final result, while the parameter δ^- controls how well co-expressed the genes should be in the S^- datasets to be excluded from the final result. Note that at the pairs $(\delta^+, 0)$ empty clusters are generated because at $\delta^- = 0$ all of the genes will be excluded from the final result.

24.4   M–N Scatter Plots Technique

Both types of UNCLES generate clusters with varying levels of wideness/tightness, which lead to largely varying sizes of such clusters. Known validation techniques are significantly affected by such variations. Therefore, a customised and sophisticated cluster-evaluation and -validation technique has been recently proposed; that is, the *M–N scatter plots* technique, where *M* refers to a modified version of the mean-square error (MSE) metric, and *N* refers to the number of genes included in the cluster, or, more specifically, the logarithm of that number (Abu-Jamous *et al.*, 2014a,b). The objective of the M–N scatter plots technique is to maximise the size of the cluster while minimising the mean-square error. This multi-objective technique suites the tunable nature of the clusters generated by the UNCLES method.

The M–N scatter plot is a 2-D plot on which the clusters are scattered, where the horizontal axis represents the MSE-based metric (MSE*) defined below, and the vertical axis represents the 10-based logarithm of the number of genes included in the cluster. The clusters closer to the top-left corner of this plot, after scaling each axis to have a unity length, are those that include more genes while maintaining low MSE* values, and are considered as better clusters based on this technique.

A sample M–N scatter plot is shown in Figure 24.3a. Each point on this plot represents one non-empty cluster where the one closest to the top left corner in Euclidean distance, after scaling the plot to have unity length at each side, is marked with a big solid circle, and is selected as the best cluster. The stars represent all of those clusters which are considered as other versions of that best cluster, and this can be identified based on the size of the overlap between the clusters. Before selecting the second best cluster, those clusters with similarity to the first best cluster are removed from the plot, and the resulting updated M–N plot, in this case, is shown in Figure 24.3b. The same step is repeated iteratively to select many clusters until the M–N plot

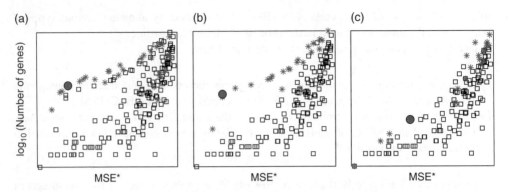

Figure 24.3   A sample of three iterations of cluster selection based on M–N scatter plots. (a) M–N scatter plot for the first iteration of cluster selection; this includes points for all of the available clusters; (b) the M–N scatter plot for the second iteration of cluster selection; this includes a subset of points of what is in (a); (c) the M–N scatter plot for the third iteration; this includes a subset of points of what is in (b)

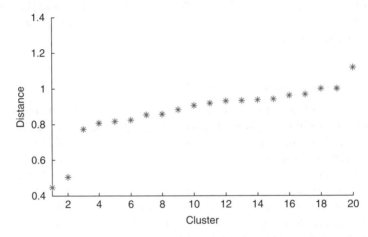

Figure 24.4   Distances of the 20 ordered clusters selected by the M–N plots from the top-left corners of those plots (shown in Figure 24.3)

has no more clusters or a specific termination criterion is met. For example, after 20 iterations, the M–N plot in Figure 24.3a becomes totally empty; the first three iterations are shown in Figure 24.3. The selected 20 clusters are ordered in quality from the closest to the top-left corner to the farthest, and those 20 distances are shown in Figure 24.4. Although 20 clusters are found in this example, the grace of having the clusters ordered allows selecting a few top clusters only. As in Figure 24.4, there is a large gap in distances between the second and the third clusters, which would lead the researcher to restrict oneself to the first two clusters only for further biological analysis.

The MSE-related metric (MSE*) is defined differently for UNCLE types A and B to meet their different objectives. For type A, the MSE* metric is the average of the MSE values for the considered cluster across all of the given datasets; whereas for type B, it is the signed difference between the average of the MSE values across the positive subset of datasets (S^+) and that average across the negative subset of datasets (S^-).

24.5   Parameter-free UNCLES with M–N Plots

The UNCLES method requires a number of parameters to run. One important parameter is K, which is a common parameter for most of the existing clustering methods. Other parameters include the tuning parameters δ, (δ^+, δ^-), and α for the different binarisation techniques of the method. The M–N scatter plots technique solves the problem of setting these parameters by its ability to evaluate the clustering results at the level of clusters rather than the level of partitions. In other words, the M–N scatter plots technique does not give a quality measure for an entire partition as many other measures do (e.g. Calinski–Harabasz (CH), Dunn index (DI), geometrical index (GI) and others); it rather evaluates each cluster individually, and then ranks them based on their relative quality. This feature can be exploited by considering multiple executions of the UNCLES method, each with a different parameter setting, and then exposing all of the generated clusters to M–N plots for final clusters selection.

The selected best few clusters might belong to different partitions; that is, they might have been generated at different K values or while considering different tightness levels. This means that the final result would not state which of the adopted K values or tuning parameters' values is the optimum; it would rather select the best clusters from the complete pool of clusters regardless of their origin. Another important note is that if different parameter settings resulted in very similar versions of the same cluster, the closest of the versions to the top-left corner will be selected only once and the rest are removed to allow for selecting the next distinct cluster.

24.6   Discussion and Summary

The UNCLES consensus clustering method has the ability to mine multiple datasets in order to identify focused subsets of genes that have consistent co-expression patterns under the given external specifications. Two main types of external specifications are presented in this chapter; type A requires the subsets of genes to be consistently co-expressed across all of the given datasets, while type B requires the subsets of genes to be consistently co-expressed in one subset of datasets while being poorly co-expressed in another subset of datasets.

Several parameters need to be set for the UNCLES methods, most notably K and the tuning parameters δ and (δ^+, δ^-). Since modifying those parameters results in different clusters with largely varying sizes, conventional cluster validation would not be suitable to evaluate the quality of such clusters in order to select the best of them. However, and as part of the UNCLES complete framework, the M–N scatter plots technique has been proposed in order to evaluate the results of UNCLES. The M–N scatter plots technique aims at minimising dispersion within the cluster (measured by the MSE), while maximising the size of the cluster. Given that, the framework of the UNCLES method and the M–N scatter plots provides complete pipelined steps for parameter-free, focused, and collective analysis of multiple expression datasets.

References

Abu-Jamous, B., Fa, R., Roberts, D.J. and Nandi, A.K. (2013a). Identification of Genes Consistently Co-Expressed in Multiple Microarrays by a Genome-Wide Approach. Proceedings of the International Conference on Acoustics, Speech, and Signal Processing (ICASSP-2013), May 2013, Vancouver, pp. 1172–1176.

Abu-Jamous, B., Fa, R., Roberts, D.J. and Nandi, A.K. (2013b). Method for the Identification of the Subsets of Genes Specifically Consistently Co-expressed in a Set of Datasets. Proceedings of the 2013 IEEE Workshop on Machine Learning for Signal Processing (MLSP-2013), September 2013, Southampton.

Abu-Jamous, B., Fa, R., Roberts, D.J. and Nandi, A.K. (2013c). Paradigm of tunable clustering using binarization of consensus partition matrices (Bi-CoPaM) for gene discovery. *Plos One*, **8**(2), e56432.

Abu-Jamous, B., Fa, R., Roberts, D.J. and Nandi, A.K. (2013d). Yeast gene CMR1 YDL156W is consistently co-expressed with genes participating in DNA-metabolic processes in a variety of stringent clustering experiments. *Journal of the Royal Society Interface*, **10**(81), 20120990.

Abu-Jamous, B., Fa, R., Roberts, D.J. and Nandi, A.K. (2014a). M-N Scatter Plots Technique for Evaluating Varying-Size Clusters and Setting the Parameters of Bi-CoPaM and UNCLES Methods. Proceedings of the International Conference on Acoustics, Speech, and Signal Processing (ICASSP-2014), May 2014, Florence, pp. 6726–6730.

Abu-Jamous, B., Fa, R., Roberts, D.J. and Nandi, A.K. (2014b). Comprehensive analysis of forty yeast microarray datasets reveals a novel subset of genes (APha-RiB) consistently negatively associated with ribosome biogenesis. *BMC Bioinformatics*, **15**, 322 (doi:10.1186/1471-2105-15-322).

Ayad, H.G. and Kamel, M.S. (2010). On voting-based consensus of cluster ensembles. *Pattern Recognition*, **43**, pp. 1943–1953.

Vega-Pons, S. and Ruiz-Shulcloper, J. (2011). A survey of clustering ensemble algorithms. *International Journal of Pattern Recognition and Artificial Intelligence*, **25**(3), pp. 337–372.

Appendix

High-throughput Data Resources

Database	Provider	Description
PubMed	NCBI	Literature citations
PMC	NCBI	Full-text journal papers
Bookshelf	NCBI	Books, reports, and documents
MeSH	NCBI/US NLM	Medical subject headings
GenBank	NCBI	DNA sequences
RefSeq	NCBI	DNA, RNA, and protein sequences (accessed by GenBank (Nucleotide) and Protein databases)
Entrez Gene	NCBI	Genes
Protein	NCBI	Proteins
GEO Series	NCBI	Microarray and NGS expression datasets
GEO Platforms	NCBI	Microarray and NGS platforms
GEO Samples	NCBI	Microarray and NGS expression samples
Taxonomy	NCBI	Organisms
HomoloGene	NCBI	Genes' homologues
SRA	NCBI	Next-generation sequencing (NGS) experiments and results
dbSNP	NCBI	Single-nucleotide polymorphisms and short sequence variations
dbVar	NCBI	Large sequence variations (insertions, deletions, translocations, and inversions)
BioProject	NCBI	Biological projects (studies)

(continued overleaf)

Integrative Cluster Analysis in Bioinformatics, First Edition. Basel Abu-Jamous, Rui Fa and Asoke K. Nandi.
© 2015 John Wiley & Sons, Ltd. Published 2015 by John Wiley & Sons, Ltd.

(*continued*)

Database	Provider	Description
BioSample	NCBI	Biological samples (e.g. expression data and epigenomics)
BioSystems	NCBI	Biological systems (e.g. metabolic and signalling pathways)
PubChem Substance	NCBI	Chemical substances submitted by researchers
PubChem Compound	NCBI	Validated chemical structures – links to PubChem Substance
PubChem Bio-Assay	NCBI	Bioactivity assays which screen chemical substances
dbGaP	NCBI	Associations of genotypes and phenotypes
MMDB	NCBI	Protein 3D structures
Genome	NCBI	Whole genomes
UniGene	NCBI	Transcript sequences that appear to come from the same transcriptional locus
Ensembl	EMBL-EBI and Wellcome Trust Sanger Institute	Genomic data for model vertebrate species and a few model non-vertebrate species (e.g. *C. elegans, D. melanogaster,* and *S. cerevisiae*)
Ensembl Genomes	EMBL-EBI	Extends Ensembl database by including thousands of genomes
EGA	EMBL-EBI	Associations of genotypes and phenotypes
ENA	EMBL-EBI	Nucleotide (DNA and RNA) sequences
GWAS	EMBL-EBI and US NHGRI	Single-nucleotide polymorphisms
ArrayExpress Archive	EMBL-EBI	Microarray and NGS expression data
Expression Atlas	EMBL-EBI	Additional layer of analysis to the data in the ArrayExpress Archive
PRIDE	EMBL-EBI	Proteomic data (e.g. protein-expression data)
MetaboLights	EMBL-EBI	Metabolomic data
Omics Archive	DDBJ	Microarray and NGS expression data
UniProt	EMBL-EBI and others	Protein sequences and information
InterPro	EMBL-EBI	Protein families, domains, functional sites, and motifs
EMDB	EMBL-EBI	Electron microscopy data
PDBe	EMBL-EBI	Protein molecular structures and cellular structures
ChEMBL	EMBL-EBI	Chemical biology data
ChEBL	EMBL-EBI	Chemical biology data
BioModels	EMBL-EBI	Computational models of biological processes
IntAct	EMBL-EBI	Molecular interactions
Reactome	EMBL-EBI	Curated human pathways
Enzyme portal	EMBL-EBI	Enzymes functions, structures, reactions, pathways, substrates, and so on.
Europe PMC	EMBL-EBI and others	Full-text literature; part of the NCBI international PMC
GO	GO Consortium	Gene ontologies (biological processes, molecular functions, ad cellular components)

(continued)

Database	Provider	Description
EFO	EMBL-EBI	An added-value GO databases annotating gene expression, genome-wide associations, and integrating genomic and disease data
EBI BioSamples	EMBL-EBI	Biological samples (gene expression, epigenomics, and others)
HMDB	University of Alberta	Chemical, clinical, and biochemical information about human metabolites
MGI/MGD	Jackson Laboratory	Mouse genomes
MGI/GXD	Jackson Laboratory	Mouse gene expression data
MGI/MTB	Jackson Laboratory	Mouse tumour data
MGI/MPD	Jackson Laboratory	Mouse phenome data
EMAP/EMA	UK MRC, Jackson Lab, and Heriot-Watt University	Mouse anatomy
EMAP/ EMAGE	UK MRC, Jackson Lab, and Heriot-Watt University	Mouse gene expression spatial data
RGD	Medical College of Wisconsin	Rat genome, genes, phenotypes, strains, diseases, physiology, and nutrition
Xenbase	International community	Frog genus *Xenopus* genomic, expression, and functional data
ZFIN	University of Oregon	Zebrafish genetic, genomic, and developmental data
WormBase	International community	Worm *C. elegans* and related roundworms genetic and genomic data
FlyBase	International community	Fruit fly genus *Drosophila* genetic and genomic data
GreenPhyl	Biodiversity International and CIRAD	Plant genomes integrative and comparative analysis
PlantDB	Munich MIPS	Plant genomes integrative and comparative analysis
TAIR	Phoenix Bioinformatics Corporation	Thale cress (*A. thaliana*) genetic, proteomic, metabolic, and other data
MaizeDB	International community	Maize crop (*Z. mais*) genetic, proteomic, metabolic, and other data
SGD	Stanford University	Baker's budding yeast (*Saccharomyces* genus) genetic, genomic, proteomic, structural, literature, and other data
CGD	Stanford University	Budding yeast (*Candida* genus) genetic, genomic, proteomic, structural, literature, and other data
MycoCosm	US Department of Energy Joint Genome Institute	Fungal genome portal for integrated fungal genomic data and promotes for the 1000 genomes project
CYGD	Munich MIPS	Baker's yeast (*S. cerevisiae*) molecular structures and functional networks
YMDB	University of Alberta, Canada	Baker's yeast (*S. cerevisiae*) metabolome database
YeastNet	Yonsei University, Korea	Baker's yeast (*S. cerevisiae*) integrated functional gene networks
MBGD	Japan Society for the Promotion of Science	Comparative genome analysis of completely sequenced microbial genomes (bacteria, archaea, and few eukaryotes)

(continued overleaf)

(*continued*)

Database	Provider	Description
PATRIC	Virginia Bioinformatics Institute	Bacterial integrated genomic, transcriptomic, proteomic, structural, and sequencing data
IMG and IMG/M	University of California	Integrated annotation, analysis, and distribution of microbial genomic and metagenomic datasets
BacDive	Leibniz Institute DSMZ	Bacterial and archaeal taxonomic, morphologic, physiologic, environmental, and molecular-biological information
COLOMBOS	International community	Cross-platform microarray datasets for organism-specific bacteria
PortEco	A consortium of US laboratories	Model bacteria *E. coli* comprehensive resource
ViralZone	Swiss SIB	Viral bioinformatics resource
Virus Variation Resource	NCBI	Viral gene and protein sequence annotations and relevant metadata

Normalisation Methods

Method	Platform/package	Function
Background correction (general)	R/Bioconductor - affy	expresso bgcorrect
Contrast normalisation	R/Bioconductor - affy	normalize.contrasts normalize.AffyBatch.contrasts
Cyclic loess normalisation	R/Bioconductor - limma	normalizeCyclicLoess
Filter low absolute-value genes	MATLAB	genelowvalfilter
Filter low-entropy genes	MATLAB	geneentropyfilter
Filter small-range genes	MATLAB	generangefilter
Filter small-variance genes	MATLAB	genevarfilter
GC-RMA	MATLAB	affygcrma gcrma gcrmabackadj
Invariant set normalisation	MATLAB	affyinvarsetnorm mainvarsetnorm
	R/Bioconductor - preprocessCore	normalize.invariantset normalize.AffyBatch.invariantset
Loess/lowess normalisation	MATLAB	malowess
	R/Bioconductor - preprocessCore	normalize.loess normalize.AffyBatch.loess
MAS 5.0	R/Bioconductor - affy	mas5
MBEI	R/Bioconductor - affy	expresso (with specific set of parameters)
Normalisation (general)	R/Bioconductor - affy	normalize

(continued)

Method	Platform/package	Function
Quantile normalisation	MATLAB	quantilenorm
	R/Bioconductor - preprocessCore	normalize.quantiles
		normalize.quantiles.robust
		normalize.AffyBatch.quantiles
RMA	MATLAB	affyrma
		rmabackadj
		rmasummary
	R/Bioconductor - affy	rma
Scaling and centring	MATLAB	manorm
	R/Bioconductor - preprocessCore	normalize.constant
Summarisation (general)	R/Bioconductor - affy	summary

Feature-selection Methods

Method	Platform/package	Function
Feature ranking	MATLAB	rankfeatures
	R/FSelector	cutoff.k
		cutoff.k.percent
		cutoff.biggest.diff
Random feature selection	MATLAB	randfeatures
PCA	MATLAB	pca
	R/stats	princomp
Genetic algorithm	MATLAB	ga
	R/genalg	rbga
		rbga.bin

Differential Expression Methods

Method	Platform/package	Function
Adjusted p-value	R	p.adjust
	R/Bioconductor - limma	topTable
ANOVA	MATLAB	anova1
		anova2
		kruskalwallis
		multcompare
ANOVA	R	aov
		anova

(continued overleaf)

(*continued*)

Method	Platform/package	Function
baySeq	R/baySeq	
BBSeq	R/BBSeq	
B-statistic	R/Bioconductor – limma	eBayes
		topTable
DESeq	R/DESeq	
edgeR	R/edgeR	
FDR for multiple hypotheses	MATLAB	mafdr
Fisher's exact test	MATLAB	By the hypergeometric cumulative distribution function:
		hygecdf
Fisher's exact test	R	fisher.test
Fold change	R/Bioconductor – limma	topTabe
Fold change	MATLAB	mavolcanoplot
Likelihood ratio test	MATLAB	lratiotest
Likelihood ratio test	R/lmtest	lrtest
MA scatter plot	MATLAB	mairplot
MA scatter plot	R/Bioconductor – limma	plotMA
Moderated t-test	R/Bioconductor – limma	topTable
q-value	MATLAB	mafdr
q-value	MATLAB	topTable
Student's t-test	MATLAB	ttest
		ttest2
		mattest
Student's t-test	R	t.test
Volcano plot	MATLAB	mavolcanoplot
Volcano plot	R/Bioconductor – limma	volcanoplot

Partitional Clustering Algorithms

Algorithm name	Year	Platform	Package	Function
k-means	1967	R/MATLAB/ JAVA	Statistics/stats/ weka	kmeans/kmeans/Simple K-Means
k-medoids	1990	R	cluster	pam
Genetic k-means	1999	R	skmeans	skmeans (method='genetic')
Spherical k-means	2001	R	skmeans	skmeans
Kernel k-means	2002	R	kernlab	kkmeans
Spherical k-medoids	2006			
Genetic k-medoids	2006			

The summary of all partitional clustering algorithms introduced in this chapter.

MATLAB Linkage Functions

Method	Description
'average'	Unweighted average distance (UPGMA)
'centroid'	Centroid distance (UPGMC), appropriate for Euclidean distances only
'complete'	Complete linkage/Furthest distance
'median'	Weighted centre of mass distance (WPGMC), appropriate for Euclidean distances only
'single'	Single linkage/shortest distance
'ward'	Minimum variance algorithm, appropriate for Euclidean distances only
'weighted'	Weighted average distance (WPGMA)/McQuitty's methods

The 'method' argument of function *linkage* in MATLAB determines the algorithm for computing distance between clusters.

Fuzzy Clustering Algorithms

Algorithm name	Platform	Package	Function
Fuzzy *c*-means	*R*/MATLAB	e1071	cmeans/fcm
FANNY	*R*	cluster	Fanny
Fuzzy *c*-shell	*R*	e1071	cshell
Fuzzy cluster indices	*R*	e1071	fclustIndex

Summary of the publicly accessible resources of fuzzy clustering algorithms.

Neural Network-based Clustering Methods

Algorithm name	Platform	Package	Function
SOM	*R*	kohonen/som/wccsom	som
	MATLAB		newsom/train/sim
	weka	Weka.clusterers	SelfOrganizingMap
ART	*R*	RSNNS	art1/art2/artmap

Collection of publicly accessible resources for neural network-based clustering.

Mixture Model-based Clustering Methods

Algorithm name	Platform	Package	Function
GMM-EM	*R*	mclust	mclust
GMM-VBEM	MATLAB	Chen, (2012)	vbgm
MFA-EM	MATLAB	Ghahramani (2000)	mfa
MFA-VBEM	MATLAB	Ghahramani (2000)	vbmfa
BHC	*R*	BHC	bhc
Bayesian nonparametric	*R*	DPpackage	

Collection of publicly accessible resources for mixture model-based clustering.

Graphs and Networks File Formats and Storage

Format	Filename Ext	Comments
GraphML	.graphml	An XML-based file format for graphs
XGMML	.xml	(The eXtensible Graph Markup and Modelling Language) is an XML application based on GML which is used for graph description
Pajek NET	.net	Pajek (Slovene word for Spider) is a program, for Windows, for analysis and visualisation of large networks.
Graphlet GML	.gml	Graphlet is GML graph data format. GML is an acronym derived from Graph Modelling Language
Graphviz DOT	.dot	DOT is the text file format of the suite GraphViz.
CSV	.csv	A comma-separated values (CSV) (also sometimes called character-separated values) file stores tabular data in plain-text form.
UCINET DL	.dl	UCINET DL format is the most common file format used by UCINET package. UCINET 6 for Windows is a software package for the analysis of social network data
GXL	.gxl	GXL (Graph eXchange Language) is designed to be a standard exchange format for graphs.
Text	.txt	Delimited text table

Summary of file formats as a storage of graphs and network.

Graph-clustering Algorithms

Name	Year	Reference	Software
Minimum-cut criterion	1993	Wu and Leahy (1993)	
METIS	1995	Karypis and Kumar (1995, 1998)	
Normalised cut	2000	Shi and Malik (2000)	
Markov clustering	2000	van Dongen (1998)	C (van Dongen, 2012)
Spectral clustering	2001	Ng, Jordan and Weiss (2001); Luxburg (2007)	
Greed method of Newman	2004	Newman (2004)	MATLAB (MIT, 2011)
Spectral modularity optimisation	2006	Newman (2006)	MATLAB (MIT, 2011)
Affinity propagation	2007	Frey and Dueck (2007)	MATLAB/R (Frey, 2011)
Directed networks	2008	Leicht and Newman (2008)	MATLAB (MIT, 2011)
Overlapping community	2013	Gopalan and Blei (2013)	

Summary of graph-clustering algorithms mentioned in Chapter 16.

Consensus Clustering Algorithms

Name	Year	Reference	Software
Voting-merging	2001	Dimitriadou, Weingessel and Hornik (2001)	
Resampling methods	2003	Monti *et al.* (2003)	
MCL (MetaClustering)	2003	Strehl and Ghosh (2003)	
CSPA			
HGPA			
Best-of-k	2004	Filkov and Skiena, (2004)	
SAOM			
BOM (Mirkin distance)			
Clustering ensembles weak	2005	Topchy, Jain and Punch, (2005)	
Evidence accumulation	2005	Fred and Jain (2005)	
Graph Consensus Clustering (GCC)	2007	Yu, Wong and Wang (2007)	
Clustering Aggregation	2007	Gionis, Mannila and Tsaparas (2007)	
Consensus Clustering	2010	Brannon *et al.* (2010); Seiler *et al.* (2010)	Python (Seiler *et al.*, 2010)

Summary of consensus clustering algorithms mentioned in Chapter 17.

Biclustering Algorithms

Bicluster method	Class	Year	Availability
δ-clustering (Hartigan, 1972)		1972	
CC (Cheng and Church, 2000)	VMB	2000	Barkow *et al.* (2006); Kaiser and Leisch (2008); Eren *et al.* (2013)
δ-jk (Califano, Stolovitzky and Tu, 2000)		2000	
CTWC (Getz, Levine and Domany, 2000)	TWC	2000	
ITWC (Tang *et al.*, 2001)	TWC	2001	
DCC (Busygin *et al.*, 2002; Busygin, Prokopyev and Pardalos, 2008)	TWC	2002	
SA (Ihmels *et al.*, 2002; Ihmels, Bergmann and Barkai, 2004)	TWC	2002	Barkow *et al.* (2006)
Plaid (Lazzeroni and Owen, 2002; Turner, Bailey and Krzanowski, 2005)	PGM	2002	Kaiser and Leisch (2008); Eren *et al.* (2013)
SAMBA (Tanay, Sharan and Shamir, 2002)	PGM	2002	
δ-Pclustering (Wang *et al.*, 2002)	VMB	2002	
Gibbs clustering (Sheng, Moreau and Moor, 2003)	PGM	2003	
ISA (Bergmann, Ihmels and Barkai, 2003)	TWC	2003	Eren *et al.* (2013)

(*continued overleaf*)

(*continued*)

Bicluster method	Class	Year	Availability
OP-clustering (Liu and Wang, 2003; Liu, Wang and Yang, 2004)	CMB	2003	
OPSM (Ben-Dor *et al.*, 2003)	CMB	2003	Barkow *et al.* (2006); Eren *et al.* (2013)
Spectral (Kluger *et al.*, 2003)	VMB	2003	Kaiser and Leisch (2008); Eren *et al.* (2013)
xMOTIF (Murali and Kasif, 2003)	VMB	2003	Barkow *et al.* (2006); Kaiser and Leisch (2008); Eren *et al.* (2013)
GEM (Wu *et al.*, 2004; Wu and Kasif, 2005)	PGM	2004	
FLOC (Yang *et al.*, 2005)	CMB	2005	
SA (Bryan, Cunningham and Bolshakova, 2005)	CMB	2005	
ZBDD (Yoon *et al.*, 2005)	VMB	2005	
ROBA (Tchagang and Twefik, 2005)	CMB	2005	
Bimax (Prelić *et al.*, 2006)	CMB	2006	Barkow *et al.* (2006)
CMonkey (Reiss, Baliga and Bonneau, 2006)	PGM	2006	
SEBI (Divina and Aguilar-Ruiz, 2006)	VMB	2006	
MOEB (Mitra and Banka, 2006)	CMB	2006	
UBCLUST (Li *et al.*, 2006)	CMB	2006	
R/MSBE (Liu and Wang, 2007)	VMB	2007	
Bayesian biclustering (Gu and Liu, 2008)	PGM	2008	Eren *et al.* (2013)
ACV (Teng and Chan, 2008)	CMB	2008	
Bayesian plaid (Caldas and Kaski, 2008)	PGM	2008	
BiMine (Ayadi, Elloumi and Hao, 2009)	CMB	2009	
CPB (Bozdağ, Parvin and Catalyurek, 2009)	CMB	2009	Eren *et al.* (2013)
COALESCE (Huttenhower *et al.*, 2009)	CMB	2009	Eren *et al.* (2013)
QUBIC (Li *et al.*, 2009)	VMB	2009	Eren *et al.* (2013)
FABIA and FABIAS (Hochreiter *et al.*, 2010)	PGM	2010	Eren *et al.* (2013)
TreeBic (Caldas and Kaski, 2010)	PGM	2010	

Summary of biclustering algorithms.

References

Ayadi, W., Elloumi, M. and Hao, J.-K. (2009). A biclustering algorithm based on a bicluster enumeration tree: application to DNA microarray data. *BioData Mining*, **2**(1), p. 9.

Barkow, S., Bleuler, S., Prelic, A. *et al.* (2006). BicAT: a biclustering analysis toolbox. *Bioinformatics*, **22**(10), pp. 1282–1283.

Ben-Dor, A., Chor, B., Karp, R. and Yakhini, Z. (2003). Discovering local structure in gene expression data: the order-preserving submatrix problem. *Journal of Computational Biology*, **10**(3–4), pp. 373–384.

Bergmann, S., Ihmels, J. and Barkai, N. (2003). Iterative signature algorithm for the analysis of large-scale gene expression data. *Physical Review E*, **67**(3), p. 031902.

Bozdağ, D., Parvin, J.D. and Catalyurek, U.V. (2009). A biclustering method to discover co-regulated genes using diverse gene expression datasets. In: *Bioinformatics and Computational Biology* (ed. S. Rajasekaran), Springer, New York, pp. 151–163.

Brannon, A.R., Reddy, A., Seiler, M. *et al.* (2010). Molecular stratification of clear cell renal cell carcinoma by consensus clustering reveals distinct subtypes and survival patterns. *Genes and Cancer*, **1**(2), pp. 152–163.

Bryan, K., Cunningham, P. and Bolshakova, N. (2005). Biclustering of Expression Data using Simulated Annealing. Proceedings of the 18th IEEE Symposium on Computer-Based Medical Systems, 23–24 June 2005, Dublin. pp. 383–388.

Busygin, S., Jacobsen, G., Krämer, E. and Ag, C. (2002). Double Conjugated Clustering Applied to Leukemia Microarray Data. 2nd SIAM ICDM, Workshop on Clustering High Dimensional Data, Arlington, VA.

Busygin, S., Prokopyev, O. and Pardalos, P.M. (2008). Biclustering in data mining. *Computers and Operations Research*, **35**(9), pp. 2964–2987.

Caldas, J. and Kaski, S. (2008). Bayesian Biclustering with the Plaid Model. IEEE Workshop on Machine Learning for Signal Processing, MLSP 2008, Cancún, *Mexico*, pp. 291–296.

Caldas, J. and Kaski, S. (2010). *Hierarchical Generative Biclustering for microRNA Expression Analysis.* Lisbon, Research in Computational Molecular Biology, pp. 65–79.

Califano, A., Stolovitzky, G. and Tu, Y. (2000). *Analysis of Gene Expression Microarrays for Phenotype Classification*, Springer, Boston, MA, pp. 75–85.

Chen, M. (2012). *Mathworks File Exchange Center,* http://www.mathworks.co.uk/matlabcentral/fileexchange/35362-variational-bayesian-inference-for-gaussian-mixture-model/content/vbgm.m (accessed 28 July 2014).

Cheng, Y. and Church, G.M. (2000). *Biclustering of Expression Data.* ISMB, Boston, MA, pp. 93–103.

Dimitriadou, E., Weingessel, A. and Hornik, K. (2001). Voting-merging: an ensemble method for clustering. In: *Artificial Neural Networks—ICANN 2001*, Springer, New York, pp. 217–224.

Divina, F. and Aguilar-Ruiz, J.S. (2006). Biclustering of expression data with evolutionary computation. *IEEE Transactions on Knowledge and Data Engineering*, **18**(5), pp. 590–602.

van Dongen, S. (1998). *A New Cluster Algorithm for Graphs,* CWI (Centre for Mathematics and Computer Science), Amsterdam.

van Dongen, S. (2012). *MCL - a Cluster Algorithm for Graphs,* http://micans.org/mcl/ (accessed 28 July 2014).

Eren, K., Deveci, M., Küçüktunç, O. and Çatalyürek, Ü.V. (2013). A comparative analysis of biclustering algorithms for gene expression data. *Briefings in Bioinformatics*, **14**(3), pp. 279–292.

Filkov, V. and Skiena, S. (2004). Integrating microarray data by consensus clustering. *International Journal on Artificial Intelligence Tools*, **13**(04), pp. 863–880.

Fred, A.L. and Jain, A.K. (2005). Combining multiple clusterings using evidence accumulation., *IEEE Transactions on Pattern Analysis and Machine Intelligence*, **27**(6), pp. 835–850.

Frey, B.J. (2011). *Frey Lab,* http://genes.toronto.edu/index.php?q=affinity%20propagation (accessed 28 July 2014).

Frey, B.J. and Dueck, D. (2007). Clustering by passing messages between data points. *Science*, **315**(5814), pp. 972–976.

Getz, G., Levine, E. and Domany, E. (2000). Coupled two-way clustering analysis of gene microarray data. *Proceedings of the National Academy of Sciences*, **97**(22), pp. 12079–12084.

Ghahramani, Z. (2000). MATLAB codes for EM for Mixture of Factor Analyzers and Variational Bayesian Mixture of Factor Analysers, Online resources, http://mlg.eng.cam.ac.uk/zoubin/software.html (accessed 28 July 2014).

Gionis, A., Mannila, H. and Tsaparas, P. (2007). Clustering aggregation. *ACM Transactions on Knowledge Discovery from Data (TKDD)*, **1**(1), Article No. 4.

Gopalan, P.K. and Blei, D.M. (2013). Efficient discovery of overlapping communities in massive networks. *Proceedings of the National Academy of Sciences*, **110**(36), pp. 14534–14539.

Gu, J. and Liu, J.S. (2008). Bayesian biclustering of gene expression data. *BMC Genomics*, **9**(Suppl 1), S4.

Hartigan, J.A. (1972). Direct clustering of a data matrix. *Journal of the American Statistical Association*, **67**(337), pp. 123–129.

Hochreiter, S., Bodenhofer, U., Heusel, M. *et al.* (2010). FABIA: factor analysis for bicluster acquisition. *Bioinformatics*, **26**(12), pp. 1520–1527.

Huttenhower, C., Mutungu, K.T., Indik, N. *et al.* (2009). Detailing regulatory networks through large scale data integration. *Bioinformatics*, **25**(24), pp. 3267–3274.

Ihmels, J., Friedlander, G., Bergmann, S. *et al.* (2002). Revealing modular organization in the yeast transcriptional network. *Nature Genetics*, **31**(4), pp. 370–377.

Ihmels, J., Bergmann, S. and Barkai, N. (2004). Defining transcription modules using large-scale gene expression data. *Bioinformatics*, **20**(13), pp. 1993–2003.

Kaiser, S. and Leisch, F. (2008). *A Toolbox for Bicluster Analysis in R*, Proceedings in Computational Statistics, Heidelberg.

Karypis, G. and Kumar, V. (1995). *Metis-Unstructured Graph Partitioning and Sparse Matrix Ordering System, Version 2.0.* Department of Computer Science, University of Minnesota, Minneapolis.

Karypis, G. and Kumar, V. (1998). A fast and high quality multilevel scheme for partitioning irregular graphs. *SIAM Journal on Scientific Computing*, **20**(1), pp. 359–392.

Kluger, Y., Basri, R., Chang, J.T. and Gerstein, M. (2003). Spectral biclustering of microarray data: coclustering genes and conditions. *Genome Research*, **13**(4), pp. 703–716.

Lazzeroni, L. and Owen, A. (2002). Plaid models for gene expression data. *Statistica Sinica*, **12**(1), pp. 61–86.

Leicht, E. and Newman, M. (2008). Community structure in directed networks. *Physical Review Letters*, **100**(11) pp. 1–5.

Li, H., Chen, X., Zhang, K. and Jiang, T. (2006). A general framework for biclustering gene expression data. *Journal of Bioinformatics and Computational Biology*, **4**(04), pp. 911–993.

Li, G., Ma, Q., Tang, H. *et al.* (2009). QUBIC: a qualitative biclustering algorithm for analyses of gene expression data. *Nucleic Acids Research*, **37**(15), e101.

Liu, J. and Wang, W. (2003). Op-cluster: Clustering by Tendency in High Dimensional Space. Third IEEE International Conference on Data Mining. ICDM 2003, Melbourne, FL, pp. 187–194.

Liu, X. and Wang, L. (2007). Computing the maximum similarity bi-clusters of gene expression data. *Bioinformatics*, **23**(1), pp. 50–56.

Liu, J., Wang, W. and Yang, J. (2004). Gene Ontology Friendly Biclustering of Expression Profiles. Proceedings of the IEEE Computational Systems Bioinformatics Conference, CSB 2004, Stanford, CA, pp. 436–447.

Luxburg, U.V. (2007). A tutorial on spectral clustering. *Statistics and Computing*, **17**(4), pp. 395–416.

MIT (2011). *Matlab Tools for Network Analysis,* http://strategic.mit.edu/downloads.php?page=matlab_networks (accessed 28 July 2014).

Mitra, S. and Banka, H. (2006). Multi-objective evolutionary biclustering of gene expression data. *Pattern Recognition*, **39**(12), pp. 2464–2477.

Monti, S., Tamayo, P., Mesirov, J. and Golub, T. (2003). Consensus clustering: a resampling-based method for class discovery and visualization of gene expression microarray data. *Machine Learning*, **52**(1–2), pp. 91–118.

Murali, T. and Kasif, S. (2003). Extracting Conserved Gene Expression Motifs from Gene Expression Data. Pacific Symposium on Biocomputing, 3–7 January 2003, Lihue, HI. World Scientific Pub Co Inc., pp. 77–88.

Newman, M.E.J. (2004). Fast algorithm for detecting community structure in networks. *Physical Review E*, **69**(6), p. 066133.

Newman, M.E.J. (2006). Modularity and community structure in networks. *Proceedings of the National Academy of Sciences of the United States of America*, **103**(23), pp. 8577–8582.

Ng, A.Y., Jordan, M.I. and Weiss, Y. (2001). On spectral clustering: analysis and an algorithm. *Proceedings of Advances in Neural Information Processing Systems* (eds T.G. Dietterich, S. Becker and Z. Ghahramani), MIT Press, Cambridge, MA, Volume **14**, pp. 849–856.

Prelić, A., Bleuler, S., Zimmermann, P. *et al.* (2006). A systematic comparison and evaluation of biclustering methods for gene expression data. *Bioinformatics*, **22**(9), pp. 1122–1129.

Reiss, D.J., Baliga, N.S. and Bonneau, R. (2006). Integrated biclustering of heterogeneous genome-wide datasets for the inference of global regulatory networks. *BMC Bioinformatics*, **7**(1), p. 280.

Seiler, M., Huang, C.C., Szalma, S. and Bhanot, G. (2010a). *Consensus Clustering Software Package*, http://code.google.com/p/consensus-cluster/ (accessed 28 July 2014).

Seiler, M., Huang, C.C., Szalma, S. and Bhanot, G. (2010b). ConsensusCluster: a software tool for unsupervised cluster discovery in numerical data. *OMICS A Journal of Integrative Biology*, **14**(1), pp. 109–113.

Sheng, Q., Moreau, Y. and Moor, B.D. (2003). Biclustering microarray data by Gibbs sampling. *Bioinformatics-Oxford*, **19**(2), pp. 196–205.

Shi, J. and Malik, J. (2000). Normalized cuts and image segmentation. *IEEE Transactions on Pattern Analysis and Machine Intelligence*, **22**(8), pp. 888–905.

Strehl, A. and Ghosh, J. (2003). Cluster ensembles – a knowledge reuse framework for combining multiple partitions. *Journal of Machine Learning Research*, **3**, pp. 583–617.

Tanay, A., Sharan, R. and Shamir, R. (2002). Discovering statistically significant biclusters in gene expression data. *Bioinformatics*, **18**(suppl 1), pp. S136–S144.

Tang, C., Zhang, L., Zhang, A. and Ramanathan, M. (2001). Interrelated Two-Way Clustering: an Unsupervised Approach for Gene Expression Data Analysis. Proceedings of the IEEE 2nd International Symposium on Bioinformatics and Bioengineering Conference, 2001, Bethesda, MD, pp. 41–48.

Tchagang, A.B. and Twefik, A. (2005). Robust Biclustering Algorithm (ROBA) for DNA Microarray Data Analysis. IEEE 13th Workshop on Statistical Signal Processing, Bordeaux, 2005, pp. 984–989.

Teng, L. and Chan, L. (2008). Discovering biclusters by iteratively sorting with weighted correlation coefficient in gene expression data. *Journal of Signal Processing Systems*, **50**(3), pp. 267–280.

Topchy, A., Jain, A.K. and Punch, W. (2005). Clustering ensembles: models of consensus and weak partitions., *IEEE Transactions on Pattern Analysis and Machine Intelligence*, **27**(12), pp. 1866–1881.

Turner, H., Bailey, T. and Krzanowski, W. (2005). Improved biclustering of microarray data demonstrated through systematic performance tests. *Computational Statistics and Data Analysis*, **48**(2), pp. 235–254.

Wang, H., Wang, W., Yang, J. and Yu, P.S. (2002). Clustering by Pattern Similarity in Large Data Sets. Proceedings of the 2002 ACM SIGMOD International Conference on Management of Data, Madison, WI, pp. 394–405.

Wu, C.-J. and Kasif, S. (2005). GEMS: a web server for biclustering analysis of expression data. *Nucleic Acids Research*, **33**(suppl 2), pp. W596–W599.

Wu, Z. and Leahy, R. (1993). An optimal graph theoretic approach to data clustering: theory and its application to image segmentation. *IEEE Transactions on Pattern Analysis and Machine Intelligence*, **15**(11), pp. 1101–1113.

Wu, C.-J., Fu, Y., Murali, T. and Kasif, S. (2004). Gene expression module discovery using Gibbs sampling. *Genome Informatics Series*, **15**(1) pp. 239–248.

Yang, J., Wang, H., Wang, W. and Yu, P.S. (2005). An improved biclustering method for analyzing gene expression profiles. *International Journal on Artificial Intelligence Tools*, **14**(05), pp. 771–789.

Yoon, S., Nardini, C., Benini, L. and Micheli, G.D. (2005). Discovering coherent biclusters from gene expression data using zero-suppressed binary decision diagrams. *IEEE/ACM Transactions on Computational Biology and Bioinformatics (TCBB)*, **2**(4), pp. 339–354.

Yu, Z., Wong, H.-S. and Wang, H. (2007). Graph-based consensus clustering for class discovery from gene expression data. *Bioinformatics*, **23**(21), pp. 2888–2896.

Index

Note: Page numbers in *italics* refer to Figures; those in **bold** to Tables

---

Integrative Cluster Analysis in Bioinformatics, First Edition. Basel Abu-Jamous, Rui Fa and Asoke K. Nandi.
© 2015 John Wiley & Sons, Ltd. Published 2015 by John Wiley & Sons, Ltd.